THE TERRESTRIAL ENVIRONMENT AND THE ORIGIN OF LAND VERTEBRATES

Proceedings of an International Symposium
held at the University of Newcastle upon Tyne

THE SYSTEMATICS ASSOCIATION
SPECIAL VOLUME NO. 15

THE TERRESTRIAL ENVIRONMENT AND THE ORIGIN OF LAND VERTEBRATES

Edited by
A. L. PANCHEN
Department of Zoology,
University of Newcastle upon Tyne

1980

Published for the
SYSTEMATICS ASSOCIATION
by
ACADEMIC PRESS
LONDON NEW YORK TORONTO SYDNEY SAN FRANCISCO

ACADEMIC PRESS INC. (LONDON) LTD.
24-28 Oval Road
London NW1 7DX

U.S. Edition published by
ACADEMIC PRESS INC.
111 Fifth Avenue
New York, New York 10003

British Library Cataloguing in Publication Data

The terrestrial environment and the origin of land vertebrates.
—(Systematics Association. Special volumes; No. 15 ISSN 0309-2593).
1. Vertebrates—Evolution—Congresses
I. Panchen, Alec II. Series
596'.03'8 QL607.5 80-40225

ISBN 0-12-544780-9

PRINTED IN GREAT BRITAIN BY
WHITSTABLE LITHO LTD, WHITSTABLE, KENT

Contributors

Carroll, Dr R. L., *Redpath Museum, McGill University, 859 Sherbrooke Street West, Montreal, PQ, Canada H3A 2K6.*

Edwards, Dr Dianne, *Department of Plant Science, University College, P.O. Box 78, Cardiff CF1 1XL, Wales.*

Gaffney, Dr E. S., *Department of Vertebrate Paleontology, The American Museum of Natural History, Central Park West at 79th Street, New York, N.Y., 10024, U.S.A.*

Gardiner, Dr B. G., *Department of Biology, Queen Elizabeth College, University of London, Campden Hill Road, London, W8 7AH, England.*

Heaton, Dr M. J., *Department of Biology, Erindale College, University of Toronto, Mississauga, Ontario, Canada L5L IC6.*

Holmes, R. B., *Redpath Museum, McGill University, 859 Sherbrooke Street West, Montreal, PQ, Canada H3A 2K6.*

Janvier, Dr P., *Laboratoire de Paléontologie des Vertébrés et de Paléontologie Humaine, Université de Paris VI, 4 Place Jussieu, 75230 Paris, Cedex 05, France.*

Johnson, Dr G. A. L., *Department of Geological Sciences, University Science Laboratories, South Road, Durham DH1 3LE, England.*

Milner, Dr Andrew R., *Department of Zoology, Birkbeck College, University of London, Malet Street, London WC1E 7HX, England.*

Milner, Dr Angela C., *Department of Palaeontology, British Museum (Natural History), Cromwell Road, London SW7 5BD, England.*

Panchen, Dr A. L., *Department of Zoology, The University, Newcastle upon Tyne, NE1 7RU, England.*

Patterson, Dr C., *Department of Palaeontology, British Museum (Natural History), Cromwell Road, London SW7 5BD, England.*

Rackoff, Dr J. S., *Department of Biology, Bucknell University, Lewisburg, Pennsylvania, 17837, U.S.A.*

Reisz, Dr R., *Department of Biology, Erindale College, University of Toronto, Mississauga, Ontario, Canada L5L IC6.*

Rolfe, Dr W. D. I., *Hunterian Museum, Glasgow University, Glasgow G12 8QQ, Scotland.*

Scott, Dr A. C., *Department of Geology, Chelsea College, University of London, 271 King Street, London W6 9LZ, England.*

Smithson, T. R., *Redpath Museum, McGill University, 859 Sherbrooke Street West, Montreal, P.Q., Canada, H3A 2K6.*

Tarling, Dr D. H., *Department of Geophysics and Planetary Physics, The University, Newcastle upon Tyne, NE1 7RU, England.*

Thomson, Dr K. S., *Peabody Museum of Natural History, Yale University, New Haven, Connecticut, 06520, U.S.A.*

Westoll, Professor T. S., F.R.S., *Department of Geology, The University, Newcastle upon Tyne, NE1 7RU, England.*

Preface

The contributions published in this volume represent the papers given at an international symposium held at the University of Newcastle upon Tyne on 18 and 19 April 1979. The original intention had been to invite colleagues to talk on specific problems of the origin of tetrapod vertebrates from their fish ancestors, and on those alone. On reflection, however, it seemed advisable to broaden the scope of the meeting so as to set the geographical, geological, floral and faunal background to one of the most important events in the history of vertebrates, and in the history of life. It was hoped, by so doing, to encourage, in both students and colleagues, the wish to see the origin of land vertebrates as an ecological as well as an anatomical and systematic problem.

We were lucky to secure the services of a distinguished group of contributors and trust that all the papers will be of lasting value. However, discussion of the immediate problem of the group of fish from which tetrapods arose, or (as some of our authors would prefer) to which they are most closely related, may be entering a period of instability. The most recent major reviews of this problem are those of Jarvik in 1942, whose detailed conclusions were supported by a series of more recent works, and Westoll in 1943, reinforced by almost simultaneous and excellent studies on two topics echoed in this volume, the origin of the tetrapod limb and the evolution of the middle ear. The wealth of meticulous anatomical detail in Jarvik's studies is beyond praise. However, his conclusion that tetrapods were at least diphyletic, based on detailed but exclusive comparisons between two groups of fossil fish and two of the three groups of extent amphibia, has not been generally accepted, at least by workers outside Scandanavia. Moreover, it received a severe blow when it was shown that the differences in the structure of the snout of living amphibia, critical to Jarvik's thesis, disappeared as diagnostic features with a comprehensive review of both amphibian groups (Jurgens, J. D. (1971). *Ann. Univ. Stellenbosch* 46A, No. 2).

The orthodox view, therefore, is of monophyletic origin of tetrapods from within the crossopterygian fishes and it is this view that is now being challenged. That challenge is represented in this volume by the papers by Patterson and Gardiner. More than facts and hypotheses are being challenged, however. During the last dozen or so years a deep division has arisen in systematic methodology and particularly in methods of phylogeny reconstruction, between orthodox systematists and advocates of the cladistic system associated with the name of W. Hennig. The resulting controversy is echoed in the pages of this volume, as is the variable degree of influence of Hennig's system on some of our authors. It is right, in a symposium sponsored by the Systematics Association, that this should be so.

At the meeting we missed the wise councils of Professor Stanley Westoll and Dr Mahala Andrews in discussion of this controversy with respect to the origin of tetrapods. Both were on a visit to the Peoples' Republic of China, but we are glad that Professor Westoll has been able to contribute an invited paper to this volume. Thanks are due to Professor Michael House (President of the Systematics Association), Dr R. L. Carroll, Professor Barry Cox and Dr Colin Patterson, who chaired sessions of the meeting, and particularly to Dr K. S. Thomson, who initiated and controlled a lively final discussion, which, while not recorded here, undoubtedly influences the final form of some of the papers. Members of the Department of Geology and Zoology and of the Hancock Museum, here in Newcastle, helped to make the meeting a success and generous publicity was given by the Palaeontological Association. Apart from the Systematics Association the meeting was subsidized by the University of Newcastle including the Zoology Department and by the British Council. A grant from the latter body gave us further help with the expenses of our overseas contributors. The indexes in this volume were compiled by Mr Andrew McDonald of the Newcastle University Library.

May 1980
Newcastle

A.L.P.

Contents

Systematics Association Publications

2. FUNCTION AND TAXONOMIC IMPORTANCE (1959)
   Edited by A. J. Cain
3. THE SPECIES CONCEPT IN PALAEONTOLOGY (1956)
   Edited by P. S. Sylvester-Bradley
4. TAXONOMY AND GEOGRAPHY (1962)
   Edited by D. Nichols
5. SPECIATION IN THE SEA (1963)
   Edited by J. P. Harding and N. Tebble
6. PHENETIC AND PHYLOGENETIC CLASSIFICATION (1964)
   Edited by V. H. Heywood and J. McNeill
7. ASPECTS OF TETHYAN BIOGEOGRAPHY (1967)
   Edited by C. G. Adams and D. V. Ager
8. THE SOIL ECOSYSTEM (1969)
   Edited by J. Sheals
9. ORGANISMS AND CONTINENTS THROUGH TIME (1973)†
   Edited by N. F. Hughes

LONDON. Published by the Association

Systematics Association Special Volumes

1. THE NEW SYSTEMATICS (1940)
   Edited by Julian Huxley (Reprinted 1971)
2. CHEMOTAXONOMY AND SEROTAXONOMY (1968)*
   Edited by J. G. Hawkes
3. DATA PROCESSING IN BIOLOGY AND GEOLOGY (1971)*
   Edited by J. L. Cutbill
4. SCANNING ELECTRON MICROSCOPY (1971)*
   Edited by V. H. Heywood
5. TAXONOMY AND ECOLOGY (1973)*
   Edited by V. H. Heywood
6. THE CHANGING FLORA AND FAUNA OF BRITAIN (1974)*
   Edited by G. L. Hawksworth

*Published by Academic Press for the Systematics Association
†Published by the Palaeontological Association in conjection with the Systematics Association

7. BIOLOGICAL IDENTIFICATION WITH COMPUTERS (1975)*
   Edited by R. J. Pankhurst
8. LICHENOLOGY: PROGRESS AND PROBLEMS (1977)*
   Edited by D. H. Brown, D. L. Hawksworth and R. H. Bailey
9. KEY WORKS (1978)*
   Edited by G. J. Kerrich, D. L. Hawksworth and R. W. Sims
10. MODERN APPROACHES TO THE TAXONOMY OF RED AND BROWN ALGAE (1978)*
    Edited by D. E. G. Irvine and J. H. Price
11. BIOLOGY AND SYSTEMATICS OF COLONIAL ORGANISMS (1979)*
    Edited by G. Larwood and B. R. Rosen
12. THE ORIGIN OF MAJOR INVERTEBRATE GROUPS (1979)*
    Edited by M. R. House
13. ADVANCES IN BRYOZOOLOGY (1979)*
    Edited by G. P. Larwood and M. B. Abbott
14. BRYOPHYTE SYSTEMATICS (1979)*
    Edited by G. C. S. Clarke and J. G. Duckett
15. THE TERRESTRIAL ENVIRONMENT AND THE ORIGIN OF LAND VERTEBRATES (1980)*
    Edited by A. L. Panchen

*Published by Academic Press for the Systematics Association

1 | Prologue: Problems of Tetrapod Origin

T. S. WESTOLL

Department of Geology, The University, Newcastle upon Tyne, England

INTRODUCTION

The great radiation of early tetrapods in the Carboniferous Period is one of the most striking phenomena documented by palaeontological studies over the last 125 years or so. The origin of tetrapods poses problems – whether of morphology and comparative anatomy, of environment and mode of life, of function and physiology, or of the time-scale of the evolutionary processes – that demand investigation of Devonian events and fossils. Such studies are doubly important if there are any reasons to suppose that different tetrapods arose from separate fish ancestors, however closely those fishes may have been related. Here we shall briefly review the fossil evidence.

STRATIGRAPHICAL TIME-SCALE

The Devonian system, like so many others, presents problems of demarcation of boundaries, both of the limits of the system and of its recognized stages. Such boundaries should, in general, be characterized in a "type" section (stratotype) with sufficiently abundant and varied fossils to permit correlation with other developments. Correlation is always more difficult the further one goes from the

Systematics Association Special Volume No. 15, "The Terrestrial Environment and the Origin of Land Vertebrates", edited by A. L. Panchen, 1980, pp. 1–10, Academic Press, London and New York.

stratotype sections, and becomes more difficult if the sedimentary facies is significantly different from the type sections. Since many of the fossils with which we shall be concerned occur in strata of the Old Red Sandstone facies, very precise correlation, such as is possible with many marine sequences, is rare. The usage outlined by Westoll (1977, 1979) is followed here.

Fortunately we are not concerned with the lower boundary of the Devonian. The upper boundary presents problems that are outlined by Westoll (1977, p.66). To a rather good approximation this concerns the Strunian and its equivalents; the "official" Devonian–Carboniferous boundary set by the Heerlen Conference of 1935 is at, or just above, the top of the Strunian, but in the U.S.S.R. the "official" base of the Carboniferous is close to the base of the Strunian stage or substage. The main importance of this for our present purpose is that the ichthyostegid tetrapod fauna from East Greenland is from an "Old Red" facies that probably includes Strunian equivalents.

THE PROBABLE TETRAPOD ANCESTORS

The ichthyostegids from East Greenland are already functional tetrapods, but were clearly mainly aquatic in habitat and preserve many fish-like characteristics lost in most Carboniferous tetrapods. It is probable that their most important functional features as tetrapods – the nature of the girdles and the pentadactyl limbs – were present in somewhat earlier ancestors, and may confidently be expected in future finds from the Famennian (s.s.). Meanwhile the search must be within groups of Devonian fishes to identify as closely as possible the group or groups that could immediately have given rise to tetrapods. We may summarily dismiss all Agnatha; and among the gnathostome fishes we may exclude Acanthodii and Chondrichthyes without comment. We are left with the bony-jawed fishes brigaded as Sub-class Osteichthyes by Moy-Thomas and Miles (1971), and among these only the Crossopterygii and Dipnoi have combinations of characters suggesting close relationship to tetrapods. But of the fossil subgroups among these, which one (or more) gave rise to tetrapods?

There are two main approaches to this problem. One is strictly palaeontological, involving detailed comparative anatomy, the consideration of variation within and between subgroups, the stratigraphical order of the fossils, and consideration of environment and mode of life so far as they can be deduced from the geological record. Comparative anatomy and functional analysis can, of course, only be done well by detailed reference to living forms. The other main approach, at present, is to rely very largely on living tetrapods and fishes, and to use the cladistic analytical methods of Hennig; these can provide a valuable tool in the study of very large and varied groups such as insects with a comparatively poor fossil record. To apply it to fossil vertebrates, where much of the material is skeletally incomplete and where even the best material gives negligible information about any structures or functions not affecting the skeleton directly (as by foramina for nerves or vessels, or muscle insertions), is clearly less precise by orders of magnitude.

Both approaches have obvious merits; but where disagreements have emerged, it is my confident belief that search for, and detailed analysis of, geologically early members is likely to produce much more reliable evidence of interrelationships of groups. "Hard-line" cladists eschew the time-scale.

The Crossopterygii and Dipnoi must therefore be considered with reference to both points of view. The interrelationships of the Dipnoi and the generally recognized orders of Class Crossopterygii, using the non-committal higher, taxonomic group-names in the sense of Westoll (1979), have been much debated and their inclusion within larger " taxa" as Choanata and Sarcopterygii is briefly discussed there. Here we shall be content to consider the five crossopterygian orders and the Dipnoi individually as possible tetrapod ancestors, with a few comments on other interrelationships as necessary.

Of the Crossopterygii, members of three orders have been, at one time or another, particularly informative in seeking the origin of tetrapod conditions. They are the Holoptychiida (Porolepiformes), the Osteolepidida (Osteolepiformes) and Rhizodontida (*sensu* Andrews and Westoll, 1970). Of these, the Rhizodontida are important mainly because of the isolated pectoral fin and girdle of *Sauripterus* (for a recent account see Andrews and Westoll, 1971), which played a large part in the search for a tetrapod limb ancestral condition. Other features of Rhizodontida so far as known (e.g. Andrews, 1973) suggest that they lie off the main line (or lines) or tetrapod ancestry; they seem to be a sister-group of Osteolepidida.

The Holoptychiida have been regarded by Jarvik as very closely related to Urodela; the view was based originally on detailed study of the snout and nasal capsules and extended to other skeletal features such as the tongue and branchial arches (Jarvik, 1942, 1962, 1963, 1968, 1975; but see, for example, Thomson, 1964, 1967, 1968 for other views). The skull structure of this order has been magnificently illustrated by Jarvik (1972); it only needs to be said that it shows less close resemblance to that of stegocephalians than does that of Osteolepidida.

It is certainly among the Osteolepidida that the closest and most significant comparison can be made with early stegocephalians. These resemblances are outlined below. Meanwhile we may briefly note that Actinistia (coelacanths) and Struniiformes (Onychodontida) are fairly clearly members of the crossopterygian complex (see comments in Westoll, 1979), but their anatomy makes them unsuitable as direct tetrapod ancestors.

The Dipnoi, from their first known appearance in the Lower Devonian (the Devonian history is briefly summarized in Westoll, 1979), are highly specialized in being autostylic with remarkable modifications of palate and jaws, but the dermal bone pattern in early members (particularly *Dipnorhynchus* – see Thomson and Campbell, 1971) is remarkably "archaic" in many respects. Miles (1977) has added enormously to our knowledge of the cranial anatomy of the group. He has demonstrated that, even in the lower Frasnian *Griphognathus*, remnants of a marginal dentition occur (but not fully organized maxillae and premaxillae), and this would indicate that the posterior (palatally placed) nasal opening is a true choana (see also Gardiner, Chapter 8 of this volume). There is nothing else to suggest that Dipnoi could have been at all closely related to stegocephalian Amphibia. I regard the autostyly, so strongly argued by some as indicating affinity with Urodela, as independently acquired; the autostyly in Holocephali is clearly also independent.

In the briefest outline, the justification for the above comments is as follows:

1. Skull Structure

The dermal bone pattern of cheek, palate and lower jaw in early stegocephalians is essentially identical with that in Osteolepidida;

Ichthyostega and *Crassigyrinus* (Panchen, 1973 and Chapter 13 of this volume) even retain the narrow preopercular of osteolepidids. The dermal skull-roof, however, presents difficulties. Westoll (1936, 1937) pointed out a remarkable change in proportions, amply borne out by the discovery of *Elpistostege* (Westoll, 1938), and discussed in more detail in Westoll (1943a). The result was to show that the names of bones in the skull-roof previously given to the osteolepid fishes did not correspond to those of their true homologues in tetrapods; thus, for example, the fish-bones previously called frontal and parietal were shown to be the tetrapod parietal and postparietal.

The equation of osteolepid "frontal" with tetrapod parietal implies that the tetrapod frontal and nasal must be sought more anteriorly, including the "nasal" series between the fish "frontal" and the premaxilla. The discovery of *Elpistostege* (which was almost certainly still a "fish") showed that a tetrapod-type frontal was developed, incorporating posterior parts of the "nasal series"; the front of the head is unknown (Westoll, 1938). Subsequently Vorobjeva (1973) described this region in *Panderichthys*, where a tetrapod – type paired frontal is followed anteriorly by three paired "nasals", together the likely equivalents of the tetrapod nasals. *Panderichthys* is probably of late Givetian age, *Elpistostege* is early Frasnian; the former retains the osteolepidid intracranial joint, the latter has lost it. These are at present the closest to the stegocephalian pattern, and suggest that true tetrapod development took place in the Upper Devonian.

This has been denied (e.g. by Jarvik, 1967 – but see rejoinder by Parrington, 1967) but is wholly in accord with very great changes in proportions of the braincase (Romer, 1941; Westoll, 1943a). The intracranial joint so characteristic of Crossopterygii is generally obliterated in tetrapods, though in *Ichthyostega* this seems only partial. Consolidation of such a feature can involve: (a) immobilization of the endocranial joint itself; (b) interlocking of dermal bones of the skull-roof so that the tetrapod parietal extends back at the expense of the postparietal; (c) fusion at progressively earlier stages of ontogeny across the dermal gap – as possibly the fish dermosphenotic and intertemporal to produce a compound supratemporo-intertemporal in *Ichthyostega*; (d) firm sutural contact between dermal cheek and skull-roof eliminating the crossopterygian kinesis; (e) backward extension of an osteolepid-type parasphenoid, either directly or by assimilation of parotic plates, under the otico-occipital cranium; or even (f) stronger bracing by a hyomandibular.

Not all of these may be necessary. Some may be regarded as temporary makeshifts, and it is quite possible that some (for example (c) above) could later be relaxed and reversed. In these respects primitive tetrapods show considerable variety in detail, and herein lies the possibility of very rapid adaptive radiation.

Of other aspects of the skull, the origin of the stapes (Romer 1941; Westoll, 1943c) is still a live problem (Carroll, Chapter 12 of this volume). The development of an occipital plate, of specialized cervical vertebrae and of the atlas - axis complex is also currently receiving much attention.

2. *Post-cranial Skeleton*

The origin of the tetrapod limb from a crossopterygian paddle has a long history, summarized by Westoll (1943a,b) and by Andrews and Westoll (1970, 1971). The pectoral skeleton of Osteolepidida (e.g. *Eusthenopteron*) gives by far the closest model for a prototetrapod foreleg of any other crossopterygians or dipnoans. It is not difficult to derive the tetrapod humerus, radius, ulna and carpus (e.g. Westoll, 1943a,b; Andrews and Westoll, 1970, 1971); but there remain problems concerning the origin of digits. The clearest difference between these fish and early tetrapods concerns the endoskeletal shoulder-girdle, which has no large scapular blade in the fishes. However, *Rhizodus* seems to show how such a structure could develop (Andrews and Westoll, 1971).

Other Osteolepidida are built on similar patterns, so far as is known. In known Rhizodontida the pectoral paddle is broader and more complex, but it is doubtful whether the more distal and marginal elements could have given rise to tetrapod-type digits.

The more slender pectoral fins of Devonian Dipnoi, and of Holoptychiida, where the internal skeleton is poorly known, seem less likely to be ancestral to the limbs of most tetrapods, though Holmgren (reviewed in Westoll, 1943b) would derive the urodele limb from a lungfish model.

The pelvic fin and girdle of *Eusthenopteron*, as shown by Andrews and Westoll (1970), offer a very good prototype for tetrapod conditions, and it is reasonable to suppose that some other Osteolepidida would have shown similar conditions, perhaps even closer (Rackoff, Chapter 11 of this volume).

The vertebral column of Crossopterygii was investigated by Andrews and Westoll (1970, 1971). Some Osteolepidida again provide the closest comparisons with stegocephalians, with essentially temnospondylous vertebrae and – most significantly – dorsal ribs. Even with the Osteolepidida there are considerable differences in the vertebral column, including the development of ring centra. It is reasonable to suppose that a considerable variety of structural patterns could develop rapidly, even within early tetrapods, from a basically temnospondylous condition.

POSSIBLE TETRAPOD POLYPHYLY

Carboniferous tetrapods already show rapid diversification. Apart from the Labyrinthodontia (usage, for example, of Panchen, 1967) which are clearly derivable from Osteolepidida, the microsaurs, lysorophids, adelospondylids, nectridians and aïstopods have skull patterns more easily derivable from that of a generalized labyrinthodont than from any non-osteolepidid fish. Their nature and interrelationships (see Carroll and Gaskill, 1978) are beyond the scope of this paper.

It will be clear that the most obvious exception to the general picture of tetrapod origins from osteolepidid fishes is the Urodela, for which a dipnoan or porolepid ancestry has been claimed. Unfortunately all living orders of Amphibia have a poor to very bad fossil record. While Anura may not unreasonably be considered as derived from small temnospondyls, the Urodela and Apoda are more difficult. If pedicellate teeth are regarded as a character uniquely linking nearly all modern Amphibia, then forms related to the modified temnospondylous Lower Permian *Doleserpeton* may indicate that all modern groups converge on its close allies. Carroll and Gaskill (1978) carefully discuss possible microsaurian ancestry for urodeles and apodans. Here we need much more fossil evidence.

Cladistic methods applied to living Amphibia and to early tetrapods and fishes have been claimed, in recent and largely unpublished discussions, to show significant connections between Urodela and Dipnoi (Gardiner, Chapter 8 of this volume); by implication this is extended to other tetrapods. Such a conclusion is firmly opposed by the vast number of features indicating the contrary for the other tetrapods. It seems, however, that Dipnoi and the various groups of Crossopterygii share a common ancestral stock. From this, with our present palaeontological knowledge, three major groups emerged – Dipnoi, Holoptychiida and Osteolepidida, of which the last two can be thought of as close "sister-groups". The Actinistia, on direct fossil evidence, seem to be a "sister-group" of Holoptychiida, the Rhizodontida of Osteolepidida; the status of Struniiformes is uncertain (see brief notes in Westoll, 1979).

In many ways the extant groups of Amphibia are specialized, as are the living coelacanth and lungfish; it is thus dangerous to use some of their characters to indicate origins or to rule out relationships. Who, for example, could have deduced the characters of Lower Devonian lungfish from the living genera?

ENVIRONMENT AND DISTRIBUTION

Nearly all the Devonian fish groups concerned have representatives in both marine and continental ("Old Red") deposits; they are often remarkably abundant in lacustrine, fluvial, flood-plain and coastal-plain deposits. The bulk of the important material comes from North America, Greenland and the Arctic, the Russo-Baltic area and Western Europe. There are, however, very significant indications of faunal relationships with Australia and Antarctica (Westoll, 1979) that suggest the need for revision of some palaeogeographical reconstruction of the Devonian continents (Tarling, Chapter 2 of this volume).

Recent Australian discoveries reinforce these views. Campbell and Bell (1977) describe a mandibular ramus as *Metaxygnathus denticulatus*, of ichthyostegalian character, from strata near Forbes, New South Wales, associated with *Soederberghia, Phyllolepis, Remigolepis* and *Bothriolepis*. In East Greenland, such an assemblage would be found only in the lowest "Remigolepis Series", which I regard as close to the Famennian (s.s.)/Strunian boundary. Young (1974) suggested that a number of Greenland genera occur distinctly earlier in Australia. While this is possible, Young's arguments are by no means decisive. Far firmer evidence is needed. The three Upper Devonian and ("most probably Frasnian") trackways from the Genoa River Beds of Victoria (Warren and Wakefield, 1972) differ from one another. How far this results from different buoyancy and locomotory methods in variably shallow waters is uncertain. Trackway I indicates a very sprawling gait, the pes obliterating part of the manus imprint. The pes had five very short free digits, and was almost laterally directed – a very primitive condition. An otherwise ichthyostegalian body-habit was deduced.

REFERENCES

Andrews, S. M. (1973). Interrelationships of crossopterygians. *In* "Interrelationships of Fishes" (P. H. Greenwood, R. S. Miles and C. Patterson, eds), pp. 137–177, Academic Press, London and New York.

Andrews, S. M. and Westoll, T. S. (1970). The postcranial skeleton of *Eusthenopteron foordi* Whiteaves. *Trans. R. Soc. Edinb.* 68, 207–329.

Andrews, S. M. and Westoll, T. S. (1971). The postcranial skeleton of rhipidistian fishes excluding *Eusthenopteron. Trans. R. Soc. Edinb.* **68**, 391–489.

Campbell, K. S. W. and Bell, M. W. (1977). A primitive amphibian from the Late Devonian of New South Wales. *Alcheringa* **1**, 369–381

Carroll, R.L. and Gaskill, P. (1978). The Order Microsauria. *Mem. Am. Phil. Soc.* No. 126, 211 pp.

Jarvik, E. (1942). On the structure of the snout of crossopterygians and lower gnathostomes in general. *Zool. Bid. Upps.* **21**, 235–675

Jarvik, E. (1962). Les prolépiformes et l'origine des urodéles. *Colloques int. Cent. natn. Rech. Scient.* **104**, 87–101.

Jarvik, E. (1963). The composition of the intermandibular division of the head in fish and tetrapods and the diphyletic origin of the tetrapod tongue. *K. svenska Vetensk Akad. Handl.* 9(4), 1–74.

Jarvik, E. (1967). The homologies of the frontal and parietal bones in fishes and tetrapods. *Colloques int. Cent. natn. Rech. Scient.* **163**, 181–213.

Jarvik, E. (1968). Aspects of vertebrate phylogeny. *Nobel Symposium* 4, 497–527.

Jarvik, E. (1972). Middle and Upper Devonian Porolepiformes from East Greenland with special reference to *Glyptolepis groenlandica* n. sp. *Meddr Grønland* **187**, 1-307.

Jarvik, E. (1975). On the saccus endolymphaticus and adjacent structures in osteolepiforms, anurans and urodeles. *Colloques int. Cent. natn. Rech. Scient.* **218**, 191–211.

Miles, R. S. (1977). Dipnoan (lungfish) skulls and relationships of the group: a study based on new species from the Devonian of Australia. *Zool. J. Linn. Soc.* **61**, 1–328.

Moy-Thomas, J. A. and Miles R. S. (1971) "Palaeozoic Fishes", 2nd Edn, 259 pp. Chapman and Hall, London.

Panchen, A. L. (1967). Amphibia. *In* "The Fossil Record" (Harland *et al.*, eds), pp. 685–694. Geological Society of London.

Panchen, A.L. (1973). On *Crassigyrinus scoticus* Watson, a primitive amphibian from the Lower Carboniferous of Scotland. *Palaeontology* **16**, 179–193.

Parrington, F. R. (1967). The identification of the dermal bones of the head. *J. Linn. Soc. (Zool.)* **47**, 231–39.

Romer, A. S. (1941). Notes on the crossopterygian hyomandibular and braincase. *J. Morph.* **69**, 141–160.
Thomson, K. S. (1964). The comparative anatomy of the snout in rhipidistian fishes. *Bull. Mus. comp. Zool. Harv.* **131**, 313–357.
Thomson, K.S. (1967). Notes on the relationship of the rhipidistian fishes and the ancestry of the tetrapods. *J. Paleont.* **41**, 660–674.
Thomson, K. S. (1968). A critical review of the diphyletic theory of tetrapod relationships. *Nobel Symposium* **4**, 285–305.
Thomson, K. S. and Campbell, K. S. W. (1971). The structure and relationships of the primitive Devonian lungfish *Dipnorhynchus sussmilchi* (Etheridge). *Bull. Peabody Mus. nat. Hist.* **38**, 1–109.
Vorobjeva, E. I. (1973). Einige Besonderheiten in Schadelbau von *Panderichthys rhombolepis* (Gross). *Palaeontographica* **143A**, 221–229.
Warren, J. W. and Wakefield, N. A. (1972). Trackways of Tetrapod Vertebrates from the Upper Devonian of Victoria, Australia. *Nature, Lond.* **238**, 469–470.
Westoll, T. S. (1936). On the structures of the dermal ethmoid shield of *Osteolepis*. *Geol. Mag.* **73**, 157–171.
Westoll, T. S. (1937). On a specimen of *Eusthenopteron* from the Old Red Sandstone of Scotland. *Geol. Mag.* **74**, 507–524.
Westoll, T. S. (1938). Ancestry of the tetrapods. *Nature, Lond.* **141**, 127–128.
Westoll, T. S. (1943a). The origin of tetrapods. *Biol. Rev.* **18**, 78–98.
Westoll, T. S. (1943b). The origin of the primitive tetrapod limb. *Proc. R. Soc.* **B131**, 373-393.
Westoll, T. S. (1943c). The hyomandibular of *Eusthenopteron* and the tetrapod middle ear. *Proc. R. Soc.* **B131**, 393–413.
Westoll, T. S. (1977). Northern Britain. *In* "A Correlation of the Devonian Rocks in the British Isles" (M. R. House, ed.), *Spec. Rep. geol. Soc. Lond.* **8**, 66–93.
Westoll, T. S. (1979). Devonian fish biostratigraphy. *In* "The Devonian System", *Spec. Pap. Palaeont.* **23**, 341–353.
Young, G. C. (1974). Stratigraphic occurrence of some placoderm fishes in the Middle and Late Devonian. *Newsl. Stratigr. Leiden* **3**, 4, 243–261.

2 | Upper Palaeozoic Continental Distributions Based on Palaeomagnetic Studies

D. H. TARLING

Department of Geophysics and Planetary Physics, The University, Newcastle upon Tyne, England

Abstract: An evaluation of Upper Palaeozoic palaeomagnetic data indicates that reliable data are few and that magnetic overprinting is commoner than normally thought. This means that the palaeomagnetic data for any particular continental block at any one period are unreliable for detailed analysis, but consideration of the total changes in palaeolatitude and orientation allows consistent reconstructions to be drawn for 50-million-year intervals from Lower Devonian to Upper Permian. These indicate Laurentia and Gondwanaland in close proximity for much of the Palaeozoic, and Manchuria–China acting as a migration route between Australia and Arctica in the Devonian. Siberia is shown as an isolated block. Subsequent palaeogeographies are controlled by clockwise rotations of both Laurentia and Gondwanaland during the Devonian. Following the Acadian Orogeny, Laurentia begins to rotate anticlockwise, resulting in a North–South American collision at the end of the Carboniferous. This results in Gondwanaland rotating anticlockwise and fracturing of both supercontinents. Siberia has gradually been moving northwards but at a slower rate than the motion of Laurentia which intersects with it during Permian times. The position of the tectonic blocks of central Asia and south-eastern Asia are not clear until Permian times, but are thought to provide partial continental linkages between Australia and Iran during most of the Upper Palaeozoic. It is emphasized that the reconstructions provided now require to be tested and modified by other palaeontological and palaeoclimatic evidence.

Systematics Association Special Volume No. 15, "The Terrestrial Environment and the Origin of Land Vertebrates", edited by A. L. Panchen, 1980, pp. 11–37, Academic Press, London and New York.

INTRODUCTION

Study of the distribution of fossils led, in the nineteenth century, to many reconstructions of world palaeogeographies, utilizing extensive land-bridge connections to allow free access between continents now separated by thousands of kilometres (Schuchert, 1910). The existence of land-bridges formed by continental rocks became increasingly doubted in the twentieth century when geophysical studies showed that the low density of continental rocks would mean that their presence would be clearly indicated on geophysical surveys of the oceans. At the same time, deeper geophysical studies were being interpreted as indicating that continental drift movements could not take place as the Earth's mantle appeared too rigid for such motions to take place. Subsequent palaeontological investigations, particularly of Triassic vertebrates (Cox, 1965), led to increasing belief in the reality of continental drift, and palaeomagnetic studies, in the 1950s, convinced many people of the reality of drift. On the palaeontological evidence, in particular, it was recognized that the "southern" continents – South America, Africa, Madagascar, India, Ceylon, Australia, New Zealand and Antarctica – had formed a single continental unit, called Gondwanaland, since late Precambrian times until the Jurassic. Similarly, North America, Greenland and Europe were known to be united as the Laurentian continent which, in the late Mesozoic, combined with Siberia to form Laurasia. It was, however, palaeomagnetic studies of the ocean floors (Mason, 1958) which led to the dating of the oceans and to the confirmation of Hess's (1962) concept of ocean-floor spreading (Vine and Matthews 1963). Palaeomagnetic techniques have therefore been crucial in the acceptance of the previous contiguity of continental rocks, and they continue to provide the main potential for determining the past latitude and orientation of continental blocks relative to the Earth's axis of rotation for all times since the Archean. (Unfortunately Archean rocks have all been subjected to later metamorphism that has re-set any earlier magnetization and it is unlikely that reliable observations will be obtained from such old rocks.) However, while there is a major potential present in such investigations, surprisingly few data are available for Palaeozoic times and there are reasons for doubting the reliability of many of the available observations. The distribution of the continents in the Upper Palaeozoic is, however, fundamental to any considerations of the origin and evolution of terrestrial flora and fauna. While sedimentological and general geological characteristics are invaluable in reconstructing local palaeogeographies, the regional pattern can only be evaluated when the more general distribution of the continents is known, in particular their latitudinal distribution and contiguity. When such reconstructions are available, it is then possible to plot on them the distribution of land and sea for particular times. (This is not attempted in this more generalized survey.) With such information, it is possible to use standard meteorological principles to predict the distribution of climatic zones, not merely in terms of equatorial, tropical and polar, but the likelihood of a monsoon climate, rain-shadow areas and so forth. Such considerations are of prime importance as the distributions of the continents themselves have a major climatic influence because of continental effects on both air pressures and oceanic circulation.

In this analysis of predominantly Upper Palaeozoic palaeomagnetic data, with subsequent palaeogeographic reconstructions, an unconventional but systematic and objective approach has been taken in the selection of those data considered to be reliable and this has necessitated, at times, the rejection of all available observations for a particular period, other than as possible indicators of the true situation. It must also be emphasized that the restrictions on palaeomagnetic interpretations also mean that the optimum approach to palaeogeographic reconstructions must come from a synthesis of palaeomagnetic, palaeoclimatic and palaeontological data.

THE PALAEOMAGNETIC METHOD

The fundamental basis of palaeomagnetic observations (Irving, 1964; Tarling, 1971; McElhinny, 1973a) is that many rocks acquire a magnetization at or very shortly after their formation and, that this magnetization reflects the direction of the geomagnetic field at that particular time. Igneous rocks mainly acquired their magnetization as they cooled from a red-hot but already solid state at 600–700°C, down to ambient temperatures. This thermal remanent magnetization process can be extremely rapid in lavas erupted into water, but can occur over several years for thick lava sequences, or over several thousand years for plutonic rocks, particularly if the plutons are predominantly cooled by thermal conduction. Sediments may have acquired a magnetization by alignments of already magnetized particles in the prevailing geomagnetic field when these particles were deposited or while the particles were surrounded by interstitial water in the gradually compacting sediment. These detrital remanent magnetization processes can therefore occur instantaneously on deposition or within a few years or centuries of deposition, depending on the rate of compaction and diagenesis. However, sediments often acquire an additional chemical magnetization as detrital particles break down into components that include iron oxides, and also as cementing fluids deposit iron oxides within interstices. These original magnetizations, in both rock types, gradually decay although they can persist in particular grain sizes for many billions of years. The magnetic particles in rocks also gradually acquire time-dependent (viscous) magnetizations along the changing directions of the geomagnetic field. Even more complex magnetizations can also be superimposed by prolonged chemical changes, such as gradual exsolution, or when the rocks cool from being heated during a metamorphic episode or deep burial.

The magnetizations observed in rocks of different ages are therefore complex and it is usually essential to examine all components of magnetization that are actually represented within each rock sample. Such analyses are generally carried out by studying how their magnetization changes during partial demagnetization in gradually incremented alternating magnetic fields or during heating. In general, the time-dependent magnetizations will be readily removed in fields of 100–150 oesteds or temperatures of 100–150°C, and the direction of magnetization of the sample often moves away from the present geomagnetic field direction at the sampling site towards an earlier magnetic direction during this demagnetization range. In rocks that have not been subjected to any later chemical or thermal events, this direction will usually be the original geomagnetic direction at the time that the rock formed and this will remain constant, within instrumental error, until the magnetization is entirely destroyed by the demagnetization procedures. In many cases the magnetic direction does not stabilize until much higher demagnetization levels or it may indicate two or more components of stable magnetization, isolated at different demagnetization levels. In such cases it is often difficult to ascertain the relative ages of the different components. It is common practice, in such circumstances, to interpret the directions isolated at the highest temperature or alternating magnetic fields as being the oldest and hence most likely to be associated with the time of formation of the rock. This interpretation is based on the fact that the high temperature or high coercivity component is least likely to have been affected by the time-dependent magnetic decay processes or by later lower temperature heatings. However, this assumption is not necessarily valid as the high temperature-high coercivity component is primarily associated with magnetic particles of a precise grain-size range (monodomains), about 1 μm in diameter. If, for example, haematite grains of this dimension were formed by some later chemical process – the exsolution of ferromagnetic grains, breakdown of iron-bearing minerals and so forth – then the magnetization of such grains would be isolated at high fields or temperatures but would relate to the direction of the geomagnetic field at the time that the haematite formed. Haematite of slightly less stable magnetization, or magnetite, may well carry an original magnetization acquired when the rocks formed, yet this magnetization will be demagnetized before that of the later chemically acquired younger magnetic components.

If a previous magnetization has been successfully isolated, then this usually reflects the direction of the geomagnetic field at some particular time, e.g. when the rock formed. In certain rocks, however, the magnetic minerals have preferred orientations, as in some flow-banded trachytes, gneisses etc., and the direction of magnetization of the rock is then a compromise between the actual direction of the geomagnetic field and the direction along which the rock is most early magnetized. In most extant studies such anisotropic effects are minimal as it is only recently that direct studies have been initiated into such rock types and most previous sampling deliberately avoided rocks that were likely to be magnetically anisotropic. In general, therefore, this effect can be neglected, although it probably does affect some observations. More serious is the difficulty of determining the original horizontal in rocks that have been tilted or folded. Again, most studies have been on rocks in which the original horizontal can be ascertained by the local tilt of the bedding, but some results are from igneous bodies that have to be assumed to be still in their original position as there is no evidence to determine whether these have been subjected to later tilting. Fortunately for this review, most data are from sediments or lavas that have good control on their original orientation.

As the geomagnetic field is continually changing direction on a scale of some 2000–8000 years, it is necessary to average out these short-term changes in order to define the average direction of the geomagnetic field. Igneous rocks in which the magnetization was acquired rapidly require extensive sampling. For example, sampling through a complete lava sequence may be necessary to achieve such averaging. In most sediments, however, the process of diagenetic acquisition of remanence is protracted and the magnetization of a single sample may, in fact, already average out all such secular changes of the geomagnetic field during the time that the magnetization was acquired. When such averaging has been achieved, the direction of magnetization very closely approximates to a geomagnetic field attributable to a single dipole magnet at the Earth's centre that is aligned along the Earth's axis of rotation (an axial geocentric dipole). Such consistency can be demonstrated by direct palaeomagnetic observations on rocks formed during the last 20 million years, and the agreement with palaeoclimatic evidence indicates that this agreement (± 5°; Tarling, 1971) persisted throughout geological time. However, the observation that the geomagnetic field also changes polarity (approximately three times every million years during the last 60 million years) also means that some palaeomagnetic observations will also reflect the brief periods (probable some 3000–5000 years) during which the Earth's magnetic field was undergoing a polarity transition.

Clearly one final problem in any palaeomagnetic interpretation is that while the magnetization defines the direction of the geomagnetic field at the sampling site at a particular time, the corresponding palaeolatitudes and orientations relative to the polar axis can only be extrapolated to areas that are part of the same tectonic block, i.e. to areas that have remained in the same relationship to the sampling site. Observations from blocks on the same continent cannot, therefore, be combined if there are reasons for thinking these blocks have undergone relative motion since the magnetization was acquired. Conveniently, the palaeomagnetic data themselves can be used to assess whether such areas were in the same relative relationship, as an exact agreement in the location of the pole calculated from two different areas for the same period of time (or preferably for a range of different times) is unlikely to occur unless their past relationship was the same as today. When considering results for the Upper Palaeozoic, it is therefore clear that areas subjected to late Caledonian or younger orogenies are likely to have magnetizations that cannot be used to define past position of now adjacent areas.

The approach adopted here has been to examine published pole positions (over 8000) for rocks of Phanerozoic age. Data from rocks of uncertain age (i.e. their age estimate ranges over two or more complete periods) were eliminated from active consideration, although they were sometimes used to indicate the agreement, or lack of it, between more precisely dated observations. As much of the data have been subjected to a variety of treatments, it was not thought practicable to eliminate all data which had not been subjected to successful demagnetization procedures, unless the original author commented that no stable direction could be isolated. This is unfortunate but any adequate criteria for "successful isolation of the stable component" could only be applied to comparatively few studies. The data from the U.S.S.R., for example, are generally comprised of lists of poles with symbols indicating the treatment applied without the actual observations or further information. It is not possible to assess properly such data. The data therefore necessarily contain observations on which no demagnetization tests have actually been carried out. It was thus necessary to find some method, preferably objective, for determining which were the most reliable observations that could be related directly to the geomagnetic field at the time that the rocks were formed.

In order to avoid tectonic complications, data from within known orogenic belts were largely excluded (Tarling, 1969). Data, from individual tectonic blocks were then examined for each period (see the Australian Devonian, discussed later). Where the data appeared to be internally consistent, they were accepted as being possibly reliable indicators of the palaeolatitude and orientation of that block for one period. If there was a moderately clear grouping of poles for that period from that block, then any aberrant poles were initially rejected. The assumption here is, of course, that "aberrant" directions may have arisen from inadequate averaging of secular variations, the trapping of polarity transitions, the presence of unsuspected anisotropy of inhomogeneity, local tectonic disturbances, and either orientation or measurement errors. The definition of aberrant poles was necessarily subjective in that the precision of grouping differs from time to time and block to block, and hence the evaluation of what was considered aberrant also varied. Any apparent grouping of "aberrant" poles was also noted and averaged. The data from stable blocks within the same continent were then compared to see which blocks appeared to have been magnetized at a similar time in approximately their present relationship. Such comparison was again subjective but it was generally quite clear when the data from different blocks agreed or differed. The data from such blocks were then combined to obtain the mean pole position for those continental areas. An apparent polar wandering curve for that continent or continental block was then produced and evaluated.

The approach, up to this stage, was not significantly dissimilar to conventional approaches. However subsequent evaluation of the apparent polar wandering curves took into consideration such factors as the unlikelihood that polar wandering paths should oscillate or diverge on a short-term basis. For example, it was thought suspicious that the Silurian pole for Africa (Krs, 1978) should suddenly revert to the present North Pole (Fig. 1) and that the Silurian pole (Krs, 1978) for Asia (east of the Urals) should have moved eastwards, but that the older poles reverted to the previous path. Naturally there is no reason why a continent should not change its direction of motion relative to the pole, but it seems intuitively unlikely that a second change should result in an earlier motion being resumed. Similarly it was decided to suspect initially data in which older poles lay near or strung towards a younger pole position. Such a situation was considered to indicate the possibility that the observed data had arisen entirely, or in part, from a later remagnetization at the time of the younger pole. This suspicion is not, of course, fully justified as a continent in purely longitudinal motion relative to the pole at that time would have an apparently stationary pole position for the duration of that motion. Nonetheless, it was felt that such movements would be exceptional and warranted closer examination. It must be emphasized that such considerations were only used to isolate periods and places where it was thought that further examination for the presence of extraneous observations should be undertaken. Data were not, at this point, eliminated.

In many cases, the suspect pole distributions were thought to indicate the presence of some later magnetic overprinting. In particular, some groups showed distinctly "strung" distributions indicating the presence of more than one magnetic component. In such

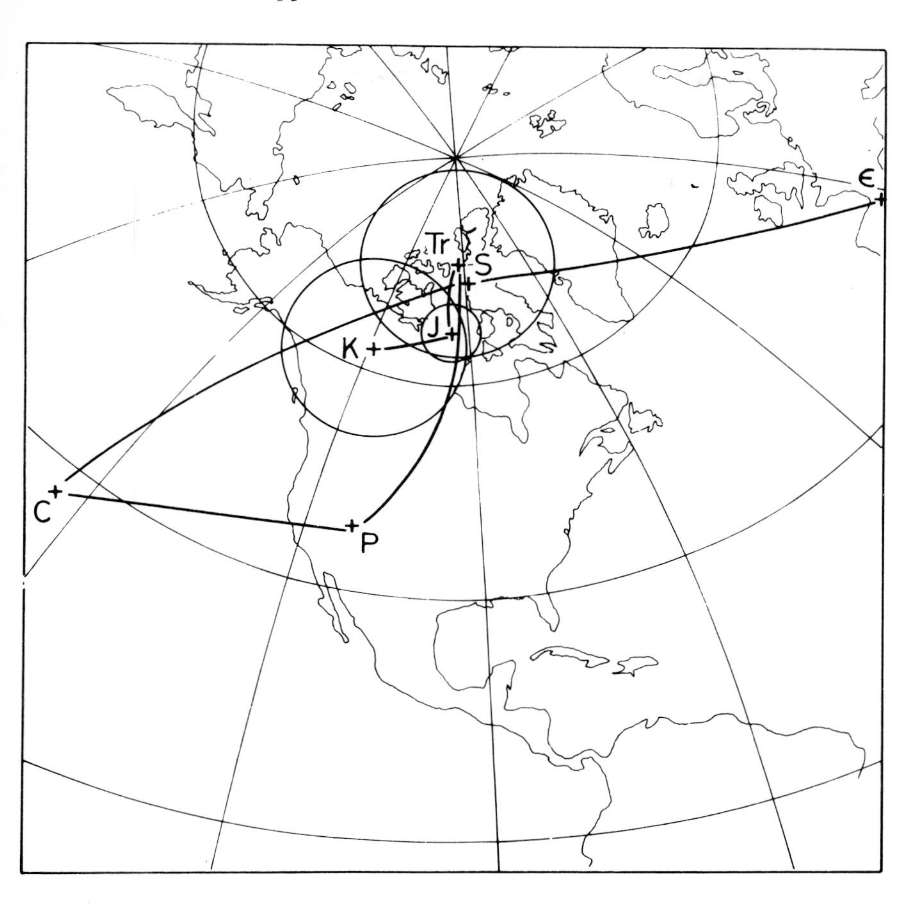

Fig. 1. The African polar wandering curve (after Krs, 1978). The "standard" method of determining a continental apparent polar wandering curve is to isolate the most magnetically stable palaeomagnetic observations and to take the statistical mean as representative of the palaeomagnetic pole for that continent for that period. An example of a recent analysis in this way is by Krs (1978) for Africa for the Cretaceous (K), Jurassic (J), Triassic (Tr), Permian (P), Carboniferous (C), Silurian (S) and Cambrian (Є) periods. No statistics are provided for the Palaeozoic pole positions as the data were too few. It is suggested here that the Palaeozoic pole positions calculated in this way can be seriously misleading as the method takes no account of possible later magnetic overprinting. (Krs's calculations were also on the basis that the Gondwanan glaciations were of Carboniferous age, and they are now known to occur within the Sakmarian stage of the Permian (Anderson and Schwyzer, 1977).)

circumstances, the one associated with a known, younger pole position was interpreted as being a component imposed later, and the polar positions relating to the geological age of the rocks were then evaluated from grouping(s) of poles away from the over-printed position or were estimated to be near the far end of the distribution. Each assessment was carried out independently for each period for each stable tectonic unit, and the estimated mean poles were then plotted to obtain an estimated apparent polar wandering curve for that continental unit.

The nature of these assessments is indicated here by consideration of the (1) Australian and (2) European Devonian data, although other specific areas and periods will be mentioned later when considering the reconstructions of Upper Palaeozoic geographies. The general assessment to determine common polar wandering paths is then illustrated (3) for the South American–African continents.

1. Australian Devonian Palaeomagnetic Data

Most of the Devonian pole positions are based on rocks sampled in south-eastern Australia, an area that was strongly affected by the Lower Carboniferous Kanimblan Orogeny, although some sites have been sampled in Queensland and in the Northern Territories (Fig. 2). The Queensland data are considered to be magnetically unreliable (Chamalaun, 1968) but all other sites are apparently magnetically stable and are generally considered to be of genuine Devonian age. The individual mean poles for each locality give very consistent palaeolatitudes (8°S. ± 11°). This suggests a southerly motion of Australia during the Devonian of some 1600 km and is consistent with the Lower Palaeozoic and later Upper Palaeozoic observations. The individual pole positions, however, are in very poor agreement, forming a distinct girdle from Rangoon to Manchuria, the Aleutians and the north-western United States. The conventional approach has been to accept the mean of these pole positions as being the best representative of the Devonian pole for Australia. The distribution of individual pole positions is not really consistent with this interpretation because the scatter in palaeolatitude is very small, but the rotations implied for the whole of Australia are very large and do not correlate with age differences of the rocks sampled. It is suggested here that the pole positions for south-eastern Australia have all been

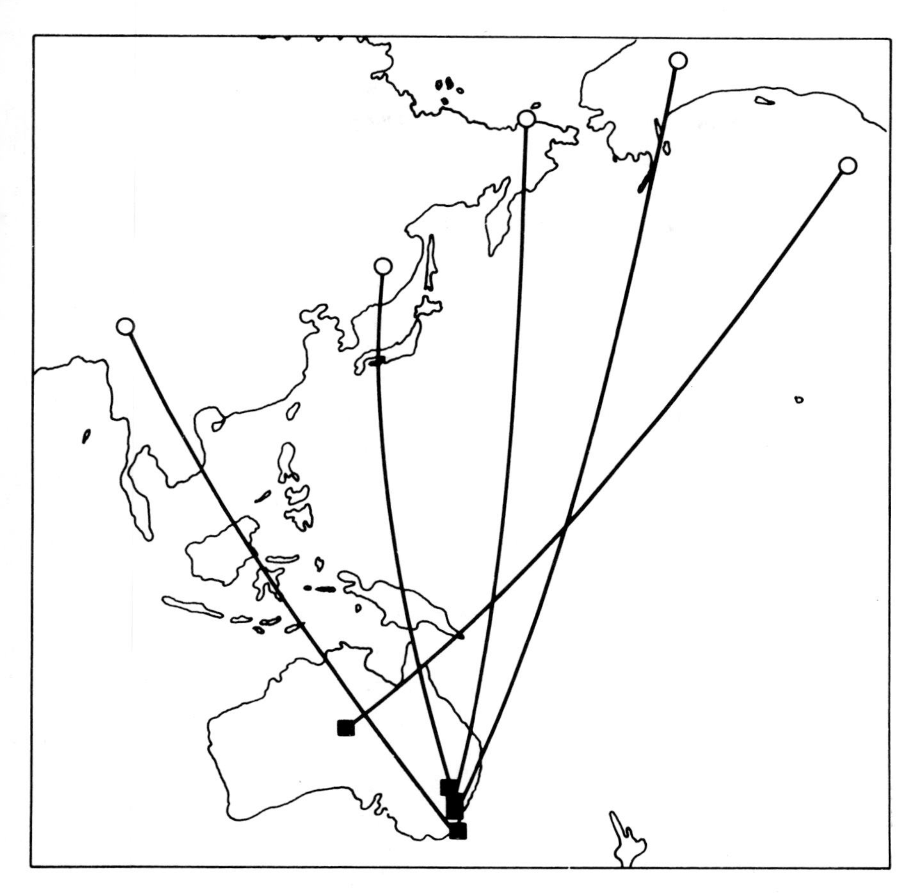

Fig. 2. Devonian poles and sites for Australia. The sites of Devonian age (squares) and their corresponding palaeomagnetic poles (circles) are shown. It is considered that the similarity in palaeolatitude (indicated by the length of the line joining the site and pole (suggests that they are all of genuine Devonian age, but that the wide sweep of pole positions indicates that the sites in south-eastern Australia have been subjected to localized post-Devonian anticlockwise rotations.

subjected to anticlockwise tectonic rotation during the Kanimblan or later orogenies so that the palaeolatitudes for these areas are valid, but their corresponding pole positions are no longer representative of the entire Australian continent. As the Queensland observations are unstable, it appears to be only the Mereenie Sandstones from central

Australia (Embleton, 1972) that provide a pole that can be considered representative of the pole position for most of the Australian block. However, even the age of this pole (42°N. 229°E.) is somewhat uncertain as these sediments could be of Silurian age and, on this basis, it is accepted, *pro tem*., as being probably of Lower Devonian age and more representative of the Devonian Australian pole position than the other data.

2. European Palaeomagnetic Data

A large quantity of data are available from Europe, even excluding areas that have been definitely affected by the Alpine and Variscan Orogenies, but the corresponding pole positions are widely scattered over most of the western Pacific (Fig. 3). There are moderately clear regional patterns within these polar distributions. Most poles from the Moscow, south Russian and Urals areas lie very close to the mean Carboniferous–Permian pole positions, which, if accepted, would indicate little or no latitudinal motion of these parts of Europe during the Devonian, Carboniferous or Permian. The pole positions for Devonian rocks from northern Russia (Timan) are quite different and lie east of New Guinea, well over 3000 km from the mean pole for the other far-eastern European poles. The north-western European data differ from both other groupings in that the majority of observations lie near to the Carboniferous-Permian grouping, but also form a broad swathe passing from it to the south and east.

It is considered here that this distribution indicates the presence of significant Permo-Carboniferous magnetic overprinting of the east European and north-western European data, as has already been suggested for certain formations (Tarling *et al*., 1976; Storetvedt, 1968) and as a distinct possibility for many other areas (Creer, 1968). On this basis, the true mean Devonian pole position is thought to lie somewhere close to the Equator, but with its longitudinal position uncertain. Consideration of Carboniferous and Lower Palaeozoic observations suggest that the average Devonian pole lies near New Guinea at about 2°N. 150°E. and that the Timan data indicate a Lower Devonian pole near 5°N. and 165°E. Such pole positions are distinctly different to the mean position yielded by an analysis of all reasonably stable Devonian data (marked "D" in Fig. 3) which corresponds with the "conventionally" adopted mean Devonian pole position for north-western Europe.

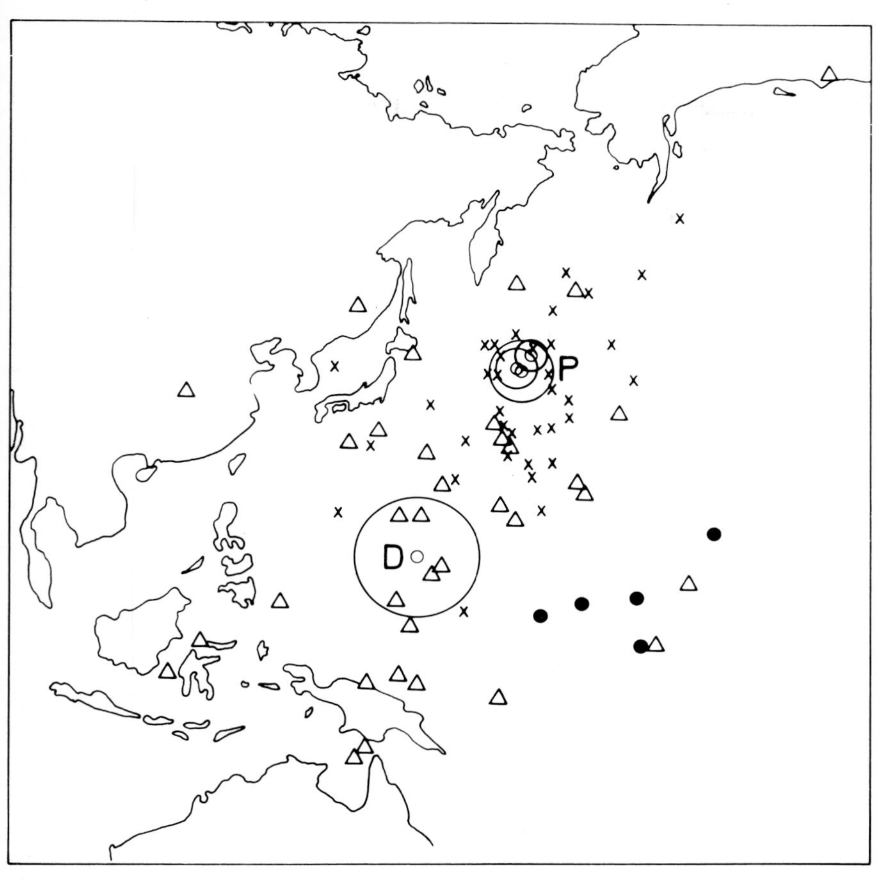

Fig. 3. Devonian poles for "Extra-Alpine" Europe. When poles from rocks in areas known to have been affected by Alpine rotations are excluded, the remaining poles show three main groupings. Those from the Moscow, Urals and southern U.S.S.R. areas (crosses) tend to group near to the mean European pole positions for Upper Carboniferous, Lower, Middle and Upper Permian times (circles, with confidence limits of 95% confidence, indicated by P). Poles from northern Russia (dots) lie near the Equator, while those from north-western Europe (triangles) cover much of the western Pacific, with a mean pole and circle of 95% confidence (D) well away from the Russian groups.

3. "Western" Gondwanan Polar Wandering Curve

As there is unanimity concerning the palaeogeographic reconstructions for South America and Africa for pre-Jurassic times, these two

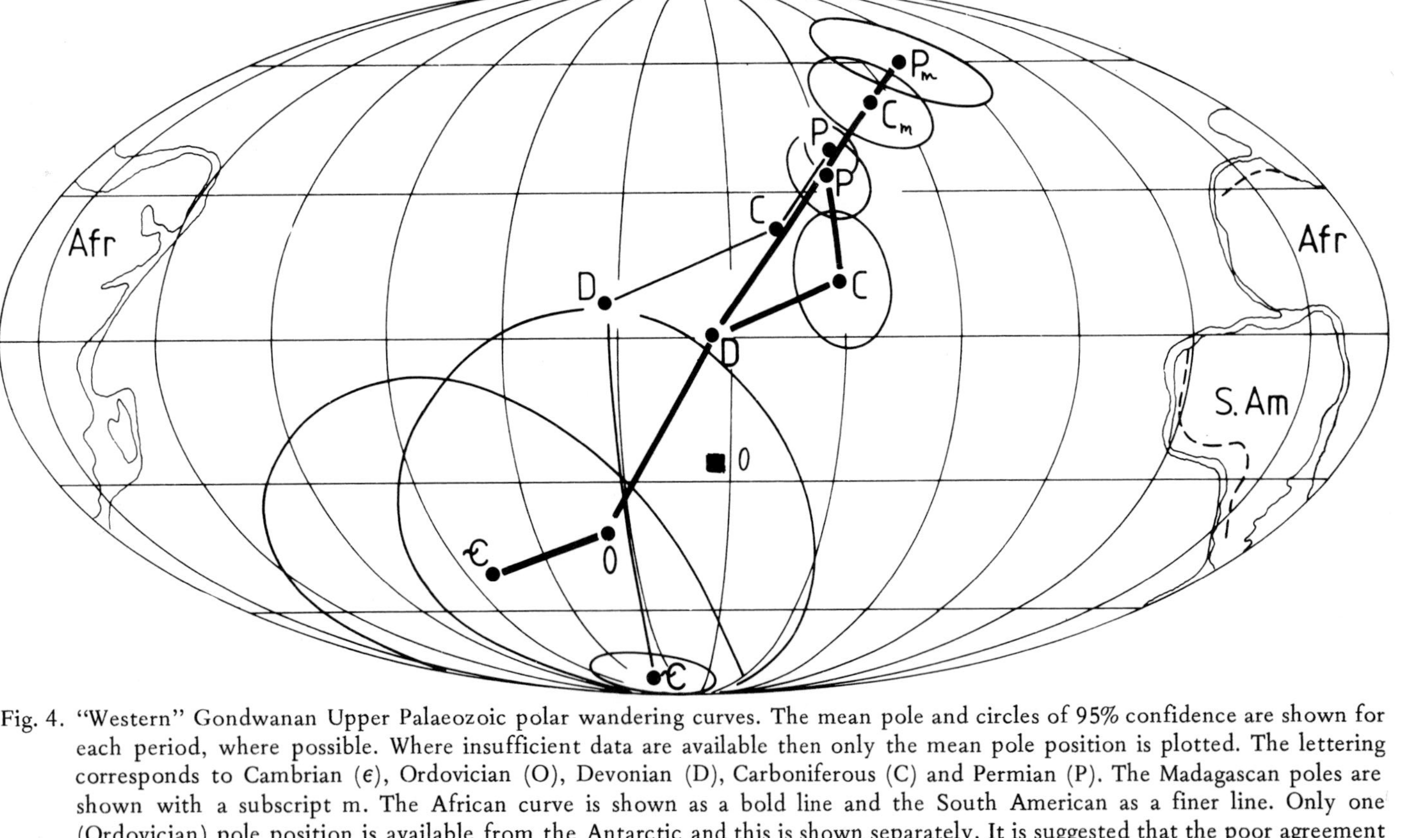

Fig. 4. "Western" Gondwanan Upper Palaeozoic polar wandering curves. The mean pole and circles of 95% confidence are shown for each period, where possible. Where insufficient data are available then only the mean pole position is plotted. The lettering corresponds to Cambrian (Є), Ordovician (O), Devonian (D), Carboniferous (C) and Permian (P). The Madagascan poles are shown with a subscript m. The African curve is shown as a bold line and the South American as a finer line. Only one (Ordovician) pole position is available from the Antarctic and this is shown separately. It is suggested that the poor agreement between these two paths for a widely accepted reconstruction indicates the need for caution in interpretation of the palaeomagnetic data, but also that a consistent path can be derived for these combined units, the reliability of which can then

continents should have an identical Palaeozoic polar wandering path when South America, and its corresponding polar wandering path, are rotated to Africa by 56·1° about a Euler pole at 43·9N. 30·1W (Bullard *et al.*, 1965). (Such a rotation does not, in fact, reproduce the actual motion of the two continents since the Jurassic, but is a summation of all such motions.) This fit provides an excellent match for both Permo-Triassic and Precambrian features on both continents (Tarling, 1980) but it is difficult to reconcile precisely with the currently determined Palaeozoic paths for the two continents (Fig. 4). The Lower Palaeozoic data from South America are entirely from western Argentina and are mostly only poorly defined both individually and as separate age groups. The Cambro-Ordovician poles, although poorly defined, lie near to the Sahara on the reconstruction and these are accepted purely on the basis that they are supported by the evidence for Ordovician ice-sheet glaciation in the Sahara. There are several observations for rocks of probably Cambrian age from Africa, some of which are better defined palaeomagnetically than the South American data, but their age is often even less certain. Positions of four of the seven poles lie loosely grouped together near Australia and are assumed to be broadly indicative of the location of the Cambrian African pole. Technically, the South American and African Cambrian poles are, in fact, identical at a 95% probability level as the South American mean pole and its circle of 95% confidence are enclosed within that of the African 95% confidence limits. However, the apparent precision of the mean South American pole position is itself misleading as its components are themselves only very poorly defined, so that the common Cambrian pole is, so far, only defined within a quadrant.

Two Devonian poles are available from South America and one from Africa, and their mean position is thus only an indication of the possible actual pole position. On the reconstructed "Western" Gondwanaland, this mean South American pole and the African Devonian pole are over 3000 km apart, but are both approximately consistent with the possible location of the palaeoclimatic South Pole in the central or west African area. The Carboniferous data from Africa are predominantly from the north and west and are of Lower to Middle Carboniferous age, while most of the South American Carboniferous data are of Upper Carboniferous or Permo-Carboniferous age. Both the African and South American mean poles are somewhat closer to each other than their Devonian counterparts, but still differ by some 2000 km. In contrast, the Permian data are very consistent with each other. In summary, therefore, while the palaeomagnetic data must certainly be considered to be more consistent with the geometric–geological fit of the two continents than with the continents in their present relationships, the actual polar paths are extremely badly defined for the Lower Palaeozoic and only poorly defined for Upper Palaeozoic times. It is also relevant that the Madagascan Permian and Carboniferous data are not consistent with those from the mainland of Africa, but the imprecision of available data inhibit objective analysis (Tarling, in preparation). The pole positions for "Western" Gondwanaland are thus still poorly defined although generally consistent with drift of the supercontinent of some 180° during this entire era. It is therefore essential that the available palaeomagnetic determinations should be supplemented by palaeoclimatic and palaeontological evidence to determine the probable location of the rotational pole at any one time and to decipher the relation of this part of Gondwanaland with other regions of the world.

UPPER PALAEOZOIC RECONSTRUCTIONS

The four reconstructions presented here are based entirely on palaeomagnetic data for North America, Europe, Siberia and Gondwanaland. However, these data have been analysed in a subjective manner and are necessarily schematic. Nonetheless, the general trends of the palaeomagnetic data are clear and reduce the available options for any one period as obviously each reconstruction must be consistent with both the previous and subsequent history of the areas concerned. If "abnormal" behaviour is assumed to be rare and fairly smooth, gradual changes in continental motion being taken to indicate the average motion during 50-million-year intervals, then the following reconstructions are thought to be reasonably accurate as indications of the major continental relationships even though the individual errors are large and other palaeoclimatic and palaeontological data are essential for a further evaluation.

1. Lower Devonian (Fig. 5)

The orientation and palaeolatitude of Gondwanaland at the start of the Devonian is quite well defined in that the position of the south rotational pole must clearly lie within or near central Africa on the

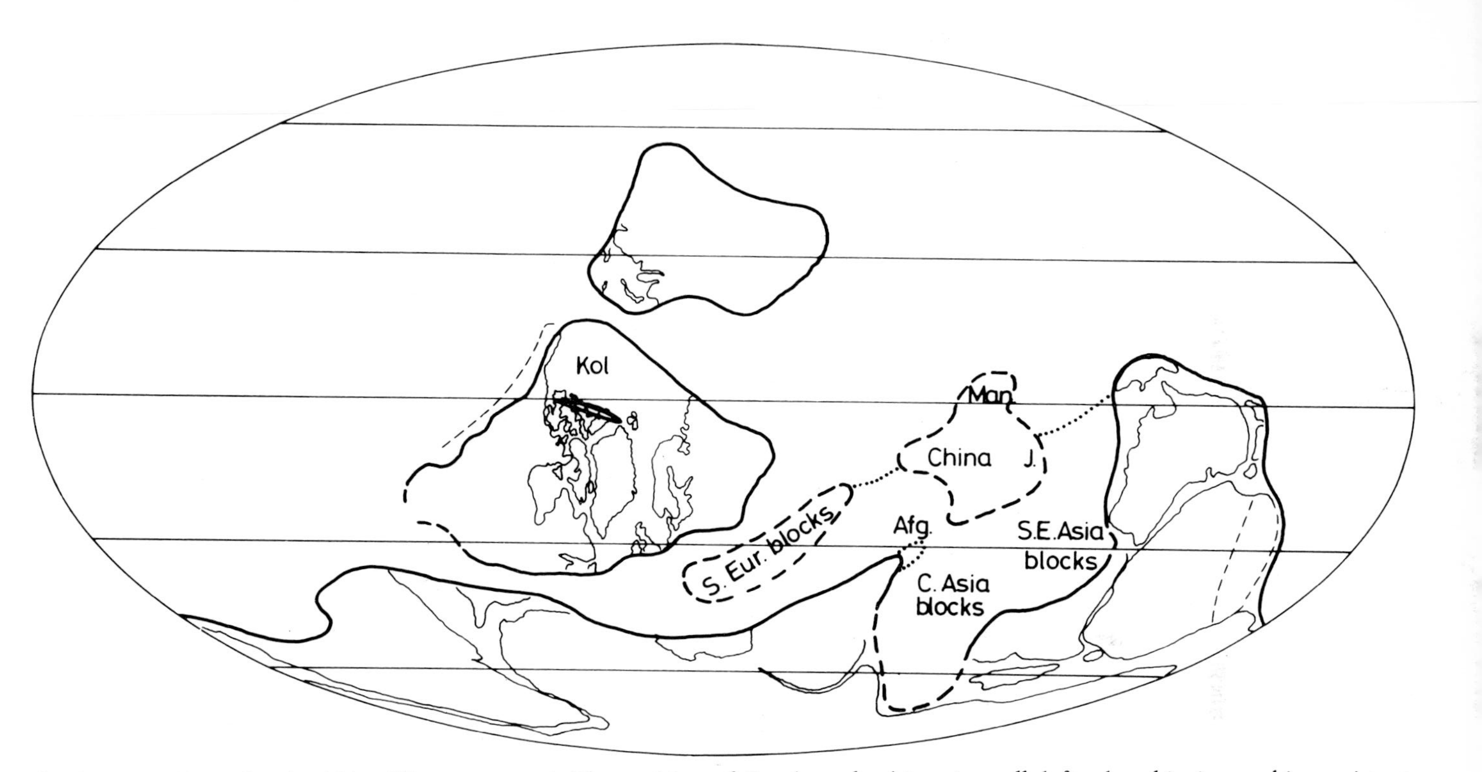

Fig. 5. Lower Devonian (*c.* 400 million years ago). The position of Gondwanaland is quite well defined at this time and is consistent with the interpretation of Laurentian palaeomagnetic data. The positions of the China–Manchuria, south European, central Asian and south-eastern Asian blocks are not clear at this time. It is thought that most of these blocks were somewhat nearer Australia than in Permian times (Fig. 8). Kol. is the Kolyma block which is now part of the far eastern U.S.S.R.; Man., China and J. are the Manchurian, north and south Chinese and Japanese tectonic blocks; Afg. is Afghanistan, but the central (C) Asian and south-eastern (S.E.) Asian blocks are undifferentiated.

basis of palaeoclimatic evidence for the Ordovician (Beuf *et al.*, 1968) and Permian (Anderson and Schwyzer, 1977) glaciations in that continent, and such a polar location is consistent with the palaeomagnetic evidence from South America, Africa, Australia and India, although the available data can only be considered to confirm the Devonian pole in that general position. However, when this position of Gondwanaland is considered in both earlier and later contexts, and with the postulated position of the Laurentian continent, the possible alternatives are few if north-western Africa and eastern North America are considered to have been in very close proximity at the end of the Caledonian and Taconic Orogenies and then to have been closed yet again during the later Acadian Orogeny (Tarling, 1979). Such a close proximity is also indicated by the distribution of the Appalachian faunal province (Boucot *et al.*, 1969) during the Lower and Middle Devonian. The data from Laurentia for this period are, at first glance, plentiful, but on closer examination the really genuine Lower Devonian pole positions are not well known for either Europe or North America. Nonetheless, assuming these continents to be united (House, 1971), their palaeomagnetic data are consistent enough to provide both orientation and palaeolatitudes of this supercontinent. The position of the south European continental block is not clear on palaeomagnetic data. It is probable that it comprised several separate blocks, some of which were partially united with Gondwanaland or Laurentia, and others that were isolated from both. It is certain that these tectonic blocks were not in their present position relative to each other at this time and they have, until further data are available, been placed in their possible Permian relationship with each other. The changing positions of these units through the Upper Palaeozoic was, in fact, critical in determining the location of the different Variscan orogenic events (Tarling, 1979) and it seems probable that their "eastern" extremities may have been crucial as Devonian migration routes from China-Manchuria-Japan to the Arctic areas of northern Europe and North America which, at the start of the Devonian, were all in similar equatorial latitudes. The Kolyma block had, at the start of the Devonian, not quite joined with the Canadian Shield, but did so during the Lower Devonian and remained in that position until the opening of the present Canadian Arctic Ocean in the Jurassic when this block moved out to collide eventually with the Siberian block in Cretaceous times, forming the Verkhoyansk Mountains.

The Siberian block is shown isolated from all other continental blocks, although its longitudinal relationship with Laurentia is not known. It could thus be moved to have partial contact with either the Kolyma block or, via island arcs etc., with Manchuria or even Australia. Its orientation and palaeolatitude are apparently well defined for this period. It is not clear, from the palaeomagnetic data, whether Kazakhstan was, at this time, a part of the Siberian block or not. The various components of the Siberian block appear to have become largely united by, or during, the Caledonian Orogeny, and Kazakhstan may also have formed part of this block at this period. Some palaeomagnetic data are available for this area, but appear to show evidence of very young, possibly even Cretaceous, overprinting. It is not felt, at this stage, that the palaeomagnetic data are adequate to evaluate its true position during most of the Devonian and Carboniferous, and it is therefore treated as part of the Siberian block.

Some Devonian palaeomagnetic data are available from Manchuria, Japan, China and south-eastern Asia, but these are mostly initial results with no demagnetization or other stability tests and the published values are close to those for Permian times. The position adopted here is based on Permian and Mesozoic data which indicate that Manchuria, China, Japan, together with the various tectonic blocks of south-eastern Asia, lay nearer to the Australian land mass than today (McElhinny, 1973b; Haile, 1978). Indeed, it is possible that these units formed a single, but loosely co-ordinated, tectonic unit that was in partial connection with Australia during much of the Palaeozoic and early Mesozoic. The positions of the central Asia blocks (Tarim, Dzungaria etc.) are completely unknown palaeomagnetically, but the geological similarities between Iran and Afghanistan for much of this period indicate that these, and possibly other blocks, were in a similar (but certainly not identical) relationship to India and Arabia as today. It is suggested here that all Asian units remained in similar positions relative to India and Australia for much of the Palaeozoic, although breaking off from Gondwanaland during the early Mesozoic and providing the units against which India initially collided in the Cretaceous, prior to further collisions with Asia in the Tertiary.

It is difficult to overstate the uncertainties in this reconstruction. The known Palaeozoic deformations in South America, western Antarctica, south-eastern Australia almost certainly caused modifications in the shape of Gondwanaland. Small Asian units probably lay off western North America, together with the southern part of Alaska. However, it is considered that this reconstruction is at least moderately consistent with the sparse palaeomagnetic data actually available and with the subsequent evolution of the larger units, at least, during the Carboniferous and Permian.

2. Lower Carboniferous (Fig. 6)

During the Devonian, Gondwanaland was rotating clockwise so that "Western" Gondwanaland was moving northward while "Eastern" Gondwanaland was moving south. Laurentia was also rotating clockwise but its main motion was northerly. These relative motions resulted in the Mid-Upper Devonian closure of the very narrow "Atlantic" Ocean that had been open since Taconic times. This closure formed the Acadian Orogeny and also destroyed the isolation of the Appalachian trilobite province. The reaction of the two supercontinents to this collision resulted in the previous slight clockwise motion of Laurentia being changed to anticlockwise in Upper Devonian times although Laurentia continued to move predominantly northwards. The Acadian collision only slightly affected the motion of Gondwanaland, which continued its clockwise rotation even after the Acadian Orogeny. The new relative motion again re-opened the "Atlantic" Ocean as a long narrow strip, possibly never more than 200–300 km wide and therefore readily bridged by island arcs or continental fragments. Indeed, it is possible that the changes in motion only ended the compressional stress along the line of the Acadian Orogeny, rather than re-opening a significant ocean gap. The tectonic effects of the relative motion between the two supercontinents thus became concentrated on blocks, such as the Iberian Massif, that had become wedged between the two main continental plates. Siberia only moved slightly further northwards during this time, with little or no rotation. It is not clear what was happening in the China–Manchuria–Japan–south-east Asia–central Asia areas. It is suggested that while they remained as a poorly coherent unit, some internal

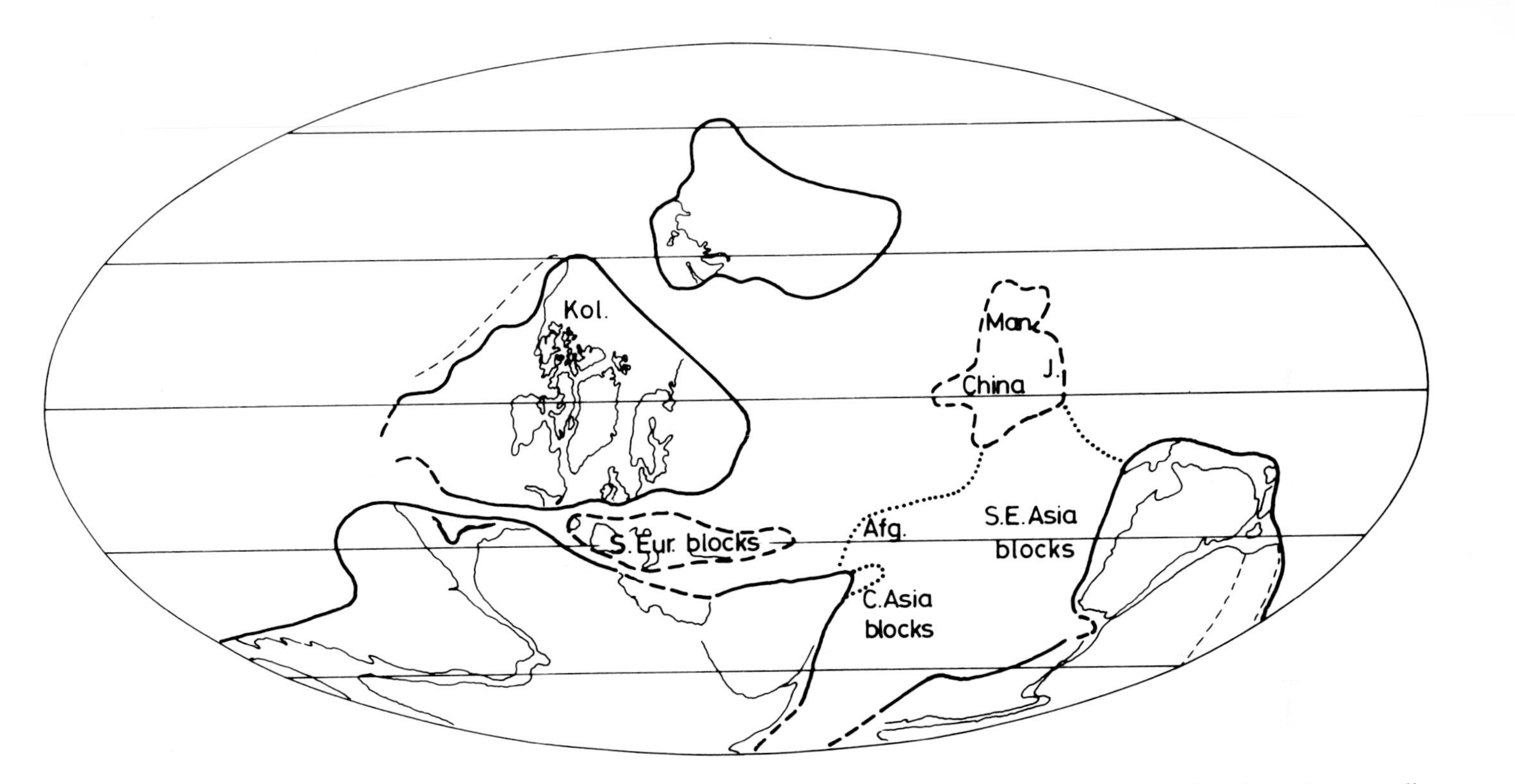

Fig. 6. Lower Carboniferous (*c.* 350 million years ago). Both the Laurentian and Gondwanan positions are thought to be mutually consistent, and the latitudinal position of Siberia is also well constrained. It is thought that the China–Manchuria blocks had not yet moved to their approximately Permian positions relative to Australia, although there are no palaeomagnetic data for this period.

motions took place, possibly extending the latitudinal span of the area and possibly creating only restricted barriers to migrations between Australia and Laurentia towards the end of the Devonian.

3. Upper Carboniferous (Fig. 7)

During all of the Carboniferous, Laurentia was reacting to the Acadian Orogeny by an anticlockwise rotation and slightly northerly motion which resulted in southern North America remaining in a similar latitude through this period, although most other parts were being carried into somewhat more northerly latitudes. Gondwanaland, however, was still rotating clockwise, carrying "Western" Gondwanaland northwards and Australia towards the South Pole. Both supercontinents were only in partial tectonic contact in the Iberian section of the south European block and thus this area was subjected to both north-south compression and right-lateral displacements. The Siberian block continued its very slow northerly motion and may have been in partial contact with Laurentia towards the end of the Carboniferous, although essentially isolated from it – however its exact lateral relationship can only be assessed by non-palaeomagnetic methods. There is no palaeomagnetic information available that can help determine the situation in China–central Asia etc., and it is assumed that these were still in essentially their same relationship with India–Australia as in Devonian times.

4. Upper Permian (Fig. 8)

At the very end of Carboniferous times, the anticlockwise rotation of Laurentia and clockwise rotation of Gondwanaland resulted in a major collision between northern South America and southern North America (King, 1975). This resulted in a drastic change in the relative motions of the two supercontinents. Although both continents, in the Mediterranean region, continued to have a northerly motion, the anticlockwise rotation of Laurentia became initially reduced and Gondwanaland began to rotate clockwise. There was thus a differential motion between Laurentia and Gondwanaland that closed the area of the Acadian Orogeny, forming the final phase of the Variscan Orogeny. At the same time, the Adriatic Promontory and associated parts of the south European blocks collided with the Iberian block, giving rise to the main Variscan events in central and

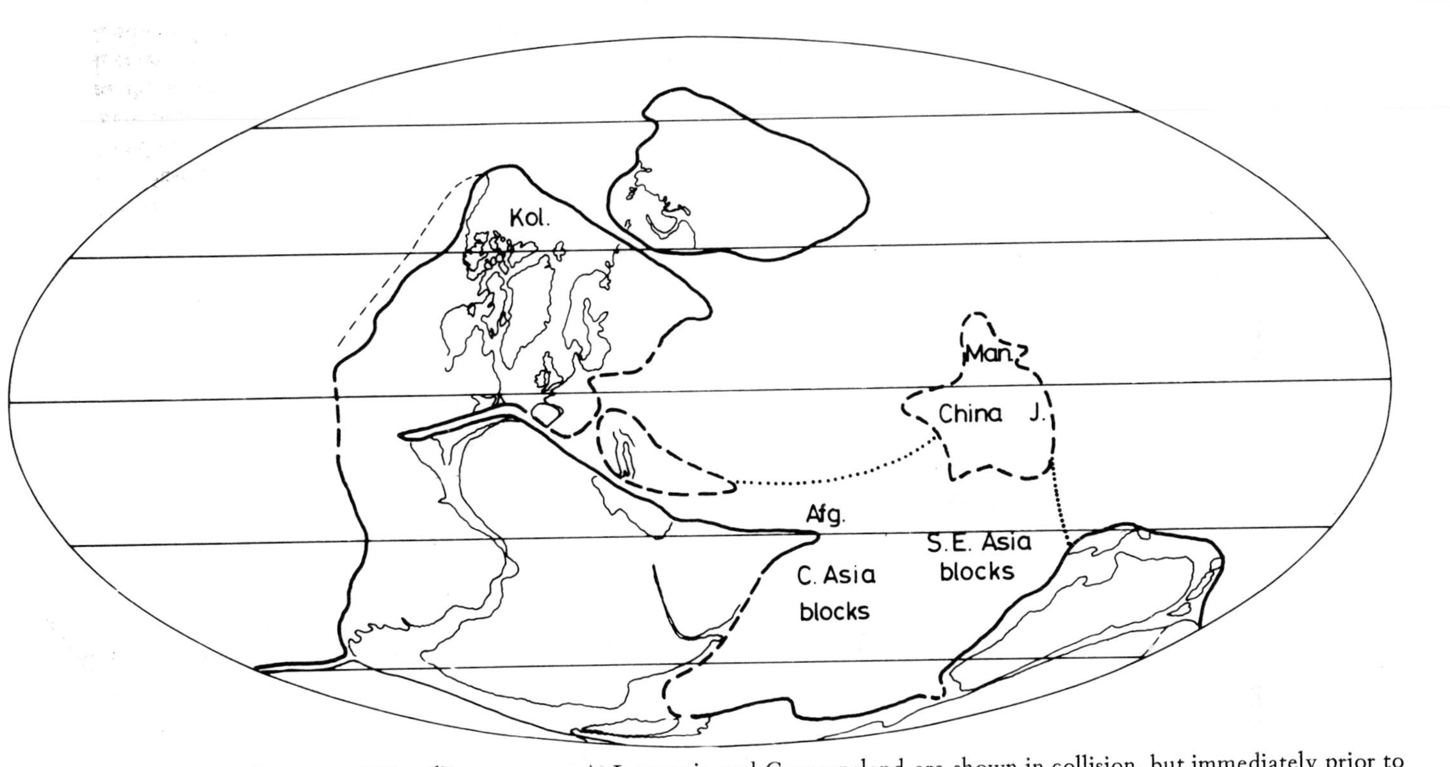

Fig. 7. Upper Carboniferous (*c.* 300 million years ago). Laurentia and Gonwanaland are shown in collision, but immediately prior to the change in rotation of Gondwanaland that was to create the main Variscan tectonic events in Europe and eastern North America. The western edges of the south European blocks are thought to have become attached between Laurentia and Gondwanaland, but the Adriatic Promontory had not yet collided with this region. The position of Siberia is still undefined in longitude but its consistent northward motion is interpreted as indicating that it had not yet been in collision with any major tectonic block.

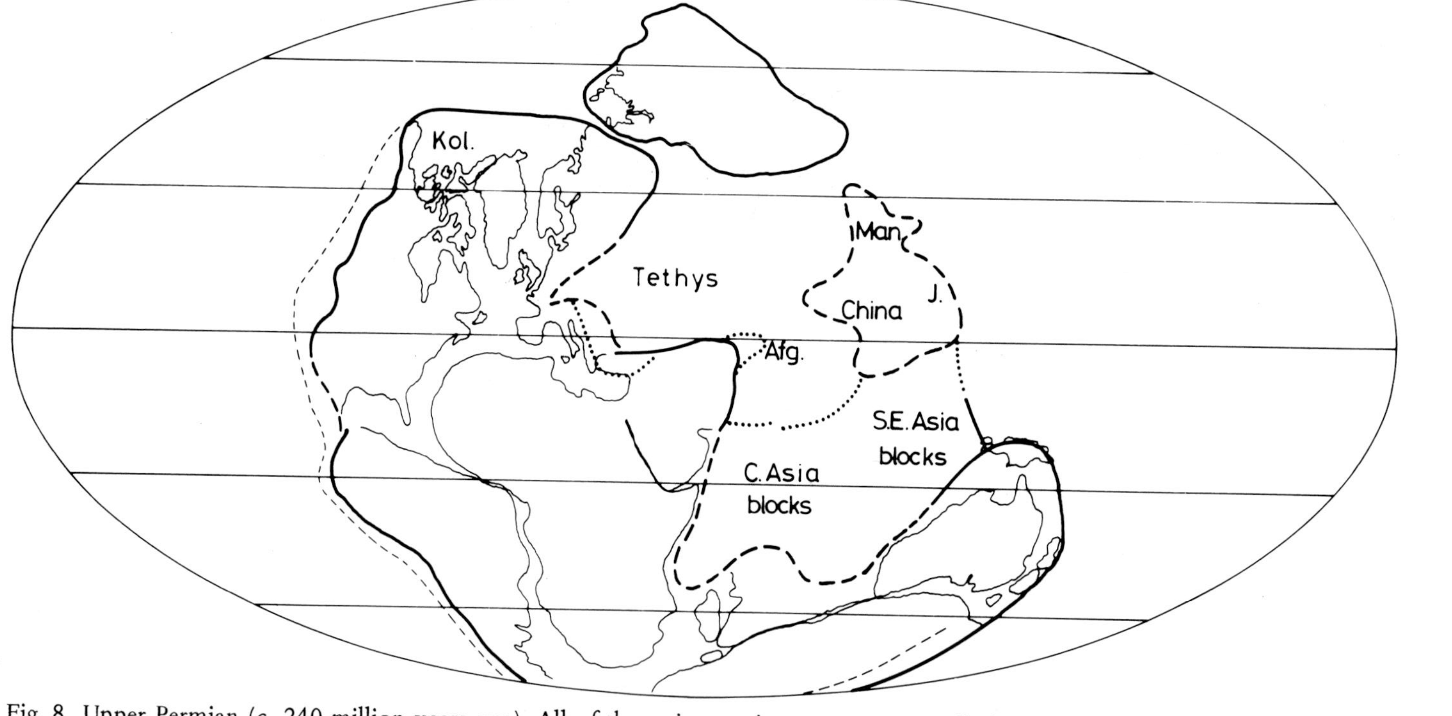

Fig. 8. Upper Permian (*c.* 240 million years ago). All of the major continents were essentially linked as Pangaea, although Siberia was only just beginning to become attached to Europe and the China–Manchuria block was probably only loosely linked with Gondwanaland via Australia. Although geographically forming one major continental block, both Laurentia and Gondwanaland were in motion relative to each other and both continents were undergoing anticlockwise rotation, creating complex tectonics in the Mediterranean region in particular. It is emphasized that while this reconstruction is thought to indicate the general location of the main continental blocks, the distribution of fauna and flora is likely to have been strongly controlled by both land–sea distributions and climatic zonations for this and the earlier reconstructions.

southern Europe. As a consequence of these motions, the two supercontinents formed the continent of Pangaea, but this was a temporary phenomenon in that the two continents continued to move relative to each other during the Permian. It seems probable, however, that the drastic changes in motion of the two supercontinents were responsible for the widespread fracturing (Kent, 1976) that was often the site of ocean formation during the Mesozoic. At the start of the Permian, Siberia may have only been in partial contact with Laurentia, although these two continents collided gradually during the Permian, forming the Laurasian continent by the end of the Permian.

It seems probable that the changes in the sense of rotation of Gondwanaland had drastic effects on the central and south-eastern Asian blocks lying between India and Australia. Although there is no palaeomagnetic evidence, it is suggested that some of these areas began to separate from Gondwanaland during the Permian, although the main Chinese-Manchurian block, at least, remained loosely connected to the Australian continent via some of the south-eastern Asia blocks.

CONCLUSIONS

The palaeomagnetic data available for evaluating the distribution of the continents for any period in the Palaeozoic are both few and unreliable. In any case, even well-dated palaeomagnetic data of high reliability must nevertheless be treated cautiously. However, while such caution is necessary for the evaluation of the situation for any one period, there is sufficient data for the entire Upper Palaeozoic to determine moderately consistent reconstructions for intervals of approximately 50 million years. These broader controls appear to be capable of explaining many of the tectonic events of the Earth's history during this time, but it is essential that other data (palaeoclimatic and palaeontological) be examined to test the reality of these constructions for particular areas at particular times. It should also be emphasized that while the general distribution of the main continental blocks is becoming clearer, slightly smaller scale features, such as land–sea distributions, climatic belts, island arcs etc., may have had a controlling influence on the detailed distribution of organisms. It is not, however, possible to evaluate the significance of these or other features until the general pattern is better defined, and that requires a synthesis of all geological and geophysical observations.

REFERENCES

Anderson, J. M. and Schwyzer, R. V. (1977). The biostratigraphy of the Permian and Triassic. *Trans. geol. Soc. S. Afr.* **80**, 211–234.

Beuf, S., Biju-Duval, B., Mauvier, A. and Legrand, P. (1968). Nouvelles observations sur le "Cambro-Ordovicien" du Bled el Mass (Sahara Central). *Sci. geol. Algerie, Bull.* **38**, 39–52.

Boucot, A. J., Johnson, J. G. and Talent, J. A. (1969). Early Devonian brachiopod zoogeography. *Geol. Soc. Am. spec. Pap.* **119**, 113.

Bullard, E. C., Everett, J. E. and Smith, A. G. (1965). The fit of continents around the Atlantic. *Phil. Trans. R. Soc.* **A258**, 41-51.

Chamalaun, F. H. (1968). The magnetization of the Dotswood Red Beds (Queensland). *Earth Planet. Sci. Lett.* **3**, 439–443.

Cox, C. B. (1965). New Triassic dicynodonts from South America, their origins and relationships. *Phil. Trans. R. Soc.* **B248**, 457–514.

Creer, K. M. (1968). Palaeozoic palaeomagnetism. *Nature, Lond.* **219**, 246–250.

Embleton, B. J. J. (1972). The palaeomagnetism of some palaeozoic sediments from central Australia. *J. Proc. R. Soc. N. S. W.* **105**, 86ff.

Haile, N. S. (1978). Reconnaissance palaeomagnetic results from Sulavesi, Indonesia, and their bearing on palaeogeographic reconstructions. *Tectonophysics* **46**, 77–85.

Hess, H. H. (1962). History of ocean basins. *In* "Petrologic Studies" (A. E. J. Engels *et a.*, eds, pp. 599–620. Geological Society of America.

House, M. R. (1971). Devonian faunal distributions. *In* "Faunal Provinces in Space and Time" (F. A. Middlemass *et al.*, eds), pp. 77–94. Seel House Press, Liverpool.

Irving, E. (1964). "Palaeomagnetism and its Application to Geological and Geophysical Problems", 399pp. Wiley, New York.

Kent, P. E. (1976). Major synchronous events in continental shelves. *Tectonophysics* **36**, 87–91.

King, P. B. (1975). The Ouachita and Appalachian orogenic belts. *In* "The Ocean Basins and Margins", Vol. 3, "The Gulf of Mexico and the Caribbean" (A. E. M. Nairn and F. G. Stehli, eds), pp. 201–241. Plenum, New York.

Krs, M. (1978). An analysis of the palaeomagnetic field. *Studia geophys. geod.* **22**, 368–389.

McElhinny, M.W. (1973a). "Palaeomagnetism and Plate Tectonics", 358pp. Cambridge University Press.

McElhinny, M. W. (1973b). Palaeomagnetism and plate tectonics of eastern Asia. *In* "The Western Pacific : Island Arcs, Marginal Seas, Geochemistry" (P. J. Coleman, ed.), pp. 407–414. University of Western Australia Press.

Mason, R. G. (1958). A magnetic survey off the West Coast of the United States between 32° and 36°N. and longitude 121° and 128°W. *Geophys. J.* **1**, 320–329.

Schuchert, C. (1910). Biologic principles of paleogeography. *Pop. Sci. Mon.* **76**, 591–600.

Storetvedt, K. M. (1968). On remagnetization problems in palaeomagnetism. *Earth Planet. Sci. Lett.* 4, 107–112.

Tarling, D. H. (1969). The palaeomagnetic evidence of displacements within continents. *In* "Time and Place in Orogeny" (P. E. Kent *et al.*, eds), pp. 95–113. Geological Society of London.

Tarling, D. H. (1971). "Principle and Applications of Palaeomagnetism", 164pp. Chapman and Hall, London.

Tarling, D. H. (1979). Palaeomagnetic reconstruction and the Variscan Orogeny. *Proc. Ussher Soc.* 4, 233–261.

Tarling, D. H. (1980). The geologic evolution of South America during the last 200 million years. *In* "Evolutionary Biology of the New World Monkeys and Continental Drift" (R. L. Ciochon and A. B. Chiarelli, eds). Plenum, New York (in press).

Tarling, D. H., Donovan, R. N., Abou-Deeb, J. M. and El-Batrouk S. I. (1976). Palaeomagnetic dating of haematite genesis in Orcadian Basin sediments. *Scott. J. Geol.* 12(2), 125–134.

Vine, F. J. and Matthews, D. H. (1963). Magnetic anomalies over oceanic ridges. *Nature, Lond.* **199**, 947–949.

3 | Carboniferous Geography and Terrestrial Migration Routes

G. A. L. JOHNSON

Department of Geological Sciences, University of Durham, England

Abstract: Convergent drift of continental plates caused a continual pattern of changes in world geography during Upper Palaeozoic times. Initially this caused accretionary growth of the Laurasian continent, but by the end of the era the supercontinent of Pangaea was formed by the rafting together of Laurasia with Asia and the southern continent of Gondwanaland. The first evolutionary radiation of the land flora, non-marine invertebrates and tetrapods took place during this period of coalescing continents and their geographical distribution is linked, to some extent, to this pattern of change. Of these groups the land flora was relatively cosmopolitan early in the Carboniferous and was differentiated into five kingdoms by the Permian. The non-marine invertebrates show more limited migration within a continent and the tetrapods were similarly restricted to Laurasia except for two records in Gondwanaland (south-east Australia). In Laurasia, Carboniferous and Permian tetrapod migration has a southerly trend following the Equator as the continent drifted northwards. Migration onto new southerly extensions to Laurasia became possible as marine barriers closed when continental plates were rafted together. This can be demonstrated in south Nova Scotia and in south Europe. Elsewhere newly uplifted mountain chains formed barriers to migration, and climatic changes, caused by drifting continents, produced progressively changing limits to the early tetrapod habitat.

INTRODUCTION

The Carboniferous world was dominated by three vast continents that straddled the climatic zones almost from pole to pole. Lying well within the tropics, Laurasia, made up of Russia west of the Urals,

---

Systematics Association Special Volume No. 15, "The Terrestrial Environment and the Origin of Land Vertebrates", edited by A. L. Panchen, 1980, pp. 39–54, Academic Press, London and New York.

Europe, Greenland and North America, had all the characteristics of an equatorial continent and remained in this position throughout the Carboniferous. Further north, Angaraland, formed of Asiatic Russia, Siberia and China lay partly in the temperate zone and stretched northwards towards the Arctic. Mainly to the south, the major continent of Gondwanaland was formed of Africa, South America, Australia, India and Antarctica. These continents were not stationary through the Carboniferous, drift and rotational movements affected all of them, causing Laurasia to drift gently northwards, Angaraland to drift and rotate southwards towards Laurasia, and Gondwanaland to drift and rotate more rapidly southwards so that part of it came to lie near to the South Pole before the end of the period. Together with these major continents there were smaller continental fragments that were also capable of drift movement. Several of these fragments lay to the south of Laurasia and were successively rafted against Laurasia during the Devonian and Carboniferous Periods. Later, in the Permian, all of these movements culminated in the formation by continental collision of the Pangaea landmass which encompassed all three of the Carboniferous continents. An interpretation of the relative positions of the Carboniferous continents is given by Tarling (Chapter 2 of this volume).

The mechanism of continental drift that caused the separation and coalescence of landmasses and the enlargement and closing of oceans has been better understood during the last two decades with the formulation of the theory of global tectonics. As it is now seen, new ocean floor is formed at central ridge systems and ancient ocean floor is consumed in down-thrusting subduction zones often near continental margins. This process can be shown to be taking place today and evidence for continental movement and collision in the past can be found in the stratigraphical record.

With the more actualistic reconstruction of changing patterns of palaeogeography it is now possible to give a better assessment of continental palaeobiogeography. The distribution of late Palaeozoic land plants has been described by Chaloner and Lacey (1973) and the tetrapods have been studied by Panchen (1973, 1977). In this paper an attempt is made to trace the migration of the land flora and fauna during the later Palaeozoic based on the fossil and stratigraphical record in the setting of mobile continents.

PALAEOZOIC TERRESTRIAL COMMUNITIES

The Devonian is the oldest fossiliferous system in which non-marine formations are extensively developed and it is significant that all the major divisions of terrestrial and freshwater plants and animals have their early representatives present. Thus the first reliable records of terrestrial plants occur in the Lower Devonian, with them being found the first insects and scorpions. Slightly higher in the Devonian the non-marine bivalves enter, freshwater fish are abundant and towards the end of the system the first tetrapods are found. The sudden appearance of diverse stocks of plants and animals apparently well adapted to life on land or in fresh water has many parallels elsewhere in the stratigraphical column. The seemingly abrupt entry of these communities is brought about by development of favourable conditions for the preservation of fossils rather than any sudden burst of evolution. The gradual development of plants and animals capable of colonizing the land environment probably took place through the Lower Palaeozoic, perhaps over an even longer period. Only a fragmentary record of these pre-Devonian forms is preserved, partly owing to the lack of preservable hard-parts and partly owing to the rare occurrence of Lower Palaeozoic terrestrial deposits. Land communities were established by the Lower Devonian and the fossil record here is relatively good. Some features of the Upper Palaeozoic land and freshwater flora and fauna are discussed below.

1. The Land Flora

According to some authorities the early land flora evolved from the green algae possibly on a world-wide scale and over a long period of time. The Devonian land flora includes representatives of the Psilopsida, Pteropsida, Lycopsida and Sphenopsida capable of producing both forest trees and ground-cover vegetation. Problems of reproduction and migration had been solved by the production of vast quantities of fertile spores from each individual. These plants colonized the equable climatic zones throughout the world to provide the early almost cosmopolitan flora of later Devonian and early Carboniferous times. Floral divergence owing to plant evolution with geographical and climatic isolation, took place during the Carboniferous and by the Permian five floral provinces were established: the Angara, Atlantic, North American, Cathaysian and Gondwana floras (Chaloner and Lacey, 1973).

2. *The Land and Freshwater Invertebrates*

Non-marine bivalves enter in the Upper Devonian and are referred to the genus *Archanodon* which is fairly widely distributed in Europe and North America – i.e. central and western Laurasia. During the Carboniferous the *Anthracosia* - *Carbonicola* group of freshwater mussels evolved in the British area where they are abundant in the Coal Measures. Though known across Europe eastwards to the Donets Basin, they are restricted to this region of Laurasia. They are absent in the Upper Carboniferous of North America and they have not been reported outside Laurasia. Freshwater myalinids on the other hand, such as *Naiadites* and *Anthraconaia*, occur in the Upper Carboniferous and Permian throughout Laurasia and some genera are cosmopolitan. The difference in capacity to migrate in these bivalves is perhaps controlled by family characteristics. The myalinids are a family with many marine forms, only one branch of which lives in brackish and freshwater. The marine branch is cosmopolitan and could have produced freshwater forms widely in the Carboniferous and Permian. The *Anthracosia* group, however, was derived from a stock which inhabited fresh water from its first appearance in the Devonian. Migration of this group would be expected to be more restricted.

Land and freshwater arthropods were well established during the Devonian. Scorpions enter during the Silurian and all the major groups of arachnids are present in the fossil record by the Carboniferous. Crustaceans, including branchiopods, ostracods and malacostricans, were present in the Devonian and the insects were rapidly evolving. The fossil record of these groups is usually limited and their distribution is only partly known. However, some genera have been recorded quite widely in central and western Laurasia, in what is now Europe and North America. Their capacity to migrate and colonize in Laurasia seems to have been good.

3. *The Freshwater and Land Vertebrates*

Freshwater fish were abundant throughout the Devonian and representatives of all the main groups of fish are present as fossils. The

Lower Devonian was characterized by the lower fish groups such as the ostracoderms and placoderms with only rare members of the higher bony fishes. This latter group, the Osteichthyes, appears in great abundance and variety suddenly in the Middle Devonian. Crossopterygians, dipnoans and actinopterygians were present in distinct groups which implies that they must have a long antecedent history of which we have a fragmentary record only. The fish faunas are widely distributed in Laurasia, Gondwanaland and possibly Angaraland; some groups become almost cosmopolitan.

The amphibians, the first vertebrates to colonize the land, are assumed to have evolved from crossopterygian (rhipidistian) fish ancestors in Devonian or possibly pre-Devonian times. These fish had lungs, and also stout, muscular fins that could have supported the body on land. The development of the tetrapod limb from these specialized fins was possibly an adaptation in a fish group for better survival during the conditions of seasonal drought that prevailed during the Devonian Period. The earliest fossil amphibians have been found in the Upper Devonian of Laurasia (East Greenland) and Gondwanaland (south-eastern Australia), but they are rare fossils until the Upper Carboniferous. There is a sudden appearance of a variety and abundance of amphibians in the Westphalian that can be correlated with the development of widespread swamp conditions (Coal Measures) on the southern margin of Laurasia. But even under these favourable living conditions preservation of fossil amphibian bones is sporadic, perhaps because of the acid swamp environment. The early reptiles emerge early in the Upper Carboniferous and lived parallel with the amphibians in the Laurasian swamplands. At the end of the Carboniferous, when the major continents fused together to form Pangaea, the well-established tetrapod fauna of Laurasia was able to migrate widely.

PRELUDE—DEVONIAN GEOGRAPHY AND MIGRATION

The broad picture of Devonian geography is given on the provisional palaeogeographical maps presented by Tarling (Fig. 5, Chapter 2 of this volume). As stressed by Tarling, the relative positions of the Devonian continental masses based on palaeomagnetic observations are still to a large extent conjectural and evidence from comparative stratigraphy and palaeontology is important. Most of the detailed studies of the Devonian come from Laurasia (Europe, Greenland and

North America). In other parts of the world more limited studies have been conducted and direct comparisons are more difficult. Laurasia was dominated by a vast continental hinterland in which post-orogenic fluviatile sediments were laid down in intermontane molasse basins. These sediments have been known since the work of the early stratigraphers as the Old Red Sandstone and they form a widespread facies throughout central Laurasia (Fig. 1). The Devonian Equator crossed Laurasia some way north of Britain and an equatorial climate produced dry desert conditions though wind-blown dune sands are not conspicuous. The Old Red Sandstone facies consists of

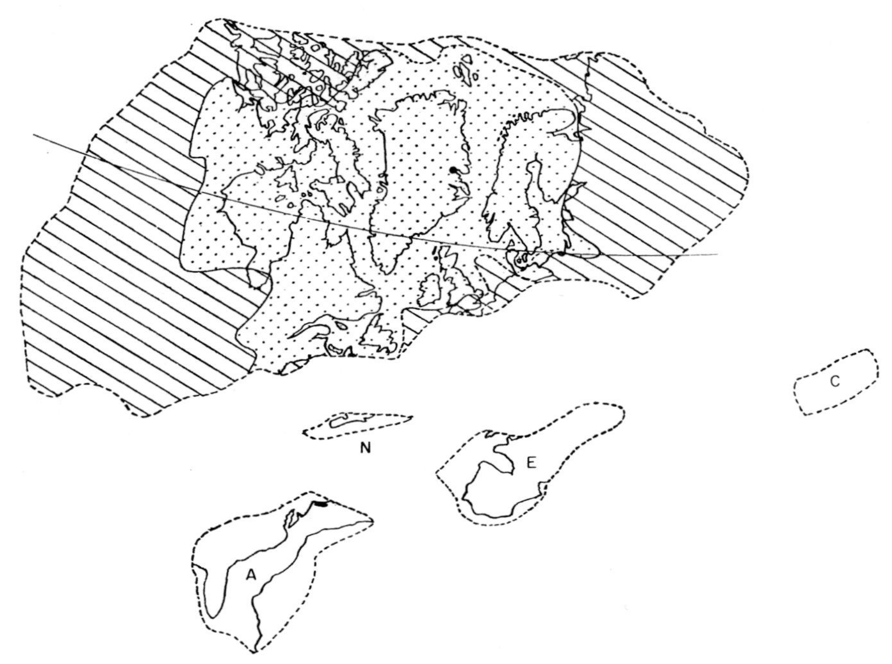

Fig. 1. Palaeogeographical interpretation of the Laurasian continent in Middle Devonian times showing the conjectural position of microcontinents to the south. Devonian Equator shown lying to the north of the British Isles. (A) Eastern North America with north-west Africa; (C) Caucasus; (E) south Europe; (N) southern Nova Scotia. Stipple indicates terrestrial hinterland, cross-hatching shelf seas and marginal marine environment. The location of the late Devonian ichthyostegid fauna of east Greenland is shown by a black spot. After many authors including Bullard *et al.* (1965), House (1968), Smith *et al.* (1973), Arthaud and Matte (1977) and Johnson (1978).

sediments laid down in estuary, river and lake environments. Soil horizons are frequent, particularly caliches, light-coloured calcareous soils, known in Britain as cornstones. Freshwater fish fossils are often well preserved in the Old Red Sandstone sediments and near the top of the sequence the first amphibians have been found. In the Devonian oceans surrounding the continents marine life flourished with abundant algae, invertebrates and fish. But the flora and fauna in many respects continue the pattern set in the previous Silurian period and there are relatively few innovations except for the rise of the marine fish. Towards the end of the Devonian there was a major crisis in marine invertebrate life that affected the coelenterates, molluscs, branchiopods and arthropods, causing mass extinction of many of the older stocks. The reason for this crisis is not fully known, but recent work has suggested that variations in the Earth's rotation rate and in the polarity of the geomagnetic field at this time may have disrupted biological ecosystems and led to mass extinctions (Whyte, 1977). Geographic and climatic changes induced by these effects may have been the main cause of extinction. It is noteworthy that dramatic changes in the fish faunas affected both freshwater and marine stocks towards the end of the Devonian. But the land flora was not affected and spread widely, becoming cosmopolitan by the end of the period. The non-marine bivalves (*Archanodon*) first appear in the Upper Devonian and are widely distributed across Laurasia, in Europe and North America, and the first fossil amphibians (ichthyostegids) also appear before the end of the period. These early amphibians are particularly interesting because they possess certain specializations that disqualify them as ancestors of any known post-Devonian tetrapods. They are specialized or evolved forms perhaps of larger size than the contemporary main tetrapod stock, of which to date we have no fossil evidence from the Devonian. Their larger size and more massive skeleton would possibly explain their appearance in the fossil record at this time. It is also possible that the evolution of the ichthyostegid stock and the non-marine bivalves at the end of the Devonian could link in with the major changes in other groups of animals at this time.

The distribution of the Devonian tetrapods in Laurasia (East Greenland) and Gondwanaland (south-east Australia), but not elsewhere, suggests that there was a migration route between these two areas, possibly used by species of which we have no fossil record.

The position of south-east Australia with respect to east Greenland is far from certain as has been discussed by Tarling (Chapter 2 of this volume). From palaeomagnetic evidence Tarling considers that south-east Australia was within the tropics during late Devonian times and he suggests that it could lie near eastern or north-eastern Laurasia. A near juxtaposition of the two areas is further supported by similar stratigraphy and structural history. In east Greenland some 7000 m of typical Old Red Sandstone sediments interbedded with acid and mafic volcanic rocks overlie a folded basement of Lower Palaeozoic and Precambrian age. The mainly fluviatile sediments were laid down, with breaks in deposition in a wide molasse basin within the Laurasian continent at the close of the Caldedonian Orogeny (Escher and Watt, 1976). The Old Red Sandstone of south-east Australia was laid down under similar conditions. A fluviatile and lacustrine sedimentary succession, with breaks in deposition and basal and interbedded acid volcanic rocks, is found, and this overlies the Lower Palaeozoic sequence with or without unconformity. The Precambrian is only exposed to the west on the edge of the Adelaide Basin, but similarities with east Greenland extend to this part of the succession with tillites developed in the Upper Precambrian in both regions (Brown *et al.*, 1968).

The Devonian tetrapods of Australia were first reported from Victoria where well-defined trackways and footprints were discovered in the Upper Devonian Genoa River Beds (Warren and Wakefield, 1972). From a similar horizon in the Cloghan Shale a single specimen of the lower jaw of an amphibian has been found near to Forbes, New South Wales, and has been named *Metaxygnathus* (Campbell and Bell, 1977). Although this genus is new to science, ichthyostegid affinities are suggested by Campbell and Bell and it is comparable to the much more abundant and diverse Upper Devonian amphibians of east Greenland. The Greenland fauna includes *Ichthyostega*, *Ichthyostegopsis* and *Acanthostega*; these differ considerably from each other and suggest a lengthy evolution of the stock prior to the Upper Devonian. This amphibian fauna has been under study in Scandinavia for many years, but has not yet been fully described. The presence of amphibia in Australia and Greenland is paralleled by remarkable similarities of the freshwater fish faunas (Ritchie, 1975). Of the fish, one dipnoan (*Soederberghia*) has only been recorded in east Greenland and in the Forbes area of south-east Australia (Campbell and Bell,

1977). There are clearly strong grounds for suggesting near juxtaposition of south-east Australia (Gondwanaland) with east Greenland (Laurasia) during the Devonian, and it seems probable that migration of freshwater and terrestrial stocks between the two regions was possible.

LOWER CARBONIFEROUS CONTINENTAL MIGRATION

The broad motion of continental plates during the Lower Carboniferous has been deduced from palaeomagnetic records and shows that Laurasia tended to drift northwards and Gondwanaland to drift southwards. Land connection, which appears to have been developed between them in the later Devonian, seems to have been lost prior to the beginning of the Carboniferous. Along the southern margin of Laurasia the Lower Carboniferous enters with a widespread marine transgression. Marine limestones dominate the early Carboniferous successions from the Donets and Russian Platform westwards to North America. Continental deposits were laid down in interior basins of Laurasia and on the northern margin of the continent marine transgression occurs later near the top of the Lower or even in the Upper Carboniferous. Extensive shallow shelf seas developed round the margins of southern Laurasia, and under a tropical climate marine life flourished and quickly evolved new faunas from ancestral stocks that survived the late Devonian crisis. From the shorelines, deltas, formed of sediment derived from eroding mountain ranges within the continent, built out to sea. The delta flats close to sea-level proved acceptable for the successful terrestrial swamp plants and luxuriant forests developed in this environment from quite early in the Carboniferous. Extensive lowlands, with mainly freshwater swamps, pools and lakes, sluggish streams and distributary channels, provided an environment in which freshwater and terrestrial animals could live. The non-marine bivalves migrated into these paralic swamplands early in the Carboniferous and evolved steadily, providing chronological succession of faunas until the end of the period. The amphibians flourished in this environment, but because of the rare chance of preservation only a tiny proportion became fossilized. Under exceptional circumstances, mainly in deposits laid down in ponds or shallow lakes, skeletal remains of small amphibians belonging to the lepospondyls are preserved. These forms are possibly nearer to the

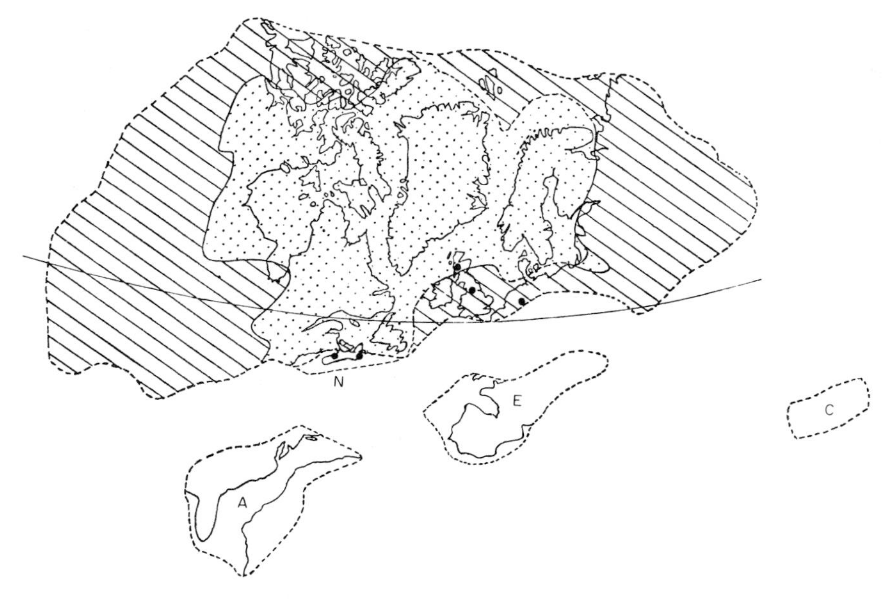

Fig. 2. Palaeogeographical interpretation of the Laurasian continent in Middle Carboniferous times showing the conjectural position of microcontinents and the south Nova Scotia plate (N) fused to Laurasia. The Carboniferous Equator is shown to the south of the British Isles. Lower and Mid-Carboniferous tetrapod localities in Britain, west Germany and south Nova Scotia are shown by black spots. Other details as for Fig. 1.

main line of amphibian evolution than the more massive labyrinthodonts that first appeared in the late Devonian. They are also present in a great diversity of forms suggesting that they had a long history of evolution prior to the occurrence of the first fossils in the Lower Carboniferous. Without doubt, the well-established amphibian stocks migrated to the Lower Carboniferous lowland coastal swamps where they could, and did, flourish.

A further event that took place in the late Devonian was the rafting together and collision of a fragment of continental crust, including what is now south Nova Scotia, with Laurasia (North America). Interpretations of the pre- and post-collision positions of south Nova Scotia are shown in Figs 1 and 2. A number of independent fragments of continental crust are believed to have been present to the south of Laurasia during the Upper Palaeozoic and, one by one, these closed with the major continent and were welded together (Johnson, 1977, 1978; Keppie, 1977b). The south Nova Scotia fragment is believed to have originated as a fragment of Gondwanaland that developed a Pacific-type margin on the west during the Silurian before finally colliding with Laurasia (Keppie, 1977a). A zone of subduction dipping southwards beneath south Nova Scotia and running parallel to the coast of Laurasia is required to explain the structure of the region and the destruction of intervening oceanic crust prior to collision.

Collision of the south Nova Scotia fragment with Laurasia made further migration of tetrapods possible on to the new southern margin of the continent. Both Lower and Upper Carboniferous amphibians have been recorded from this region in rocks laid down in a lowland paralic swamp environment (Panchen, 1973, 1977; Rayner, 1971). The successful migration of amphibians into the south Nova Scotia area by early Carboniferous times confirms the continental link and fits with the structural and stratigraphical evidence.

UPPER CARBONIFEROUS CONTINENTAL MIGRATION

By Upper Carboniferous times southerly drift and rotation had taken the Gondwana Devonian amphibian-bearing region of south-east Australia out of the tropics and into the influence of the south polar zone. The earliest evidence of glacial deposits comes from south-east Australia (New South Wales) and is dated as low in the Upper Carboniferous (Namurian). These deposits include glacial marine sediments, indicating the presence of ice at sea-level even at this early date. The glaciation continued through the rest of the Upper Carboniferous Westphalian and Stephanian, and the whole of the Australian continent, except for the north-eastern corner, appears to have been affected by ice (Brown *et al.*, 1968). Now, the Carboniferous amphibians and the early reptiles that evolved from them during the Upper Carboniferous were cold-blooded. Like the majority of extant forms, the Carboniferous tetrapods were probably adapted to a relatively high and constant atmospheric temperature. The middle and late Carboniferous glaciation of Australia (Gondwanaland) must have caused the indigenous tetrapods to migrate northwards, and ultimately it appears to have led to their extinction on this continent.

On the southern margin of Laurasia, however, the tropical lowland paralic swamps were at their maximum development in the Coal Measures and Pennsylvanian, Upper Carboniferous. The non-marine swamp and pool environment would be expected to support amphibian life and there are widespread records of fossils in Laurasia from Europe through to central North America (Panchen, 1977). The fossil record is mainly remarkable for its rarity. The geological succession is perhaps better known and exposed than most other divisions owing to the extraction of coal, but fossil amphibians are exceedingly rare elements of a very abundant and diverse fossil flora and fauna. This may be due to the difficulty of preservation of small bones in an acid swamp environment, because the tetrapods would be expected to live there in abundance.

At the end of the Carboniferous another fragment of continental crust, including south Europe, was rafted against southern Laurasia (Johnson, 1978). This was a repetition of very much the same series of events as those that took place when the south Nova Scotia plate approached and collided with Laurasia in the Devonian. Again, a zone of subduction developed parallel to the coast of Laurasia and the intervening oceanic crust was consumed by down-thrusting beneath south Europe (Fig. 2). An active Pacific-type continental margin developed adjacent to the subduction zone in the early Upper Carboniferous and formed the east–west Hercynian chains of mountains across Europe. During the Carboniferous the seaway between Laurasia and south Europe was relatively narrow so that the marine faunas on either side of the sea were very similar; at least the pelagic larval stages of these creatures were able to migrate across the sea. As time progressed, the sea narrowed and the east to west marine passage is believed to have been lost in mid-Carboniferous times (Ramsbottom, 1971; Johnson, 1973). The sea across Europe finally closed, with a mild collision orogeny, at the end of the Carboniferous forming the late Hercynian fold-belt. The line of suture between north and south Europe can be traced eastwards from south-west England to where it disappears against the Carpathian Mountains. Fragments of oceanic crust (ophiolites) are believed to have been preserved along the suture at the Lizard (Bromley, 1975) and in the Harz Mountains (Anderson, 1975).

With a mid-Carboniferous connection between north and south Europe the possibility exists for the migration of tetrapods onto the south Europe plate. The first tetrapods recorded here are of Westphalian D age, late in the Carboniferous, where they are found in intermountain lake basins in Czechoslovakia (Panchen, 1977). The marine barrier of the Mid-European Sea seems to have been effective in preventing the southward migration of the amphibians, but when a land-bridge was formed by closing the seaway, they were quickly able to colonize the freshwater lakes that developed in deep valleys between the young Hercynian mountains.

At the end of the Carboniferous another continental collision took place in western Laurasia. Here a further fragment of continental crust, including the eastern seaboard of the U.S.A. and Florida, probably attached to north-west Africa, was rafted against North America to regenerate the southern Appalachian mountains (Figs 2 and 3). No Carboniferous tetrapod records are known from the southern Appalachians and it is possible that this new mountain chain formed a barrier to migration. Alternatively it could well be either that amphibian bones have not been preserved in this area or that their fossil remains have yet to be discovered.

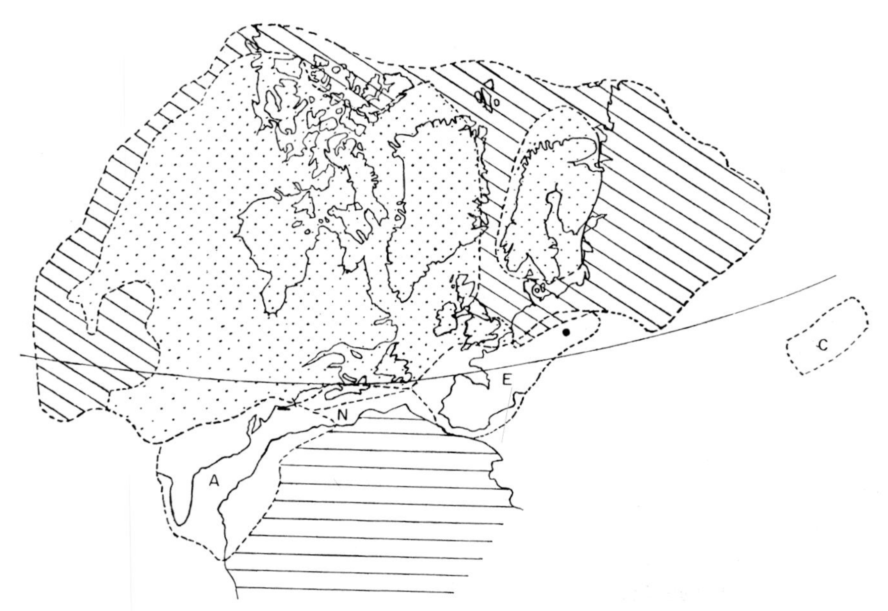

Fig. 3 Palaeogeographical interpretation of the Laurasian continent in Permian times showing the Pangaea fit of southern microcontinents and Africa, Gondwanaland (horizontal hatching). The Permian Equator is shown passing through southern Europe. Late Carboniferous tetrapod locality in Czechoslovakia, south Europe plate, is shown by a black spot. Other details as for Fig. 1.

EPILOGUE – PERMIAN COSMOPOLITAN MIGRATION

The progressive closing of continents witnessed during the Carboniferous culminated in the Permian with Gondwanaland drifting and rotating against North America and, later, Angaraland closing with east Laurasia along the line of the Urals. In this way the three continents coalesced to form the supercontinent of Pangaea. The exact fit of the continents and the detailed chronology of events is uncertain at present; several interpretations have been suggested (Smith *et al.*, 1973; Bullard *et al.*, 1965; Irving, 1977), but for our purposes it is significant that by the end of the Permian the existence of land connections between each of the continents had become probable and tetrapod migration made possible over vast new territories. The Appalachian and Ural mountain barriers did not check migration for long. Permian amphibians and early reptiles are known outside Laurasia in Gondwanaland, early in the period in Africa, India and South America; later in the period they are also recorded in Australia. Permian tetrapod migration was clearly rapid and successful. Similarly, in the early Trias, after the closing of the marine barrier between Angaraland and Laurasia, tetrapod migration eastwards was equally successful. Thus, in the early Mesozoic, cosmopolitan tetrapod faunas became established.

Another major crisis in marine life at the close of the Permian may have had an effect on tetrapod migration. This crisis caused widespread extinction among invertebrates and fishes which has led to the suggestion that it could have been produced by an exceptional physical situation. It could have been caused partly by exceptional emergence of the continents and consequent reduction of shallow shelf seas, and partly by the surface waters of the seas becoming brackish at this time (Rudwick, 1970). Though speculative, each of these possibilities would have much assisted tetrapod migration. In brackish water the possibility exists that even amphibians could have taken to the seas, as one group, the trematosaurs, are known to have done in the Trias.

REFERENCES

Anderson, T. A. (1975). Carboniferous subduction complex in the Harz Mountains, Germany. *Bull. geol. Soc. Am.* **86**, 77–82.

Arthaud, F. and Matte, P. (1977). Late Palaeozoic strike-slip faulting in southern

Europe and northern Africa: Result of a right-lateral shear zone between the Appalachians and the Urals. *Bull. geol. Soc. Am.* 88, 1305–1320.

Bromley, A. V. (1975). Tin mineralization of Western Europe: is it related to crustal subduction? *Trans. Instn Min. Metall.* B84, B28–30.

Brown, D. A., Campbell, K. S. W. and Cook, K. A. W. (1968). "The Geological Evolution of Australia and New Zealand." Pergamon, Oxford.

Bullard, E. C., Everett, J. E. and Smith, A. G. (1965). The fit of the continents around the Atlantic. *In* "A Symposium on Continental Drift", *Phil. Trans. R. Soc.* A258, 41–51.

Campbell, K. S. W. and Bell, M. W. (1977). A primitive amphibian from the late Devonian of New South Wales. *Alcheringa* **1**, 369–381.

Chaloner, W. G. and Lacey, W. S. (1973). The distribution of late Palaeozoic floras. *In* "Organisms and Continents through Time" (N. F. Hughes, ed.), *Spec. Pap. Palaeont.* No. 12, 271–289.

Escher, A. and Watt, W. S. (1976). "Geology of Greenland". Geological Survey of Greenland, Copenhagen.

House, M. R. (1968). "Continental Drift and the Devonian System", Inaugural Lecture Series, University of Hull, pp. 1–24.

Irving, E. (1977). Drift of major continental blocks since the Devonian. *Nature, Lond.* **270**, 304–309.

Johnson, G. A. L. (1973). Closing of the Carboniferous Sea in Western Europe. *In* "Implications of Continental Drift to the Earth Sciences" (D. H. Tarling and S. K. Runcorn, eds), Vol. 2, pp. 845–850. Academic Press, London and New York.

Johnson, G. A. L. (1977). Palaeozoic accretion of western Europe. *Annls Soc. geol. N.* **96**, 347–352.

Johnson, G. A. L. (1978). European plate movements during the Carboniferous. *In* "Evolution of the Earth's Crust" (D. H. Tarling, ed.), pp. 343–360. Academic Press, London and New York.

Keppie, J. D. (1977a). Tectonics of southern Nova Scotia. *Nova Scotia Dept Mines* Paper 77-1, 1–34.

Keppie, J. D. (1977b). Plate tectonic interpretation of Palaeozoic world maps (with emphasis on circum-Atlantic orogens and southern Nova Scotia). *Nova Scotia Dept. Mines* Paper 77-3, 1–45.

Panchen, A. L. (1973). Carboniferous tetrapods. *In* "Atlas of Palaeobiogeography" (A. Hallam, ed.), pp. 117–125. Elsevier, Amsterdam.

Panchen, A. L. (1977). Geographical and ecological distribution of the earliest tetrapods. *In* "Major Patterns in Vertebrate Evolution" (M. K. Hecht, P. C. Goody and B. M. Hecht, eds), pp. 723–728. Plenum, New York.

Ramsbottom, W. H. C. (1971). Palaeogeography and goniatite distribution in the Namurian and early Westphalian. *C.r. 6th Congr. int. Strat. Geol. Carbonif., Sheffield, 1967* 4, 1385–1400.

Rayner, D. H. (1971). Data on the environment and preservation of late palaeozoic tetrapods. *Proc. Yorks. geol. Soc.* 38, 437–495.

Ritchie, A. (1975). *Groenlandaspis* in Antarctica, Australia and Europe. *Nature, Lond.* **254**, 569-573.

Rudwick, M. J. S. (1970). "Living and Fossil Brachiopods." Hutchinson, London.

Smith, A. G., Briden, J. C. and Drewry, G. E. (1973). Phanerozoic world maps. *In* "Organisms and Continents through Time" (N. F. Hughes, ed.), *Spec. Pap. Palaeont.* No. 12, 1–42.

Warren, J. W. and Wakefield, N. A. (1972). Trackways of tetrapod vertebrates from the Upper Devonian of Victoria, Aust. *Nature, Lond.* 238, 469–470.

Whyte, M. A. (1977). Turning points in Phanerozoic history. *Nature, Lond.* 267, 679–682.

4 | Early Land Floras

DIANNE EDWARDS

Department of Plant Science, University College, Cardiff, Wales

Abstract: The main objects of this review are to give some idea of the general appearance, habit and affinities of early vascular plants and to present the meagre evidence pertinent to the reconstruction of terrestrial vegetation in late Silurian and Devonian times. The description of vascular plants in the late Silurian (Ludlow Series) provides unequivocal evidence for the existence of terrestrial vegetation. By early Pridoli, floras of remarkable uniform composition were widespread. These comprised plants showing great simplicity of organization with a predominance of rhyniophytes and centred on the genus *Cooksonia*. Later Gedinnian floras indicate the beginnings of diversification with the appearance of zosterophyllophytes, but it is in the later Lower Devonian that extensive assemblages of very varied composition are first recorded. These include the first lycopods as well as the trimerophytes, which are considered intermediate in organization between the earliest vascular plants and the Middle Devonian ferns, "presphenopsids" and progymnosperms. Although recent anatomical and morphological studies on Silurian and Devonian macroplants have permitted such an evolutionary overview and have resulted in major revisions of their classification, it is concluded that little progress has been made in elucidating details of terrestrial vegetation.

INTRODUCTION

The two recent major reviews of Devonian floras by Chaloner and Sheerin (1979) and by Banks (1980b) more than adequately reflect the intensity of interest and activity in the study of early land plants. Banks' attempt at a stratigraphy of late Silurian (Pridoli) and Devonian sediments based on macroplants, in which he distinguishes seven generic assemblage zones, is the outcome of the

Systematics Association Special Volume No. 15, "The Terrestrial Environment and the Origin of Land Vertebrates", edited by A. L. Panchen, 1980, pp. 55–85, Academic Press, London and New York.

accumulation of a wealth of data, both morphological and anatomical, over the past 60 years. Such data have also made possible important revisions in the classification of early vascular plants: the "psilophytes", for example, have been replaced by three new subdivisions, Rhyniophytina, Zosterophyllophytina and Trimerophytina (Banks, 1968a, 1975a), and Beck (1960) established a new class, the Progymnospermopsida, comprising plants with pteridophyte reproduction and gymnospermous anatomy. Both of the recent reviews emphasize the late Silurian and Devonian as a period of major evolutionary innovation, a time of diversification following the initial colonization of the land.

In general appearance, this early terrestrial vegetation would be strikingly different from that of the present day which is dominated by a wide variety of different forms of leaf. Apart from lycopods and more enigmatic plants with fan-shaped leaves, plants consisted of collections of stem-like structures, variously branched. Thus, instead of presenting yet another review of early land floras, I propose to illustrate some of the commonest genera to give some idea of the appearance of individual plants. Although it would be highly desirable to be able to group these plants into communities and then to speculate on their palaeoecology, there is unfortunately less direct evidence for the reconstruction of vegetation in the Siluro-Devonian than at any other time in the history of plant life on land.

Exactly when plants first colonized land surfaces remains a matter of considerable conjecture (e.g. Banks, 1972, 1975b,c; Gray and Boucot,1977; Edwards *et al.*, 1979). In this account I shall concentrate on vascular plants, which, as Raven (1977) in his novel physiological approach to the early evolution of vascular plants has pointed out, are able to regulate their rate of water loss because they possess a gas-impermeable cuticle and functional stomata connected to an intercellular space system. They thus remain hydrated even when their water supply is limited. Such homoiohydric plants are the true conquerors of the land and such a level of anatomical organization is seen in the probably mid-Siegenian Rhynie Chert plants (Kidston and Lang, 1917. Earlier coalified compression fossils of vascular plants lack the superb cellular preservation found in the Chert and so it is impossible to determine whether or not plants such as *Cooksonia* had attained this degree of water control. Nevertheless the demonstration of vascularized axes and fertile plants assignable to *Cooksonia* in the Ludlow Series of south Wales (Edwards and Davies, 1976; Edwards and Rogerson, 1979; Edwards *et al.*, 1979) is a firm indication that plant life existed on land in the late Silurian.

Some workers (e.g. Gray and Boucot, 1977; Pratt *et al.*, 1978) have argued very forcefully that pre-Ludlow microfossil evidence indicates the presence of terrestrial vegetation prior to the appearance of vascular plant macrofossils, although the affinities of these plants remain obscure. Such microfossils include resistant spores with triradiate marks (Gray and Boucot, 1971, 1974), cuticle-like films bearing the outlines of cells, sheets of cells and tubes with thickenings paralleling those seen in tracheids (Pratt *et al.*, 1978).

Fragments of similar type are found in the same sediments as early vascular plants (see, for example, Lang, 1937) as indeed are non-vascular plant macrofossils such as *Parka, Pachytheca* and *Prototaxites*. Whether or not these grew on the surface of the ground or were immersed in fresh water cannot be determined with certainty. Niklas (1976a) has compared the organization of *Parka* with that of an extant green alga *Coleochaete*: affinities with the Chlorophyta being further supported by chemical analysis (Niklas, 1976b). The presence of a thin, cuticle-like covering led Niklas to suggest that *Parka* grew as an epiphyte in shallow freshwater pools, subject to periods of desiccation. The affinities of *Pachytheca* have been considered with the blue-green (Lang, 1945) and green algae (Niklas, 1976b). Indeed Niklas suggests the possibility that *Parka* and *Pachytheca* were related on the basis of similar patterns of growth as well as chemical analyses. *Prototaxites* is also difficult to assign to an extant algal group, although often considered to be brown (Johnson and Konishi, 1958; Schmid, 1976). Niklas (1976c) has isolated cutin and suberin derivatives from coalified compressions which indicate that *Prototaxites* too could have survived on land. Indeed if *Prototaxites* was similar to extant algae in lacking an intercellular space system, it is difficult to understand how, with a trunk 1 m in diameter and more than 2 m long (records from the Lower Devonian of Canada—Schmid, 1976), it could have survived completely immersed in water without the central region becoming anaerobic (Raven, 1977 and personal communication).

SILURIAN PLANTS

1. *Assemblage Zone I :* Cooksonia *(mid-Ludlow to end of Pridoli; Fig. 1)*

Following the recent discovery of *Cooksonia* Lang in Bringewoodian strata in central Wales (*incipiens* Zone; Ludlow Series) Banks' Assemblage Zone I is extended back to mid-Ludlow sediments. Although there are reports of pre-Pridoli vascular plants from Libya (Boureau *et al.*, 1978) and Australia (Garratt, 1979), it is only in Wales that macrofossils of vascular plant aspect occur in sediments in which the stratigraphical dating is supported unequivocally by independent

---

Fig. 1. (opposite). Silurian plants A–D: (A) *Hostinella* Stur – smooth dichotomously branching axis; *incipiens* Zone (Ludlow Series), Wales (× 3·2). (B) *Cooksonia pertoni* Lang – smooth forking axis terminating in oval sporangia; *incipiens* Zone (Ludlow Series), Wales (× 3·1). (C) *Psilophytites* Høeg – axis with triangular spines (enations); Pridoli (Downton), Wales (× 8). (D) *Steganotheca striata* Edwards – smooth dichotomously branching axes with elongate terminal sporangia; Pridoli (Downton), Wales (× 1·2).

Devonian plants E–N: (E) *Zosterophyllum myretonianum* Penhallow – spike of spirally inserted lateral sporangia; ?early Siegenian (Dittonian), Scotland (× 1·2). (F) *Hedeia corymbosa* Cookson – cluster of elongate sporangia terminating smooth axes; ?Siegenian–Emsian, Australia (× 3·3). (G) *Yarravia oblonga* Lang and Cookson – cluster of elongate sporangia fused in basal region; ?Siegenian–Emsian, Australia (× 3·7). (H) *Gosslingia breconensis* Heard – smooth axes with scattered reniform, lateral sporangia; Siegenian, Wales (× 4·2). (J) *Gosslingia breconensis* Heard – sterile branching specimen showing pseudomonopodial branching; Siegenian, Wales (× 0·6). (K) *Zosterophyllum australianum* Lang and Cookson – sterile smooth branching axis showing K-configuration; ?Siegenian–Emsian, Australia (× 2·3). (L) *Baragwanathia longifolia* Lang and Cookson – stem covered with elongate narrow leaves (microphylls); ?Siegenian-Emsian, Australia (× 0·4). (M) *Drepanophycus spinaeformis* Goeppert – stem with falcate leaves (microphylls); Emsian, Gaspé, Canada (× 0·7). (N) *Sawdonia ornata* Hueber – sterile axis with pointed spines; Emsian, Canada (× 0·8).

(A,B) from Edwards *et al.*, 1979; (C) from Edwards, 1979; (D) from Edwards, 1970a; (E) from Edwards, 1975; (F,K) from Cookson, 1935; (G) from Lang and Cookson, 1935; (H,J) from Edwards, 1970b; (K,L) from negative loaned by J. Tims; (M) from Stubblefield and Banks, 1978; (N) from collection of author.

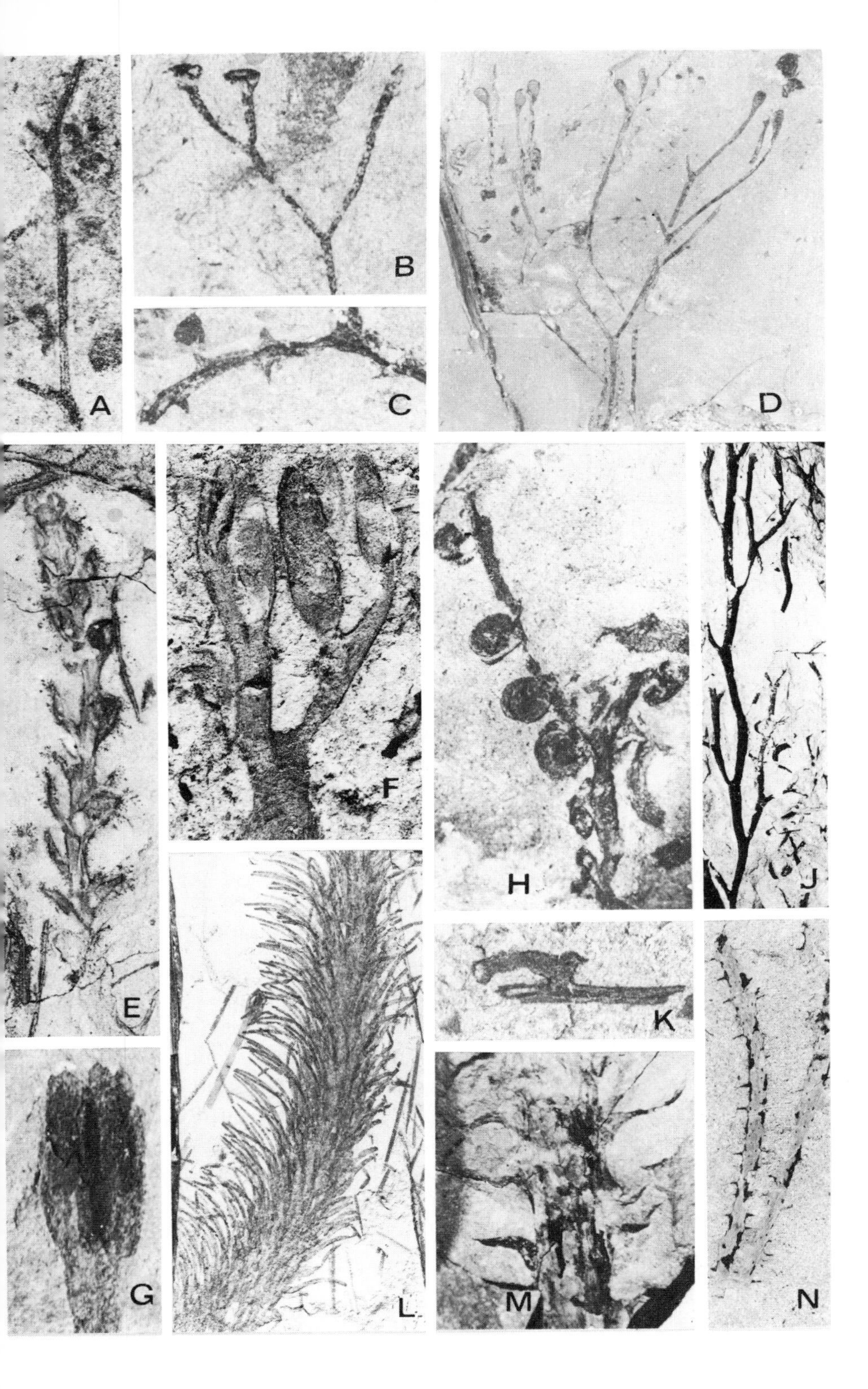
A
B
C
D
E
F
G
H
J
K
L
M
N

faunal content (Banks, 1980b; Chaloner and Sheerin, 1979; Edwards *et al.*, 1979).

By early Pridoli times, *Cooksonia* dominated floras were widespread, occurring in Bohemia (Obrhel, 1962), New York State (Banks, 1973), the U.S.S.R. (Ischenko, 1975), southern Britain (Lang, 1937; Edwards, 1979; Edwards and Rogerson, 1979) and possibly north Africa (Daber, 1971). *Cooksonia* itself (Fig. 1B) was a small plant, not more than a few centimetres high, consisting of a tuft of smooth, forking axes which terminated in spherical or ellipsoidal sporangia. It is known only from the compression fossils. *Steganotheca* (Fig. 1D) looks very similar but has elongate sporangia, each with a lens-shaped apical thickening (Edwards, 1970a; Edwards and Rogerson, 1979). By far the commonest vascular plant fossils in these Ludlow–Pridoli sediments are smooth, dichotomously branching sterile axes (Fig. 1A) which are placed in the form genus, *Hostinella* (Banks, 1968b). In addition, Obrhel records unbranched axes with a narrow central line as ?*Taeniocrada*, while in an extensive flora from south Wales (Edwards, 1979) I have described axes with spines (Fig. 1C) as well as smoother axes with more complex branching.

DEVONIAN PLANTS

1. *Assemblage Zone II :* Zosterophyllum *(Gedinnian to mid-Siegenian; Fig. 1)*

This zone is marked by the appearance of a second type of vascular plant organization. In *Zosterophyllum* (Fig. 1E) sporangia are borne laterally on smooth axes and are aggregated into terminal spikes (Lang, 1927). Dichotomous branching is still seen, but *Zosterophyllum* also has branches resembling the letters K and H (Fig. 1K). These are concentrated near the base of the plant so that it has a tufted appearance (Walton, 1964). If Richardson is correct in assigning, on palynological evidence, an early Siegenian age to the Scottish Dittonian floras (Westoll, in House *et al.*, 1977; Edwards, 1980) the oldest fertile *Zosterophyllum* is late Gedinnian or early Siegenian (Scotland – Lang, 1927; Belgium – Leclercq, 1942; Welsh Borderland – Edwards and Richardson, 1974). *Cooksonia* is recorded from the Gedinnian of Czechoslovakia (Obrhel, 1968) and the Dittonian of Scotland (Edwards, 1970a).

2. *Assemblage Zone III :* Psilophyton *(mid-Siegenian to end of Emsian; Figs 1 and 2)*

This zone has a much larger number of records of vascular plant assemblages, which show an overall increase in number of taxa and a greater diversity of forms. The discovery and description of various kinds of petrifaction (e.g. silica–Kidston and Lang, 1917; calcium carbonate–Banks *et al.*, 1975; iron sulphide–Edwards, 1970b) have yielded important information on three-dimensional anatomy.

Cooksonia is still represented in these floras (Croft and Lang, 1942), but far more complex rhyniophytes had already evolved. These include *Rhynia* and *Horneophyton* (Kidston and Lang, 1917) from the ?mid-Siegenian Rhynie Chert and *Yarravia* (Fig. 1G) and *Hedeia* (Fig. 1F) which have terminal clusters of elongate sporangia from Australia (Lang and Cookson, 1935). The genus *Zosterophyllum* has a far greater geographical distribution than in Zone II, being represented by several species (Australia – Lang and Cookson, 1935; Germany–Kräusel and Weyland, 1935; Russia–Ananiev, 1960; ?South Africa–Plumstead, 1977; China–Lee and Tsai, 1978).

The description of *Sawdonia ornata* (Fig. 1N), a zosterophyll with spiny axes from Siegenian sediments of Arctic Canada (Hueber, 1971) marks the earliest appearance of a plant which had a world-wide distribution in the Emsian (see references in Banks, 1975c) and which persisted into the Frasnian (Hueber and Grierson, 1961).

Gosslingia, a zosterophyll characterized by scattered lateral sporangia (Fig. 1H) on smooth axes, has been reported from Wales, Germany and Russia (Edwards, 1970b). The sterile branching system (Fig. 1J) of this plant may be used to illustrate an important general trend in the growth of erect axes. It is described as pseudo-monopodial, one branch of a fork growing more strongly and over-topping the other. Such branching which results in the distribution of a main axis and a number of lateral branches is the commonest type seen in later Emsian floras and is characteristic of the majority of species of the trimerophyte, *Psilophyton*, which first appeared in the Siegenian (see, for example, records of *Dawsonites*, a form genus for isolated trusses of sporangia probably belonging to *Psilophyton*, in Croft and Lang (1942).

A general impression of the growth habit of these early vascular plants may be gained from the extant *Psilotum*, but none of the

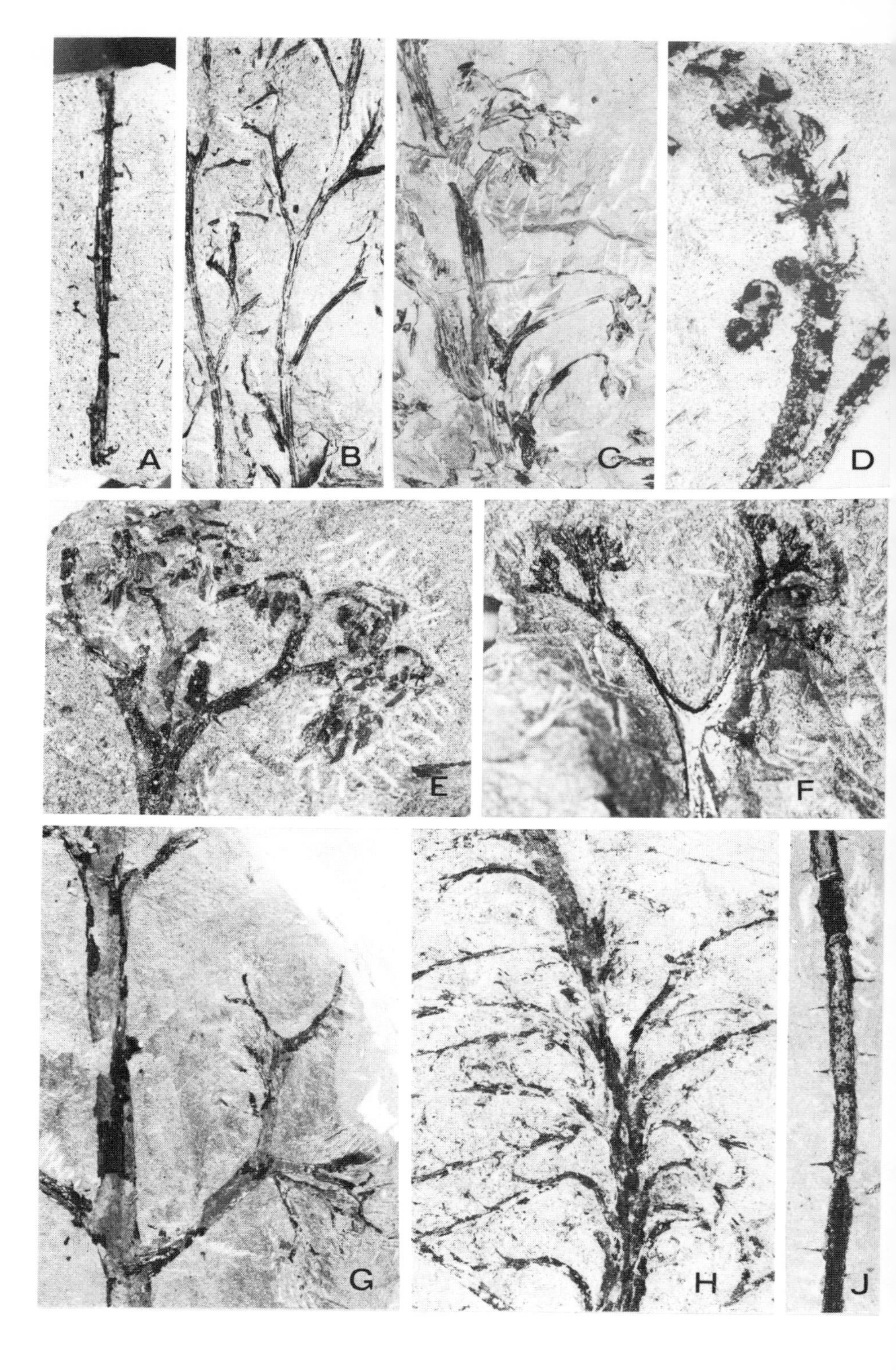
A
B
C
D
E
F
G
H
J

pteridophytes I have described so far can be directly related to extant groups (see Bierhorst (1971) and Gensel (1977) for discussion on the relationships between extant Psilotales and early vascular plants). It was in the Siegenian, however, that the lycopods (club-mosses) first appeared. By far the most convincing representative is *Baragwanathia* (Fig. 1L) from Australia (Lang and Cookson, 1935) which has undivided elongate leaves (microphylls), sporangia which appear to be axillary and a characteristically shaped, stellate xylem strand. The excellently preserved *Asteroxylon* in the Rhynie Chert (Kidston and Lang, 1920) is best described as a pre-lycopod as it possesses a level of organization intermediate between zosterophylls and true lycopods. Similarly *Drepanophycus spinaeformis* (Fig. 1M) which first occurred in the Siegenian and extended into the Upper Devonian (Stubblefield and Banks, 1978), has lycopod anatomy and microphylls but lacks the typical axillary sporangia, while *Kaulangiophyton* (Gensel *et al.*, 1969) from the late Lower Devonian has the overall appearance of a lycopod, stalked sporangia but lacks anatomy.

Some of the most important of late Devonian floras are found in north-eastern North America, in the Gaspé, New Brunswick and Maine. Although zosterophylls (e.g. *Crenaticaulis* and *Sawdonia* (Fig. 1D) and rhyniophytes (e.g. *Renalia*) are still present, these floras are

---

Fig. 2. (opposite). Late Devonian plants A–J: (A) *Psilophyton princeps* Dawson – sterile axis with blunt spines; Emsian, Canada (× 0·8). (B) *P. forbesii* Andrews – sterile branching axes showing longitudinal ribbing; Emsian, Canada (× 0·5). (C) *P. forbesii* – fertile specimen with cluster of recurved sporangia on lateral branches; Emsian, Canada (× 0·8). (D) *Sawdonia akanthotheca* Gensel *et al.* – loose spike of lateral sporangia terminating spiny axes; ?Emsian, Canada (× 1·8). (E) *P. charientos* Gensel – fertile lateral branch bearing numerous sporangia; Emsian, Canada (× 2·4). (F) *Pertica varia* Granoff *et al.* – fertile specimen with fan-shaped clusters of erect sporangia; Emsian, Canada (× 1·3). (G) *Pertica varia* – sterile pseudomonopodial main axis with dichotomously branching lateral systems; Emsian, Canada (× 0·45). (H) *Chaleuria cirrosa* Andrews *et al.* – erect main axis with spirally arranged lateral branches; ?Emsian, Canada (× 0·7). (J) *P. charientos* – axis with delicate spines; Emsian, Canada (× 2·2).

(A) From author's collection; (B,C,E,J) from Gensel, 1979; (D) from Gensel *et al.*, 1975; (F,G) from Granoff *et al.*, 1976; (H) from Andrews *et al.*, 1974.

dominated by Trimerophytina, a group of enormous evolutionary significance in that they are believed to have given rise to the Middle Devonian "fern" complex (Banks, 1980a). *Psilophyton* (Hueber, 1968) is considered to be the most primitive genus in the subdivision (Fig. 2). Six of its seven fertile species are recorded in this area. Branching may be dichotomous, but it more usually pseudomonopodial with lateral branches arranged spirally or distichously. Gensel (1979) suggests that the latter represents an intermediate stage in the evolution of a stem bearing leaves, and that major axes will eventually evolve into stems and the lateral branches into leaves. Some of these lateral branches terminate in trusses of numerous, elongate sporangia (as many as 128 in some species). Surface characteristics of axes are important in delimiting species. *P. dawsonii*, for example, has smooth axes. In *P. forbesii* (Fig. 2B,C) they are heavily ridged. *P. princeps* (Fig. 2A) has truncated peg-like spines, while in *P. charientos* (Fig. 2E,F) spines are pointed.

In *Pertica* (Fig. 2G,H), there is a greater distinction between robust main axes and compacted sterile lateral branch systems; sporangia are organized into spherical masses. Kasper and Andrews

---

Fig. 3. (opposite) Middle Devonian plants A–H: (A) *Hicklingia edwardii* – spike of lateral sporangia terminating smooth axis (?zosterophyll); Middle Old Red Sandstone, Scotland (× 2·5). Lycopsida B–F: (B) *Archaeosigillaria vanuxemi* (Goeppert) Kidston – stem showing small leaves along the edges and leaf bases on surface near base; Givetian, U.S.A. (× 1·43). (C) *Colpodexylon trifurcutum* Banks – stem showing leaf cushions on surface: long, needle-like leaves are visible at the margin, but do not show trifurcations; Eifelian, U.S.A. (× 1·2). (D) *Leclercqia complexa* Banks *et al.* – stem with some fertile leaves (arrowed): note a single sporangium is borne on the upper surface of a leaf (sporophyll); Givetian, U.S.A. (× 1·6). (E) *L. complexa* – leafy stem with dichotomous branch: in some cases, the two of the five tips of the leaf can be seen (arrowed); (× 1·1). (F) *Bergeria mimerensis* Høeg – part of a small branch with diamond leaf cushions: leafy shoots are called *Protolepidodendropsis pulchra;* Givetian, Spitsbergen (× 0·9). (G) *Duisbergia mirabilis* Kräusel and Weyland – part of a branch with a row of partly covered fan-shaped leaves on right-hand side; Eifelian, Germany (× 0·4). (H) *D. mirabilis* – stem with elongate, partly dissected leaves; Eifelian, Germany (× 0·9).

(A) From Edwards, 1976; (B) from Grierson and Banks, 1963; (C) negative from H. P. Banks; (D,E) from Banks *et al.*, 1972; (F) from Schweitzer, 1965; (G,H) negatives loaned by H. Mustafa.

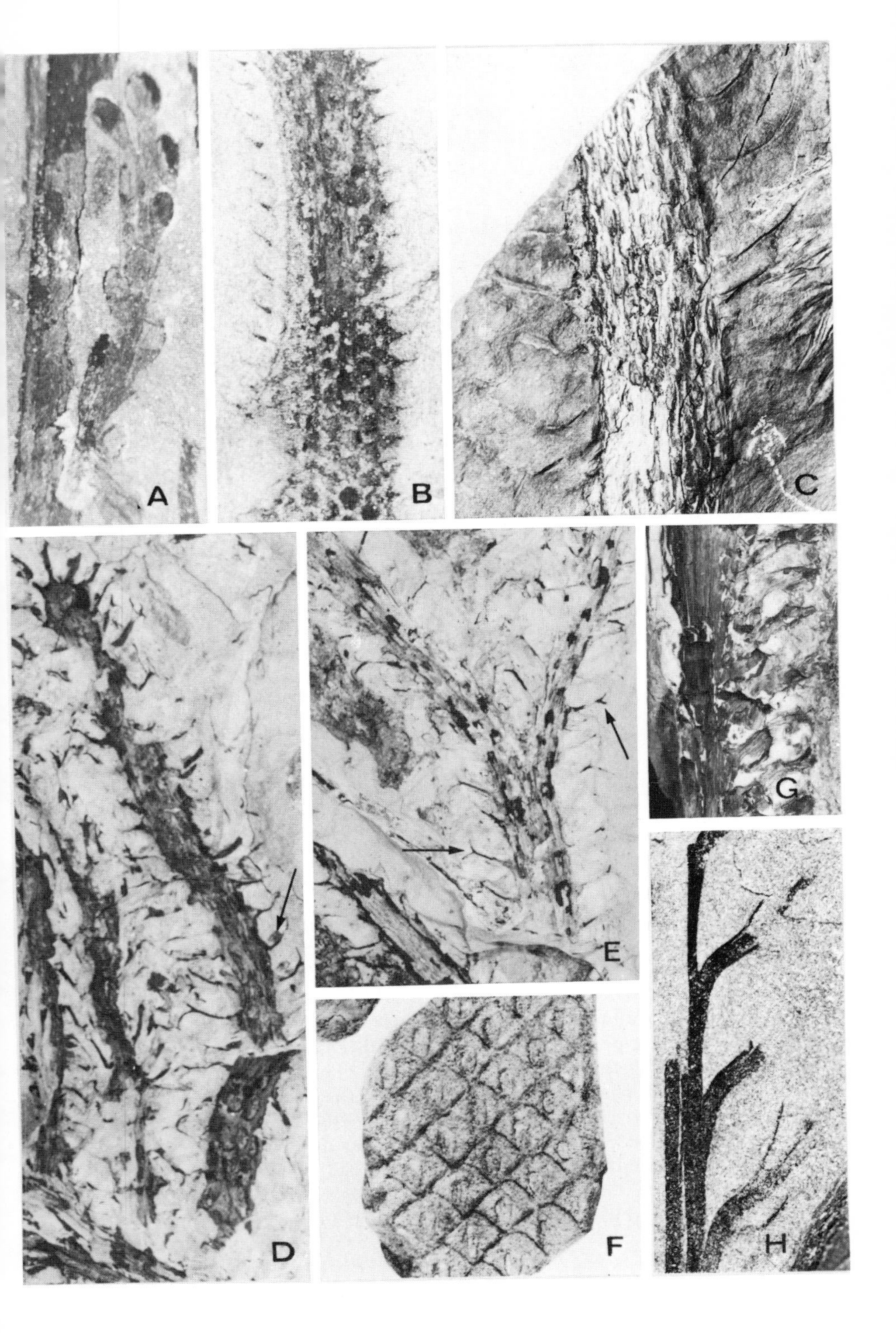
A
B
C
D
E
F
G
H

(1972) consider this increased regularity of branching and higher degree of organization an advance on the condition seen in *Psilophyton* and postulate that the highly branched three-dimensional sterile lateral system may be considered the starting point in the evolution of the fern frond (megaphyll).

A further major evolutionary advance recorded in the later Lower Devonian of New Brunswick is incipient heterospory (Andrews *et al*., 1974; Banks, in press for dating). All the plants I have described so far have been homosporous; i.e. producers of large numbers of small spores of approximately the same size, their life-cycles presumably similar, at least in outline, to the vast majority of living ferns (Edwards, 1980). *Chaleuria cirrosa* (Fig. 2J) had spores of two sizes which Andrews *et al.* speculate developed into two kinds of free-living gametophytes, a life-cycle intermediate between that of a homosporous pteridophyte and a heterosporous one such as in living *Selaginella*. In the absence of anatomy, the classification of *Chaleuria* presented problems. The authors reached the tentative conclusion that it was a primitive member of the progymnosperms and certainly more highly advanced than the trimerophytes.

3. Assemblage Zone IV : Hyenia *(Eifelian into base of Givetian). Assemblage Zone V :* Svalbardia *(much of Givetian). Figs 3 and 4.*

I propose to consider these together as Banks has indicated that Zone V is characterized by "the abundance and the wide geographic

---

Fig. 4. (opposite). Middle Devonian progymnosperms A–E: (A) *Rellimia thomsonii* Bonamo – the major axes bearing fertile organs with masses of sporangia attached adaxially; Givetian, Belgium (× 0·9). (B) *Tetraxylopteris schmidtii* Beck – fertile specimen in which paired trusses of sporangia are opposite and decussate (arrow indicates region where axis and rock have been removed to reveal sporangia beneath); *n.b.* Frasnian example, U.S.A. (× 0·8). (C) *Aneurophyton germanicum* Kräusel and Weyland – smooth sterile axes showing branching; Eifelian, Germany (× 0·7). (D) *A. germanicum* – fertile specimen showing clusters of elongate sporangia; Eifelian, Germany (× 3·0). (E) *A. germanicum* – dichotomously branching ultimate appendage interpreted as a leaf (arrowed); Eifelian, Germany (× 6·0).

(A) From Leclercq and Bonamo, 1971; (B) from Bonamo and Banks, 1967; (C) from Mustafa, 1975; (D,E) from author's collection.

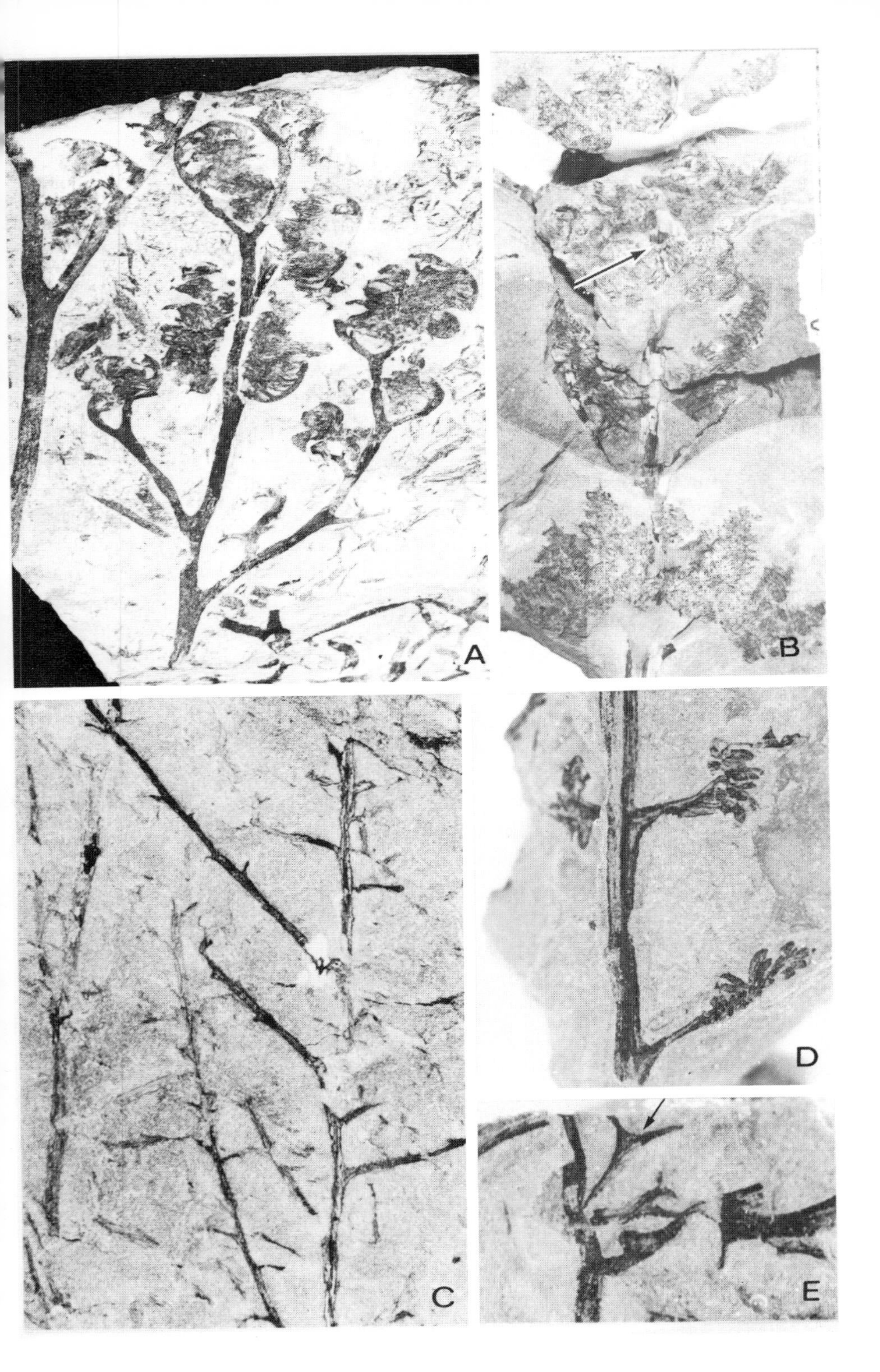
A
B
C
D
E

range of genera that had already appeared in Zone IV" as well as by appearance of more advanced progymnosperms such as *Svalbardia*.

As in the Lower Devonian, it is the lycopods which are most readily identified. Indeed similarities of Lower Devonian representatives such as *Baragwanathia* and the abundance of herbaceous lycopods in the Middle Devonian led to the idea of a relatively unchanged line of herbaceous lycopods from the Devonian to the present day (see, for example, Grierson and Banks, 1963). However, the recent discovery of a ligule in *Leclercqia complexa* (Grierson and Bonamo, 1979), the most intensively studied and best known herbaceous lycopod (Banks *et al.*, 1972; Grierson, 1976) indicates a combination of characters (ligule and homospory) hitherto unknown in the lycopods and not present in modern homosporous *Lycopodium*. Middle Devonian herbaceous lycopods also exhibit a diversity and complexity of leaf morphology not found in recent lycopods, where the leaf is a simple undivided microphyll. *Leclercqia* (Fig. 3D,E) for example, has leaves with five segments; *Colpodexylon trifurcatum* (Fig. 3C) has trifid leaves (Banks, 1944); *Protolepidodendron gilboense* (Grierson and Banks, 1963) has bifurcating leaves, although in species of *Drepanophycus* the leaves are undivided (Grierson and Banks, 1963); and *Archaeosigillaria* (Fig. 3B) has leaves with a pronounced lamina (Fairon-Demaret and Banks, 1978).

Even in the Givetian there are indications that larger lycopods, the stems of which bear the characteristic patterning of later Carboniferous forms, had already evolved. Schweitzer (1965) presents a reconstruction of *Protolepidodendropsis pulchra* (Fig. 3F) from Spitsbergen which had a small trunk approximately 1·5 m high terminating in a sparsely branched crown of leafy axes. Grierson and Banks (1963) report poorly preserved decorticated axes, up to 6·7 cm in diameter from the United States, which they named *Amphidoxodendron*, while from Kazakhstan, Yurina (1968) has described some very well preserved stem petrifactions. Their anatomy is similar to that of Carboniferous lepidodendroids in that an exarch protostele, secondary wood and well-developed bark are all present. Yurina referred these axes to the genus *Lepidodendropsis*, an assignation questioned by Banks (1980b).

Of the two remaining major groups of free-sporing vascular plants, ferns, directly related to and having the appearance of extant forms (i.e. showing the large planated fern frond and abaxial sporangia), did not evolve until the Carboniferous, while the earliest unequivocal sphenopsid is recorded in the Upper Devonian. Middle Devonian representatives are plants consisting of three-dimensional systems of branching axes usually lacking planation or webbing and with sporangia borne on stem-like structures (Scheckler, 1974), In this plexus, however, can be seen the beginnings of more familiar types of organization, e.g. an approach to the whorled arrangement of the leaves of sphenopsids, gymnospermous secondary wood and the anatomy of coenopterid ferns.

Two of the best known Middle Devonian genera, *Rellimia (Protopteridium)* (Fig. 4A) and *Aneurophyton* (Fig. 3C–E), are now considered to be members of the Progymnospermopsida, a class erected for plants with gymnospermous anatomy and pteridophytic (i.e. free-sporing) reproduction (Bonamo, 1975; Beck, 1976). Leclercq and Bonamo (1971) interpret *Rellimia* as a bushy woody shrub in which crowded branches, leaves (the dichotomously branched ultimate appendages) and fertile organs are all borne spirally. These fertile organs are far more highly organized than those of the trimerophytes, the sporangium-bearing system being branched once dichotomously then three to four times pinnately and curved adaxially so that the sporangia (250–400) are inwardly orientated and form overlapping masses (Bonamo, 1977). *Aneurophyton* also has spiral branching (Fig. 4C) but its fructifications are smaller and less branched (Fig. 4D,E). A fertile axis branches just once, with the two arms curving towards each other forming a lyre shape, each arm bearing two rows of sporangia (Serlin and Banks, 1978; Mustafa, 1975). A closely related genus *Tetraxylopteris* (Fig. 4B), first described from the Frasnian of the United States (Beck, 1957; Bonamo and Banks, 1967), has recently been recorded from the Middle Devonian (Givetian) of Germany (Mustafa, 1975). The Upper Devonian representatives have well-preserved secondary phloem, but the earliest recorded occurrence of this tissue is in another progymnosperm, *Triloboxylon hallii* from the Givetian of the United States (Scheckler and Banks, 1971). This plant also produced periderm or cork around the edge of the plant. In living plants periderm is a protective tissue which replaces the outermost layers of stem or root destroyed as a result of extensive secondary production of xylem and phloem.

The evolution of a vascular cambium and hence the capacity to increase in girth by the production of large amounts of additional wood (secondary xylem), also increases the potential for growth in height. Although some of the Lower Devonian trimerophytes reached a height of at least 1 m, their axes lacked secondary thickening and rarely exceeded 1 cm in diameter. Andrews *et al.* (1977) suggest that mutual support came from growing in dense stands. The overall size attained by the Middle Devonian progymnosperms is difficult to estimate because of the fragmentary nature of the fossil record, but they are usually interpreted as bushy shrubs or small trees. Calculations based on the casts of the stumps of tree trunks (*Eospermatopteris*) from the Givetian of New York State suggest that trees at least 13 m high evolved by the end of the Middle Devonian.

Chaloner and Sheerin (1979) discuss the wider biological implications of this increase in height with respect both to the development of more complex plant communities displaying ecological "stratification" and to the evolution of heterospory and the seed habit.

The remaining Middle Devonian plants are united in their lack of secondary wood. One of the best known is *Pseudosporochnus* (Leclercq and Banks, 1962; Leclercq and Lele, 1968), a plant estimated to be 2–3m high with a small trunk terminating in a crown of branches. These divide dichotomously and then bear spirally arranged leaves, which divide two to three times in one plane (Fig. 5A,B). The xylem consists of a number of longitudinally running plates and terete strands, anatomy typical of the cladoxylalean ferns.

Fig. 5. (opposite). Middle Devonian plants A–F: (A) *Pseudosporochnus nodosus* Leclercq and Banks – two branches with numerous leaves : note characteristic patterning on stems; Givetian, Belgium (× 0·7). (B) *P. nodosus* – narrower branch bearing several much-divided leaves; Givetian, Belgium (× 2·4). (C) *Calamophyton primaevum* Kräusel and Weyland – fertile axis bearing numerous sporangiophores (examples of sporangia are indicated by arrows); Eifelian, Germany (× 3·0). (D) *C. primaevum* – sterile specimen illustrating branching pattern and forked leaves; Eifelian, Germany (× 1·0). (E) *C. primaevum* – fertile axis: note the well-developed sterile tips of the sporangiophotes, sporangia being borne on the recurved segments (arrowed); Eifelian, Germany (× 2·5). (F) *Hyenia elegans* Kräusel and Weyland – upright fertile axis with sporangiophores: an attached sporangium is arrowed; Eifelian, Germany (× 2·2). (G) *H. elegans* – upright sterile axis bearing dichotomously branching leaves; Eifelian, Germany (× 3·0).

(A,B) From negatives from M. Fairon-Demaret; (C,D) from H. Mustafa; (E) from Schweitzer, 1973; (F,G) from Schweitzer, 1972.

A
B
C
D
E
F
G

Hyenia (Fig. 5E–G), the Zone IV indicator fossil, has long been considered to be a precursor of the sphenopsids. A rhizome up to 2 m long bears erect aerial branches up to 50 cm high. These produce sterile as well as branched stem-like structures bearing recurved sporangia (sporangiophores). These appendages were originally thought to have been borne in whorls, although now the arrangement is considered to have been spiral (Schweitzer, 1972). *Calamophyton* (Fig. 5C,D) has similar fertile and sterile lateral organs, but lacks a rhizome. Following Leclercq and Schweitzer's description of the anatomy of *Calamophyton* as cladoxylalean, the taxonomic status of *Hyenia* as a pre-sphenopsid was questioned (Leclercq and Schweitzer, 1965; Chaloner and Sheerin, 1979). Banks (1968a) however considered that evidence based on reproductive characters should outweigh that from anatomy and retained *Hyenia* in a group called the Protoarticulatae (see also Skog and Banks, 1973).

This is just a glimpse of the kind of problem encountered in discussing the affinities of these Middle Devonian pteridophytes. I have made no attempt to include the numerous compression fossils from the U.S.S.R. (e.g. Petrosian, 1968) or to describe the wide variety of petrified axes from the United States, some of which exhibit anatomy comparable to that of the coenopterid ferns (see Banks, 1968a; Scheckler, 1974).

Finally, although the typical fern frond had not appeared in the Middle Devonian, there are a few records of large, fan-shaped or flabelliform leaves. Some of these were reviewed by Høeg (1942) in his excellent, if somewhat inconclusive, discussion on the systematic position of *Enigmophyton*, an example from the uppermost Givetian of Spitsbergen. Illustrated here is *Duisbergia mirabilis* (Fig. 3G,H) which occurs in the Middle Devonian of Germany (Kräusel and Weyland, 1934). The authors reconstructed the plant as having a tapering trunk up to 3 m high with the top half covered by spirally inserted leaf-like structures and suggested affinity with the lycopods.

SILURO-DEVONIAN LANDSCAPES AND VEGETATION: AN ATTEMPT AT SYNTHESIS

The many difficulties encountered in reconstructing whole plants, in elucidating their affinities, in constructing a plant-based stratigraphy and in deducing their geographical distribution are well documented elsewhere (Banks, 1980b; Chaloner and Sheerin, 1979; Edwards, 1973). Nevertheless we are beginning to reach a better understanding of the composition of Siluro-Devonian floras, of their stratigraphy and their palaeogeography. In bringing together the few pieces of evidence relevant to their ecology, it will become apparent how lamentably little is known about vegetation as a whole, about plant communities and ecosystems.

I shall begin with the one exception. The ?mid-Siegenian Rhynie Chert of Scotland consists of successive layers of peat interbedded with thin layers of silicified sandstone (Kidston and Lang, 1921; Tasch, 1957). The identifiable plant remains in the peat indicate that the peat-formers included species of *Rhynia, Asteroxylon, Horneophyton* and *Nothia*, but the only plant preserved in growth position is *Rhynia gwynne-vaughanii*. Although it is possible that the plants grew in an environment saturated with siliceous water and therefore formed a specialized and highly adapted type of vegetation, it seems more likely that they grew on a bog which periodically became flooded following the overflow of nearby ponds fed by hot siliceous springs. Tasch believed that the abundance of crustaceans and, to a lesser extent, algae at certain horizons resulted from such episodes of flooding. (The arthropods found in the peat are reviewed by Rolfe, Chapter 6 of this volume). It is considered unlikely that the water-table was permanently high. Kidston and Lang pointed out that stomata occurred near the bases of erect axes, while the abundance of rhizoids on rhizomes suggests that the surface peat was not waterlogged for extensive periods (Rosene and Bartlett, 1950). Whether or not such vegetation was typical of the wetter areas of the intermontane basins of the "Old Red Continent" is impossible to determine. There are fragmentary compression fossils in the more typical Old Red Sandstone sediments below the Chert, but these are unidentifiable.

Scott in his paper on late Palaeozoic vegetation in this volume (Chapter 5) and elsewhere (e.g. Scott, 1977), emphasizes the importance of considering fossil plant assemblages in relation to various lines of sedimentological evidence in the reconstruction of vegetation. In 1973, I made a tentative attempt at this kind of approach in recording the geographical distribution of Devonian floras together with the sedimentary facies in which they are preserved (continental, both external and internal, and marine). Surprisingly (and disappointingly from a palaeoecological viewpoint) many of the floras occur in marine sediments, in both near-shore and deeper water facies. This is the case for all the floras of Pridoli age as well as those in the Ludlow Series, so that the habitats of the earliest vascular plants are highly conjectural (e.g. Edwards, 1980) and also for the extensive floras of the German Lower Devonian. Størmer (1977) discussed the fauna of an Emsian black shale from Alken in the Moselle Valley which was thought to have been deposited in a lagoon. The plants were originally described by Kräusel and Weyland (1935, 1962, 1968) and the reconstruction of a mangrove-type vegetation by Solle (1970) in his sedimentological analysis of the area. Some of the plants (e.g. *Chaetocladus*) were indeed algal and perhaps grew in the shallow waters of the lagoon. There is, however, no direct evidence that the vascular plants were aquatic, as is indicated in reconstructions of *Zosterophyllum rhenanum* and *Protobarinophyton* (Kräusel and Weyland, 1935; Ananiev, 1960) in which only the fertile regions grew above water. A new reconstruction of *Z. rhenanum* (H-J. Schweitzer, personal communication) shows it to have been a terrestrial plant with much-branched rhizomes. It seems more likely that the vascular plants were washed into the lagoon having originally colonized drier coastal areas. Schaarschmidt (1974) has described a putative vascular plant from the same locality, which he interpreted as a halophyte on the basis of swellings at the bases of the main axes and lateral branches. This raises the interesting question as to whether salt-marsh vegetation existed in early Devonian times. Salt-marsh environments today are exceedingly specialized ones, particularly if subaerial exposure is prolonged, and the higher plants (usually angiosperms) which colonize them are highly adapted both anatomically and physiologically to survive periods of severe water stress. Present day pteridophytes are rarely found on salt-marshes although some live in brackish environments and there is no direct evidence for salt-marsh vegetation in the early Devonian (see references in Waisel, 1972; Reimold and Queen, 1974).

Almost all of the plants I have described above grew on "Old Red Land", a landmass which straddled the Equator in Devonian times and which has been the subject of extensive sedimentological, palaeogeographical and palaeoclimatic studies (e.g. Allen *et al.*, 1968; Allen, 1974a; Allen and Friend, 1968; Woodrow *et al.*, 1973). Such studies have provided information on the types of environments available for colonization. Plants are usually considered to have lived along rivers, around lakes or in coastal areas (Andrews *et al.*, 1977; Woodrow *et al.*, 1973; Edwards, 1980). My own research interest is in the Anglo-Welsh area (Allen, 1974a,b; Allen and Tarlo, 1963). In the Brecon Beacons, plant remains are locally abundant in Lower Old Red Sandstone fluviatile desposits, the better preserved ones occurring in grey fine-grained sediments. *Gosslingia*, for example is found throughout a layer several centimetres thick at the Brecon Beacons Quarry (Edwards, 1970b), its axes showing parallel alignment and, on any one bedding plane, with the sporangial zone at the same level on all the axes. Although the alignment itself is produced by currents, the relative positions of the sporangia and the excellent preservation of the plants led me to conclude that they were swept over and rapidly covered with sediment when they were still attached to the ground, and they grew in pure, dense stands near the stream or river and were killed during periods of flooding.

Parallel alignment and vast quantities of axes extending through some thickness of sediment have also been reported from Lower Devonian localities in northern Maine. Andrews *et al.* (1977) interpret the environment as a brackish-water or freshwater coastal swamp, with the plants, e.g. *Psilophyton* sp., growing on drier, slightly elevated (a few metres high) areas usually in monospecific stands, but just occasionally in a mixture of two or three species. *Kaulangiophyton* and *Taeniocrada* are thought to have grown in "small pockets" in the marshland area, possibly occupying quite different ecological niches.

In both these examples plants were growing in areas where the chances of preservation were high and, in the case of the Welsh plants in which anatomy is known, possessed abundant thick-walled tissue which again might have increased the potential for fossilization. This makes it difficult to estimate how representative of the coastal-plain vegetation such preserved plants are. I recently suggested (Edwards, 1980), for example, that plants with predominantly thin-walled tissues such as *Rhynia* would not have been preserved in more typical Old Red Sandstone sediments and that the variety of spores and dispersed cuticular fragments recovered on bulk maceration of sandstones and shales is a further indication of the incompleteness of the macrofossil record.

There were parts of the alluvial flood-plains which were free from sedimentation for long periods, being far away from the influence of rivers (Allen, 1974a), and where conditions for the greater part of the year would probably have been very unhospitable for plant growth. From a comparison of the abundant carbonate units in the Lower Old Red Sandstone of the Anglo-Welsh area with Quaternary soil carbonates (calcretes) of low latitudes, Allen concluded that the climate was warm to hot (mean annual temperature 16°–20°C) and marked by a low seasonal rainfall (100–500 mm). These areas would perhaps have had a surface covering of algae (Edwards *et al.*, 1979) in wetter periods, but it is doubted that such conditions would have been suitable for the growth of higher plants.

There is no direct evidence for upland vegetation in early Devonian times. Comparison of the composition of two Dittonian floras, one preserved in the external (Edwards and Richardson, 1974) and the other in the internal facies of the Old Red Sandstone (Edwards, 1970a, 1975), suggests that by the early Siegenian there was similar vegetation in the intermontane basins and on the coastal plains. It seems not unlikely that plants would have migrated further up the streams into the uplands where the climate might have been more equable. Indeed plants with the type of rhizome system present in *Zosterophyllum* might well have begun to stabilize soils.

Some of the most important research which has resulted in a better understanding of the structure and relationships of Middle Devonian plants has been based on the extensive floras of New York State (e.g. Banks, 1966; Banks *et al.*, 1972; Bonamo, 1977; Matten, 1974, and reference cited therein), but as yet there have been few attempts at palaeoecological analysis even though information is available on the sedimentology, palaeogeography and palaeoclimates (Allen and Friend, 1968; Woodrow *et al.*, 1973) of the Catskill region. Matten (1974) compared the composition of the more-or-less contemporaneous floras from Cairo and Gilboa in New York State. Apart from *Eospermatopteris–Aneurophyton* he found them to be quite different, the former being dominated by ferns and progymnosperms and the latter by lycopods and sphenopsids, which comparison with later Carboniferous floras suggested a swampy environment. In addition, the swollen bases of *Eospermatopteris* trunks appeared to have been preserved *in situ*. Matten concluded that it was perhaps premature to speculate on the detailed ecology of such assemblages, but suggested that their differences in composition indicate that vegetation was not uniform with "populations of different plants growing in different places, forming different communities" (Matten, 1974, p.315).

This review has been concerned with plant macrofossils, but there is also some palynological evidence to support ecological diversity in the Middle and Upper Devonian (Richardson, 1969). In a preliminary comparison of the composition of spore assemblages from a number of depositional environments, Richardson found that assemblages from marine or lower flood-plain-marginal deltaic facies are dominated by the thick-walled spore *Geminospora*, and that *Ancyrospora* with bifurcating spines and the pseudosaccate *Rhabdosporites* are commonest in sediments deposited in fresh water. He speculated that the plants producing *Geminospora* were halophytes either growing on delta swamps or in backswamps or levels in the lower flood-plain, but that the *Rhabdosporites*- and *Ancyrospora*-producing plants colonized the upper flood-plain or nearby freshwater lakes. Recent advances in elucidating the identity of these parent plants are summarized by Allen (1980). *Rhabdosporites* has been isolated from *Tetraxylopteris* (Bonamo and Banks, 1967) and from *Rellimia* (Leclercq and Bonamo, 1971). *Geminospora*-type spores have been found in other progymnosperms such as *Aneurophyton* (Streel, 1964). In similar analyses of Lower and Middle Devonian spore assemblages from Belgium and elsewhere, Streel (1964, 1967) also recognized a correlation between composition of spore assemblage and environment of deposition which led him to propose the existence of several different types of vegetation in the Devonian.

CONCLUSIONS

Although it has been my main concern to provide information on the kinds of vascular plants which colonized the land in late Silurian and early Devonian times, it is a disappointment to have failed to organize such plants into communities and to place these in their habitats with any confidence. There is some evidence to suggest that Lower Devonian communities were simple in composition, often comprising a dense stand of a single species, although the areas covered by such stands and what factors controlled the distribution of individual species remain unknown. In later communities, although usually impossible to identify individual members, it is possible to consider community structure in terms of evolutionary attainment as increasing complexity in morphology and anatomy resulted in changes in growth and reproductive strategies.

We know nothing about the life-span of the earliest vascular plants, but we can postulate that growth in the aerial axes in *Cooksonia*-like plants was strictly determinate, with the production of sporangia curtailing any further appreciable increase in height. (There is the possibility that a rhizome system acted as a perennating organ and continually produced erect axes, but we know nothing about the basal parts of *Cooksonia*.) The pseudomonopodial type of branching seen in some zosterophylls and trimerophytes indicates that the capacity for continued vegetative growth over longer periods and for the production of larger numbers of sporangia had evolved. Such plants did reach heights of over 2 m (e.g. *Pertica varia*), but remained upright by mutual support derived from growing in dense stands. Although these communities would have provided shade and cover for animals, the amount of light for photosynthesis would also have been cut down.

The evolution of a vascular cambium by the Middle Devonian allowed increase in girth, and hence height, and would have resulted in the stratification of communities. The earliest progymnosperms were probably shrubs, but because they lacked planated webbed leaves, there may have been sufficient light reaching the ground to allow the growth of a herbaceous layer of vegetation. By the end of the Middle Devonian there is some evidence that small progymnospermous trees with well-defined trunks had evolved; the arborescent habit also being exhibited by ferns such as *Pseudosporochnus* (with the growth form of a modern tree fern) and by the lycopods, e.g. *Lepidodendropsis* (Yurina, 1969). Chaloner and Sheerin (1979) relate this increase in size of individual plants and the resulting complexity of communities to the evolution of heterospory and the seed habit. In addition this change in reproductive strategy would have permitted the colonization of much drier environments. In the majority of extant homosporous pteridophytes, the gametophyte, an independent plant, requires abundant moisture for both vegetative growth and reproduction (Edwards, 1980). It is therefore restricted to damp, humid environments and, because the sporophyte develops directly on the gametophyte following fertilization, this in turn governs the distribution of the sporophyte. The appearance of heterospory in the early Devonian marked the beginning of the elimination of the independent free-living gametophyte. This level of organization was attained in several lines of pteridophytes, but it is the progymnosperms which are considered to have given rise to the earliest seed plants. The first seed is recorded in the Upper Devonian, the evolutionary climax to a period of extensive diversification in the history of vascular plants.

ACKNOWLEDGEMENTS

It is a great pleasure to record my gratitude to many colleagues for their generosity: to Professors H. P. Banks and W. G. Chaloner for access to their unpublished manuscripts, and to Professors H. P. Banks, P. B. Bonamo, P. G. Gensel, J. D. Grierson and H-J. Schweitzer, Drs. M. Fairon-Demaret and H. Mustafa, and Ms J. Tims for entrusting me with their negatives.

REFERENCES

Allen, K. C. (1980). A review of *in situ* late Silurian and Devonian spores. *Rev. Palaeobot. Palynol.* **29**, 253–270.

Allen, J. R. L. (1974a). Studies in fluviatile sedimentation: implications of pedogenic carbonate units, Lower Old Red Sandstone, Anglo-Welsh outcrop. *Geol. J.* **9**, 181–208.

Allen, J. R. L. (1974b). The Devonian rocks of Wales and the Welsh Borderland. *In* "The Upper Palaeozoic and Post-Palaeozoic Rocks of Wales" (T. R. Owen, ed.), pp. 47–84. University of Wales Press, Cardiff.

Allen, J. R. L. and Friend, P. F. (1968). Deposition of the Catskill Facies, Appalachian Region: with notes on some other Old Red Sandstone basins. *Spec. Pap. geol. Soc. Am.* **106**, 21–74.

Allen, J. R. L. and Tarlo, L. B. (1963). The Downtonian and Dittonian facies of the Welsh Borderland. *Geol. Mag.* **100**, 129–155.

Allen, J. R. L., Dineley, D. L. and Friend, P. F. (1968). Old Red Sandstone basins of North America and Northwest Europe. *In* "International Symposium on the Devonian System, Calgary, 1967" (D. H. Oswald, ed.), Vol. 1, pp. 69–98. Alberta Society of Petroleum Geologists, Calgary.

Ananiev, A. R. (1960). Plants. *Biostratigrafiya paleozoya Sayano-Altaiskii gornoi oblasti srednii palaeozoi* 301–320. (Transl. N. L. L. RTS 5009.)

Andrews, H. N., Gensel, P. G. and Forbes, W. H. (1974). An apparently heterosporous plant from the Middle Devonian of New Brunswick. *Palaeontology* **17**, 387–408.

Andrews, H. N., Kasper, A. E., Forbes, W. H., Gensel, P. G. and Chaloner, W. G. (1977). Early Devonian flora of the Trout Valley Formation of northern Maine. *Rev. Palaeobot. Palynol.* 23, 255–285.

Banks, H. P. (1944). A new Devonian lycopod genus from southeastern New York. *Am. J. Bot.* **31**, 649–659.

Banks, H. P. (1966). Devonian flora of New York State. *Empire State Geogram.* 4, 10–24.

Banks, H. P. (1968a). The early history of land plants. *In* "Evolution and Environment" (E. T. Drake, ed.), pp. 73–107. Yale University Press, New Haven and London.

Banks, H. P. (1968b). Anatomy and affinities of a Devonian *Hostinella. Phytomorphology* **17**, 321–330.

Banks, H. P. (1972). The stratigraphic occurrence of early land plants. *Palaeontology* **15**, 365–377.

Banks, H. P. (1973). Occurrence of *Cooksonia*, the oldest vascular land plant macrofossil in the Upper Silurian of New York State. *J. Indian bot. Soc.* Golden Jubilee Volume **50A**, 227–235.

Banks, H. P. (1975a) Reclassification of Psilophyta. *Taxon* **24**, 401–413.

Banks, H. P. (1975b). The oldest vascular plants: a note of caution. *Rev. Palaeobot. Palynol.* **20**, 13–25.

Banks, H. P. (1975c). Palaeogeographic implications of some Silurian–early Devonian floras. *In* "Gondwana Geology" (K. S. W. Campbell, ed.), pp. 71–97, Australian National University Press, Canberra.

Banks, H. P. (1980a). The role of *Psilophyton* in the evolution of vascular plants. *Rev. Palaeobot. Palynol.* **29**, 165–176.

Banks, H. P. (1980b). Floral assemblages in the Siluro-Devonian. *In* "Biostratigraphy of Fossil Plants: Successional and Palaeoecological Analysis" (D. Dilcher and T. N. Taylor, eds). Dowden, Hutchinson and Ross, Pennsylvania.

Banks, H. P., Bonamo, P. B. and Grierson, J. D. (1972). *Leclercqia complexa* gen. et sp. nov., a new lycopod from the late Middle Devonian of eastern New York. *Rev. Palaeobot. Palynol.* **14**, 19–40.

Banks, H. P., Leclercq, S. and Hueber, F. M. (1975). Anatomy and morphology of *Psilophyton dawsonii* sp.n. from the late Lower Devonian of Quebec (Gaspé), and Ontario, Canada. *Palaeontogr. am.* 8, 77–116.

Beck, C. B. (1957). *Tetraxylopteris schmidtii* gen. et sp. nov., a probable pteridosperm precursor from the Devonian of New York. *Am. J. Bot.* **44**, 350–367.

Beck, C. B. (1960). The identity of *Archaeopteris* and *Callixylon. Brittonia* **72**, 351–368.

Beck, C. B. (1976). Current status of the Progymnospermopsida. *Rev. Palaeobot. Palynol.* **21**, 5–23.

Bierhorst, D. W. (1971). "Morphology of Vascular Plants." Macmillan, New York.

Bonamo, P. M. (1975). The Progymnospermopsida: building a concept. *Taxon* **24**, 569–579.

Bonamo, P. M. (1977). *Rellimia thomsonii* (Progymnospermopsida) from the Middle Devonian of New York State. *Am. J. Bot.* **64**, 1272–1285.

Bonamo, P. M. and Banks, H. P. (1967). *Tetraxylopteris schmidtii*: its fertile parts and its relationships within the Aneurophytales. *Am. J. Bot.* **54**, 755–768.

Boureau, E., Lejal-Nicol, A. and Massa, D. (1978). À propos du Silurien et du Dévonien en Libye. Il faut reporter au Silurien la date d'apparition des plantes vasculaires. *C. r. hebd. Séanc. Acad. Sci., Paris* **286D**, 1567–1571.

Chaloner, W. G. (1970). The rise of the first land plants. *Biol. Rev.* **45**, 353–377.

Chaloner, W. G. and Sheerin, A. (1979). Devonian macrofloras. *In* "The Devonian System" (M. R. House, C. T. Scrutton and M. G. Bassett, eds), *Spec. Pap. Palaeont.* **23**, 145–161.

Cookson, I. (1935). On plant remains from the Silurian of Victoria, Australia, that extend and connect floras hitherto described. *Phil. Trans. R. Soc.* **B225**, 127–148.

Croft, W. N. and Lang, W. H. (1942). The Lower Devonian flora of the Senni Beds of Monmouthshire and Breconshire. *Phil. Trans. R. Soc.* **B231**, 131–163.

Daber, R. (1971). *Cooksonia* – one of the most ancient psilophytes – widely distributed, but rare *Botanique* **2**, 35–39.

Edwards, D. (1970a). Fertile Rhyniophytina from the Lower Devonian of Britain. *Palaeontology* **13**, 451–461.

Edwards, D. (1970b). Further observations on the Lower Devonian plant, *Gosslingia breconsensis* Heard. *Phil. Trans. R. Soc.* **B258**, 225–243.

Edwards, D. (1973). Devonian floras. *In* "Atlas of Palaeobiogeography" (A. Hallam, ed.), pp. 105–115. Elsevier, Amsterdam.

Edwards, D. (1975). Some observations on the fertile parts of *Zosterophyllum myretonianum* Penhallow from the Lower Old Red Sandstone of Scotland. *Trans. R. Soc. Edinb.* **69**, 251–265.

Edwards, D. (1976). The systematic position of *Hicklingia edwardii* Kidston and Lang. *New Phytol.* **76**, 178–181.

Edwards, D. (1979). A late Silurian flora from the Lower Old Red Sandstone of south-west Dyfed. *Palaeontology* **22**, 23–52.

Edwards, D. (1980). The early history of vascular plants based on late Silurian and Lower Devonian floras of the British Isles. *In* "The Caledonides of the British Isles – reviewed" (B. F. Leake, A. L. Harris and C. H. Holland, eds). Geological Society of London.

Edwards, D. and Davies, E. C. W. (1976). Oldest recorded *in situ* tracheids. *Nature, Lond.* **263**, 494–495.

Edwards, D. and Richardson, J. B. (1974). Lower Devonian (Dittonian) plants from the Welsh Borderland. *Palaeontology* **17**, 311-324.

Edwards, D. and Rogerson, E. C. W. (1979). New records of fertile Rhyniophytina from the late Silurian of Wales. *Geol. Mag.* **116**, 93–98.

Edwards, D., Bassett, M. G. and Rogerson, E. C. W. (1979). The earliest vascular land plants: continuing the search for proof. *Lethaia* **12**, 313–324.

Fairon-Demaret, M. and Banks, H. P. (1978). Leaves of *Archaeosigillaria vanuxemii*, a Devonian lycopod from New York. *Am. J. Bot.* **65**, 246–249.

Garratt, M. J. (1979). New evidence for the Silurian (Ludlow) age for the earliest *Baragwanathia* flora. *Alcheringa* **2**, 217–224.

Gensel, P. G. (1977). Morphologic and taxonomic relationships of the Psilotaceae relative to evolutionary lines in early land vascular plants. *Brittonia* **29**, 14–29.

Gensel, P. G. (1979). Two *Psilophyton* species from the Lower Devonian of eastern Canada with a discussion of morphological variation within the genus. *Palaeontographica* **168B**, 81–146.

Gensel, P., Kasper, A. and Andrews, H. N. (1969). *Kaulangiophyton*, a new genus of plants from the Devonian of Maine. *Bull. Torrey bot. Club* **96**, 265–276.

Gensel, P. G., Andrews, H. N. and Forbes, W. H. (1975). A new species of *Sawdonia* with notes on the origin of microphylls and lateral sporangia *Bot. Gaz.* **136**, 50–62.

Granoff, J. A., Gensel, P. G. and Andrews, H. N. (1976). A new species of *Pertica* from the Devonian of eastern Canada. *Palaeontographica* **155B**, 119–128.

Gray, J. and Boucot, A. J. (1971). Early Silurian spore tetrads from New York: earliest New World evidence for vascular plants? *Science* **173**, 918–921.

Gray, J. and Boucot, A. J. (1977). Early vascular land plants: proof and conjecture. *Lethaia* **10**, 145–147.

Gray, J., Laufeld, S. and Boucot, A. J. (1974). Silurian trilete spores and spore tetrads from Gotland: their implications for land plant evolution. *Science* **185**, 260–263.

Grierson, J. D. (1976). *Leclercqia complexa* (Lycopsida, Middle Devonian): its anatomy, and the interpretation of pyrite petrifactions. *Am. J. Bot.* **63**, 1184–1202.

Grierson, J. D. and Banks, H. P. (1963). Lycopods of the Devonian of New York State. *Palaeontogr. am.* **4**(31), 221–295.

Grierson, J. D. and Bonamo, P. M. (1979). *Leclercqia complexa*: earliest ligulate lycopod (Middle Devonian). *Am. J. Bot.* **66**, 474–476.

Høeg, O. A. (1942). The Downtonian and Dittonian flora of Spitsbergen. *Norges Svalb. Ishavs-Unders. Skrifter* **83**, 1–228.

House, M. R., Richardson, J. B., Chaloner, W. G., Allen, J. R. L., Holland, C. H. and Westoll, T. S. (1977). A correlation of the Devonian rocks in the British Isles. *Spec. Rep. geol. Soc. Lond.* **8**, 1–110.

Hueber, F. M. (1968). *Psilophyton:* the genus and the concept. *In* "International Symposium on the Devonian System, Calgary, 1967" (D. H. Oswald, ed.), Vol. 1, pp. 815–822. Alberta Society of Petroleum Geologists, Calgary.

Hueber, F. M. (1971). Early Devonian plants from Bathurst Island, District of Franklin. *Geol. Surv. Pap. Can.* 71-28, 1–17.

Hueber, F. M. and Grierson, J. D. (1961). On the occurrence of *Psilophyton princeps* in the early Upper Devonian of New York. *Am. J. Bot.* **48**, 473–479.

Ischenko, T. A. (1975). "The late Silurian Flora of Podolia." Akademiya Nauk

Ukraniskoi SSR, Institut Geologicheskikh Nauk, Naukova Dumka, Kiev. (In Russian.)
Johnson, H. J. and Konishi, K. (1958). Studies on Devonian algae. *Colo. Sch. Mines.* Q53, 22–33.
Kasper, A. E., Jr and Andrews, H. N. (1972). *Pertica*, a new genus of Devonian plants from northern Maine. *Am. J. Bot.* **59**, 897–911.
Kidston, R. and Lang, W. H. (1917). On Old Red Sandstone plants showing structure from the Rhynia Chert bed, Aberdeenshire. Part I. *Rhynia gwynne-vaughani. Trans. R. Soc. Edinb.* **51**, 761–784.
Kidston, R. and Lang, W. H. (1920). Ibid. Part III. *Asteroxylon mackiei*, Kidston and Lang. *Trans. R. Soc. Edinb.* **52**, 643–680.
Kidston, R. and Lang, W. H. (1921). Ibid. Part V. The Thallophyta occurring in the peat bed; the succession of the plants throughout a vertical section of the bed, and the conditions of accumulation and preservation of the deposit. *Trans. R. Soc. Edinb.* **52**, 855–902.
Kräusel, R. and Weyland, H. (1934). Pfalanzenreste aus dem Devon. VI. *Duisbergia mirabilis* Kr. et Weyl. VIII. Pflanzenreste vom Korzert bei Elberfeld. *Senckenbergiana* **16**, 161–175.
Kräusel, R. and Weyland, H. (1935). Neue Pflanzenfunde im rheinischen Unterdevon. *Palaeontographica* **80B**, 171–190.
Kräusel, R. and Weyland, H. (1962). Algen und Psilophyten aus dem Unterdevon von Alken an der Mosel. *Senckenberg. leth.* **43**, 249–282.
Kräusel, R. and Weyland, H. (1968). Eine weitere Psilophytale aus dem Unterdevon von Alken an der Mosel, ein Beitrag zur Gattungsgruppe *Protobarinophyton-Pectinophyton-Barinophyton. Senckenberg. leth.* **49**, 241–249.
Lang, W. H. (1927). Contributions to the study of the Old Red Sandstone flora of Scotland. VI. On *Zosterophyllum myretonianum*, Penh., and some other plant remains from the Carmyllie Beds of the Lower Old Red Sandstone. *Trans. R. Soc. Edinb.* **55**, 443–452.
Lang, W. H. (1937). On the plant-remains from the Downtonian of England and Wales. *Phil. Trans. R. Soc.* **B227**, 245–291.
Lang, W. H. (1945). *Pachytheca* and some anomalous early plants *(Prototaxites, Nematothallus, Parka, Foerstia, Orvillea* n.g.). *Bot. J. Linn. Soc.* **53**, 535–552.
Lang, W. H. and Cookson, I. C. (1935). On a flora, including vascular land plants, associated with *Monograptus,* in rocks of Silurian age, from Victoria, Australia. *Phil. Trans. R. Soc.* **B224**, 421–449.
Leclercq, S. (1942). Quelque plantes fossiles receuillies dans le Dévonien inférieur des environs de Nonceveux (Bordure orientale du Bassin de Dinant). *Bull. Soc. géol. Belg.* **65B**, 193–211.
Leclercq. S. and Banks, H. P. (1962). *Pseudosporochnus nodosus* sp. nov., a Middle Devonian plant with cladoxylalean affinities. *Palaeontographica* **110B**, 1–34.
Leclercq, S. and Bonamo, P. M. (1971). A study of the fructification of *Milleria* (*Protopteridium*) *thomsonii* Lang from the Middle Devonian of Belgium. *Palaeontographica* **136B**, 83–114.
Leclercq, S. and Lele, K. M. (1968). Further investigation on the vascular system

of *Pseudosporochnus nodosus* Leclercq and Banks. *Palaeontographica* **123B**, 97–112.

Leclercq, S. and Schweitzer, H-J. (1965). *Calamophyton* is not a sphenopsid. *Bull. Acad. r. Belg. Cl. Sci.* Ser. 5, **51**, 1395–1403.

Lee, H-H. and Tsai, C-Y. (1978). Early Devonian *Zosterophyllum* remains from southwest China. *Acta palaeont. sin.* **16**, 12–34 (Abstract in English.)

Matten, L. C. (1974). The Givetian flora from Cairo, New York: *Rhacophyton, Triloboxylon* and *Cladoxylon. Bot. J. Linn. Soc.* **68**, 303–318.

Mustafa, H. (1975). Beiträge zur Devonflora I. *Argumenta Palaeobotanica* **4**, 101–133.

Niklas, K. J. (1976a). Morphological and ontogenetic reconstructions of *Parka decipiens* Fleming and *Pachytheca* Hooker from the Lower Old Red Sandstone, Scotland. *Trans. R. Soc. Edinb.* **69**, 483–499.

Niklas, K. J. (1976b). The chemotaxonomy of *Parka decipiens* from the Lower Old Red Sandstone, Scotland (U. K.). *Rev. Palaeobot. Palynol.* **21**, 205–217.

Niklas, K. J. (1976c). Chemotaxonomy of *Prototaxites* and evidence for possible terrestrial adaptation. *Rev. Palaeobot. Palynol.* **22**, 1–17.

Obrhel, J. (1962). Die Flora der Pridoli-Schichten (Budnany-Stufe) des mittelböhmischen Silurs. *Geologie* **11**, 83–97.

Obrhel, J. (1968). Die Silur- und Devonflora des Barrandiums. *Paläont. Abh. Berl.* **2B**, 635–793.

Petrosian, N. M. (1968). Stratigraphic importance of the Devonian flora of the U.S.S.R. *In* "International Symposium on the Devonian System, Calgary, 1967" (D. H. Oswald. ed.), Vol. 2, pp. 579–586. Association of Petroleum Geologists, Calgary.

Plumstead, E. P. (1977). A new phytogeographical Devonian zone in southern Africa which includes the first record of *Zosterophyllum. Trans. geol. Soc. S. Afr.* **80**, 267-277.

Pratt, L. M., Phillips, T. L. and Dennison, J. M. (1978). Evidence of non-vascular land plants from the early Silurian (Llandoverian) of Virginia, U.S.A. *Rev. Palaeobot. Palynol.* **25**, 121–149.

Raven, J. A. (1977). The evolution of vascular land plants in relation to supracellular transport processes. *In* "Advances in Botanical Research" (H. W. Woolhouse, ed.), Vol. 5, pp. 154–219. Academic Press, London and New York.

Reimold, R. J. and Queen, W. H. eds (1974). "Ecology of Halophytes." Academic Press, New York and London.

Richardson, J. B. (1969). Devonian spores. *In* "Aspects of Palynology" (R. H. Tschudy and R. A. Scott, eds), pp. 193–222. Wiley, New York.

Rosene, H. F. and Bartlett, L. E. (1950). Effect of Anoxia on water influx of individual radish root hair cells. *J. Cell. comp. Physiol.* **36**, 83–96.

Schaarschmidt, F. (1974). *Mosellophyton hefter* n.g., n.sp. (?Psilophyta) ein sukkulenta Halophyt aus dem Unterdevon von Alken an der Mosel. *Paläont. Z.* **48**, 188–204.

Scheckler, S. E. (1974). Systematic characters of Devonian ferns. *Ann. Mo. bot. Gdn* **61**, 462–473.

Scheckler, S. E. and Banks, H. P. (1971). Anatomy and relationships of some Devonian progymnosperms from New York. *Am. J. Bot.* **58**, 737–751.

Scheckler, S. E. and Banks, H. P. (1974). Periderm in some Devonian plants. *In* "Advances in Plant Morphology" (Y. S. Murty, B. M. Johri, H. Y. Ram Mohan and T. M. Vargehese, eds), pp. 58–64. Meerut City University Press, India.

Schmid, R. (1976). Septal pores in *Prototaxites*, an enigmatic Devonian plant. *Science* **191**, 287–288.

Schweitzer, H-J. (1965). Über *Bergeria mimerensis* und *Protolepidodendropsis pulchra* aus dem Devon Westspitzbergens. *Palaeontographica* **115B**, 117–138.

Schweitzer, H-J. (1972). Die Mitteldevon-Flora von Lindlar (Rheinland) 3. Filicinae – *Hyenia elegans* Kräusel & Weyland. *Palaeontographica* **137B**, 154–175.

Schweitzer, H-J. (1973). Die Mitteldevon-Flora von Lindlar (Rheinland) 4. Filicinae – *Calamophyton primaevum* Kräusel & Weyland. *Palaeontographica* **140B**, 117–150.

Scott, A. C. (1977). A review of the ecology of Upper Carboniferous plant assemblages, with new data from Strathclyde. *Palaeontology* **20**, 447–473.

Serlin, B. S. and Banks, H. P. (1978). Morphology and anatomy of *Aneurophyton,* a progymnosperm from the late Devonian of New York. *Palaeontogr. am.* 8 (51), 343-359.

Skog, J. E. and Banks, H. P. (1973). *Ibyka amphikoma* gen. et sp. nov., a new protoarticulate precursor from the late Middle Devonian of New York State. *Am. J. Bot.* **60**, 366–380.

Solle, G. (1970). Die Hunsrück-Insel im oberen Unterdevon. *Notizbl. hess. Landesamt. Bodenforsch. Wiesbaden* **98**, 50–80.

Størmer, L. (1977). Arthropod invasion of land during late Silurian and Devonian times. *Science* **197**, 1362–1364.

Streel, M. (1964). Une association de spores du Givétian inférieur de la Vesdra à Goé (Belgique). *Ann. Soc. géol. Belg.* **87**, 1–29.

Streel, M. (1967). Associations de spores du Dévonien inférieur Belge et leur signification stratigraphique. *Ann. Soc. géol. Belg.* **90**, 11–54.

Stubblefield, S. and Banks, H. P. (1978). The cuticle of *Drepanophycus spinaeformis*, a long-ranging Devonian lycopod from New York and eastern Canada. *Am. J. Bot.* **65**, 110–118.

Tasch, P. (1957). Flora and fauna of the Rhynie Chert: a palaeoecological evaluation of published evidence. *Univ. Stud. Munic. Univ. Wichita* **32**, 3–24.

Waisel, Y. (1972). "The Biology of Halophytes." Academic Press, New York and London.

Walton, J. (1964). On the morphology of *Zosterophyllum* and some early Devonian plants. *Phytomorphology* **14**, 155–160.

Woodrow, D. L. Fletcher, F. W. and Ahrnsbank, W. F. (1973). Palaeogeography and palaeoclimate at the depostion site of the Devonian Catskill and Old Red facies. *Bull. geol. Soc. Am.* **84**, 3051–3064.

Yurina, A. L. (1969). Devonian floras of central Kazakhstan. *Materielў po geologii Shentral 'nogo Kazakhstana* **8**, 1–208. *(In Russian).*

5 | The Ecology of Some Upper Palaeozoic Floras

ANDREW C. SCOTT

Geology Department, Chelsea College (University of London), 271 King Street, London W6, England

Abstract: The ecology of some Upper Devonian, Carboniferous and Permian floras is briefly reviewed. A number of specific examples of combined sedimentological and palaeontological studies are taken to emphasize the need for interdisciplinary analysis in fossil plant ecology. Ecological changes in the late Devonian are important with, for instance, the occurrence of trees causing a structure in plant communities not present before that time. The basic pattern in late Palaeozoic plant ecology appears to have been set late in the Devonian or early in the Carboniferous with various lowland floras being associated with specific ecological conditions. Sphenopsids and lycopods often appear to dominate lake margins and "swamp forests", whilst pteridosperms appear to dominate drier flood-plain environments. This pattern persists throughout the Carboniferous and in some regions into the earliest Permian. Upland floras are first recognized at the beginning of the Upper Carboniferous and are dominated by various gymnosperms, conifers and cordaites in particular. These floras spread into the lowland areas during the Permian. The general patterns of change in the plant ecology are thought to be controlled in part by broad climatic changes. Relationships of climate, vegetation and erosion as well as evidence for animal/plant interactions are also considered.

INTRODUCTION

This paper serves only as one to "set the scene" of the terrestrial floras during the later part of the Devonian, Carboniferous and

Systematics Association Special Volume No. 15, "The Terrestrial Environment and the Origin of Land Vertebrates", edited by A. L. Panchen, 1980, pp. 87-115, Academic Press, London and New York.

Permian. I shall discuss various aspects of preservation, distribution and ecology, as well as animal/plant interactions in the broadest sense. It is not intended to be a comprehensive review of Palaeozoic plant ecology since the Carboniferous has already been adequately treated (Scott, 1977b), while for the Devonian our evidence is very incomplete. However, I hold the view of Seward (1924) who wrote: "A useful purpose will have been served if, by premature expression of opinion based on incomplete data, attention is directed to the need for fuller information and to the importance of taking stock of such knowledge as we already possess".

As few fossil plants are found *in situ*, their distribution within rock strata is controlled by transport and depositional processes. In general, fossil plant assemblages are controlled by the interactions of their original ecology with sedimentary processes. The reconstruction of original plant communities is dependent upon the understanding of transport history and depositional environments interpreted from the sedimentological evidence (Scott, 1977b). In addition not all the plants have survived the fossilization process. We probably know less than half of the plants at any one moment in time, and maybe as few as 10%. Preservational environment is as important with fossil plants as it is with vertebrates. Fossil plant material is often best preserved in acidic or anaerobic conditions, but bone is best preserved in alkaline environments (Schopf, 1975; Rayner, 1971). This fact can influence our ideas concerning the distribution of these fossils. This problem of preservational environments must, therefore, be considered in any attempt to understand terrestrial communities. Most of the famous vertebrate finds in the Carboniferous have been found in lacustrine or oxbow lake deposits (Boy, 1977). This might cause an overrepresentation of aquatic, semi-aquatic or marginal-aquatic vertebrates. Those animals living away from such environments (just as with upland floras) may not be preserved, living in erosional rather than depositional sites (Boy, 1977). In addition, vertebrates living in peat swamps, i.e. preserved in coal seams, are often found as "ghosts", the original bone not surviving the acidic conditions (Rayner, 1971).

A number of changes in the floras and plant communities from the late Devonian to the Permian may be related to climatic and evolutionary factors, and may result from the adaptation of biological strategies of the plants to these differing conditions. In this paper I shall try and trace some of these changes and try to relate them to the evolution of tetrapods. I shall, however, restrict my comments to the "Euramerian" palaeobiological province, i.e. Europe and North America, as it is from this area that virtually all important vertebrates are found; it is also the area for which we have the fullest knowledge of late Palaeozoic ecology (Chaloner and Meyen, 1973; Panchen, 1973).

SEDIMENTARY ENVIRONMENTS

I am concerned in this paper with terrestrial vegetation, ecology and ecosystems and it is a truism to state that most of our information concerning the distribution of fossil plants comes from terrestrial sediments, although a few floras are known from near-shore marine deposits. Very often, however, either palaeobotanists are not interested in the sedimentology, or sedimentologists ignore the plants, so that our knowledge of their relationships is often scanty. As many plants live in erosional rather than depositional areas, information about the former may be difficult to obtain whereas we may tend to overemphasize the latter. Our knowledge of upland floras is scanty and until recently evidence of these floras has been generally overlooked.

The first step in reconstructing the vegetation is to observe the distribution of plant remains in various sedimentary facies. The next step involves detailed interpretation of the depositional environments and likely transport history of the plants. Only then can hypotheses be made concerning possible plant communities, which may be subsequently tested by the acquisition of new data. Such reconstructions are hazardous, but are offered in the spirit of encouraging further work and critical comment.

The Devonian continents, when reassembled to their pre-drift positions, may be shown to have spanned the Equator (Smith *et al.*, 1973), with a central "Old Red Sandstone" continent surrounded by tropical seas (House, 1968, 1974). Allen *et al.* (1967) have shown that the continental sediments were laid down under fluvial, lagoonal or lacustrine conditions. In addition they consider that equatorially there was a variable climate which was locally warm-humid. The "Old Red Sandstone" deposits may also be considered as syn- or post-orogenic and have been shown to interfinger with marine deposits (Allen *et al.*, 1967). Evidence of rainfall (its amount and variation) is difficult to assess. The common presence of concretions of calichelike cornstones suggest periods of low water-table (Read and Johnson, 1967). Fossil plant material in Upper Devonian strata is often found in flood-plain or near-shore marine clastic facies. Some of the characters of the Old Red Sandstone which are typical of present-day arid or semi-arid regions may have resulted from a lack of vegetation rather than a lack of rainfall (Schumm, 1968; Glennie, 1970). It is likely that a land flora first colonized near-shore or lowland alluvial areas, and this would mean that the delivery of sediment from source areas was not inhibited and sheet-like alluvial deposits could still form (Schumm, 1968). Schumm also observes that "with increased plant cover, alluvial deposits were stabilized, but large floods caused periodic flushing of sediment from the system, thereby creating cyclic sedimentary deposits".

The Carboniferous began with widespread marine transgressions which spread over the eroded remnants of the "Old Red Sandstone" continents (Ramsbottom, 1973), but it has been clearly demonstrated that this was not exactly at the Devonian/Carboniferous boundary (Sleeman, 1977; Clayton *et al.*, 1977). Consequently non-marine Lower Carboniferous rocks are of smaller extent than those of the preceding Devonian and our knowledge of the distribution of plant material in terrestrial and near-shore marine rocks is sparse. The late Carboniferous, in contrast, is characterized by large tracts of coal "swamps" spread over wide continental areas (Schopf, 1974). These coal-forming belts, which straddled the Equator (Frederiksen, 1972) belonged to a single (Euramerian) palaeobotanical province (Chaloner and Lacey, 1973). The sediments were laid down in marginal marine or alluvial swamps, on flood-plains by various types of alluvial systems, or in lakes (Wanless *et al.*, 1969; Kelling, 1974; Scott, 1978). In addition, deltaic sediments were deposited in marine or freshwater bodies. The vegetation was at least luxuriant, being responsible for forming the thick coal seams (Teichmüller, 1952, 1962), but plant material is also abundant in many of the associated sediments (Wanless, 1959; Robertson, 1952). We are able to recognize upland floras for the first time in the Upper Carboniferous (Leary, 1977), and they must have played a major role in stabilizing erosional areas just as the often dense lowland vegetation must have helped stabilize many of the alluvial systems (Robertson, 1952; Schumm, 1968).

The disappearance of the "coal swamps" at the end of the Carboniferous was probably due to a change in climate of which evidence can be seen even in Stephanian times (Frederiksen, 1972; Phillips *et al.*, 1974). The Permian is often characterized by red beds which are generally believed to have been formed in arid or semi-arid desert conditions (Glennie, 1970; Waugh, 1973). Permian terrestrial sediments often consist of piedmont breccias, wadi-fan and dune complexes, alluvial fan and flood-plain sediments (Brookfield, 1978), deposition often taking place in a series of cuvettes which may have opened into larger basins (Laming, 1966). There is also some evidence, at least for the Lower Permian in Britain, of some moderate relief (Smith *et al.*, 1975). There appears to have only been a generally thin plant cover which declined because of diminishing rainfall.

The Upper Permian began with a widespread marine transgression (Smith *et al.*, 1975). Plant material occasionally occurs in these sediments (Stonely, 1958) and suggests a sparse conifer forest surrounding the new marine basins. These floras take on more of the character of the succeeding Mesozoic flora (Frederiksen, 1972).

CLIMATE AND WEATHER

Climatic changes and to a lesser extent weather patterns played a major role in the distribution and ecology of late Palaeozoic floras, as with those of the present day. Throughout the late Palaeozoic the Euramerian belt straddled the Equator with a slight northward drift (Smith *et al.*, 1973). The climate in Europe and North America has generally thought to have been warm, locally humid, if not hot, with red beds particularly common in the Upper Devonian (Old Red Sandstone facies) (Allen *et al.*, 1967). These continental deposits seem to show evidence of a sporadic rainfall which may be contrasted with the Carboniferous.

A more humid, but not necessarily cooler climate appears to have been extensive during the later Carboniferous and, with regular precipitation, provided an environment for luxuriant growth of vegetation and the formation of thick peat (coal) (Schopf, 1974). Evidence from sediments and from fossil charcoal suggests that, although the climate may have been fairly stable, there were variations in weather, shown by periods of drought accompanied by forest fires and by periods of storms and floods. The regular banding in some Coal Measure lacustrine deposits may suggest in addition a more regular weather cycle not recorded by woody trees in the form of growth rings (Chaloner and Creber, 1974).

At the end of the Carboniferous period, rainfall appears to have declined again. The British Permian seems to be typical of the northern continents and Smith *et al.* (1975) write that "with the decline in rainfall came a depletion of plant cover, and movement of finer sediment by prevailing easterly to east north-easterly winds became the major sedimentary process in the central areas of most depositional basins". Most of the familiar lowland habitats of the late Carboniferous seem to disappear gradually during the early Permian, so that following the mid- to late Permian transgressions those floras and faunas best adapted to drier conditions thrived when new continental areas became available.

Although it is evident that broad climatic shifts were responsible for changing sedimentological and ecological conditions, rainfall must have been a very important controlling factor. Published evidence suggests that rainfall was sporadic during late Devonian times, although it is difficult to assess the effect of no vegetation on upland areas. Vegetation was probably more uniform in the Euramerian belt during the Carboniferous, although Frederiksen (1972) suggests that "the uplands were probably cooler, less rainy, and had greater seasonal variations in rainfall than most tropical uplands do now". The increased rainfall caused swamping of lowland areas, as well as the formation of large lakes. Rainfall appears to have declined during the Permian, allowing the onset of more arid conditions. Temperatures may have remained fairly constant at the Equator at this time.

PLANT COMMUNITIES

We may consider, within the Upper Palaeozoic, two major "explosions" in land plant evolution. The first was the rise of the earliest vascular land plants and the evolution of the psilophytes in the late Silurian and early Devonian: this is, the diversification from a *Cooksonia*-type to the development of the lycopods via the zosterophylls, and the development of the trimerophytes from the rhyniophytes (Banks, 1975; Gensel, 1977). The second explosion represents the diversification of the trimerophytes and the evolution of the "pteropsids", progymnosperms and sphenopsids during the Middle and Upper Devonian (Scheckler, 1974; Gensel, 1977). By the end of the Devonian, therefore, most of the basic botanical strategies that are seen during the rest of the Palaeozoic had evolved, including gymnospermous structure, arborescent habit, true leaves and seeds (Chaloner, 1970;

Beck, 1970; Pettit, 1970; Kevan *et al.*, 1975). Further diversification and evolution was, therefore, geared to the filling of new ecological niches made available by the changing climate and sedimentary environments.

Data concerning the original ecology may be derived from the distribution of plants within sediments, the interpretation of sedimentary environments and the transport history of the plants. Alternatively, there is what may be termed the structuralist approach; that is, using the structures of the plants to deduce their ecology. This approach has been well illustrated by Weiss (1924). More recently there has been the tendency to ascribe fossil plant species to pseudoecological groupings under such terms as hygrophile biotype, mesophile biotype etc. (Remy and Remy, 1977), but perhaps this kind of compartmentalization reflects as much our uncertainty of the true ecological status of the plants rather than anything else. We are plagued by the problem of whether a plant, for instance, may have been a true xerophyte, or merely a pathological xerophyte (Wartman, 1969). Chaloner and Collinson (1975) have demonstrated that the interpretation of ecology from botanical structures can be very misleading and perhaps even dangerous. There is no doubt that the evolution of certain botanical structures may have been in response to particular environmental pressures, but it does not mean that plants possessing these structures must always be found in the same environment. The evolution of the seed may be said to enable the plant to reproduce without reliance on water being widely available, but this does not prevent the extant seed-producing conifer *Taxodium distichum* from standing with its roots and even its trunk in open water.

1. *Floras of the Devonian/Carboniferous Transition*

Our knowledge concerning the ecology of Upper Devonian floras is very sparse, but a few general comments may be made before taking examples from which general conclusions may be made.

Within the Euramerian area during the late Devonian there were wide continental margins (bounded to the north by shield areas) and, according to Allen *et al.* (1967) "their coastlines, complex river systems, alluvial flood plains, lagoons and lakes were ideal for the development of a wide variety of non-marine habitats for the vertebrates". In addition this presented land plants with a diversity of niches to fill. The mid- to late Devonian diversification of land plants may have been a response to this situation.

A number of biological strategies adopted by land plants would have a direct influence on all types of animal life. The evolution of leaves (megaphylls) and the development of heterospory and the seed, for example, may relate to the evolution of herbivory in different animal groups. Perhaps the most important innovation as far as whole-ecosystem dynamics was concerned was the evolution of secondary growth which enabled plants to attain an arborescent habit. Until the early Middle Devonian most land plants were no bigger than shrubs, perhaps a metre or so high. The acquisition of secondary growth enabled some plants to increase girth, grow taller and give stability to these larger plants, now truly trees. With forests, therefore, a two-level "structure" to the ecology can be discerned for the first time: a division into a tree-top ecosystem and a ground-cover ecosystem (see Gosz *et al.*, 1978). It has been argued that this acquisition of an arborescent habit by plants stimulated the evolution of flying arthropods (Kevan *et al.*, 1975). Theoretically this structuring of plant communities would have given great stimulus to both animal and plant evolution, providing a wide range of new ecological niches to be filled.

Most known Upper Devonian floras come from continental or near-shore marine deposits. In the British Isles our only well-known flora comes from the Devonian/Carboniferous boundary at Kiltorcan, County Kilkenny, Ireland. More diverse and widespread floras are known from North America and it is these that I shall describe first.

The sediments of the Middle and Upper Devonian of the Appalachian region, including New York, are rather similar to the British Old Red Sandstone. These continental rocks, known as the Catskill facies, interfinger westwards with marine rocks (Allen and Friend, 1968). According to these authors, commenting on the similarity of some European and American facies: "Assuming continental drift, a rational Devonian palaeogeography is developed for the North Atlantic region that involves the symmetrical occurrence of marine influenced coastal plains in relation to the unified Caledonide fold belt, a number of intermontane basins apparently unconnected with the sea". These authors have shown that sediments of braided streams (Pocono facies), meandering streams and tidal flats (Catskill facies) and coastal barrier and shelf (Chemung facies) may be encountered.

Banks (1966) has described the stratigraphic distribution of the Middle and Upper Devonian floras of New York State. A few comments concerning the sedimentological distribution of some of these Upper Devonian floras may be made. Banks' list gives a good indication of those plants occurring together at one locality. *Archaeopteris* is abundant in the Oneonta Formation in the Catskill facies and is associated with *Tetraxylopteris, Aneurophyton, Schizopodium, Drepanophycus* sp. and *Colodexylon* sp. (At other localities *Callixylon* spp., the wood of *Archaeopteris*, are also present.) Slightly higher in the succession in the Walton Formation, in the same facies, a number of species of *Archaeopteris* plus *Callixylon* are present. In the Chemung facies, *Taenocradia, Knorria* and *Lepidosigillaria* are present. Within the Catskills there is also some ecological diversity illustrated by the palynomorphs, with *Geminospora* dominating spore assemblages of the tidal Catskill facies, together with other plants, fish and the bivalve *Archanodon*, while *Ancyrospora* dominates sediments of the flood-plain (Catskill, interior facies) associated with fish and plants (Berg, 1977; J. Richardson, personal communication).

Similar palynological results have been obtained from Belgium where hystrichospores (e.g. *Ancyrospora*) are abundant in coarse sediments and are presumed to "originate with fluvial sediments from 'uplands'" (Streel, in Becker *et al.*, 1974). Where the sediments are fine, however, the spore flora is dominated by *Aneurospora greggsii* which possibly represents the local (*Archaeopteris*?) Frasnian vegetation (Streel, ibid.).

Beck (1964) has already commented on the predominance of *Archaeopteris* in the Upper Devonian flora of the Western Catskills, in the near-shore or tidal facies. From its abundance and good preservation and rarity of other species, Beck concludes that the coast of this "Upper Devonian inland sea, in the region studied, was probably densely forested by woody plants of which *Archaeopteris* was the most common".

Evidence from British Old Red Sandstone deposits is very sparse. Most of the Upper Devonian rocks consist of red sandstones and conglomerates, and these facies persist in many areas until well into the Tournaisian (Carboniferous). The one area from which we have any detailed knowledge of the stratigraphy, sedimentology and flora is in County Kilkenny, south-eastern Ireland. The flora from this region has long been realized to be of importance (Haughton, 1855). This flora is best known from the Kiltorcan Formation, approximately equivalent to the "Yellow Sandstone Series" of the old surveyors (Holland, 1977). The flora and fauna are best known from the type locality on Kiltorcan Hill and are less evident elsewhere (Holland, 1977; Colthurst, 1978). The classic old quarry, from which most of the famous Kiltorcan flora was obtained, has yielded almost exclusively an assemblage of the progymnosperm *Archaeopteris* and the lycopod *Cyclostigma* (Chaloner, 1968). In addition, however, specimens of the seed *Spermolithus devonicus* have been recorded during the recent excavations by W. G. Chaloner (Chaloner *et al.*, 1977). Rarer specimens of *Ginkgophyllum kiltorkense* and *Sphenopteris hookeri* have also been recorded (Johnson, 1914).

A new quarry at Kiltorcan Hill exposes fossiliferous strata approximately 3 m above the level in the "classic" quarry (Colthurst, 1978). Recent collecting by Chaloner and the author has yielded abundant *Lepidodendroposis*, leaf fragments cf. *Rhacopteris* and seeds very similar to *Spermolithus devonicus*. Both of these assemblages are presently being studied by W. G. Chaloner and W. S. Lacey. These floras are from approximately the Devonian/Carboniferous boundary, and Chaloner (see Holland, 1977) has commented on the close similarity of the lower assemblage to the *Cyclostigma*/*Archaeopteris* flora from Bear Island (Kaiser, 1970) which has an associated spore flora of Tournaisian (Tn1a) age (= PL Zone; K. Higgs, personal communication). Recent palynological work in the Kiltorcan area has shown that rocks a little way above the plant-bearing horizons are of Tournaisian (Tn1b–Tn2) age, (VI zone; Clayton *et al.*, 1977). New unpublished data show that rocks stratigraphically below this horizon elsewhere yield a PL Zone spore assemblage (K. Higgs, personal communication), the position of the PL/VI boundary not being known.

The sediments in the "classic" quarry consist of fine greenish-yellow sandstones and siltstones overlying coarse flaggy cross-bedded sandstone (Colthurst, 1978). The plant material consists either of branches of *Archaeopteris*, occasionally fertile, or stem fragments (leaves and sporophylls) of *Cyclostigma*. These are preserved in a chloritoid which is replacing a coalified compression (Chaloner, 1968; Schopf, 1975). Colthurst (1978) has interpreted these sediments as representing a bar-tail in a meandering channel. In addition to the flora an interesting fauna has been recorded, consisting of the bivalve *Archanodon jukesii*, fish including *Coccosteus* and crustaceans

(Forbes , 1853; Harper, in Gill, 1956). No bivalves or fish, however, were found in the most recent excavations of the classic quarry. The plants occur in fairly coarse silts and sands as well as in finer silts. Although often occurring together, *Archaeopteris* is more often found in coarser sediment than *Cyclostigma*. It is interesting to note that no pieces of *Archaeopteris* trunk, known as *Callixylon*, have been recorded.

The plant material in the newer, or council, quarry is found near the top of the section in laminated sandy silts which overlie green-, brown- and red-coloured mudstones. The plants, according to Colthurst (1978), occur in finer grained laminae within the more silty units. Colthurst interprets these sediments as being deposited during flash floods in small, shallow ephemeral lakes or on near-coastal flood-plains.

The main problem concerning these two floras is to know whether they are more or less contemporaneous and whether their distribution is ecological rather than stratigraphic. The fauna may have been washed into a small abandoned meander of only local occurrence. Berg (1977) has shown that American species of *Archanodon* may have lived in flowing rivers and were in fact burrowers. No burrows have been recorded from Kiltorcan. With such little data the reconstruction of the plant ecology is difficult.

The Kiltorcan beds span at least two spore zones (PL and VI) and rocks of similar age are present in England in the Avon Gorge. Utting and Neves (1970) record a lower assemblage of spores, within the Old Red Sandstone, which may be placed in the PL Zone (K. Higgs, personal communication). The transition beds to the Lower Limestone Shales yield a macroflora containing abundant *Rhacophyton*. VI Zone spores are recorded from a sample above this horizon within the lowest part of the Lower Limestone Shales. The macroflora may be considered as possibly being of a similar age to the Kiltorcan flora and the distribution of these genera may be ecological rather than stratigraphic. Three separate plant assemblages occur in the uppermost Devonian/Lower Carboniferous rocks: an *Archaeopteris/Cyclostigma* assemblage; a *Lepidodendropsis* assemblage; and a *Rhacophyton* assemblage. All three are world-wide in distribution (Zalessky, 1931; Iurina and Lemoigne, 1975; Andrews and Phillips, 1968; Nathorst, 1902), with the *Lepidodendropsis* assemblage dominating Lower Carboniferous floras world-wide (Chaloner and Lacey, 1973).

2. Lower Carboniferous Floras

As with the previous floras, we have little data on those of the lowest Carboniferous (Tournaisian). Most of these floras have been recovered from either Old Red Sandstone facies (Matten *et al.*, 1975) or from marine rocks (Hoskins and Cross, 1941). Some of the well-known Tournaisian floras (e.g. Barnard and Long, 1973) now appear to be Lower Viséan in age (Scott, in preparation). In Britain we have a number of important Viséan compression and petrified floras from near-shore marine and non-marine sequences. When, however, floral lists are compared, few species are found in common. Most of the detailed work has been done on the petrified floras and, although our knowledge of the stratigraphic distribution of compression floras is great (Crookall, 1932), we have little knowledge of their ecological distribution. This being so I will briefly mention two localities which I am at present studying.

One of the best known Lower Carboniferous floras is that of the Loch Humphrey Burn, north of Glasgow, studied by J. Walton and his student D. L. Smith (e.g. Walton *et al.*, 1938; Smith, 1964). These authors described a wide range of plants preserved both as petrifactions and as compressions. Recent field investigation has shown that there is more than one plant bed. The lowest part of the section consists of green silts with upright *Archaeocalamites*. Its cone *Pothocites* is also present. This bed is overlain by a fusain-rich drifted coal followed by yellow–green siltstones and mudstones containing a pteridosperm-rich flora. Siltstone blocks at the base of this bed contain contorted fronds of *Rhacopteris* and *Spathulopteris*. The sediments, which have been dated palynologically as Lower Viséan in age (G. Clayton, personal communication), may be interpreted as marginal-lake and alluvial flood-plain sediments and have some analogy, both sedimentologically and palaeoecologically, with some Coal Measure sediments (Scott, 1978). The *in situ Archaeocalamites* may have an analogy with the marginal-lake reed-swamp *Calamites* of the later Westphalian. The upper part of the sequence may represent flood-plain and river deposits, the siltstone blocks being reworked channel bank collapse deposits (Scott, 1978) which may have incorporated plants living on the river banks. Not surprisingly these data give us a picture of the rich pteridosperm flora being associated with the river banks and flood-plain. Smith (1964) considered that the flora from Glenarbuck, nearby, which consisted mainly of lycopods, may be of the same age but represented a different ecological situation. As yet, however, there are no data as to the precise age of this deposit so no further comment can be made.

The second example is from Lower Carboniferous sediments (Lower Viséan: Chadian) of north-west Ireland. Here, in near-shore marine deposits (Shalwy beds) of the Donegal Syncline (George and Oswald, 1957), there are a number of fusain bands. The bands, up to 50 in all, occur in carbonate-rich sandstones which also contain a shallow marine fauna—gastropods, ostracods, brachiopods, foraminifera, nautiloids and fish teeth—and a flora of algae. The fusain bands contain an almost monotypic flora or a herbaceous lycopod which is extremely well preserved and has been studied using SEM (Scott and Collinson, 1978). Although other plants are present they are rare in comparison with the lycopod. As yet this plant has not been identified, but is still under study. The rock is black with the fusain which consists of 1–3 cm long stelar fragments and 1 cm cubes and irregular fragments of other tissues. The occurrence of such fusain in thick (up to 1 cm) numerous bands in near-shore marine deposits may be the result of storms washing in debris from coastal areas following extensive forest fires over a lycopod scrubland. This deposit, together with a similar deposit from County Mayo, is still being studied and further results may help in interpretation.

Daber (1959, 1964) recognized in Middle Viséan rocks of Germany four plant associations: *Asterocalamites* association; *Sphenophyllum* association; a lycopod coal seam association; and a general pteridosperm assemblage.

Finally, there is the Upper Viséan Pettycur flora from Scotland. This flora was first described as a whole by Gordon (1909). The flora, although dominated by ferns and pteridosperms, also contains lycopods and sphenopsids. Further investigation of the sedimentology and field relations of this deposit may give some idea of the palaeoecology of this interesting flora.

3. *Upper Carboniferous Floras*

As with the Upper Devonian and Lower Carboniferous, our knowledge of the ecology of early Upper Carboniferous floras (Namurian) is incomplete. We have some idea of the range of plants present from well-known compression floras, but know little of their structure, except in a few isolated cases (Hass, 1975).

Recent work by Leary (1974, 1975, 1977) and Leary and Pfefferkorn (1977) has for the first time given us a detailed picture of Namurian upland, lowland and mixed plant assemblages (see also Havlena, 1961). Both of the floras described by Leary contain the enigmatic plants *Megalopteris, Lesleya* and *Lacoea*, and have high percentages of pteridosperms and Noeggerathiales, and a low percentage of ferns and lycopods which contrasts with their dominance in lowland or swamp communities. There can be little doubt that these plants were associated with upland environments. This work by Leary and others is of special importance as it gives us, for the first time, a chance to separate two distinctive floras (upland and lowland/swamp) which can be seen in the succeeding stages, and an assessment can be made concerning the evolutionary diversification of floras in both. It is often considered that the climatic change at the end of the Carboniferous enabled upland floras to spread into the lowland basins (Frederiksen, 1972), thus proving a very different setting for later tetrapod evolution. In addition, upland plant cover may have had a major role in stabilizing erosional areas (Schumm, 1968).

As has been previously mentioned, the distribution of floras is controlled sedimentologically as well as ecologically. I have shown that in the Westphalian of Britain lacustrine sequences yield a sparse flora which is species-poor but which often yields an abundant and sometimes diverse fauna. Many of the best Namurian vertebrate localities occur in sequences which are dominantly lacustrine, often associated with fish and non-marine bivalves (Andrews *et al.*, 1977). Plants are rarely recorded in such sediments. It is often difficult, therefore, to assess vertebrate/plant interactions (but see below).

By early Westphalian times a variety of upland and lowland plant associations were well developed. Evidence for this may be derived from macro- as well as from microfloral data (Sabry and Neves, 1971; Scott, 1977b). Sabry and Neves (1971) concluded from palynological evidence that two principle plant associations were established at this time in southern Scotland, namely a lycopod–*Calamites* swamp flora and a *Cordaites*–conifer backswamp and/or upland flora. It is interesting to note that from the same area (Sanquhar Coalfield) evidence suggests that the upland flora disappeared later in the Westphalian because of an overstepping and possible erosion of the upland area (Scott, 1976).

The ecology of Westphalian floras has been well documented and I will only summarize the main results here (for fuller reviews see Havlena, 1971; Scott, 1977b). Most recent work on coal-ball floras suggests that coals, particularly in the Pennsylvanian of North America, are dominated (in terms of biomass) by arborescent lycopods, but occasionally by other plants (Darrah, 1941; Phillips *et al*., 1977). These floras have also been shown to change through time to become dominated by large tree ferns such as *Psaronius*, probably as a result of "broad climatic shifts" at the close of the Carboniferous (Phillips *et al*., 1974).

Peppers and Pfefferkorn (1970), working with data from compression floras and palynology, recognized a variety of lowland associations: a wet swampy association with lycopods, ferns and rare *Cordaites*; dry swampy with ferns, lycopods and sphenopsids; levees and flood-plains with pteridosperms, ferns and sphenopsids; and upland with pteriodsperms, *Cordaites* and rare Noeggerathiales. European workers have basically come to similar conclusions (Daber, 1964; Oshurkova, 1967; Havlena, 1971).

Recent work on Westphalian floras in Britain has suggested that the peat- (coal-) forming vegetation, or "swamp", was dominated by various groups of lycopods; the flood-plain areas supported a flora dominated by pteridosperms with some ferns, sphenopsids and lycopods; levee banks of meandering rivers supported various pteridosperms; whilst *Calamites* grew around the lakes and on point bars (Scott, 1978, 1979). It has also been suggested that the upland supported a cordaite–conifer forest (Chaloner, 1958).

Evidence from America indicates an extensive uppermost Carboniferous upland flora dominated by *Taeniopteris, Walchia* and *Dichophyllum* (Cridland and Morris, 1963).

4. *Permian Floras*

Most of our knowledge of the ecology of Permian floras comes from the work of German authors (Gothan and Gimm, 1930; Gothan and Remy, 1957; Barthel, 1976) on floras from the Rotliegende (Lower Permian). Gothan and Gimm (1930) investigated the distribution of plants occurring in different rock types from a number of sections in Thuringia. These authors (and later Gothan and Remy, 1957) recognized two main plant assemblages: a *Pecopteris–Calamites* assemblage associated with lowland and coal-forming environments, and a *Callipteris–Walchia* assemblage associated with an upland environment. This was one of the first papers to detail upland floras in the Palaeozoic, although their presence has been suspected from the time of Witham (1833).

Barthel (1976), in an extensive monograph on the Rotliegende flora from Saxony, considered that five plant communities could be recognized:

(1) A hygrophile community with *Cordaites, Psaronius, Eucalamites*, and sphenopterids, locally with *Nemejcopteris feminaeformis* which lived in swamp (coal-forming) forests.
(2) A hygrophile community with *Stylocalamites gigas* (sporadically with other *Calamites*) in pure lake-side stands.
(3) A hygrophile community with *Nemejcopteris feminaeformis* and *Sphenophyllum oblongifolium* (sporadically with *Botryopteris burgkensis, Taeniopteris jejunata, Eucalamites*) in large stands, but not coal-forming, forming a reed-swamp surrounding short-lived lakes.
(4) A species-rich mesophile pteridosperm community (with other gymnosperms, sporadically with ferns) accumulated in rivers or remote lakes and on sandbanks within swamps.
(5) A xerophitic community with conifers, poacordaites and other gymnosperms in erosional areas with remote underground water.

Frederiksen (1972) has pointed out that during the early Permian plants were already migrating from the uplands down into the basins. This migration was a result of a climatic change which caused the diverse swamp and lowland habitats to disappear to be replaced by drier habitats. This major climatic change, which has already been discussed, caused the extinction of many of the familiar Upper Carboniferous plants such as the arborescent lycopods (Chaloner, 1967).

Later Permian floras (and subsequent Mesozoic floras) were dominated by gymnosperms (Stoneley, 1958; Frederiksen, 1972). It is interesting to note finally that the late Palaeozoic was a time for the palaeogeographic diversification of floras (Chaloner and Meyen, 1973).

ANIMAL/PLANT INTERRELATIONSHIPS

Kevan *et al.* (1975) have shown how, from the early Devonian, there were close relationships between arthropods, vascular plants and fungi. There is no doubt that these interrelationships continued and indeed diversified during the later part of the Palaeozoic. These authors have also pointed out that arborescence and structures on the stems of plants preceded flying insects and may have acted as an evolutionary stimulus. The evolution of trees, as already mentioned, must have caused some structuring of ecosystems not seen before that time, i.e. into an upperstorey, or tree-top, ecosystem and an understorey ecosystem, the one affecting the other. One of the results may have been the shading of the undergrowth and the development of micro-climates which may have stimulated the evolution of tetrapods adapted to such new ecological niches.

In general two major, but related, ecosystems may be considered; the lacustrine and the terrestrial systems (Fig. 1).

1. Lacustrine Systems

These, being the sites for sediment deposition, are often those from which most data concerning faunal associations may be obtained. Commonly, most lacustrine deposits in the Coal Measures yield non-marine bivalves and, in addition, ostracods and fish (Pollard, 1973; Bless and Pollard, 1973). Less frequently, but perhaps more importantly, these deposits yield vertebrate faunas often rich in species (Milner and Panchen, 1973). The relative rarity of such faunas must reflect, as Rayner (1971) has pointed out, unfavourable conditions of preservation rather than their absence from the system (Boy, 1977). A. R. Milner (Chapter 17 of this volume), in particular, has shown how data from assemblages of vertebrates may be used to build up a reasonable picture of the food web.

Most of the amphibians are thought to have been carnivorous, feeding on smaller vertebrates or arthropods. Fish probably fed on non-marine bivalves, ostracods, other arthropods and other invertebrates. These other animals probably fed either on each other or on the accumulation of organic detritus, which consisted of vascular plant material washed into the system, algae or decaying animal material (Bless and Pollard, 1973; Moore, 1968; Schram, 1976). Gensel and Skog (1977) have recently commented: "Also it is

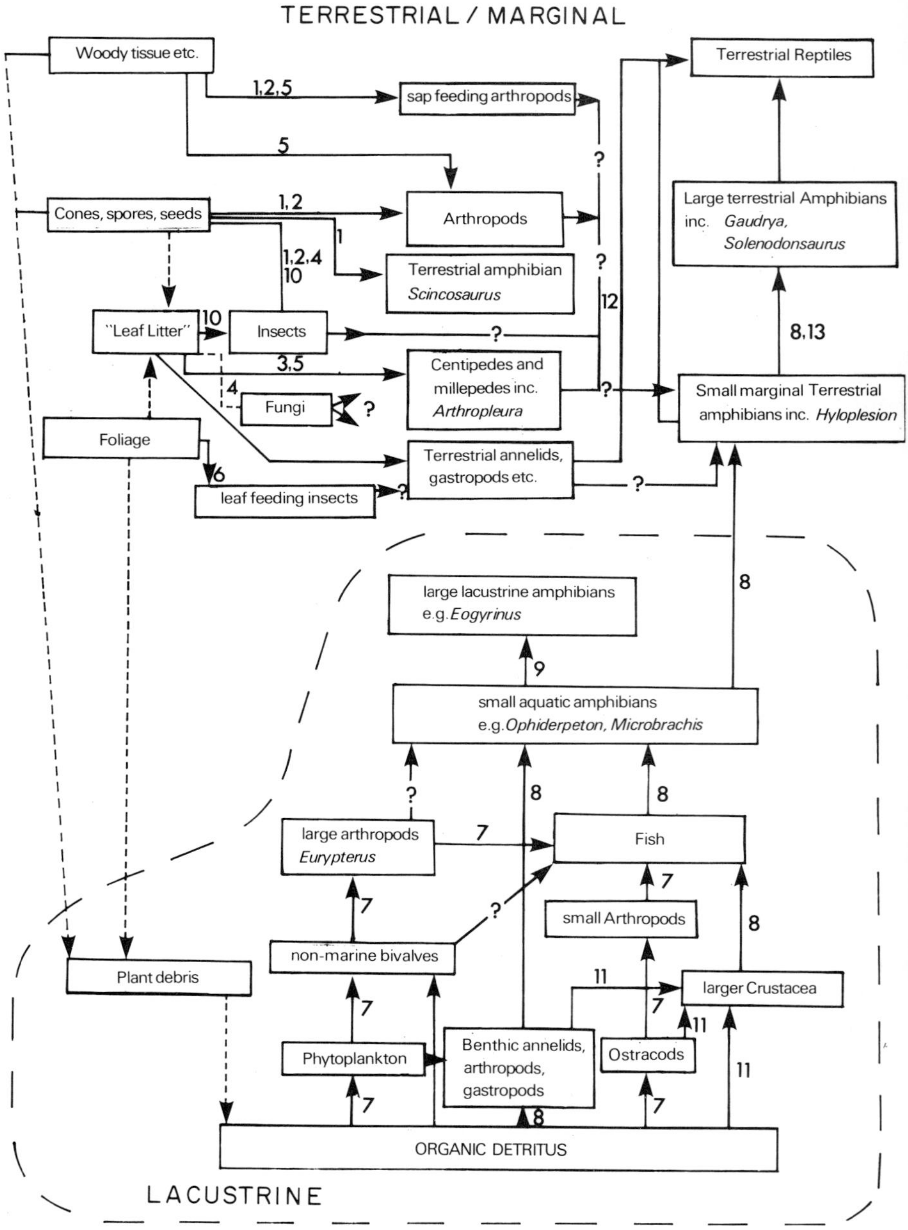

TERRESTRIAL / MARGINAL

Woody tissue etc.
1,2,5
sap feeding arthropods
5
Terrestrial Reptiles
?
Cones, spores, seeds
1,2
Arthropods
1
1,2,4
10
Terrestrial amphibian *Scincosaurus*
Large terrestrial Amphibians inc. *Gaudrya*, *Solenodonsaurus*
?
12
"Leaf Litter"
10
Insects
?
8,13
3,5
Centipedes and millepedes inc. *Arthropleura*
4
Fungi
?
?
Small marginal Terrestrial amphibians inc. *Hyloplesion*
Foliage
Terrestrial annelids, gastropods etc.
6
leaf feeding insects
?
?
8
large lacustrine amphibians e.g. *Eogyrinus*
9
small aquatic amphibians e.g. *Ophiderpeton*, *Microbrachis*
?
8
8
large arthropods *Eurypterus*
7
Fish
7
7
?
small Arthropods
8
non-marine bivalves
Plant debris
11
larger Crustacea
7
7
11
Phytoplankton
Benthic annelids, arthropods, gastropods
Ostracods
11
7
8
7
ORGANIC DETRITUS

LACUSTRINE

quite possible that tetrapods or crossopterygians of the Mississippian swamps snatched the seeds from the water surface as a potential food source and further dispersed the seeds . . .".

Vascular plants may be considered as sedimentary filters as well as being the source of organic detritus. I have suggested that through the late Palaeozoic the Coal Measure lakes were surrounded either by large lycopod swamp forests or else by "reed-swamps" formed by large sphenopsids such as *Calamites*. These may have been used by some of the larger vertebrates as "hideouts" from which to ambush their prey!

It is interesting to note that it is this environment which persists unchanged in its basic ecological structure throughout the later Carboniferous and into the Lower Permian, perhaps accounting for the conservatism in many of the vertebrate faunas of this age (Milner and Panchen, 1973).

2. *Terrestrial Systems*

Data on true terrestrial systems is somewhat sparser. As has been discussed earlier we have been able to build up a fairly detailed vegetational picture, but data on the faunas and their relationship to the floras are lacking. Good terrestrial faunas, both of vertebrates and arthropods, are sparse, and data have been obtained from only a few localities. I have discussed elsewhere data concerning plants as food (Scott, 1977a). Most workers seem to agree that the majority of late Palaeozoic vertebrates were carnivorous, feeding on smaller vertebrates and arthropods, especially insects (Olson, 1976). Olson also considers that late in the early Permian "coincident with the co-evolution of reptiles and insects, however, semi-aquatic reptilian herbivores emerged (best known being *Edaphosaurus*) presumably feeding on the vegetation in and around freshwater lakes and streams".

Fig. 1. A simplified hypothetical Coal Measures food web. Solid arrows show the direction of energy flow, dashed arrows indicate decay products. Data from: (1) Scott, 1977a; (2) Smart and Hughes, 1973; (3) Rolfe and Ingham, 1967; (4) Kevan *et al.*, 1975; (5) Hoffman, 1969; (6) Van Amerom, 1966; (7) Bless and Pollard, 1973; (8) Milner, Chapter 17 of this volume and personal communication; (9) Panchen, 1970; (10) North, 1931; (11) Schram, 1976; (12) Dawson, 1860; (13) Carroll, 1967.

It is interesting to note that this emergence of reptilian herbivores occurs in the early Permian when there is a marked change in Euramerian floras probably caused by a change in climate. There would also have been an environmental crisis with many of the vertebrate inhabitants vanishing. This crisis may have stimulated the evolution of new types, mainly reptiles which could exploit the new environments including using the vegetation as food—a strategy not tried to any great extent previously. Panchen (1973) has pointed out, however, that *Edaphosaurus* appears in the late Carboniferous and is likely to have herbivorous forebears. Other tetrapods such as *Diadectes* may have been herbivorous so that the picture of the change in floras stimulating herbivory in tetrapods may not be completely accurate.

The luxuriant vegetation of the Carboniferous, little exploited by the vertebrates as food, would have provided a variety of arthropods with a range of diet. Published evidence suggests litter feeders, sap feeders, and cone- and seed-feeding insects, myriapods and other arthropods (Smart and Hughes, 1973; Scott, 1977a). Finds of insects and other terrestrial arthropods in the Coal Measures are as infrequent as those of tetrapods, the two rarely occurring together or with plants, probably because each has different preservational requirements (Rayner, 1971; Olson, 1976; Boy, 1977). Many of the well-known insect faunas have, however, been recorded from outside the Euramerian area (Riek, 1970, 1973).

One of the most interesting sites which yields data for the understanding of terrestrial faunas and floras must still be that of Joggins, Nova Scotia. Here, in mid-Westphalian rocks, a rich fauna is preserved inside hollow trunks of trees (Dawson, 1860; Carroll, 1967). The trunks of the trees were apparently hollow, caused by the centres of the stumps rotting away (Carroll, 1967). These hollow stumps served as traps for animals living on the land surface and Carroll (1967) considers that they survived for some time within the trees as there is a considerable amount of coprolitic material associated. A diverse vertebrate fauna (reptiles and amphibians) as well as invertebrates (myriapods, terrestrial gastropods etc.) are preserved (Dawson, 1860; Carroll, 1967). Dawson (1860) records fossil charcoal from the base of some of the Joggins trees and it may well be that some of these lycopods in fact represent burnt-out stumps following forest fires (Scott and Collinson, 1978). It is clear that this deposit gives us the most complete picture of a terrestrial fauna and illustrates the general problem of inadequate data from other sites because of the nature of the preservational environment.

The interactions of plants and tetrapods, therefore, would appear to be less of one of diet rather than one of shelter and protection. Likewise with the insects, although more are thought to be herbivorous (phytophagous); North (1931) has commented that

> We have still less to go upon in connection with the feeding habits of the flying insects, but such evidence as is available concerning their mouthparts, and comparison with their nearest living representatives suggest that many of them were carnivorous. They probably preyed upon the blattoids and it has been suggested that the similarity of the blattoid wing and a pinnule of *Neuropteris* may have afforded some measure of protection to the insect, and if that is the case it is a very early instance of mimicry.

CONCLUSIONS

Our knowledge concerning the ecology of late Palaeozoic floras, in particular, of the Euramerian area, has increased rapidly over the last decade. Work in both Europe and North America has indicated that a wide variety of "plant communities" flourished, each under different ecological conditions. These may be divided into three main groups: the floras of the lowland swamps; the floras of the drier lowland environments; and upland floras. Importance must be placed on the close integration of sedimentological and palaeontological studies.

Most of the biological strategies adopted by vascular land plants in the late Palaeozoic had evolved by the end of the Devonian. Two of these, arborescence and the production of seeds, must have had a dramatic effect on the evolving ecosystems, enabling plants to spread into new ecological niches as well as causing a two-level community structure in forest ecosystems which may have had a profound influence on tetrapod evolution. Although data are sparse, evidence suggests that the basic ecological associations of plants adopted in the early Carboniferous continued with some evolution and diversification throughout the rest of the Carboniferous and into the early Permian. In particular, there appears to have been a greater diversification in the presumed upland floras. The spread of these floras may have caused the stabilization of erosional areas and affected the patterns of sedimentation (Table I). Floras of the lowlands and the coal swamps in particular appear to have been more conservative.

Table I. Some factors affecting Upper Palaeozoic ecosystems

ECOLOGICAL FACTOR / AGE	EARLY DEVONIAN	LATE DEVONIAN	EARLY CARBONIFEROUS	LATE CARBONIFEROUS	EARLY PERMIAN
PLANT COVER	Small herbaceous plants in lowland alluvial floodplain and coastal areas but mainly near water.	Herbaceous plants, shrubs and trees living in lowland areas. Coastal forests.	Diverse lowland plant communities including forests.	Extensive plant cover of both lowland and upland areas with diversification of floral communities.	Demise of lowland swamp environments with extinction of some plant groups. Upland floras spread into lowland and coastal areas.
CLIMATE AND INFLUENCE OF WEATHER	Warm - humid plant growth restricted to areas with good water supply.	Warm - humid. Two level forest ecosystem gives rise to forest micro-climate. Difficulty in interpreting climatic regime because of lack of plant cover on upland areas.	Warm - humid with wet periods and peat formation. During dry periods lightning strikes cause extensive shrub fires.	Generally wet-humid with extensive peat formation but also dry periods with frequent forest fires.	More arid climate with swamp areas drying out.
ANIMALS	Fish confined to channel systems, land arthropods.	Diversification of terrestrial arthropods and possibly evolution of flying insects.	Arthropods dominate land fauna. Early amphibians.	Arthropods including insects dominate the land fauna but amphibians an important element.	Amphibians, reptiles and mammal-like reptiles all important elements of the land fauna.
SEDIMENTOLOGY	Channel stabilisation. Total run-off in upland areas.	Increased floodplain and channel stabilisation but still upland run-off.	Generally stable lowland areas but forest fire causes increased erosion.	Stabilisation of upland areas in part by plants. Influence of plants upon sedimentation.	Climate causes depletion in plant cover, increased erosion and wind blown sediments.
OTHER COMMENTS	Problem of facies identification. Earliest evidence of animal/plant interactions.	Delay factor - i.e. in the evolution of trees and the occurrence of flying insects.	Problems of stratigraphic correlation and facies identification.	Diversification of plant communities may have stimulated the evolution of vertebrates to inhabit new ecological niches.	The major climatic change enable plants well adapted to the new conditions to survive and hence diversify.

Changes that did occur were probably due to broad climatic changes at the end of the Carboniferous and in the Permian. By the mid-Permian most of the lowland swamp areas had dried out allowing the spread of typical upland floras dominated by conifers to spread down into the arid lowland basins. This change in the distribution of the floras and their ecological setting would have had a marked effect on the various groups of tetrapods present in these different environments.

The vegetation of the late Palaeozoic may not have been a major food source for the vertebrates, but rather provided shelter and protection. This apparently changed with the ecological crisis in the Permian.

Finally, it may be reiterated that any attempt at ecosystem analysis is extremely hazardous, but I hope that this review may stimulate further investigations.

ACKNOWLEDGEMENTS

I thank Professor W. G. Chaloner, F.R.S. for providing many stimulating comments during the course of this work, also Dr A. R. Milner and Dr W. D. I. Rolfe for their continued interest in the wider aspects of Palaeozoic terrestrial ecology. I thank Dr K. Higgs for information concerning the dating of Devonian/Carboniferous sequences. I thank Professor W. G. Chaloner, Professor H. Banks, Dr M. E. Collinson, Dr A. L. Panchen and Dr A. R. Milner for kindly reading and commenting on the manuscript. This work was initiated during the tenure of an N.E.R.C. Studentship and completed during the tenure of an Irish Department of Education Fellowship. Both awards are gratefully acknowledged.

REFERENCES

Allen, J. R. L. and Friend, P. F. (1968). Deposition of the Catskill Facies Appalachian Region: With notes on some other Old Red Sandstone Basins. *Geol. Soc. Am. spec. pap.* **106**, 21–74.

Allen, J. R. L., Dineley, D. L. and Friend, P. F. (1967). Old Red Sandstone basins of North America and northwest Europe. *In* "International Symposium on the Devonian System, Calgary, 1967" (D. H. Oswald, ed.), Vol. 1, pp. 69–98. Alberta Society of Petroleum Geologists.

Andrews, H. N. and Phillips, T. L. (1968). *Rhacophyton* from the Upper Devonian of West Virginia. *J. Linn. Soc. (Bot.)* **61**, 37–64.

Andrews, S. M., Browne, M. A. E., Panchen, A. L. and Wood, S. P. (1977) Discovery of amphibians in the Namurian (Upper Carboniferous) of Fife. *Nature, Lond.* **265**, 529–532.

Banks, H. P. (1966). Devonian flora of New York State. *Empire State Geogram.* 4(3), 10–24.

Banks, H. P. (1975). Early vascular land plants: proof and conjecture. *Bioscience* 25, 730–737.

Barnard, P. D. W. and Long, A. G. (1973). On the structure of a petrified stem and some associated seeds from the Lower Carboniferous rocks of East Lothian, Scotland. *Trans. R. Soc. Edinb.* 69, 91–108.

Barthel M. (1976). Die Rotliegendflora Sachens. *Abh. Staatl. Mus. Mineral. Geol.* **24**, 1–190.

Beck, C. B. (1964). Predominance of *Archaeopteris* in Upper Devonian flora of Western Catskills and adjacent Pennsylvania. *Bot. Gaz.* **125**, 126–128.

Beck, C. B. (1970). The appearance of gymnospermous structure. *Biol. Rev.* **45**, 379–400.

Becker, G., Bless, M. J. M., Streel, M. and Thorez, J. (1974). Palynology and ostracode distribution in the Upper Devonian and basal Dinantian of Belgium and their dependence on sedimentary facies. *Meded. Rijks. geol. Dienst. N. S.* **25**, 9–99.

Berg, T. M. (1977). Bivalve burrow structures in the Bellvale Sandstone, New Jersey and New York. *Bull. N. J. Acad. Sci.* **22**(2), 1–5.

Bless, M. J. M. and Pollard, J. E. (1973). Palaeoecology and ostracod faunas of Westphalian Ostracod bands from Limburg, the Netherlands, and Lancashire, Great Britain. *Med. Rijks. geol. Dienst. N.S.* **24**, 1–22.

Boy, J. A. (1977). Typen and genese jungpaläozoischer Tetrapoden-lagerstätten. *Palaeontographica* **156A**, 111–167.

Brookfield, M. E. (1978). Revision of the stratigraphy of Permian and supposed Permian rocks of Southern Scotland. *Geol. Rund.* **67**, 110–149.

Carroll, R. L. (1967). Labyrinthodonts from the Joggins Formation. *J. Paleont.* **41**, 111–142.

Chaloner, W. G. (1958). The Carboniferous upland flora. *Geol. Mag.* **95**, 261–262.

Chaloner, W. G. (1967). Les lycophytes. *In* Traité de Palaeobotanique" (E. Boureau, ed.), Vol. 2, pp. 433–880. Masson, Paris.

Chaloner, W. G. (1968). The cone of *Cyclostigma kiltorkense* Haughton from the Upper Devonian of Ireland. *J. Linn. Soc. (Bot.)* **61**, 25–36.

Chaloner, W. G. (1970). The rise of the first land plants. *Biol. Rev.* 45, 353–377.

Chaloner, W. G. and Collinson, M. E. (1975). Application of S.E.M. to a sigillarian impression fossil. *Rev. Palaeobot. Palynol.* **20**, 85–101.

Chaloner, W. G. and Creber, G. T. (1974). Growth rings in fossil woods as evidence of past climates. *In* "Implications of Continental Drift to the Earth Sciences" (D. H. Tarling and S. K. Runcorn, eds), Vol. 1, 425–437. Academic Press, London and New York.

Chaloner, W. G. and Lacey, W. S. (1973). The distribution of late Palaeozoic floras. *In* "Organisms and Continents through Time" (N. F. Hughes, ed.), *Spec. Pap. Palaeont.* **12**, 271–289.

Chaloner, W. G. and Meyen, S. V. (1973). Carboniferous and Permian floras of the northern continents. *In* "Atlas of Palaeobiogeography" (A. Hallam, ed.), pp. 169–186. Elsevier, Amsterdam.

Chaloner, W. G., Hill, A. and Lacey, W. S. (1977). First Devonian platyspermic seed and its implication in gymnosperm evolution. *Nature, Lond.* 265, 233–235.

Clayton, G., Colthurst, J. R. J., Higgs, K., Jones, G. L. L. and Keegan, J. B. (1977). Tournaisian miospores and conodonts from County Kilkenny. *Geol. Surv. Ireland Bull.* 2, 99–106.

Colthurst, J. R. J. (1978). The geology of the Old Red Sandstone rocks surrounding the Slievenamon Inliner, Counties Tipperary and Kilkenny. *J. Earth Sci. R. Dubl. Soc.* 1, 77–103.

Cridland, A. A. and Morris, J. E. (1963). *Taeniopteris, Walchia* and *Dichophyllum* in the Pennsylvanian system of Kansas. *Univ. Kansas Bull.* 44, 71–85.

Crookall, R. (1932). The stratigraphical distribution of British Lower Carboniferous plants. *Summ. Prog. geol. Surv.* 2, 70–104.

Daber, R. (1959). Die Mittel-Vise-Flora der Tiefbohrungen von Doberlug-Kirchhain. *Geologie* 26, 1–83.

Daber, R. (1964). Fazies and Palaeobotanik. *Ber. geol. Ges. D.D.R.* 9, 141–147.

Darrah, W. C. (1941). Studies of American coal-balls. *Am. J. Sci.* 239, 33–53.

Dawson, J. W. (1860). On a terrestrial mollusc and a chilognathous myriapod and some new species of reptile from the coal formation of Nova Scotia. *Q. Jl geol. Soc. Lond.* 16, 268–277.

Forbes, E. (1853). On the fossils of the Yellow Sandstone of the south of Ireland. *Rep. Br. Ass. Adv. Sci.* Belfast 1852, 43.

Frederiksen, N. O. (1972). The rise of the Mosphytic flora. *Geosci. Man.* 4, 17–28.

George, T. N. and Oswald, D. H. (1957). The Carboniferous rocks of the Donegal Syncline. *Q. Jl geol. Soc. Lond.* 113, 137–179.

Gensel, P. G. (1977). Morphologic and taxonomic relationships of the Psilotaceae relative to evolutionary lines in early land vascular plants. *Brittonia* 29, 14–29.

Gensel, P. G. and Skog, J. E. (1977). Two early Mississippian seeds from the Price Formation of southwestern Virgina. *Brittonia* 29, 332–351.

Glennie, K. W. (1970). "Developments in Sedimentology", Vol. 14, "Desert Sedimentary Environments", 222pp. Elsevier, Amsterdam.

Gordon, W. T. (1909). On the nature and occurrence of the plant bearing rocks of Pettycur, Fife. *Trans. Edin. geol. Soc.* 9, 353–360.

Gosz, J. R., Holmes, R. T., Likens, G. E. and Bormann, F. H. (1978). The flow of energy in a forest ecosystem. *Scient. Am.* 238, 93–102.

Gothan, W. and Gimm, O. (1930). Neuere beobachtungen und betrachtungen über die flora des Rotliegenden von Thüringen. *Arb. Inst. Paläobot.* 2, 39–74.

Gothan, W. and Remy, W. (1957). "Steinkohlenpflazen", 248pp. Glückauf, Essen.

Harper, J. C. (1956). *In* "Lexique Stratigraphique International", "Europe", Vol. 1 (W. D. Gill, ed.), Part 3b, "Irlande".

Hass, H. (1975). *Arthroxylon werdensis.* n.sp. – Ein Calamit aus dem Namur C des Ruhrkarbons mit vollständig erhaltenen geweben. *Argumenta Palaeobotanica* 4, 139–154.

Haughton, S. (1855). On the evidence afforded by fossil plants, as to the boundary line between the Devonian and Carboniferous rocks. *Jl geol. Soc. Dublin* 6, 227–241.

Havlena V. (1961) Die Flöznahe und Flözfermde flora des oberschlesischen. Namurs A und B. *Palaeontographica* **B108**, 22–38.

Havlena, V. (1971). Die Zeitgleichen floren des europäischen oberkarbons und die mesophile flora des Ostraukarwiner steinkohlenreviers. *Rev. Palaeobot. Palynol.* **12**, 245–270.

Holland, C. H. (1977). Ireland. *In* "A Correlation of the Devonian Rocks of the British Isles" (M. R. House *et al.*, eds), *Spec. Rep. geol. Soc. Lond.* **7**, 54–66.

Hoskins, J. H. and Cross, A. T. (1952). The petrification flora of the Devonian - Mississippian black shale. *Palaeobotanist* **1**, 215–238.

House, M. R. (1968). "Continental Drift and the Devonian System", 24pp. University of Hull.

House, M. R. (1975). Facies and time in Devonian tropical areas. *Proc. Yorks. geol. Soc.* **40**, 233–288.

Iurina, A. and Lemoine, Y. (1975). Anatomical characters of the axes of arborescent Lepidophytes of the Devonian, referred to *Lepidodendropsis kazachstanica* Senkevitsch, 1961. *Palaeontographica* **B150**, 162–168.

Johnson, T. (1914). *Ginkgophyllum kiltorkense.* sp. nov. *Sci. Proc. R. Dublin Soc.* **41**, 169–177.

Kaiser, H. (1970). Die oberdevon-flora der Bareninsel. 3 microflora des Hohoren oberdevons und des Unterkarbons. *Palaeontographica* **B129**, 72–124.

Kelling, G. (1974). Upper Carboniferous sedimentation in South Wales. *In* "The Upper Palaeozoic and Post-Palaeozoic Rocks of Wales" (T. R. Owen, ed.), pp. 185–224. University of Wales Press, Cardiff.

Kevan, P. G., Chaloner, W. G. and Saville, D. B. O. (1975). Interrelationships of early terrestrial arthropods and plants. *Palaeontology* **19**, 391-417.

Laming, D. J. C. (1966). Imbrication, palaeocurrents and other sedimentary features in the Lower New Red Sandstone, Devonshire, England. *J. sedim. Petrol.* **36**, 940–959.

Leary, R. (1974) Two early Pennsylvanian floras of Western Illinois. *Trans. Ill. State Acad. Sci.* **67**, 430–440.

Leary, R. (1975). Early Pennsylvanian palaeogeography of an upland area, Western Illinois, U.S.A. *Bull. Soc. geol. Belg.* **84**, 19–31.

Leary, R. (1977). Palaeobotanical and geological interpretations of palaeoenvironments of the eastern interior basin. *In* "Geobotany" (R. C. Romans, ed.), pp. 157–164. Plenum, New York.

Leary, R. L. and Pfefferkorn, H. W. (1977). An early Pennsylvanian flora with *Megalopteris* and Noeggerathiales from West - Central Illinois. *Ill. State. geol. Surv. Circ.* **500**, 1–77.

Matten, L. C., Lacey, W. S. and Edwards, D. (1975). Discovery of one of the oldest Gymnosperm floras containing cupulate seeds. *Phytologia*. 32, 299–303.

Milner, A. R. and Panchen, A.L. (1973). Geographic variation in the tetrapod faunas of the Upper Carboniferous and Lower Permian. *In* "Implications of Continental Drift to the Earth Sciences" (D. H. Tarling and S. K. Runcorn, eds), Vol. 1, pp. 353–368. Academic Press, London and New York.

Moore, L. R. (1968). Cannel coals, bogheads and oil shales. *In* "Coal and Coal-bearing Strata" (D. G. Murchison and T. S. Westoll, eds), pp. 19–30. Oliver and Boyd, Edinburgh.

Nathorst, A. G. (1902). Zur Oberdevonischen flora der baren - insel. *K. svenska Vetensk Akad. Handl.* 36(3), 1–60.

North, F. J. (1931). Insect-life in the Coal forests, with special reference to South Wales. *Trans. Cardiff Nat. Soc.* 62, 16–44.

Olson, E. C. (1976). The exploitation of land by early tetrapods. *Linn. Soc. Symp. Ser.* 3, 1–30.

Oshurkova, M. V. (1967). "Paleophytological Validation of Stratigraphy of the Upper Suites of the Carboniferous Deposits in the Karaganda Basin", 150pp. Izd-vo Nauka Leningrad. (In Russian.)

Oshurkova, M. V. (1978). Palaeophytocoenogenesis as the basis of a detailed stratigraphy with special reference to the Carboniferous of the Karaganda Basin. *Rev. Palaeobot. Palynol.* 25, 181–187.

Panchen, A. (1970). "Teil 5a Anthracosauria. Handbuch der Paläoherpetologie". Fischer, Stuttgart.

Panchen, A. L. (1973). Carboniferous tetrapods. *In* "Atlas of Palaeobiogeography" (A. Hallam, ed.), pp. 117–125. Elsevier, Amsterdam.

Peppers, R. A. and Pfefferkorn, H. W. (1970). A comparison of the floras of the Colchester (No. 2) Coal and Francis Creek Shale. *In* "Depositional Environments in Parts of the Carbondale Formation – Western and Northern Illinois" (W. H. Smith *et al.*, eds), *Ill. State geol. Surv. Guide Ser.* 8, 61–74.

Pettitt, J. (1970). Heterospory and the origin of the seed habit. *Biol. Rev.* 45, 401–415.

Phillips, T. L., Kunz, A. B. and Mickish, D. J. (1977). Paleobotany of permineralised peat (coal balls) from the Herrin (No. 6) Coal Member of the Illinois Basin. *Geol. Soc. Am. Microform Publ.* 7, 18–49.

Phillips, T. L., Peppers, R. A., Avcin, M. J. and Laughnan, P. F. (1974). Fossil plants and coal: Patterns and change in Pennsylvanian coal swamps of the Illinois Basin. *Science* 154, 1367–1369.

Pollard, J. E. (1969). Three Ostracod-mussel bands in the Coal Measures (Westphalian) of Northumberland and Durham. *Proc. Yorks. geol. Soc.* 37, 239–276.

Ramsbottom, W. H. C. (1973). Transgressions and regressions in the Dinantian; a new synthesis of British Dinantian stratigraphy. *Proc. Yorks. geol. Soc.* 39, 567–607.

Rayner, D. H. (1971). Data on the environment and preservation of late Palaeozoic tetrapods. *Proc. Yorks. geol. Soc.* 38, 437–495.

Read, W. A. and Johnson, S. R. H. (1967). The sedimentology of sandstone formations within the Upper Old Red Sandstone and lowest Calciferous Sandstone Measures, West of Stirling, Scotland. *Scott. J. Geol.* **3**, 242–267.

Remy, W. and Remy, R. (1977). "Die Floren des Erdaltertums", 468 pp. Glückhauf, Essen.

Riek, E. F. (1970). Origin of the Australian insect fauna. *2nd Symp. Gondwana, S. Africa* 593–598.

Riek, E. F. (1973). Fossil insects from the Upper Permian of Natal, South Africa. *Ann. Natal Mus.* **21**, 513–532.

Robertson, T. (1952). Plant control in rhythmic sedimentation. *C.r. 3^e Congr. Adv. Etud. Strat. Geol. Carbonif., Heerlen 1951* **2**, 515–521.

Rolfe, W. D. I. and Ingham, J. K. (1967). Limb structure, affinity and diet of the Carboniferous "centipede". *Arthropleura. Scott. J. Geol.* **3**, 118–124.

Sabry, H. and Neves, R. (1971). Palynological evidence concerning the unconformable Carboniferous basal measures in the Sanquhar Coalfield, Dumfriesshire, Scotland. *C. r. 6^e Congr. Int. Strat. Geol. Carbonif., Sheffield 1967* **4**, 1441–1458.

Scheckler, S. E. (1974). Systematic characters of Devonian ferns. *Ann. Mo. bot. Gdn* **61**, 462–473.

Schopf, J. M. (1974). Coal, climate and global tectonics. *In* "Implications of Continental Drift to the Earth Sciences" (D. H. Tarling and S. K. Runcorn, eds), Vol. 1, pp. 609–622. Academic Press, London and New York.

Schopf, J. M. (1975). Modes of fossil preservation. *Rev. Palaeobot. Palynol.* **20**, 27–53.

Schram, F. R. (1976). Crustacean assemblage from the Pennsylvanian Linton vertebrate beds of Ohio. *Palaeontology* **19**, 411–412.

Schumm, S. A. (1968). Speculations concerning paleohydrological controls of terrestrial sedimentation.*Geol. Soc. Am. Bull.* **79**, 1573–1588.

Scott, A. C. (1976). "Environmental Control of Westphalian Plant Assemblages from Northern Britain", Vols 1 and 2. Ph.D. Thesis, University of London.

Scott, A. C. (1977a). Coprolites containing plant material from the Carboniferous of Britain, *Palaeontology* **20**, 59–68.

Scott, A. C. (1977b). A review of the ecology of Upper Carboniferous plant assemblages, with new data from Strathclyde. *Palaeontology* **20**, 447–473.

Scott, A. C. (1978). Sedimentological and ecological control of Westphalian B plant assemblages from West Yorkshire. *Proc. Yorks. geol. Soc.* **41**, 461–508.

Scott, A. C. (1979). The ecology of Coal Measure Floras from northern Britain. *Proc. geol. Ass.* **90**, 97–116.

Scott, A. C. and Collinson, M. E. (1978). Organic sedimentary particles: Results from SEM studies of fragmentary plant material. *In* "SEM in the Study of Sediments" (W. B. Whalley, ed.), pp. 137–167. Geo Abstracts, Norwich.

Seaward, A. C. (1924). The later records of plant life. *Q. Jl geol. Soc. Lond.* **80**, Lxi–xCvii.

Seward, A. C. (1935). "Forests of Coal Age", Abbott Memorial Lecture, 1935, 13pp. University College, Nottingham.

Sleeman, A. G. (1977). Lower Carboniferous transgressive facies in the Porter's

Gate Formation at Hook Head, County Wexford. *Proc. R. Irish Acad.* **77B**, 269–284.

Smart, J. and Hughes, N. F. (1973). The insect and the plant: progressive palaeoecological integration. *In* "Insect/Plant Relationships" (Van Emden, ed.), *Symp. R. Ent. Soc. Lond.* 6, 143–155.

Smith, A. G., Briden, J. C. and Drewery, G. E. (1973). Phanerozoic world maps. *In* "Organisms and Continents Through Time" (N. F. Hughes, ed.), *Spec. Pap. Palaeont.* **12**, 1–42.

Smith, D. B. *et al.* (1975). A correlation of Permian rocks in the British Isles. *Geol. Soc. Lond. Spec. Rep.* **5**, 1–45.

Smith, D. L. (1964). Two Scottish Lower Carboniferous floras. *Trans. Bot. Soc. Edinb.* **39**, 460–466.

Stoneley, H. M. (1958). The Upper Permian flora of England. *Bull. Br. Mus. nat. Hist., Geol.* 3, 295–337.

Teichmüller, M. (1952). Vergleichende mikroskopische Untersuchungen versteinerter Torfe des Ruhr-Karbons und der daraus enstanden en Steinkohlen. *C. r. 3^e Congr. Adv. Etud. Strat. Geol. Carbonif., Heerlen 1951* **2**, 607–613.

Teichmüller, M. (1962). Die Genese der Kohle. *C. r. 4^e Congr. Adv. Etud. Strat. Geol. Carbonif., Heerlen 1958, 3*, 699–722.

Utting, J. and Neves, R. (1970). Palynology of the Lower Limestone Shale Group (basal Carboniferous Limestone Series), and Portishead beds (Upper Old Red Sandstone) of the Avon Gorge, Bristol, England. *Congr. Collog. Univ. Liege* 55, 411–422.

Walton, J., Weir, J. and Leitch, D. (1938). A summary of Scottish Carboniferous stratigraphy and palaeontology. *C. r. 2^e Congr. Adv. Etude. Strat. Geol. Carbonif., Heerlen 1935* 1343–1356.

Wanless, H. R. (1950). Late Palaeozoic cycles of sedimentation in the United States. *Rep. 18th Int. Geol. Congr. G. B.* **IV**, 17–28.

Wanless, H. R., Baroffio, J. R. and Trescott, P. C. (1969). Conditions of deposition of Pennsylvanian Coal Beds. *In* "Environments of Coal Deposition" (E. C. Dapples and M. E. Hopkins, eds), *Geol. Soc. Am. Spec. Pap.* **114**, 105–142.

Wartmann, R. (1969). Studie über die papillen-förmigen Verdickingen auf der kutikule bei *Cordaites* an Material aus dem Westfal C des Saar-Karbons. *Argumenta Palaeobotanica* 3, 199–207.

Waugh, B. (1973). The distribution and formation of Permian - Triassic red beds. *In* "The Permian and Triassic Systems and their Mutual Boundary" (A. Logan and L. V. Hills, eds), *Can. Soc. petrol. Geol., Mem.* No. 2, 678–693.

Weiss, F. E. (1925). Plant structure and environment with special reference to fossil plants. *J. Ecol.* **13**, 301–313.

Witham, H. T. M. (1833). "The Internal Structure of Fossil Vegetables Found in Carboniferous Oolitic Deposits of Great Britain, Described and Illustrated". A. and C. Black, Edinburgh.

Zalessky, M. D. (1931). Végétaux nouveaux du Devonien superieur du Bassin du Donetz. *Bull. Acad. Sci. URSS 557–588.*

6 | Early Invertebrate Terrestrial Faunas

W. D. IAN ROLFE

Hunterian Museum, University of Glasgow, Scotland

Abstract: The few faunas known indicate that their dominant members, the arthropods, were amphibious (Xiphosura, Eurypterida) or fully terrestrial (Arachnida, Myriapoda) by the early Devonian, and thus were available to tetrapod predators. Carnivorous arachnids in these faunas imply the availability of unfossilizable small metazoan and protozoan prey.

Primary producers (plants) probably preceded primary consumers (myriapods) onto land in the Silurian (the Ludlow *Necrogammarus* may be a giant myriapod). Arthropod aeolian feeders and cryptobiotic, aerially dispersed microarthropods could even have preceded some plant pioneers, and aided colonization by "leap-frogging". Secondary and higher trophic level consumers probably succeeded each other in time (trophic level succession or web) as terrestrial invaders, inducing increased variety and biomass at lower levels of the trophic pyramid. Many early arthropods were important decomposers, building up early humus levels and soil structures via coprolites. Some of these arthropods were large, a possible adaptation for reducing water loss.

Trace fossils, important in being autochthonous, hint at the presence of a diversity of Devonian terrestrial arthropods. Experimental neoichnology with millipedes produces different traces under different conditions. Until variables are explored it will remain difficult to identify arthropod groups responsible for Devonian ichnocoenoses.

By the Upper Carboniferous, terrestrial arthropods had evolved explosively. Insectivorous reptiles probably played a larger part in the co-evolution of insects than of arachnids. The latter have shown little evolution since the Carboniferous, and many groups today are similarly confined to humid niches.

Nematode oligochaete and nemertean worms are known from the Carboniferous, but they are all aquatic. Several genera of small land snails are known from the Carboniferous; these probably inhabited tree-trunk base and forest-floor litter.

Systematics Association Special Volume No. 15, "The Terrestrial Environment and the Origin of Land Vertebrates", edited by A. L. Panchen, 1980, pp. 117–157, Academic Press, London and New York.

INTRODUCTION

There are few early invertebrate terrestrial faunas in the sense of complete fossil assemblages from the time of the origin of tetrapods – presumably Upper Devonian (Panchen, 1977, p.729). Such Devonian faunas as do exist will be reviewed here, together with observations on single fossils that may be relevant to the terrestrialization of invertebrate groups. More abundant faunas occur in the Upper Carboniferous, but to review these adequately would be beyond the required remit. Instead, observations on certain Carboniferous faunas are presented, with generalizations about the major groups present. Unlike plants and vertebrates, very little modern work has been done on fossils of fully terrestrial members of the most relevant invertebrate group – the arthropods.

TERRESTRIALIZATION OF ARTHROPODS

1. Origin

Terrestrial arthropods were probably derived independently from several aquatic stocks, although opinion differs on the extent of sclerotization of those stocks prior to their emergence (Størmer, 1976, p.134; cf. Manton, 1977, pp. 25–26). Such sclerotization may well have occurred "more than once during the evolution of arthropods, taking place independently on the sea and on the land" (Manton, 1970, p.29; 1977, p.31). *Aysheaia*, a lobopod from the Middle Cambrian, is the best example of one of these ancestral stocks (Whittington, 1978). It scarcely differs from living onychophorans – terrestrial animals highly adapted to squeeze through the narrowest of crevices (Manton, 1977, p.284) – in having a terminal mouth but lacking jaws and a projecting posterior. From such a lobopod group probably descended both the living Onychophora and the Tardigrada (Whittington, 1978, p.195; Manton, 1977, pp. 288, 498).

2. Terrestrialization

Tardigrades are unknown as fossils before the Cretaceous, probably on account of their small size – they do not exceed 1·2 mm long (Kaestner, 1968, p.16). Their significance lies in their ability to undergo cryptobiosis – the ability to survive almost total dehydration and remain in suspended animation for years, until revitalized by rehydration (Crowe and Cooper, 1971; Hinton, 1971b, 1977). Such an ability would obviously prove of great significance in distributing early microarthropods, since the desiccated tuns can be dispersed aerially like plant spores, their small size permitting them to fall without injury (Hinton, 1977, p.71). Tardigrades are mainly herbivores, sucking out the content of algal and other plant cells (Kaestner, 1968, p.24). Such microarthropods could therefore have accompanied the "greening of the landscape" that occurred with the spread of algae over land surfaces during the early Palaeozoic (Edwards, Chapter 4 of this volume), once ultraviolet levels had fallen sufficiently. Before that time, such ancestral forms may have scavenged on the shoreline, at night, when the ultraviolet radiation was cut out (Smart, 1971, p.305).

Arthropods could probably only evolve from this miniaturized phase (although small size is an important attribute of land arthropods – Edney, 1977, pp. 2–3) when sufficient plant debris had accumulated to provide protective litter and soil layers. Although this may have occurred with seaweeds rotting on beaches to form soil (Lawrence, 1953, p.367), no great spread can have occurred until the first land plants evolved and colonized the land during the Ludlovian (Edwards, Chapter 4 of this volume). As in present-day tropical and subtropical forests, leafmould probably merged with algal debris of the supralittoral (Bliss and Mantel, 1968, p.674), permitting the ready exploitation by arthropods of the new environment. Such early arthropods probably played a part in assisting the distribution of early plants (Kevan *et al.*, 1975).

Some living terrestrial arthropods can survive above the height at which plants can grow, and feed off wind-blown plant detritus in what has been termed the aeolian zone (Hutchinson, 1965, p.5). Similar habits occur in arthropods of extreme regions such as the Namib Desert. There are no autotrophs in these situations, which may provide an analogy with Palaeozoic continental interiors. Such ecological pioneers could assist the spread of early plants, by pelletizing spores in coprolite substrates, after passage through the gut (Kevan *et al.*, 1975, p. 406; Chaloner, 1976, pp. 4–5).

Most workers accept a marine origin for terrestrial arthropods, but Hinton (1977, pp. 72–73) has proposed that the cryptobiotic common ancestor of arthropods lived in a saturated terrestrial environment or just beneath the soil surface. Cryptobiosis might have existed in other lobopod stocks (as it does today in brine shrimp eggs and, secondarily, in midge larvae) to give rise to various groups of arthropods. Other lines, e.g. the myriapods, could be more conventionally derived, but again from a lobopod ancestor. In this sense, the Arthropoda are monophyletic (Patterson, 1978; cf. Manton, 1977).

3. Routes Ashore

Arthropods probably terrestrialized by different routes, as is reflected in their divergent habits and associated functional morphology to cope with the new habitats (Manton, 1973, pp. 316–317; 1977, pp. 25–36). Some may have ventured "across sandy beaches or the rocky intertidal zone, through mangrove swamps, or by way of freshwater lakes and streams" (Bliss and Mantel, 1968, p. 674). As Rapoport and Tschapek (1967, pp. 6–7) point out, access via the littoral zone is more difficult than via estuaries or swamps, since all the problems of terrestrialization exist at the same time – especially those of terrestrial locomotion, aerial respiration and temperature fluctuation. One of the few groups that may have followed this route is the isopod crustaceans, whose evaporative water cooling reflects the large, rapid temperature fluctuations of the littoral zone where water is nevertheless abundant (Edney, 1977, p. 95).

Manton (1973, p. 316) and Størmer (1976, p. 134) therefore favour "emergence from quiet estuaries, where terrestrial vegetation and cover exist right to the water's edge . . . The early inhabitants of the land probably gained shelter wherever it could be found, under larger bits of the substratum or plant matter, and by shallow penetration into soft soil without pushing and unaided by any morphological facilitations". Subsequent rapid differentiation of habits, following uniramian arrival on land (Manton, 1977, p. 489), led to the separation of Onychophora and the four myriapod and five hexapod classes, as explained by Manton (1974, p. 164).

There has been much "experimentation", but little consequent radiation within the Merostomata (Eldredge, 1974, p. 38). Although many "eurypteroids" probably attempted to come ashore, only two lines met with success – the scorpions and the arachnids. Many of the latter owe their survival to the adoption of a cryptozoic habit (Savory, 1977b, p. 312).

THE EARLIEST TERRESTRIAL FOSSILS

Many separate records indicate that the myriapods are the oldest terrestrial fossils. Little is known about this material, although current research under Professor H. B. Whittington's supervision at Cambridge should improve this position. The oldest record is that of *Archidesmus loganensis* from the Llandovery/Wenlock of Lesmahagow (Peach, 1899, pp. 123–125). This specimen (Fig. 4B; I.G.S. Edinburgh 5974) is very poorly preserved, and there is nothing to prove that it is a myriapod. The specimen may have deteriorated since its collection (see also Størmer, 1976, p. 151), but it now resembles some of the plant debris which occurs in the same *Jamoytius* horizon, as first pointed out by Ritchie (1963, p. 125). Ritchie collected a new specimen (Royal Scottish Museum 1970.2) from a higher horizon, the Fish Bed Formation fish bed (Walton, 1965, p. 196) of Wenlock/Ludlow age (Westoll, 1977, p. 72). This (Fig. 4A) resembles a juliform diplopod in having *c.* 20 similar somites with an ornament of terrace-lines, but it is impossible to be certain of this fossil's true affinity.

Necrogammarus salweyi Woodward, from the channel-fill Ludlow of Leintwardine (Whitaker, 1962) is even more problematical (Fig. 1). Originally compared with amphipod crustaceans by Huxley and Salter (1859), this unique specimen (B.M. (N.H.) In 43786) was later referred to the Diplopoda by Peach (1899, p. 126). The two body somites show a lateral sulcus such as is present on Old Red diplopods, and this suggests diplosomites are present. Two limbs are appended to one diplosomite, and one to the other, which is also a diplopodan character. Presupposing 11 somites were originally present, a body length of at least 275 mm is indicated. There are no characters to suggest that this large arthropod must have been terrestrial and the beast may have been an aquatic precursor of terrestrial forms (see Bergström, 1978, pp. 19–25). It *could* have been washed down-channel from a terrestrial situation, and the uniqueness of the specimen may be seen as evidence of such an origin, although

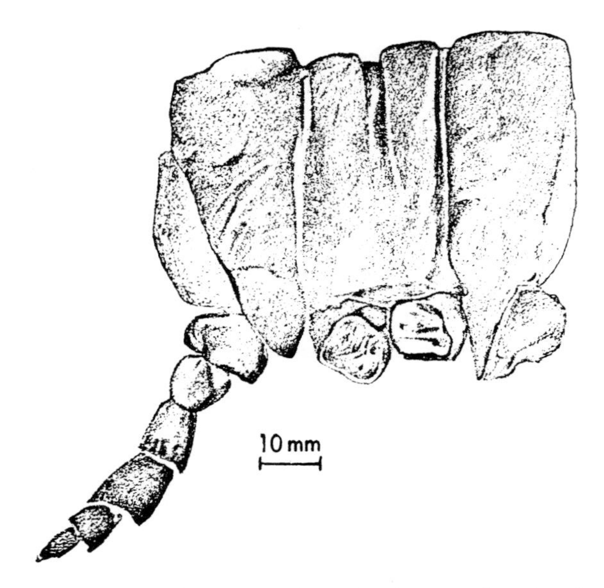

Fig. 1. *Necrogammarus salweyi* Woodward, a large millipede? from the Ludlow of Leintwardine, Herefordshire (after Huxley and Salter, 1859).

the shoreline was many miles distant. Local shallowing did occur and islands may have formed (J. D. Lawson, personal communication), which could give a more immediate origin for such a fossil – if it were terrestrial.

1. Old Red Sandstone Myriapods

More material is available here, but revision is also required. *Archidesmus* and *Kampecaris* (but not *Anthracodesmus* and *Pattonia* – cf. Hoffman, 1969, p.R. 578) are known from much of the Lower Old Red Sandstone of Scotland, probably of Gedinnian age (Westoll, 1977, pp. 72,76), as is also *Kampecaris* from England (Clarke, 1951; Allen, 1977, p. 45). Specimens as old as "Infra-Gedinnian" occur at Stonehaven and as young as Siegenian in Ayrshire (Westoll, 1977, pp. 72, 73). Many of the specimens are decalcified (e.g. *Kampecaris obanensis,* Fig. 4D), and it is not clear whether this is due to pre-moult autolysis, or to diagenesis. Only exceptionally is the heavily calcified exoskeleton preserved as in *Archidesmus macnicoli* (R.S.M. 1891-92-72), and this reveals it to have large paratergal lobes. "Such 'flat-backed' millipedes are adepts at widening crevices which tend to split open along one plane, such as under bark or in layered decaying leaves" (Manton, 1977, p. 357). Others may be rounder-bodied juliform types, e.g. *Kampecaris forfarensis* (R.S.M. 1969.16 – Fig. 4E). Such millipedes burrow into soil and decaying wood by head-on pushing (Manton, 1977, pp. 352–356), rather than by wedging. Both millipedes imply the presence of decaying vegetation and perhaps soils well inland by the early Devonian. Specimens are often found on slabs accompanied by much plant debris.

Their comparative abundance as fossils may be due to the habit of some millipedes of forming subterranean moulting chambers: "hollows in the soil, of material which has passed through the gut, smoothed internally by rocking movements performed by the animal". Others construct subterranean tents of mineral-rich soil, stuck together to form continuous sheets (Heath *et al.*, 1974, pp. 455, 462). The moulting process may take up to three weeks (Kaestner, 1968, p.409), and the chamber serves both as a protection from cannibalistic members of its own species (Evans, 1910, pp. 289–290) and to prevent desiccation. Tropical forms may remain throughout the dry season in such moulting chambers (O'Neill, 1969; Lewis, 1974). Against this habit favouring preservation must be set the fact that most millipedes eat their moults (Cloudsley-Thompson, 1968, p. 45), as well as their own dead (Cloudsley-Thompson, 1949, p. 137).

2. *The Rhynie Chert Fauna*

This fauna, of Siegenian age (Westoll, 1977, p. 76), has been reviewed by Kevan *et al.* (1975, pp. 392–396), and will only be summarized here. A unique fauna, largely of microarthropods, is known, due to the exceptional preservation of this silicified peat bog. The terrestrial fauna comprises the collembolan *Rhyniella*, a mite *Protacarus*, a spider *Palaeocteniza*, and the trigonotarbid arachnids *Palaeocharinus* and *Palaeocharinoides* (fig. 2). An additional arthropod, *Heterocrania rhyniensis* has been described but is too fragmentary to be interpreted as a chelicerate. Kevan *et al.* (1975, p. 394) note that collembolans are generally herbivorous or saprophagous, and suggest *Rhyniella* fed on soil micro-organisms and spores. Like *Protacarus, Rhyniella* might have obtained plant juices through the puncture wounds which disfigure so many *Rhynia* axes. The other arthropods are carnivorous, although Kevan *et al.* (1975, p. 395) were misled by the literature into

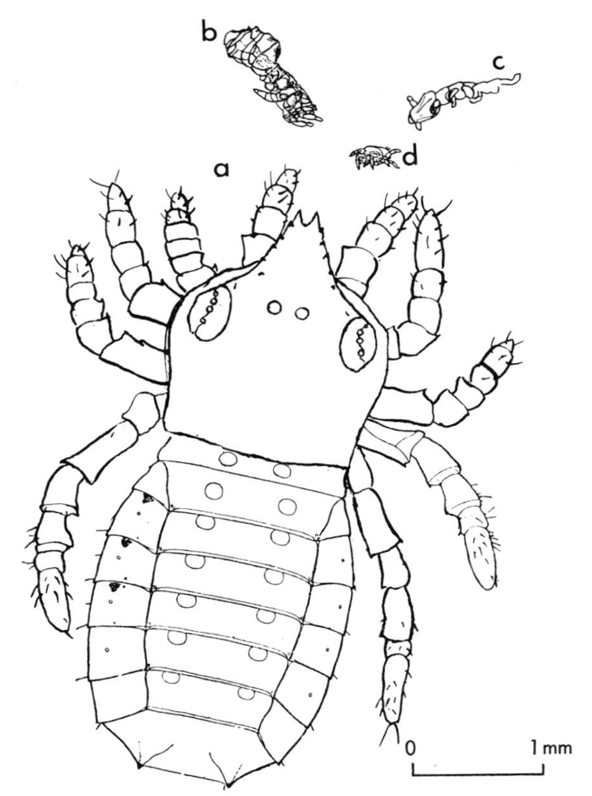

Fig. 2. Representative terrestrial arthropods of the Rhynie Chert to show relative sizes of predator and prey. All to the same scale (× 17 – after Hirst, 1922). (a) A small individual trigonotarbid arachnid *Palaeocharinus* sp. (new reconstruction based on several specimens, with limbs of *Palaeocharinoides* type, eye detail from Hirst and Maulik (1926)). The largest individual known would be twice the size of this page, to the same scale. (b) Possible liphistiid spider *Palaeocteniza crassipes* Hirst. (c) Collembolan *Rhyniella praecursor* Hirst & Maulik. (d) Pachygnathid mite *Protacarus crani* Hirst.

thinking that they were all small – less than 3 mm long. They suggested that these small predators fed on an unfossilized element of the food web – soft-bodied protozoans and small metazoans. The Hunterian Museum Kidston collection of thin sections shows several vertical sections (Kid. 2405, 2406) across palaeocharinid opisthosomata 6·5 mm broad, implying a total body length of at least 14 mm.

A similar cross-section was illustrated by Størmer (1976, Pl. 9, Fig. 1 – inverted), showing the median ventral projection of the anterior opisthosomal sternite. Such sections also reveal the book-lungs, the earliest unequivocal evidence of a truly terrestrial arthropod. That arachnids of this size existed in the Rhynie Chert is confirmed by a series of almost horizontal sections preserved in the Entomology Department of the British Museum (Natural History), one of which (In 24691) shows practically an entire body measuring 14 mm long. It seems most unlikely that such a large arachnid was even a facultative herbivore (cf. Kevan *et al.*, 1975, pp. 395–396) and it must have preyed upon the Rhynie microarthropods. A normal-sized palaeocharinid is shown to scale in Fig. 2a.

Kevan *et al.*, (1975, p. 396) and Chaloner (1976) suggested that the palaeocharinids found within hollow stems and sporangia (Rolfe, in Kevan *et al.*, p. 396, Pl. 56) were spore-feeders, and associated such feeding with the predatory habit and evolution therefrom. It is possible that the small arthropods were inhabiting these damp, rotted hollows in order to conserve their body moisture – always a problem with terrestrial arthropods – much as do many living land arthropods. They could readily prey upon other microarthropods seeking refuge in such a humid niche, and also evade their larger predatory relatives outside. It is tempting to imagine such small arachnids actually lurking in dehisced sporangia while the latter were borne upright upon the parent plant (although cf. Kevan *et al.*, 1975, p. 406). Rather than snipping spores, they might have preferred to await the arrival of food in the form of early insect, or other arthropod, visitors to the sporangium for spore-feeding and dispersal (Kevan *et al.*, 1975, pp. 402–406). The large numbers of dismembered fragments seen in such hollows (one such hollow is 14 mm long – B.M. (N.H.) In 24704) may represent the sucked-dry husks of former victims of the dominant preying arachnids. Edwards (1973 and personal communication) has examined *c.* 15 dehisced sporangia interpreted as being in growth position, both from the alignment of the parent stems (roughly perpendicular to the soil layer) and from the presence of spores in the basal portion only of the sporangium. None of these contains arthropod fragments, which argues against the above interpretation. The relationship between the palaeocharinid spirit level infill and the axis of the *Rhynia* shown by Rolfe (in Kevan *et al.*, 1975, Pl. 56, Fig. 1) indicates that the stem was far from vertical.

Together with the disrupted nature of such Rhynie "straws", this suggests the stems only became occupied by arthropod fragments after they had fallen. That some are also *post mortem* assemblages is shown by B.M. In 38236, a fungus-covered arthropod with its body cavity filled with arthropod fragments, as well as a large fungal resting phase. Finally, the small arachnids so regularly found within stems may only be "spiderlings". Complete individuals occasionally closely accompany one another inside stems (e.g. B.M. In 27756 – Hirst and Maulik, 1926, Fig. 1), and were presumably instantly killed by silicification in this biocoenosis. It suggests that they were juveniles since "only during their first instar do the spiderlings of most species stay together". Their behaviour is fundamentally different then, since they are absorbing embryonic yolk filling the mid-gut. By their first moult, they "are able to eat other food and would kill each other, as they do in captivity, but for the new and powerful inner urge to disperse" (Petrunkevitch, 1952, p. 109).

Although difficult to interpret, and known only from the holotype, *Palaeocteniza* appears to be a mygalomorph spider, a group known also from the Carboniferous. This group includes the trapdoor spiders today, and the living *Liphistius,* thought to be a "living fossil". *Liphistius* burrows have about eight long, straight threads radiating downwards from the rim of the burrow, which serve as trip-wires indicating the passage of crawling invertebrates (Bristowe, 1975, p. 115). Pit-fall studies indicate how abundant such prey is (F. R. Wanless, personal communication), and the spider has only to rush out and overpower the victim. Predation by such primitive hunting spiders (and also early amphibians and insects? – Smart 1967, p. 116) upon early wingless insects may have been one of the main factors leading to the evolution of insect wings. "When their prey took to the air to escape, spiders evolved aerial webs as a means of trapping it in flight" (Cloudsley-Thompson, 1968, p. 186; 1975, p. 192). This escape by flight may well have taken place under the impetus given by the advent of late Devonian trees, which not only provided a launching pad, but also a new, tree-top ecosystem (Kevan *et al.*, 1975; Scott, Chapter 5 of this volume). It is worth recalling that the high canopy of modern tropical rain-forests contains a little-known but large biomass not only of insects, but also of large mygalomorph and web spiders, as well as millipedes (Elton, 1973, p. 91). It seems likely that many arachnid predators pursued their insectan prey up trees as they evolved, and that a high-canopy ecosystem evolved during the Upper Devonian–Carboniferous. This may have been simpler and less sophisticated than the present tropical forest ecosystem, which is the fragile end-product of evolutionary interaction between predator and prey (Elton, 1973).

Crowson (1970 and personal communication) believes that both the mite and collembolan of the Rhynie Chert are so indistinguishable from living forms, that they must be Tertiary or later contaminants. According to Zakhvatkin (1952, p. 24) the Rhynie mite *Protacarus* can be placed in one of the modern families Alycidae, Alicorhagiidae or Nanorchestidae. Other workers have commented on this remarkably modern aspect of the collembolan, although the mite is often said to be "primitive" (Kevan *et al.*, p. 394). Crowson believes that this view of the mite is based on circular reasoning since the "primitiveness" of the families of pachygnathid mites, with which the Rhynie fossil is compared, are regarded as such by acarologists since they are known to occur in the Rhynie Chert! A primitive mite should resemble the segmented *Opilioacarus* instead of the unsegmented Rhynie form. It is worth recording that Crowson came to this view as a result of a section of Rhynie Chert referred to him by the late John Walton. It contained a thysanopteran insect nymph which Crowson was able to identify with a living form, that must have crawled into a crevice in the chert. He believes it is significant that the collembolan and mite are small habitual crevice seekers. Rapid examination of relevant material at the British Museum (Natural History) shows little optical discontinuity between the silica containing the mite and its matrix. It is conceivable that mobilization of silica could lead to such contaminants being preserved, although one would expect to be able to see evidence of such diagenesis in the rock. Some acarologists do consider this group of mites to be "primitive" on other grounds, however. Thus Hirst (1922, p. 458) noted the group was regarded as primitive in his original description of the Rhynie form, and D. Macfarlane (personal communication) points out that the setation and other characters are truly "primitive". The ancient origin of the group and its extremely conservative nature is also borne out by the discovery of a Jurassic mite attributable to the *living* genus *Hydrozetes* (Sivhed and Wallwork, 1978). Furthermore "nothing like *Protacarus* exists at present" according to Vitzthum (Petrunkevitch, 1953, p. 57).

Dubinin (1962, pp. 462–464) has created an additional two genera to accommodate the variety of form shown by the Rhynie mites. All the families to which these mites are referred characteristically inhabit environments subject to alternate wetting and drying (J. B. Kethley, personal communication). Such mites today are fungivores or algal feeders, as far as is known, but Kethley points out that the mouthparts may not be diagnostic of feeding habit.

3. The Alken Fauna

This fauna, from Alken an der Mosel in Germany, is of Lower Emsian age (Størmer, 1970–1976). In addition to a marine fauna, the locality yields brackish, amphibious and terrestrial forms. An extensive Hunsrück Island, fringed by tidal flats with occasional lagoons, existed in the Lower Devonian of the Mosel–Rhine area. Alken represents one of these elongate, land-locked lagoons, communicating by channels with the sea at high tide. The lagoon was surrounded by psilophytes and other land plants, although the suggestion that some of them were partially submerged to create a "mangrove" ecosystem (Størmer, 1976, pp. 127–133) is no longer accepted by palaeobotanists (Edwards, Chapter 4 of this volume). Such a situation was suitable for the preservation of terrestrial and amphibious faunal elements, and these may be summarized as follows.

There are at least three terrestrial forms at Alken – the arthropleurid myriapod *Eoarthropleura*, and the trigonotarbid arachnids *Alkenia* (Fig. 3b) and *Archaeomartus*. An additional arthropod fragment was doubtfully referred to the Carboniferous spider *Archaeometa* (Størmer, 1976, pp. 115–116).

Eoarthropleura is the earliest arthropleurid known (the Siegenian *Bundenbachiellus* is now thought to be based on distorted specimens of *Cheloniellon* – Stürmer and Bergström, 1978). It probably lived among and fed off decaying and damp plant litter in the psilophyte groves. With its relatively longer limbs and greater trunk flexibility coupled with smaller size, *Eoarthropleura* was probably more agile than the Carboniferous *Arthropleura*. It may have been able to twist its way through the dense psilophyte stands, rather than forcing a passage like *Arthropleura* (Briggs *et al.*, 1979, p. 288). *Alkenia* and the little known *Archaeomartus* approach the size of the largest palaeo-

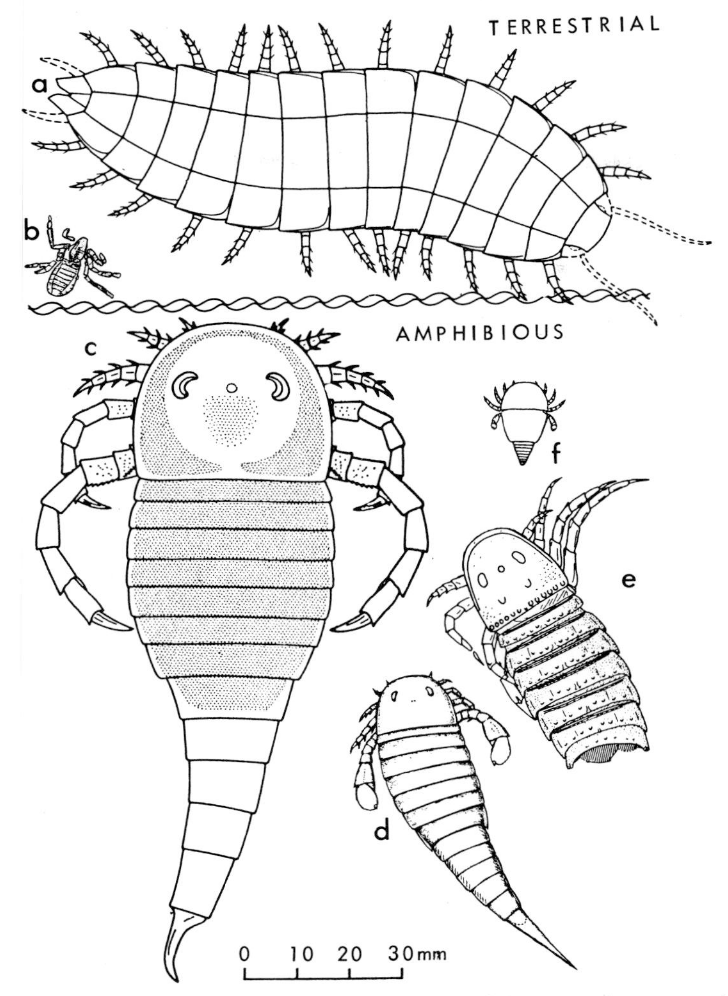

Fig. 3. Representative elements of the Alken amphibious and terrestrial arthropod fauna.

(a) Arthropleurid myriapod *Eoarthropleura devonica* Størmer (limb phasing not corrected – *v.* Briggs *et al.*, 1979).

(b) Trigonotarbid arachnid *Alkenia mirabilis* Størmer.

(c-e) Eurypterids. (c) Drepanopteroid *Moselopterus ancylotelson* Størmer (d) Hughmillerioid *Parahughmilleria hefteri* Størmer. (e) Rhenopteroid *Rhenopterus macrotuberculatus* Stømer (body of *R. diensti* restored to suggest appearance).

(f) Chasmataspidid xiphosuran *Diploaspis casteri* Størmer.

Natural size, after Størmer (1936, 1969, 1970, 1972, 1973, 1974, 1976), with permission of the author and the *Senckenbergische Naturforschende Gesellschaft*.

charinids mentioned above from the Rhynie Chert. They probably led a similar, predatory mode of life.

Many other arthropods from Alken have been suggested by Størmer to be more or less amphibious (Fig. 3c–f), since they are now known to be pseudotracheate. Many of the eurypterids may have been able to leave the water and move into the Alken swampy hinterland. Even those such as *Parahughmilleria* that were well adapted for swimming may have used their paddles for crawling along, as the existence of a Silurian eurypterid trail suggests (Hanken and Størmer, 1975). Forms without paddles, such as *Drepanopterus, Alkenopterus, Moselopterus, Willwerathia* and *Rhenopterus* would be better adapted for such excursions. They were probably more terrestrial than the living *Limulus*, which can spend a considerable time on shore or on the river banks (Størmer, 1976, pp. 137, 143). The small chasmataspidid xiphosurans *Diploaspis* and *Heteroaspis* (Fig. 3f) had a ventral shield protecting the gills from desiccation. They too may have crawled about in the psilophyte thickets bordering the lagoon. The Alken scorpion *Waeringoscorpio* possessed gills and is considered to have been completely aquatic (Størmer, 1976, pp. 129, 151).

4. Other Devonian Faunas

Faunas probably exist elsewhere in the geological record, but have yet to yield sufficient material for description. The early Upper Devonian Escuminac Formation, for example, contains "a fragment tentatively identified as *Arthropleura*", in addition to many fish and plants, conchostracans and a scorpion (Schultze, in Carroll *et al.*, 1972, p.94). Work with early fossil plants has shown how successful the study of such fragmentary fossils can be, and it is therefore important that all such scraps be retained in collections for future investigation. A third Devonian terrestrial arthropod fauna has recently been discovered in the Givetian of New York State by P. Bonamo and D. Grierson.

5. Trophic Level Succession.

With the emergence of plants in the Silurian, the first trophic level, producers, was established on land. The second trophic level, primary consumers (herbivores) probably emerged immediately thereafter, as shown by Gedinnian myriapods. It is possible that third and fourth trophic level consumers (carnivores and predators) succeeded each other in time, producing a trophic level succession with time – and this had probably happened by the Siegenian, as the Rhynie fauna indicates. This model is probably too simple to be very fruitful, although it can indicate the absence of members of particular food-chains from fossil assemblages. It has already been suggested above, for example, that the first primary consumers accompanied or even locally preceded the producers inland.

TRACE FOSSILS

Trace fossils yield potentially important evidence of early terrestrial life, since they are known to be autochthonous, unlike many body fossils. They also give direct evidence of habit and behaviour of early terrestrial organisms, and indirect evidence of their morphology. Thus, the Devonian trail *Paleohelcura (= Beaconichnus) antarcticum* (Gevers) may provide evidence for the amphibious nature of early large scorpions (Briggs *et al.*, 1979, p. 279) such as *Brontoscorpio* (Kjellesvig-Waering, 1972) or *Praearcturus* (Rolfe, 1969). This also exemplifies the ambiguity of trace fossil evidence, since *Paleohelcura*-type trackways are also left by jumping thysanuran insects (Manton, 1977, p. 332, Fig. 7.13h).

1. *Ichnocoenoses*

Müller (1975) has reviewed aspects of the ichnology of terrestrial arthropods. Several Palaeozoic assemblages are known, e.g. the Orcadian Middle Devonian (N. H. Trewin, personal communication) and Rotliegenden (Boy, 1976), but little interpretation is possible in the absence of relevant neoichnology. One promising assemblage comprising many hundreds of specimens has been collected from the Lower Old Red Sandstone of Dunure, Ayrshire (Smith, 1909), and is being re-studied by Dr J. Pollard and myself.

The siltstone and mudstone of this deposit occur as baked, laminated and occasionally penecontemporaneously upended sediments, which require investigation: they are either aeolian or from lava-surface wash. Most of the trace fossils of this deposit are of arthropod origin,

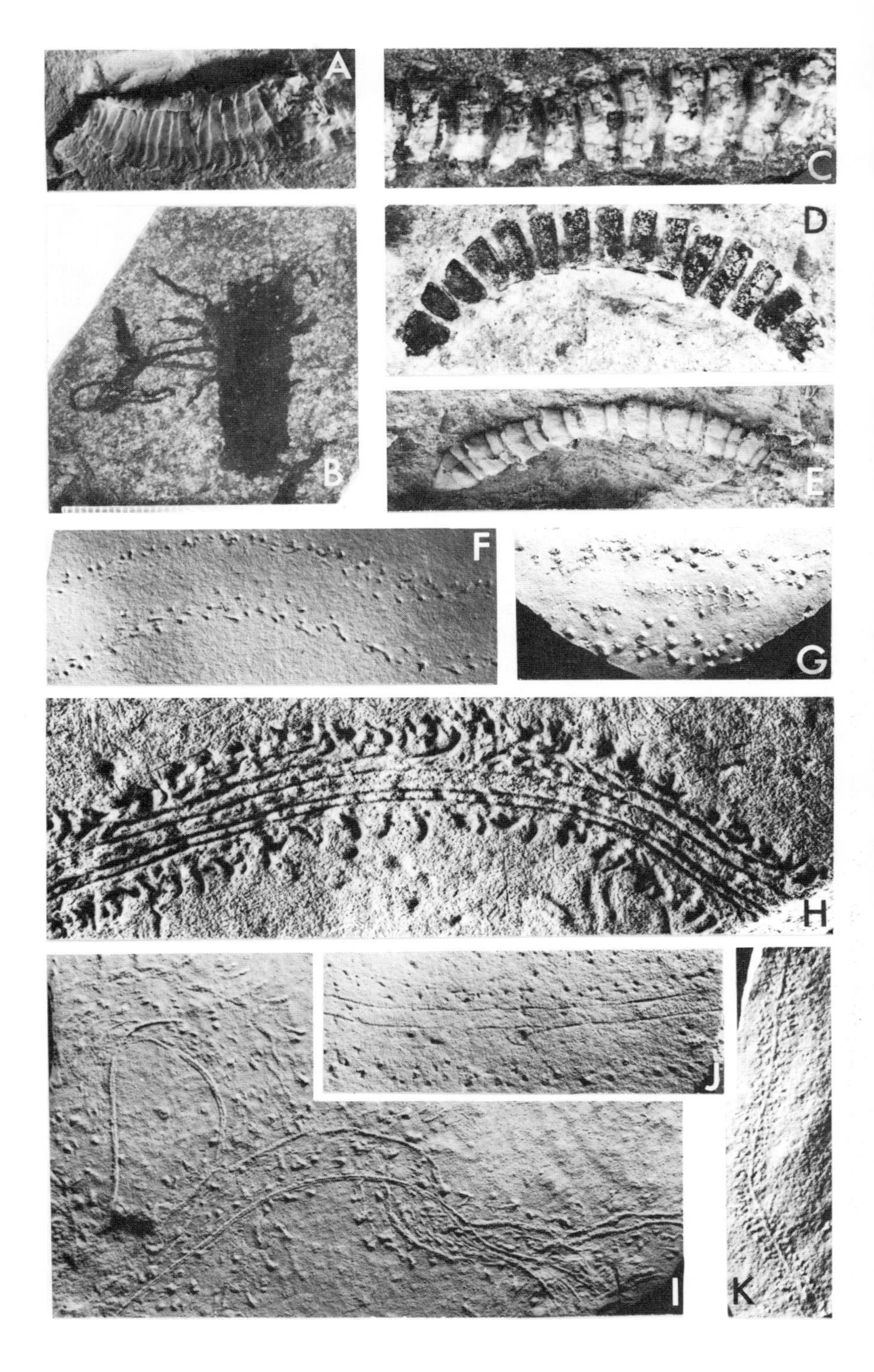

A
C
B
D
E
F
G
H
J
I
K

and millipede traces do appear to be present (Fig. 4). The unique specimen of the body fossil *Kampecaris tuberculata* Brade-Birks (1923) comes from this deposit, formed in the Old Red temporarily desert environment (Bluck, 1978).

2. Experimental Neoichnology

Manton (1977) has done much work on locomotion of arthropods, but her arthropods left traces on smoked plates, and these records are not readily comparable with traces left in soft substrates. A few references are available (Chamberlain, 1975, p. 436; Demoor, 1890, 1891; Fiori *et al.*, 1966; McKee, 1947) but emphasis on indigenous non-marine aquatic traces meant that few terrestrial arthropods were dealt

---

Fig. 4. Siluro-Devonian millipedes and trackways. (A) Possible millipede from Fish Bed Formation, Wenlock/Ludlow, Hagshaw Hills, Lanarkshire. R.S.M. 1970.2. × 2·0. (B) Holotype of *Archidesmus loganensis* Peach, *Jamoytius* horizon, Llandovery/Wenlock, Lesmahagow. I.G.S. (E) 5974. × 2·0. (C) Calcareous, flat-backed millipede, *Anthracodesmus macconochiei* Peach, Lower Carboniferous, Lennel Braes, Coldstream (resembles the Lower Old Red Sandstone *Archidesmus macnicoli* Peach). I.G.S.(E) 2176. × 6·5. (D) Decalcified millipede, *Kampecaris obanensis* Peach, L.O.R.S., Kerrera, Oban, Argyll. I.G.S.(E) 10385. × 4·0. (E) Round-bodied? millipede, *Kampecaris forfarensis*? Peach, Cairnconnan Series, L.O.R.S., Mirestone Quarry, Angus. R.S.M. 1969.16. × 3·0.

(F–K) Lower Old Red Sandstone millipede? trackways from Dunure, Ayrshire, all × 1·0. (F) *Stiaria simplex* Smith, a typical, simple millipede trackway. I.G.S.(E) 13480. (G) Holotype of *Narunia lunanova* Smith, showing median structures possibly left by sternites and eversible vesicles. I.G.S.(E) 13478. (H) Holotype of *Keircalia multipedia* Smith, showing multiple grooves caused by "miring" of posterior limbs, which are then dragged behind body. I.G.S.(E) 13463. (I) *Stiallia/Danstairia* trackway with grooves possibly made by terminal locomotor? filaments, postulated to exist in both early myriapods and hexapods (Müller, 1975, p. 81; Sharov, 1966). H.M. 1082A. (J) *Danstairia vagusa*? Smith, trackway with posterior limbs only occasionally touching substrate. Such behaviour recalls the "posterior antennal" role of the last pair of legs, which are trailed backwards on the ground, in the blind geophilid centipedes. Such animals living in cramped surroundings are able to move backwards as well as forwards, without turning the body as a whole (Lawrence, 1953, pp. 87, 89). I.G.S.(E) 13479. (K) *Danstairia kennediea* Smith, trackway with central groove, perhaps made by head (cf. Fig. 5F,G). I.G.S.(E) 13481.

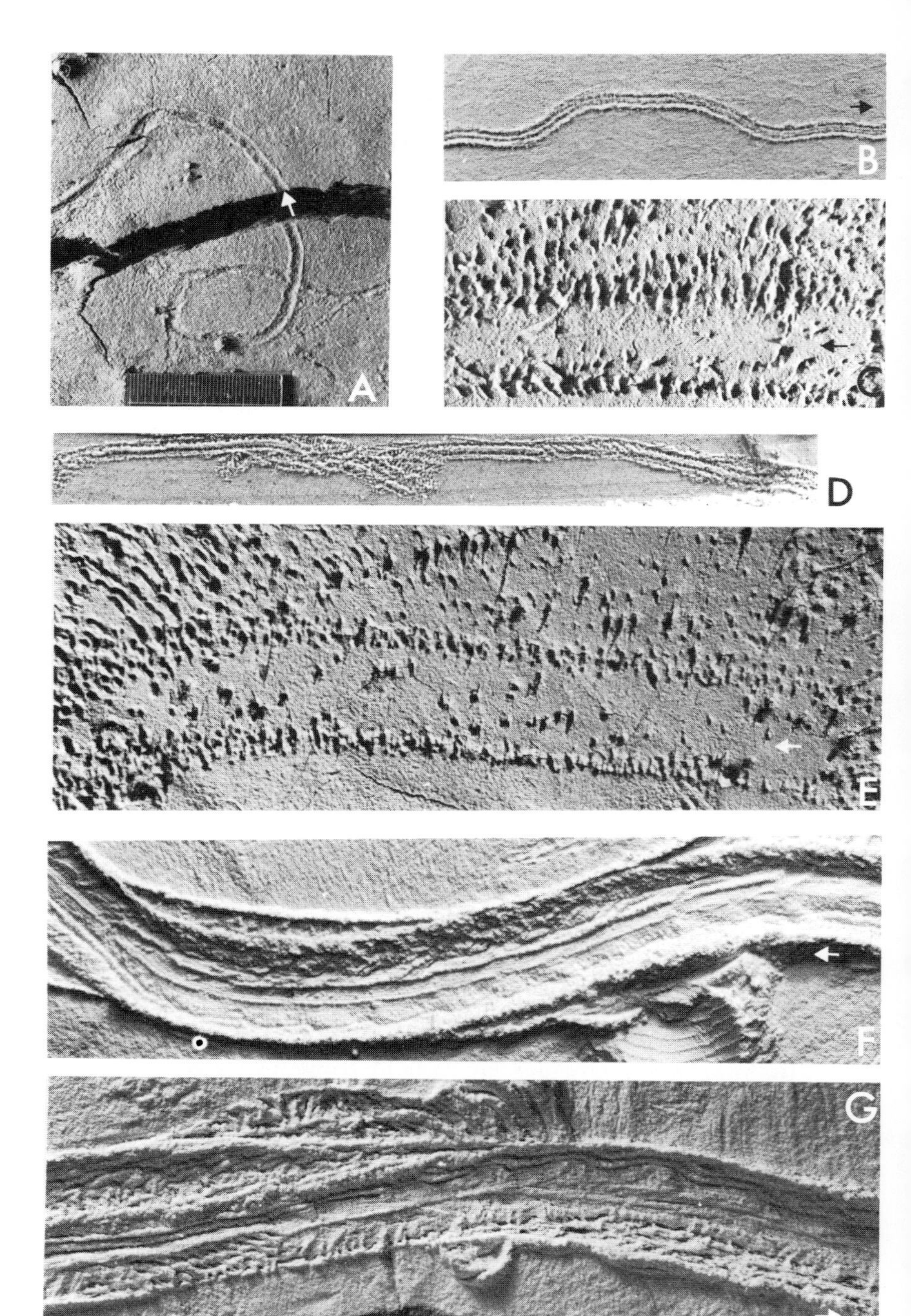

A
B
C
D
E
F
G

with in the recent volume by Frey (1975), and millipedes scarcely warrant a mention. Since the millipedes are the earliest known terrestrial animal fossils, their trace fossils are likely to be widespread.

Despite much recent work, relatively little is known about the kinds of traces left by many living arthropods, particularly the nature, limits and origin of their variability. A few experiments recently conducted by Miss Elaine Walker and myself at Glasgow University only serve to reveal the extent of the problem. As expected (Müller, 1975, p. 82) any one arthropod's trace will vary with many factors: type and size of organism, habit and gait, substrate grain-size, inclination and moisture content, and temperature (Müller, 1975, p. 81).

3. Traces of Modern Millipedes

On a wet mud depth of unconsolidated substrate the small millipede *Julus* leaves a broad simple groove (Fig. 5A) caused by body drag, which would yield a trace fossil resembling *Gordius* (cf. Boy, 1976, Fig. 41b). On a dry muddy silt substrate, a trail of two or three distinctly incised grooves is left (Fig. 5B; cf. McKee, 1947, p.27), comparable with *Aulichnus* or *Beaconichnus darwinum* (Gevers). The large (25 cm long) burrowing millipede *Scaphiostreptus* – morphologically very like *Julus* – does not leave a simple trail in wet mud, like the small *Julus*, but a complex trackway (Fig. 5C,E). These may be compared with the many myriapod ichnogenera listed by Briggs *et al.* (1979, p. 288) and with *Beaconichnus gouldi* (Gevers). On a dry substrate the head of *Scaphiostreptus* bulldozes a deep median groove (Fig. 5F) in advance of its trackway – which differs from that left in the wet sediment. It is not known whether this groove is made deliberately, to create a tunnel to ensure the limbs and spiracles are not clogged by sediment. It may only be inadvertent since the

---

Fig. 5. Trackways of living myriapods (see text for detail). All natural size, juliform millipedes, except (D) which is a centipede. Arrows indicate direction of travel. (A,B) *Julus sp. c.* 25 mm long; (A) wet mud substrate, (B) dry powder substrate. (C, E–G) *Scaphiostreptus seychellarum, c.* 250 mm long; (C, E) wet mud substrate, (F,G) dry powder substrate, (F) thick substrate, (G) thin substrate, but head groove present; individual tracks discernible on right of trackway, but obliterated on left side by trailing limbs (cf. Fig. 4K). (D) *Geophilus* sp., dry powder substrate.

darkling beetle *Blaps* also makes a groove, but with the tip of its abdomen, in dry sediment (Müller, 1975, Taf. 3, Fig. 1), which it does not do under other conditions (Graber, in Chamberlain, 1975, p. 436, Fig. 19.1). When the unconsolidated substrate is kept thin, so that *Scaphiostreptus* can support the head and body well above the substrate during locomotion, no median groove results (Fig. 5G).

4. *Beaconites*

Müller (1975, pp. 82–83, Pl. 1, Fig. 3) has described Recent surface trails superficially resembling the smaller Devonian trails attributed to *Beaconites*. These were made by beetles climbing up steep, blown sand banks. Gravitational slumping of individual transverse tracks produces a structure resembling the supposed stuffed-burrow menisci of *Beaconites*. It is worth speculating whether longer *Beaconites* originate from desert-inhabiting arthropods which construct a burrow system, rather than from early amphibians (Pollard, 1976). Menisci might then be formed by passive slumping, at intervals determined by the arthropod's footfalls. Such slumping should only occur in steeply inclined regions of the burrow system, and existing evidence from *Beaconites* (Pollard, 1976) may be sufficient to rule out this hypothesis. It is a tempting notion, however, since arthropods (like many other animals) are known to burrow in the desert seeking increased relative humidity and avoiding extremes of temperature, only emerging at optimal temperatures (Cloudsley-Thompson, 1975a, p. 80; Müller, 1975, p. 81). Proof could come from the tracing of the burrow structure into a surface trace, a situation observed in the Recent desert cockroach *Arenivaga* (Edney, 1977, pp. 198–200, Fig. 91).

CARBONIFEROUS FAUNAS

By the Upper Carboniferous, terrestrial faunas, at least of arthropods, had evolved explosively. It would be beyond the requirements of this volume to review these in detail, and very little modern work is available to permit such a review. Coal Measures invertebrate faunas were reviewed by Calver (1968), although the emphasis there is on acquatic forms. The evidence will be briefly reviewed on a systematic basis, omitting some major arthropod groups which are dealt with subsequently. Only the most terrestrial *in situ* fauna, that of Joggins, will be examined closely.

1. Insects

The earliest insects are known from two specimens in the Namurian of Poland and Pennsylvania (Orders Protorthoptera and Miomoptera – although one of these records may rather be Palaeoptera – Carpenter, 1977, pp. 65–66). Very many orders are known from the Upper Carboniferous, although part of the previous picture of this radiation was a monographic burst. Carpenter (1977, p. 63) has stated that of the 52 extinct, mainly Palaeozoic, orders named, only nine are acceptable in yielding the minimal criteria of fore and hind wings, heads and mouthparts: the remainder must remain order unknown. Nevertheless such insects show highly specialized adaptations by that time (Carpenter, 1971): more orders than in the existing fauna had mouthparts for piercing and sucking up liquid food (see also Kukalova-Peck, 1980); many had ovipositors for depositing eggs in particular substrates, and even cockroaches must have laid eggs singly at a considerable depth in soil, decaying vegetation or soft plant tissue (Carpenter, 1971, p. 1244). Of particular interest is the discovery of eye-spots on the Upper Carboniferous insect *Protodiamphipnoa* (Carpenter, 1971, p. 1250, Figs 13–15). This is probably a warning response, since the insect has raptorial front legs and "when one is poisonous, noxious tasting or otherwise disagreeable or dangerous, it pays to advertise . . . It seems clear that [such] aposematic coloration evolved in response to vertebrate predators" (Matthews and Matthews, 1978, pp. 322–323). Insectivorous reptiles, as well as arachnids (see the section on the Rhynie Chert), probably played a large part in the coevolution of insects (Olson, 1976, pp. 6–9).

2. Arachnids

The arachnids show a similar explosion in the Carboniferous. Most of the existing arachnid orders are known from this period, and their fundamental evolution was completed by the Carboniferous – there has been little since (Petrunkevitch, 1953, p. 115). Of 16 orders known from the Carboniferous, 11 survive to the present day (Petrunkevitch, 1955; Savory, 1977a, p. 97).

"Arachnids today comprise the dominant invertebrate carnivores on land" (Manton, 1977, p. 15), and the same probably applied in the Carboniferous as well. The most familiar group of arachnids is the spiders, which are preyed upon today not only by each other, but also by "... toads and frogs, birds, shrews, wasps and centipedes. Invertebrate enemies are very much more numerous, and probably destroy larger numbers of spiders than do vertebrates" (Cloudsley-Thompson, 1968, p. 210). Carboniferous spiders are broadly divisible into the older groups of liphistiomorph and mygalomorph spiders, and the more modern arachnomorphs. The habits of the first group have already been referred to in the discussion of the Rhynie Chert spider. It is probable that the Coal Measure forms wove simple silk traps between aerial fronds of plants, as well as trapping early wingless insects and nymphs on the ground. Details of the spinnerets of the Carboniferous arachnomorph spiders are unknown (Petrunkevitch, 1953, p. 100), and spiders with the ability to spin the more efficient orb-webs for trapping insects are not known until the Tertiary. This may partially account for the instant success of the insects in their new milieu – they attain their greatest diversity in the Permian (Carpenter, 1977), and aerial vertebrate predators do not appear before the Mesozoic, as gliding lizards in the Triassic.

Turning to other groups of arachnids, the modern vinegaroon, practically unchanged from the Carboniferous *Prothelyphonus*, eats insects, centipedes, woodlice, worms and slugs (Cloudsley-Thompson, 1968, p. 146). Living solpugid arachnids kill and eat insects, spiders, scorpions and lizards using "the most formidable jaws in the animal world" (Savory, 1977a, p. 236). In turn they are preyed upon by lizards and other reptiles, as well as by birds and mammals (Cloudsley-Thompson, 1968, p. 116). They are the most active of living arachnids, and the most independent of atmospheric moisture (Lawrence, 1953, p. 139). One fossil solpugid is known, *Protosolpuga* from the Westphalian of Mazon Creek.

Almost nothing is known of the enemies of "living fossil" arachnids such as the Ricinulei. Doubtless they are preyed upon today by frogs, toads, lizards, small mammals, spiders, centipedes and various insects, as are most living arachnids, including the Opiliones (Cloudsley-Thompson, 1968, p. 176) which also occur in the Carboniferous.

Amphibians and reptiles figure prominently among arachnid predators, as they probably did in Carboniferous times. Such prey is often inconspicuous and retiring, and hence may not have occupied more than a small proportion of tetrapods' diet.

Plump ricinuleid arachnids live today much as their Carboniferous ancestors must have done, skulking motionless under rotten logs, behind leaf fronds on wet mud and in wet leaf-axils (laminate leaf-bases are known since the Devonian *Archaeopteris*), emerging only to catch and eat living spider and insect prey (Cooke 1967; Pollock, 1966, 1967). They illustrate very well how such terrestrial arthropods take advantage of plants to control their environment for them. By remaining in the shade of large leaves – the laminate leaf evolved in the Carboniferous – such arachnids manage to keep their temperature down in tropical climates. Plants act as wind-breaks, with locally humid niches, so arachnids are able to conserve their moisture which otherwise would be lost by transpiration across their cuticles – they lack the wax layer typical of many terrestrial arthropods.

3. Mazon Creek

Much work remains to be done on the palaeoecology of the arthropod fauna at this Westphalian D locality. Of 170 invertebrate genera listed from Mazon Creek by Horowitz and Richardson's 1978 computer print-out, no less than 137 are terrestrial arthropods (21 of them non-scorpionid arachnids, 99 insects). At species level, too, 85% of the Braidwood Mazon Creek fauna has been found to be terrestrial (Johnson and Richardson, 1966; Richardson and Johnson, 1971). All four arachnid orders found at the contemporaneous locality of Nýřany, Czechoslovakia, also occur at Mazon Creek, along with seven other orders. Two of the eight Nýřany genera also occur at Mazon Creek. This is only weak evidence supporting the theory of commonality of terrestrial faunas of this time, derived from study of the tetrapods (Milner and Panchen, 1973; Panchen, 1977). The absence of *Arthropleura* from Nýřany, although it is widespread elsewhere in the Euramerian flora belt, suggests that Nýřany is insufficiently rich in terrestrial arthropods to permit adequate comparison of Europe with North America. A comparison of Mazon Creek with the more arachnid-rich British Midlands localities might be more profitable. At lower taxonomic levels, Petrunkevitch (1913, pp. 23–

26), believed the North American fauna was distinct from that of Europe, and developed on different lines.

A variety of myriapod taxa is present at Mazon Creek.

4. Joggins

This celebrated Westphalian B locality is important in preserving a terrestrial fauna *in situ* within *Calamites* and *Sigillaria* stumps. It has been authoritatively reviewed by Carroll *et al.* (1972, pp. 64–80) and the arthropods have been discussed by Briggs *et al.* (1979, pp. 286–287). The hollow stumps were buried, creating pit-falls which are thought to have sampled the truly terrestrial fauna. "With the exception of one or two coelacanth scales (which may have been someone's dinner) no fish are known from the stumps, nor are there any strictly aquatic amphibians" (Carroll *et al.*, 1972, p. 67). Besides the land snail *Dendropupa* (see Section 5 "Snails"), the remaining invertebrates are all arthropods. Eight millipede species have been recorded as abundant in the trunks, and millipedes may well have sought out tree stumps both as a damp refuge, and as food substrate. In turn, they were preyed upon to some extent (their abundance suggests not greatly – R. L. Carroll, personal communication) by the tetrapods, since millipede somites (and other cuticles) are recorded from their coprolites (Dawson, in Scudder, 1895). Two arachnids are known, the anthracomartid *Coryphomartus triangularis* (Petrunkevitch) and the whip-spider *Graeophonus carbonarius* Scudder (Petrunkevitch, 1913, pp. 69, 101; 1953, pp. 60, 99). These doubtless came from tree-trunks (R. L. Carroll, personal communication). Many fragments of eurypterid cuticle occur, and these can be compared with *Dunsopterus, Hibbertopterus* and *Vernonopterus* (C. D. Waterston, personal communication). Dr J. Dalingwater has confirmed the eurypterid nature of the cuticle by SEM, and compares the cuticular microstructure with that of *Mycterops*? (Dalingwater, 1975). How did these eurypterid fragments enter the hollow stumps? Were they carried there by predatory tetrapods, like the coelacanth scales? Or did eurypterids fall into, or seek out, the pits whilst crawling about the new land surface? All the eurypterids mentioned are morphologically unusual forms which characteristically occur incomplete elsewhere, and are practically confined to non-marine localities. Or, finally, were they washed in by flooding of this "bayou" country, giving mixed aquatic and terrestrial biotae, in death if not in life, as K. E. Caster suggests (personal communication)? In favour of this explanation is Woodward's (1918, p. 464) point, that the stumps are occasionally found with *Spirorbis* ("more characteristic of swamp facies"—Calver, 1968, p. 163) attached both to outside and inside of the "bark". Without the most detailed palaeoecological study of a stump, it is impossible to decide this point, but the overwhelmingly terrestrial ecology of the rest of the fauna, stressed by Carroll *et al.* (1972), leads one to favour the second explanation here.

5. *Snails*

The earliest land snails are known from the Carboniferous, although this had been denied (Knight *et al.*, 1960). They have been reviewed by Solem and Yochelson (1979): *Dendropupa* from Joggins and Poland, *Anthracopupa* from the Little Captina Limestone of the Dunkard Basin (Yochelson, 1975) and the British Stephanian (Calver, 1968, p. 155), and *Dawsonella* from Illinois. Other genera are known (e.g. *Maturipupa),* but there is less agreement about their pulmonate nature and this also applies to earlier records. Solem and Yochelson suggest that *Dendropupa* lived in the tree-trunk base debris of the Joggins forest. In modern forests similar snails inhabit the buttress and tangled root litter accumulations of large trees. The elongate form of the shells is consistent with their being climbers on such vertical surfaces, rather than creepers along horizontal substrates (Cain, 1977). Such snails would doubtless have formed part of the diet of the contemporary arachnids and other carnivorous arthropods.

6. *Worms*

No truly terrestrial worms are known from the Palaeozoic (Clark, 1969). Aquatic oligochaetes probably preceded terrestrial ones (S. Conway Morris, personal communication) and the earliest record is the Carboniferous *Pronaidites*. A possible terrestrial oligochaete from the Carboniferous has been illustrated by Zangerl and Richardson (1963, p. 128, Pl. 21). Fossil nematodes infest the body of the Lower Carboniferous *Gigantoscorpio* (Størmer, 1963) but these are probably aquatic.

WATER BALANCE AND SIZE

"The arthropod level of organisation is highly successful for land life, and a large part of this success may be put down to the arthropods' effective water balance mechanisms which permit rigorous conservation while allowing for rapid elimination of excess water when this is necessary" (Edney, 1977, p. 244). There is no simple correlation between arthropod group and cuticular permeability to water: the correlation is rather with ecology – desert scorpions show the lowest permeability of any arthropod.

Size is another important factor, and animals such as desert scorpions that combine low permeability with large size are well adapted to withstand desiccating conditions for long periods (Edney, 1977, pp. 57–60). It is tempting to suggest that the several parallel and iterative trends toward gigantism in fossil arthropods – in scorpions, eurypterids, millipedes, arthropleurids and insects – are manifestations of this relationship. Some of them may reflect successive attempts at terrestrialization by various arthropod stocks. Some of these arthropods may have had thicker cuticles than the fossil record suggests and this may have been one way of reducing cuticular permeability. Thus, a Westphalian arthropod cuticle found in a borehole core by Dr M. Calver superficially resembles bone, is 1·3 mm thick, but locally reaches as much as *c*. 8 mm thick.

A similar relationship between size and cuticular permeability has been noted by Carroll (1977, p. 419) in Palaeozoic amphibians, which are large and covered with overlapping scales. Modern amphibians which depend on cutaneous gas exchange are of small size, and restricted to damp environments. The latter also applies to many terrestrial arthropods. As Hinton (1977, p. 73) has pointed out, since the oxygen molecule is larger than the water molecule, any membrane for oxygen transfer will leak water. Respiratory membranes are wet because water leaks through them, not because a water film assists respiration. On the contrary, such a film greatly inhibits the rate of oxygen diffusion. Insects and some arachnids are the only arthropods to deal successfully with the incompatible demands of dry environments. They do this by oxygen- and water-proofing their cuticles with a continuous fatty-acid layer, and at the same time invaginate an enormous surface area as tracheae for respiration.

EURYPTERIDS

Recent work on the respiratory structures of eurypterids by Størmer (1976), Waterston (1975) and Wills (1965) has culminated in the view that the "eurypterid gill tract might have acted both as gill and pseudotrachea" (Størmer, 1976, p. 143). This discovery was made on one of the most aquatically adapted eurypterids, *Baltoeurypterus*, although similar structures have also been found in the larger more amphibious-looking *Tarsopterella*. It implies that "several eurypterids at least might have been able to leave the water and stay on land for shorter or longer periods. They may have been able to remain above the supra-littoral zone, and thus be more terrestrial in their habits than the Recent *Limulus*" (Størmer, 1976, p. 143). Such amphibious eurypterids probably lived on sandy and muddy beaches, and in the early plant thickets on the land beyond. Like modern *Limulus* they probably burrowed, only shallowly, for worms and molluscan prey (Manton, 1964, p. 35) and, like Recent land hermit crabs, fed on seaweed and other plant debris, such carrion as was available and sick members of their own species (Kaestner, 1970, pp. 329–330).

1. Gill-tracts

Størmer (1976, p. 142) noted the similarity of the microstructures of the *Baltoeurypterus* gill-tract to similar structures in some insects and land isopods. He suggested that they might serve to trap air or prevent clogging of the fine respiratory passages in insects, isopods and thus eurypterids. The remarkable similarity of these structures to those of the plastron – a gaseous gill of insects and mites, formed by the trapping of a thin layer of air over areas of the cuticle (Edney, 1977, p. 215) – has been pointed out elsewhere (Rolfe, 1980). Such a plastron could conceivably have formed the normal eurypterid respiratory structure, although in modern aquatic arthropods it is only used as such by small forms. It is commoner in living terrestrial arthropods, to overcome the risk of drowning: "To be submerged in water for several hours or even days is not a rare or isolated event, but a normal hazard of the environment" (Hinton, 1971a, p. 1185; 1977, p. 72). This would imply that the eurypterids were habitual air breathers. In other forms, however, the plastron is hydrophilic to combat desiccation, and this may have been its purpose in eurypterids.

2. *Mesosomal Enlargement*

In the light of this new, amphibious view of eurypterids, the enlargement of the first two mesosomites in *Woodwardopterus*, and of at least the first (genital) mesosomite in *Mycterops*, may be significant. It may represent an adaptation to accommodate larger book-lungs or pseudotracheae, to enable better respiration on land (Rolfe, 1980).

3. *Limbs and Size*

There seems to have been a trend in the Old Red Sandstone for various eurypterids to increase their size (e.g. *Tarsopterella*, *Ctenopterus*, *Pterygotus anglicus*). Such a size increase would serve to reduce desiccation loss, by increasing the surface to volume ratio (see the section "Water Balance and Size"), and would have been selectively advantageous to those eurypterids that were to survive the periodic droughts of that environment. As with crossopterygians (Romer, 1966, p. 86; Størmer, 1976, p. 143), individuals that survived may have evolved adaptations such as the stylonuroid walking limbs in order to migrate to the nearest available waterhole, rather than for terrestrial ambitions. With the advent of the less rigorous conditions of the Carboniferous, the need for such an adaptation would have waned, but large stocks may then have proved pre-adapted to move onto land. Such forms are only known from fragments at present – as their palaeoecology would demand – and include e.g. *Vernonopterus*, *Dunsopterus*, *Hibbertopterus* and *Woodwardopterus*, and probably *Mycterops* (see Waterston, 1957, 1968). These forms all have relatively thick cuticles (e.g. Dalingwater, 1975), an adaptation which would confer great strength to the limbs.

Several of the presumed amphibious eurypterids had long, stylonuroid-like legs. Størmer (1974, p. 399) has suggested that the longer posterior legs of *Moselopterus* may indicate that the opisthosoma was held up at quite a high angle, as in scorpions, so that the telson could be used as a weapon to kill prey held by the anterior prosomal appendages. Such long legs would also enable their owner to "stilt", like some modern scorpions. In this behaviour pattern the legs are straightened to lift the body clear of the ground. Increased circulation of air around the stilting animal enables it to maintain a constant body temperature, whilst the environmental temperature rises rapidly (Alexander and Ewer, 1958). Above a certain temperature range, stilting no longer protects the animal, and the animal would need to seek cover: in the case of Joggins, perhaps a hollow tree stump refuge? It may be objected that, as in the giant Japanese crab *Macrocheira*, such legs would be insufficient to support the body weight in air. This is doubtless true of an extreme form such as *Ctenopterus*, although functional morphological investigation is required. *Hibbertopterus,* however, has short, stubby limb segments and is noticeably hexapodous: it would seem well adapted for movement on land despite its bulky opisthosoma. Its anterior limbs are modified as sensory appendages (Waterston, 1957). The limb form of most other Carboniferous genera is unknown in detail. *Woodwardopterus* has remarkable *Limulus*-like pushing spines (Manton, 1964; 1977, p. 48) on its limbs, useful for crawling across, or burrowing shallowly in, a moist substrate. Such locomotion as there was might well have involved an ungainly, crawling gait. Certainly, however, the apparent massiveness of *Pterygotus anglicus* might have prevented its locomotion on land (Waterston and Størmer, in Størmer, 1976, p. 143). This bulk may be deceptive, however, as it is in the burrowing-adapted *Limulus* (Manton, 1964), and it is doubtful if the whole body cavity was packed with weighty muscle. With the (co-evolutionary?) rise of predatory tetrapods, such feeble attempts to terrestrialize would be doomed to failure, and might explain the demise of these large eurypterids with the Carboniferous, as of the large pterygotids earlier in the Devonian (Waterston, 1967).

4. Claspers

The presence of a particular kind of clasper in some eurypterids may be additional evidence of their at least partially terrestrial habit. Many groups of terrestrial arthropods develop clasping structures on the limbs to hold the female during courtship and especially to manoeuvre her over a previously deposited spermatophore (Lawrence, 1953, pp. 223–239; Kaestner, 1968). Subaquatic arthropods also develop claspers, however, for copulation or spermatophore transfer (Kaestner, 1970, p. 37 *et seq.*) and it is not therefore a simple matter of equating claspers with terrestriality. But there may be claspers and claspers. Størmer and Kjellesvig-Waering (1969, p. 209, Fig. 2a) have found it difficult to understand from the morphology of the scimitar lobes of *Eurypterus* (Wills, 1965, Ppl. 1, Figs 2–6) how these could have served as body claspers, and have suggested they were used for scooping an egg-laying hollow. It seems more likely that those structures are an immobilizing type of clasper, used to immobilize fangs (as in living spiders – Lawrence, 1953, Fig. 92A) during courtship and to prevent biting by the hostile mate, or to contain the stinging response (as in the scorpion – Lawrence, 1953, Fig. 104; Kaestner, 1968, pp. 105–106). Cloudsley-Thompson (1968, p. 219) has suggested that the function of such "courtship is to provide releaser stimuli for the mating instinct which at the same time block hunger drives". That such cannibalism existed in the early Palaeozoic eurypterids is suggested by the long coprolite of *Megalograptus*, containing undigested *Megalograptus* cuticle fragments (Caster and Kjellesvig-Waering, 1964, p. 337, Pl. 51, Fig. 4). This type of clasper (present also in *Mixopterus*?), presumably implying a terrestrially deposited spermatophore, might then prove to be an index of terrestrial forms.

SCORPIONS

The early Siluro-Devonian scorpions are now regarded as aquatic, and the earliest unequivocal terrestrial scorpion, with stigmata, is *Palaeopisthacanthus*, from the Westphalian D of Mazon Creek. Aquatic scorpions continued throughout the Carboniferous, however, and possibly even into the Triassic. Conversely, some of the earliest Palaeozoic forms may have been amphibious.

Confirmation of the long-suspected aquatic nature of at least some of the early forms comes from the Emsian *Waeringoscorpio* from Alken, which has filamentous gills (Størmer, 1970; 1976, p. 151). Other criteria used to assess the degree of terrestriality of scorpions are: the walking legs – whether digitigrade or plantigrade; the nature of sensory setae; the degree of resemblance to eurypterids;

and the nature of the faunal assemblage (Størmer, 1970, pp. 350–351). Several of these lines of evidence can prove equivocal, however.

Three large fossil scorpions are known, between half a metre and one metre long. Two of them are Gedinnian – *Brontoscorpio* (Kjellesvig-Waering, 1972) and *Praearcturus* (Rolfe, 1969, p. R622). The latter is only here recognized as a scorpion, independently by L. Størmer (personal communication, *c.* 1974) and by E. N. Kjellesvig-Waering (personal communication, 1978) from published figures (Rolfe, 1969, Fig. 395 – cephalothorax inverted!). The narrowness of the sternite and forward swing of the coxae in *Praearcturus* suggests a preoral chamber was formed, to permit the external digestion of prey. In Recent scorpions, food is chewed in this chamber by the pedipalpal coxae, while digestive juice from the mouth is poured over the food and alternately sucked back. Chewing may be executed for as much as three hours (Manton, 1964, p. 29), and such external digestion is the rule in arachnids (Petrunkevitch, 1955, p. P53). It "is not a suitable mode of feeding in the water where dissipation of digestive juices must take place" (Manton, 1977, p. 265). Størmer has pointed out the analogy with the evolution of a digestive oral cavity in terrestrial vertebrates. It is not impossible, however, that early scorpions developed the habit of eating their prey whilst holding it above water, as do some Recent insect larvae (R. A. Crowson, personal communication). This would then be another pre-adaptation for life on land. Respiratory structures are unknown from these Gedinnian genera, but, if the trace fossil evidence discussed above is acceptable, it suggests such scorpions may have been at least partly terrestrial. *Gigantoscorpio*, from the Tournaisian of Scotland, is another large scorpion, *c.* 500 mm long (Størmer, 1976, p. 153). It has been suggested to be amphibious from its plantigrade feet, the presence of setae rather than trichobothria and the nature of the ventral plates concealing gills (Størmer, 1963).

All these scorpions might provide suitable prey for early tetrapods, and vice versa. Recent scorpion enemies include centipedes, arachnids, lizards, snakes and birds; baboons tear off the tails of scorpions before devouring the rest of the body (Cloudsley-Thompson, 1968, p. 93). Romer (1958, p. 367) has objected that scorpions "do not appear to be too nourishing a base on which to found a flourishing terrestrial vertebrate fauna", and it is likely that they only formed a supplementary item of diet.

The characteristic pectines of Recent scorpions are probably a sense organ for testing the dryness or humidity of the air, for determining a suitable substrate for spermatophore deposition, and for detecting vibrations of the ground, giving warning of the approach of enemies or prey (Kaestner, 1968, p. 103; Savory, 1977, p. 119). There would appear to be little use for such organs in aquatic scorpions, and it may be significant that none have yet been confirmed in Silurian scorpions (Størmer, 1963, p. 94). Størmer has suggested the function of the organs might have changed from their aquatic ancestors. Huge pectine-like structures are found on the enigmatic chelicerate *Cyrtoctenus*, which has been interpreted as an aquatic animal simply from its large size (Størmer and Waterston, 1968). It serves to emphasize that there are many Palaeozoic arthropods, some of them amphibious or terrestrial, yet to be recognized.

The scorpions are unusual among arachnids in having experienced much extinction: on one classification, of nine Palaeozoic superfamilies only one survives today (Savory, 1977, p. 97). Characters used in classification reflect the nature of the respiratory organs and the preoral chamber, i.e. the degree of terrestrialization. Such extinctions may therefore indicate a sequence of unsuccessful attempts to terrestrialize.

MYRIAPODS, INCLUDING ARTHROPLEURIDS

1. *Habits*

The variety of morphology shown by living millipedes is largely governed by their habits of burrowing or splitting their way through the decomposing plant substrates off which they feed (Manton, 1977, pp. 352–368). Carboniferous millipedes show a similar variety of morphology, implying similar habits (Burke, 1973, 1979; Kraus, 1974; Rolfe, 1980). Other forms, such as *Euphoberia*, *Myriacantherpestes ferox* and *Acantherpestes major* differ from most living millipedes in having an armament of long spines. Such spines were clearly defensive, since they are occasionally crushed or broken off, presumably the result of exchanges with would-be predators (Rolfe, 1980). This spinosity may be combined with large size, as in *A. major* which reached up to *c.* 300 mm long (Scudder, 1882, p. 151), and with the presence of large compound eyes. As Kraus (1974) has pointed out, these forms must have been surface dwellers, which climbed over plants as shown in Scudder's (1882, Pl. 10) evocative reconstruction. They are no longer thought to be amphibious, however, as that reconstruction suggests (Burke, 1973, pp. 20–22; Kraus, 1974, pp. 21, 22).

2. Predators

Repugnatorial gland openings have been reported on some Carboniferous forms, although Hoffman (1969, p. R582) has been unable to confirm this. The caustic repugnatorial fluids of recent millipedes will deter, or even blind, predators. Lizards will turn away an inch or so from *Tachypodoiulus*, and one which persisted eventually rejected the prey and rubbed its mouth in sand (Cloudsley-Thompson, 1949, p. 139). Amphibians and birds are their most effective predators today. They are a constant article of diet of the American toad – 77 having been found in one's stomach, and 10% by bulk of the food of this species is composed of millipedes. Nevertheless, enemies do not play a large part in the ecology of myriapods: their numbers are chiefly governed by the physical conditions of the environment (Cloudsley-Thompson, 1949).

3. Arthropleurids

Whether or not the giant 1·8 m long Carboniferous *Arthropleura* is a myriapod (Rolfe, 1969) or represents a separate group within the Uniramia (Manton, 1977, p. 27) is uncertain, but its habits must have resembled those of living polydesmoid millipedes. Such "flat-backed" forms split their way through litter which, in the case of *Arthropleura* was probably of branches and logs, rather than leaves, which would be sparse on such a tropical forest floor. The gut contents of tracheids (Rolfe and Ingham, 1967) suggest *Arthropleura* fed off the woody central stele of lycopods (A. C. Scott, personal communication), and such large animals may have found food, shelter and a natural moulting chamber within the rotted pith cavity of *Lepidodendron* trunks. That *Arthropleura* could leave these humid confines is shown by trace fossils from the Namurian of Scotland and the Westphalian of Joggins (Briggs *et al.*,

1979). These trackways suggest that *Arthropleura* traversed sand-plugged distributory channels on delta surfaces.

In such an open environment, *Arthropleura* would doubtless be more vulnerable to a wider range of predators than in the forest. The leathery cuticle and sheer size of a fully grown *Arthropleura* would protect it from much predation, although younger instars would doubtless succumb, along with other myriapods. Some Coal Measure tetrapods probably behaved like the modern coati, which "spends the greater part of its effort searching among ground litter and debris for invertebrates, as well as digging them out of the soil, using its sense of smell to locate them. Coatis also tear apart rotting logs to get at sheltering invertebrates . . . they will systematically dissect such logs for frogs, snails, centipedes and the like (mostly sheltering nocturnal species)" (Elton, 1973, p. 94).

4. Decomposition

Terrestrial ecosystems contrast with marine ecosystems in that the greatest proportion of primary production passes through the decomposer chain. Myriapods and *Arthropleura* played a vital role as decomposers in Coal Measure and earlier forests, a topic which is enlarged upon elsewhere (Rolfe, 1980). Mite droppings are known from tunnelled plant fragments in coal-balls, and mites comminuted further the plant debris left by millipedes, to contribute to soil formation. This is further evidence of the close interdependence that had developed between plants and arthropods by Carboniferous times.

5. Centipedes

Despite the doubt cast upon the existence of Palaeozoic centipedes and onychophorans by Hoffman (1969, p. R598), recent work has shown that both a scutigeromorph and a cryptopid are present in the Westphalian D of Mazon Creek, and suggests that a Palaeozoic geophilomorph is known from Canada (Mundel, 1979). This restores an important group of invertebrate carnivores to the Palaeozoic record. Knowledge of this publication, as of that recording onychophorans from the same deposit (Jones and Thompson, 1980), was received too late to permit fuller discussion here.

CONCLUSIONS

The early terrestrial invertebrate fossil record is made up of two sparse faunas, with only one arachnid order in common, supplemented by many records of myriapods. Trace fossil assemblages hint at the presence of a diversity of terrestrial arthropods during the Devonian, but neoichnology is required before progress will be possible in evaluating them. Later Carboniferous faunas could yield much palaeoecological and palaeobiogeographical information, but these are all after the event of tetrapod origin. They represent well-established communities, of which the arachnids survive today with little change. Devonian and particularly Carboniferous communities were doubtless governed, like their living counterparts, by water requirements, and are only known from humid regions of the palaeotropics. Other groups such as eurypterids and scorpions were amphibious during the Devonian and were available to tetrapod predators. Decomposer organisms played a significant part in building up early humus levels and soil structures, and fossils of these animals or their products have recently been recognized. The present record is clearly grossly incomplete and yields only a distorted picture of the situation that must have existed.

ACKNOWLEDGEMENTS

I am indebted to the following for much help and guidance during the preparation of this paper: P. Bonamo, P. Brand, D. E. G. Briggs, R. L. Carroll, W. G. Chaloner, S. Conway Morris, R. Crowson, J. Dalingwater, D. S. Edwards, Dianne Edwards, D. Grierson, J. B. Kethley, E. N. Kjellesvig-Waering, D. Macfarlane, D. Maclean, S. Morris, J. Pollard, Julia Rolfe, A. Scott, L. Størmer, E. Walker, F. Wanless, C. D. Waterston and R. B. Wilson.

REFERENCES

Alexander, A. J. and Ewer, D. W. (1958). Temperature adaptive behaviour in the scorpion *Opisthophthalmus latimanus* Koch. *J. exp. Biol.* 35, 349–359.

Allen, J. R. L. (1977). Wales and the Welsh Borders. *In* "A Correlation of the Devonian Rocks of the British Isles" (M. R. House *et al.*, eds), *Geol. Soc. Lond. Spec. Rep.* 8, 40–54.

Bergström, J. (1978). Morphology of fossil arthropods as a guide to phylogenetic relationships. *In* "Arthropod Phylogeny" (A. P. Gupta, ed.), pp. 3–56. Van Nostrand Reinhold, New York.

Bliss, D. E. and Mantel, L. H. (1968). Adaptions of crustaceans to land: a summary and analysis of new findings. *Am. Zool.* **8**, 673–685.
Bluck, B. J. (1978). Sedimentation in a late orogenic basin: the Old Red Sandstone of the Midland Valley of Scotland. *In* "Crustal Evolution in Northwestern Britain and Adjacent Regions" (D. R. Bowes and B. E. Leake, eds), *Geol. J. Spec. Issue* **10**, 249–278.
Boy, J. A. (1976). Überblick über die Fauna des saarpfälzischen Rotliegenden (Unter-Perm). *Mainzer geowiss. Mitt.* **5**, 13–85.
Brade-Birks, S. G. (1923). Notes on Myriapoda, 28. *Kampecaris tuberculata*, n.sp., from the Old Red Sandstone of Ayrshire. *Proc. R. phys. Soc. Edinb.* **20**, 277–280.
Briggs, D. E. G., Rolfe, W. D. I. and Brannan, J. (1979). A giant myriapod trail from the Namurian of Arran, Scotland. *Palaeontology* **22**, 273–291.
Bristowe, W. S. (1975). A family of living fossil spiders. *Endeavour* **34**, 115–117.
Burke, J. J. (1973). Notes on the morphology of *Acantherpestes* (Myriapoda, Archipolypoda). *Kirtlandia* **17**, 1–24.
Burke, J. J. (1979). A new millipede genus *Myriacantherpestes* (Diplopoda, Archipolypoda). *Kirtlandia* **30**, 1–24.
Cain, A. J. (1977). Variation in the spire index of some coiled gastropod shells, and its evolutionary significance. *Phil. Trans. R. Soc.* **B277**, 377–428.
Calver, M. (1968). Coal Measures invertebrate faunas. *In* "Coal and Coal-bearing strata" (D. Murchison and T. S. Westoll, eds), pp. 147–177. Oliver and Boyd, Edinburgh.
Carpenter, F. M. (1971). Adaptions among Paleozoic insects. *In* "Proceedings of the North American Paleontological Convention" (E. L. Yochelson, ed.), Vol. 2, pp. 1236–1251. Allen Press, Lawrence, Kansas.
Carpenter, F. M. (1977). Geological history and evolution of the insects. *Proc. 15th int. Congr. Entom., Washington, 1976* 63–70.
Carroll, R. L., Belt, E. S., Dineley, D. L., Baird, D. and McGregor, D. C. (1972). Vertebrate paleontology of Eastern Canada. *Guidebook, Excursion A59, 24th Int. Geol. Congr., Montreal* 1–113.
Caster, K. E. and Kjellesvig-Waering, E. N. (1964). Upper Devonian eurypterids of Ohio. *Palaeontogr. am.* **4**, 297–358.
Chaloner, W. G. (1976). The evolution of adaptive features in fossil exines. *In* "The Evolutionary Significance of the Exine" (I. K. Ferguson and J. Muller, eds), *Linn. Soc. Symp. Ser.* **1**, 1–14.
Chamberlain, C. K. (1975). Recent Lebensspuren in nonmarine aquatic environments. *In* "The Study of Trace Fossils" (R. W. Frey, ed.), pp. 431–458. Springer, Berlin.
Clark, R. B. (1969). Systematics and phylogeny: Annelida, Echiura, Sipuncula. *In* "Chemical Zoology" (M. Florkin and B. T. Scheer, eds), Vol. 4, pp. 1–68. Academic Press, New York and London.
Clarke, B. B. (1951). The geology of Dinmore Hill, Herefordshire, with a description of a new myriapod from the Dittonian rocks there. *Trans. Woolhope Nat. Fld Club* **33**, 222–236.
Cloudsley-Thompson, J. L. (1949). The enemies of myriapods. *Naturalist* **831**, 137–141.

Cloudsley-Thompson, J. L. (1968). "Spiders, Scorpions, Centipedes and Mites." Pergamon, Oxford.
Cloudsley-Thompson, J. L. (1975a). "Terrestrial Environments." Croom Helm, London.
Cloudsley-Thompson, J. L. (1975b). Adaptations of Arthropoda to arid environments. *A. Rev. Ent.* **20**, 261–283.
Crowe, J. H. and Cooper, A. F. (1971). Cryptobiosis. *Scient. Am.* 225(6), 30–36.
Crowson, R. A. (1970). "Classification and Biology." Heinemann Educ., London.
Dalingwater, J. E. (1975). Further observations on eurypterid cuticles. *Fossils Strata* **4**, 271–279.
Dawson, J. W. (1891). "Acadian Geology". 4th edn. Macmillan, London.
Demoor, J. (1890). Recherche sur la marche des insectes et des arachnides. *Archs Biol.* **10**, 567–608.
Demoor, J. (1891). Recherches sur la marche des crustacés. *Archs Zool. exp.* **9**, 477–502.
Dubinin, V. B. (1962) Klass Acaromorpha. *In* "Chlenistonogie, Trakheinye i Khelitserovye" (B. B. Rodendorf, ed.), pp. 447–473. *In* "Osnovy Paleontologii" (Yu. A. Orlov, ed.), Akad. nauk S.S.R., Moscow.
Edney, E. B. (1968). Transition from water to land in isopod crustaceans. *Am. Zool.* **8**, 309–326.
Edney, E. B. (1977). "Water Balance in Land Arthropods." Springer, Berlin.
Edwards, D. S. (1973). "Studies on the Flora of the Rhynie Chert." Ph. D. Thesis, University of Wales.
Eldredge, N. S. (1974). Revision of the Suborder Synziphosurina (Chelicerata, Merostomata), with remarks on merostome phylogeny. *Am. Mus. Novit.* **2543**, 1–41.
Elton, C. S. (1973). The structure of invertebrate populations inside neotropical rain forest. *J. Anim. Ecol.* **42**, 55–104.
Evans, T. J. (1910). Bionomical observations on some British millipedes. *Ann. Mag. nat. Hist.* (8)**6**, 284–291.
Fiori, G., Mellini, E. and Crovetti, A. (1966). Brevi considerazione sulle orme lasciate sulla sabbia da alcuni insetti subdeserticoli e deserticoli. *Studi Sassaresi* **3**, 170–190.
Frey, R. W., ed. (1975). "The Study of Trace Fossils." Springer, Berlin.
Hanken, N-M. and Størmer, L. (1975). The trail of a large Silurian eurypterid. *Fossils Strata* **4**, 255–270.
Heath, J., Bocock, K. L. and Mountford, M. D. (1974). The life history of the millipede *Glomeris marginata* (Villers) in north-west England. *In* "Myriapoda" (J. G. Blower, ed.), *Symp. zool. Soc. Lond.* **32**, 433–462.
Hinton, H. E. (1971a). Plastron respiration in the mite *Platyseius italicus*. *J. Insect Physiol.* **17**, 1185–1199.
Hinton, H. E. (1971b). Reversible suspension of metabolism. *Proc. 2nd int. Conf. theoret. Physics Biol. (Versailles)* 69–89.
Hinton, H. E. (1977). Enabling mechanisms. *Proc. 15th int. Congr. Entom., Washington, 1976* 71–83.
Hirst, S. (1922). On some arachnid remains from the Old Red Sandstone (Rhynie

Chert Bed, Aberdeenshire). *Ann. Mag. nat. Hist.* (9)**12**, 455–474.

Hirst, S. and Maulik, S. (1926). On some arthropod remains from the Rhynie Chert (Old Red Sandstone). *Geol. Mag.* **63**, 69–71.

Hoffman, R. L. (1969). Myriapoda, exclusive of Insecta. *In* "Treatise on Invertebrate Paleontology" (R. C. Moore, ed.), Part R, pp. R572–606. Geological Society of America.

Hutchinson, G. E. (1965). "The Ecological Theater and the Evolutionary Play." Yale University Press, New Haven.

Huxley, T. H. and Salter, J. W. (1859). On the anatomy and affinities of the genus *Pterygotus* (Huxley). *Mem. geol. Surv. U.K., Monogr.* **1**, 1–105.

Johnson, R. G. and Richardson, E. S. (1966). A remarkable Pennsylvanian fauna from the Mazon Creek area, Illinois. *J. Geol.* **75**, 626–631.

Jones, D. and Thompson, I. (1980). Onychophora of the Carboniferous of Mazon Creek, Illinois. *J. Paleont.* (in press).

Kaestner, A. (1968, 1970). "Invertebrate Zoology", Vols 2 and 3. Interscience, New York.

Kevan, P. G., Chaloner, W. G. and Saville, D. B. O. (1975). Interrelationships of early terrestrial arthropods and plants. *Palaeontology* **18**, 391–417.

Kjellesvig-Waering, E. N. (1972). *Brontoscorpio anglicus*: a gigantic lower Paleozoic scorpion from central England. *J. Paleont.* **46**, 39–42.

Knight, J. B., Batten, R. L., Yochelson, E. L. and Cox, L. R. (1960). Supplement. *In* "Treatise on Invertebrate Paleontology" (R. C. Moore, ed.), Part I, pp. I.310–330. Geological Society of America.

Kraus, O. (1974). On the morphology of Palaeozoic diplopods. *In* "Myriapoda" (J. G. Blower, ed.), *Symp. zool. Soc. Lond.* **32**, 13–22.

Kukalova-Peck, J. (1980). Primitive Carboniferous insects and their convergence, parallelism and homology with Recent insects. *Proc. 9th int. Congr. Carb. Strat. Geol.* (in press).

Lawrence, R. F. (1953). "The Biology of the Cryptic Fauna of Forests." Balkema, Cape Town and Amsterdam.

Lewis, J. G. E. (1974). The ecology of centipedes and millipedes in northern Nigeria. *In* "Myriapoda" (J. G. Blower, ed.), *Symp. zool. Soc. Lond.* **32**, 423–431.

Loomis, H. F. and Hoffman, R. L. (1962). A remarkable new family of spined polydesmoid Diplopoda, including a species lacking gonopods in the male sex. *Proc. biol. Soc. Wash.* **75**, 145–158.

McKee, E. D. (1947). Experiments on the development of tracks in fine cross-bedded sand. *J. sedim. Petrol.* **17**, 23–28.

Manton, S. M. (1964). Mandibular mechanisms and the evolution of arthropods. *Phil. Trans. R. Soc. Lond.* **B247**, 1–183.

Manton, S. M. (1970). Arthropods: introduction. *In* "Chemical Zoology" (M. Florkin and B. J. Scheer, eds), Vol. 5, pp. 1–34. Academic Press, New York and London.

Manton, S. M. (1973). The evolution of arthropodan locomotory mechanisms, Part II. *J. Linn. Soc. (Zool.)* **53**, 257–375.

Manton, S. M. (1974). Segregation in Symphyla, Chilopoda and Pauropoda in relation to phylogeny. *In* "Myriapoda" (J. G. Blower, ed), *Symp. zool. Soc. Lond.* **32**, 163–190.

Manton, S. M. (1977). "The Arthropoda: Habits, Functional Morphology and Evolution." Clarendon Press, Oxford.

Matthews, R. W. and Matthews, J. R. (1978). "Insect Behaviour." John Wiley, New York.

Milner, A. R. and Panchen, A. L. (1973). Geographical variation in the tetrapod faunas of the Upper Carboniferous and Lower Permian. *In* "Implications of Continental Drift to the Earth Sciences" (D. H. Tarling and S. K. Runcorn, eds), Vol. 1, pp. 353–367. Academic Press, London and New York.

Moore, S. J. (1976). Some spider organs as seen by the S.E.M., with special reference to the book-lung. *Bull. Br. arachnol. Soc.* **3**, 177–187.

Müller, A. H. (1975). Zur Ichnologie limnisch-terrestrischer Sedimentationsräume mit Bemerkungen über Ichnia landbewohnender Arthropoden aus Gegenwart und geologischer Vergangenheit. *Freiberg. Forschungsh.* **C304**, 79–87.

Mundel, P. (1979). The centipedes (Chilopoda) of the Mazon Creek. *In* "Mazon Creek fossils" (M. H. Nitecki, ed.), pp. 361–378. Academic Press, New York and London.

Olson, E. C. (1976). The exploitation of land by early tetrapods. *In* "Morphology and Biology of Reptiles" (A.d'A. Bellairs and C. B. Cox, eds), *Linn. Soc. Symp. Ser.* **3**, 1–30.

O'Neill, R. V. (1969). Adaptive responses to desiccation in the millipede *Narceus americanus* (Beauvois). *Am. Midl. Nat.* **81**, 578–583.

Panchen, A. L. (1977). Geographical and ecological distribution of the earliest tetrapods. *In* "Major Patterns in Vertebrate Evolution" (M. K. Hecht, P. C. Goody and B. M. Hecht, eds), pp. 723–738. Plenum, New York.

Patterson, C. (1978). Arthropods and ancestors. *Antenna, Bull. R. Ent. Soc. Lond.* **2**, 99–103.

Peach, B. N. (1899). On some new myriapods from the Palaeozoic rocks of Scotland. *Proc. R. phys. Soc. Edinb.* **14**, 113–126.

Petrunkevitch, A. (1913). A monograph of the terrestrial Paleozoic Arachnida of North America. *Trans. Conn. Acad. Arts Sci.* **18**, 1–137.

Petrunkevitch, A. (1952). Macroevolution and the fossil record of Arachnida. *Am. Scient.* **40**, 99–122.

Petrunkevitch, A. (1953). Paleozoic and Mesozoic Arachnida of Europe. *Mem. geol. Soc. Amer.* **53**, 128 pp.

Petrunkevitch, A. (1955). Arachnida. *In* "Treatise on Invertebrate Paleontology", (R. C. Moore, ed.), Part P, pp. P.42–162. Geological Society of America.

Pollard, J. E. (1976). A problematic trace fossil from the Tor Bay Breccias of South Devon. *Proc. Geol. Ass.* **87**, 105–107.

Rapoport, E. H. and Tschapek, M. (1967). Soil water and soil fauna. *Rev. Ecol. Biol. Sol.* **4**, 1–58.

Richardson, E. S. and Johnson, R. G. (1971). The Mazon Creek faunas. *In* "Proceedings of the North American Paleontological Convention" (E. L. Yochelson, ed.), Vol. 2, pp. 1222–1235. Allen Press, Lawrence, Kansas.

Ritchie, A. (1963). "Palaeontological Studies on Scottish Silurian Fish Beds." Ph.D. Thesis, University of Edinburgh.

Rolfe, W. D. I. (1969). Arthropleurida and Arthropoda *incertae sedis. In* "Treatise on Invertebrate Paleontology", (R. C. Moore, ed.), Part R, pp. R.607-625. Geological Society of America.

Rolfe, W. D. I. (1980). Aspects of the Carboniferous terrestrial arthropod community. *Proc. 9th int. Congr. Carb. Strat. Geol., Urbana* (in press).

Rolfe, W. D. I. and Ingham, J. K. (1967). Limb structure, affinity and diet of the Carboniferous "centipede" *Arthropleura. Scott. J. Geol.* **3**, 118–124.

Romer, A. S. (1958). Tetrapod limbs and early tetrapod life. *Evolution* **12**, 365–369.

Romer, A. S. (1966). "Vertebrate Paleontology", 3rd ed. University Press, Chicago.

Savory, T. (1977a). "Arachnida", 2nd ed. Academic Press, London and New York.

Savory, T. (1977b). Cyphophthalmi: the case for promotion. *Biol. J. Linn. Soc.* **9**, 299–304.

Scudder, S. H. (1882). Archipolypoda, a subordinal type of spined myriapods from the Carboniferous Formation. *Mem. Boston Soc. nat. Hist.* **3**, 143–182.

Scudder, S. H. (1895). Notes upon myriapods and arachnids found in sigillarian stumps in the Nova Scotia coal field. *Geol. Surv. Canada, Contrib. Can. Palaeont.* **2**, 57–66.

Sharov, A. G. (1966). "Basic Arthropodan Stock." Pergamon, Oxford.

Sivhed, U. and Wallwork, A. (1978). An early Jurassic oribatid mite from southern Sweden. *Geol. For. Stockh. Förh.* **100**, 65–70.

Smart, J. (1967). Origin of insect wings. *In* "The Fossil Record" (W. B. Harland *et al.*, eds), *Geol. Soc. Lond.* p.116.

Smart, J. (1971). Palaeoecological factors affecting the origin of winged insects. *Proc. 13th int. Congr. Ent. Moscow* **1**, 304–306.

Smith, J. (1909). "Upland Fauna of the Old Red Sandstone Formation of Carrick, Ayrshire." Cross, Kilwinning.

Solem, A. and Yochelson, E. L. (1979). "North American Paleozoic Land Snails". *Prof. Pap. U.S. geol. Surv.* **1072**, 42pp.

Størmer, L. (1963). *Gigantoscorpio willsi*, a new scorpion from the Lower Carboniferous of Scotland. *Skr. Norske Vidensk. -Akad. Oslo* I. *Mat. Nat. Kl.* N. S. **8**, 1–171.

Størmer, L. (1970, 1972, 1973, 1974, 1976). Arthropods from the Lower Devonian (Lower Emsian) of Alken an der Mosel, Germany. Parts 1-5. *Senckenberg. leth.* **51**, 335–369; 53, 1–29; **54**, 119–205; 359–451; **57**, 87–183.

Størmer, L. and Kjellesvig-Waering, E. (1969). Sexual dimorphism in eurypterids. *In* "Sexual Dimorphism in fossil Metazoa" (G. E. G. Westermann, ed.), I.U.G.S. A1, 201–214. Schweitzerbart'sche, Stuttgart.

Størmer, L. and Waterston, C. D. (1968). *Cyrtoctenus* gen. nov., a large late Palaeozoic arthropod with pectinate appendages. *Trans. R. Soc. Edinb.* **68**, 63–104.

Stürmer, W. and Bergström, J. (1978). The arthropod *Cheloniellon* from the Devonian Hunsrück Shale. *Paläont. Z.* **52**, 57–81.

Walton, E. K. (1965). Lower Palaeozoic rocks – stratigraphy. *In* "The Geology of Scotland" (G. Y. Craig, ed.), pp. 161–200. Oliver and Boyd, Edinburgh.

Waterston, C. D. (1957). The Scottish Carboniferous Eurypterida. *Trans. R. Soc. Edinb.* **63**, 265–288.

Waterston, C. D. (1967). Eurypterida. *In* "The Fossil Record" (W. B. Harland *et al.*, eds), *Geol. Soc. Lond.* pp. 499–502.

Waterston, C. D. (1968). Further observations on Scottish Carboniferous eurypterids. *Trans. R. Soc. Edinb.* **68**, 1–20.

Waterston, C. D. (1975). Gill structures in the lower Devonian eurypterid *Tarsopterella scotica. Fossils Strata* **4**, 241–254.

Westoll, T. S. (1977). Northern Britain. *In* "A Correlation of the Devonian Rocks of the British Isles" (M. R. House *et al.*, eds), *Geol. Soc. Lond. Spec. Rep.* **8**, 66–93.

Whitaker, J. H. McD. (1962). The geology of the area around Leintwardine, Herefordshire. *Q. Jl geol. Soc. Lond.* **118**, 319–351.

Whittington, H. B. (1978). The lobopod animal *Aysheaia pedunculata* Walcott, Middle Cambrian, Burgess Shale, British Columbia. *Phil. Trans. R. Soc. Lond.* **B284**, 165–197.

Wills, L. J. (1965). A supplement to Gerhard Holm's Über die Organisation des *Eurypterus fischeri* Eichw. *Ark. Zool.* (2)**18**, 93–145.

Woodward, H. (1918). Fossil arthropods from the Carboniferous rocks of Cape Breton. *Geol. Mag.* **55**, 462–471.

Yochelson, E. L. (1975). Monongahela and Dunkard nonmarine gastropods. *In* "The Age of the Dunkard" (J. A. Barlow, ed.), *Proc. 1st. I. C. White Memorial Symp.* pp. 249–262.

Zakhvatkin, A. A. (1952). Razdeleniye kleshchei (Acarina) na otryady i ikh polozheniye v sistemye Chelicerata. *Parazit. sborn. Zool. Inst. Akad. Nauk SSSR* **14**, 5–46.

Zangerl, R. and Richardson, E. S. (1963). The paleoecological history of two Pennsylvanian black shales. *Fieldiana, Geol. Mem.* **4**, 352 pp.

7 | Origin of Tetrapods: Historical Introduction to the Problem

COLIN PATTERSON

British Museum (Natural History), Cromwell Road, London SW7, England

Abstract: The problem of tetrapod origins is treated here as one that arose before Darwin, when the discovery of living lungfishes showed that the distinction between fishes and tetrapods was blurred. Rather than concentrating on how that difficulty might be resolved, early evolutionists welcomed it, and projected the indistinct boundary back into the Palaeozoic. Current thought on the origin of tetrapods centres on the gap between osteolepiform rhipidistians and Palaeozoic amphibians. These seem both to be paraphyletic groups, and contain many poorly known or fragmentary fossils, so that their morphotypes may readily be conflated in the mind. A fresh approach to the problem might be to return to the pre-Darwinian question of how tetrapods are characterized. This could fruitfully be attacked through a phylogeny or synapomorphy scheme of Recent tetrapods. Introducing fossils, and monophyletic groups of fossils, into that scheme should test and amplify it, and produce clearer answers to such questions as the monophyly of tetrapods and their nearest relatives amongst fishes.

This is an abbreviated version of the paper presented at the symposium, which was intended as a historical review emphasizing method rather than facts. A more detailed historical review is to be published elsewhere (Rosen, Forey, Gardiner and Patterson, in press). My approach is influenced by a belief that since Darwin we have been too eager to think in terms of origins, and to view problems such as how vertebrates got on land as the province of the palaeon-

Systematics Association Special Volume No. 15, "The Terrestrial Environment and the Origin of Land Vertebrates", edited by A. L. Panchen, 1980, pp. 159-175, Academic Press, London and New York.

tologist, soluble through discovery of fossils intermediate between extant groups. Such fossils have been found, and the mosaics of tetrapod-like and fish-like characters observed or restored in *Ichthyostega* and *Eusthenopteron*, for example, are generally seen as a vindication of this method. I take a different line, that the problem surfaced before Darwin, when living lungfishes were first discovered; that problem was how to distinguish tetrapods and fishes. It was still unsolved when acceptance of evolution directed attention elsewhere, and has remained so since (but see Gaffney, 1979a,b).

In my view, the problem of tetrapod origins first arose in 1837. In that year, South American and African lungfishes first came to Europe. The South American specimens went to Germany, were named *Lepidosiren* by Fitzinger (1837), and were fully described by Bischoff (1840a,b). Bischoff and Fitzinger thought *Lepidosiren* was an amphibian. Bischoff's reasons for this opinion were the presence of five amphibian-like characters: internal nostrils; large lungs connected to the oesophagus by a glottis; gills and their vascular supply reduced, so that most blood bypasses them; heart with two auricles, one receiving blood from the lungs; and valves in the conus separating branchial and pulmonary blood. He gave greater weight to these than to several characters found in *Lepidosiren* and one or more groups of fishes: scales, sensory canals, no ossified vertebrae, opercular bones, no external ear, labial cartilages. Bischoff's argument is readily converted into cladistic terms. It implies that there is a monophyletic group comprising lungfishes and tetrapods, characterized by synapomorphies of the respiratory and vascular systems; and that within this group there is a monophyletic group comprising tetrapods, characterized by synapomorphies of the ear/opercular region and lateralis system.

The African lungfishes came to London, were named *Protopterus* by Owen, and fully described by him in 1841. Owen thought they were fishes, and justified his opinion by a review of every system in the body, seeking the essential difference between fishes and tetrapods. He concluded that only one character would serve, the presence of internal nostrils in tetrapods, and their absence in fishes. Unfortunately, perhaps through defective preservation, Owen failed to find the internal nostril in his specimens, and so rated lungfishes as fish. Owen later admitted his mistake, but having committed himself in this way, never altered his opinion on the systematic position of lungfishes. During the next 25 years there followed a controversy over whether lungfishes should be classed as fishes or as amphibians (for a review see Duméril, 1870). Owen's authority surely played a part in this controversy, and other big guns—Müller, Agassiz, Oken—took the same view, that they were fishes. Unlike Bischoff's, the arguments of these authorities are not favourably converted into cladistic terms; in grouping lungfishes with other fishes they were giving weight to primitive characters, and seeking to maintain a paraphyletic group.

From this controversy there emerged a consensus view that lungfishes were not amphibians, but the most amphibian-like of fishes. Their transitional position was emphasized by Owen (1859), who wrote that *Lepidosiren* "conducts the march of development" towards the perennibranchiate amphibians, and M'Donnell (1860), who recommended that lungfishes be classified as an order uniting fishes with amphibians. As for fossils, they were first seriously brought into this argument by Huxley (1861), who named the Crossopterygidae for *Polypterus* and an assortment of mainly Palaeozoic fishes, including what we now call rhipidistians, lungfishes and coelacanths. Huxley failed to recognize the relationship between living and Palaeozoic lungfishes, but he called living lungfishes "next of kin" to crossopterygians, a term that is open to various interpretations (parent, sibling, child etc.).

Such opinions were made to measure for evolutionists. Haeckel, who produced the first evolutionary trees (1866), placed lungfishes as the sister-group of tetrapods, and took the latter back to the Devonian, where they joined with "Dipnoi ignota". As with other groups (Patterson, 1977, p. 591), Haeckel merely took outdated pre-Darwinian taxonomy (for instance, he seems to have been ignorant of Huxley's work) and converted it into tree form, treating uncertainty about the position of groups like lungfishes as evidence of their transitional or ancestral status. With acceptance of evolution, and of a taxonomy of ancestry and descent, the problem that Owen and Bischoff had argued about—how tetrapods are characterized, and whether lungfishes belong with them—simply disappeared. If evolution were true, then naturally tetrapods and fishes would not be easy to separate, and animals such as lungfishes, in some ways transitional between the two, were favourable to the theory.

In 1870, the Australian lungfish was discovered. As with protopterids, *Neoceratodus* was first reported as an amphibian (Krefft, 1870). It was fully described by Günther (1872), a staunch opponent of Darwinism. Günther argued that lungfishes were nothing to do with amphibians, but were ganoid fishes. He was the first to recognize that Devonian fishes such as *Dipterus*, included by Huxley (1861) in the Crossopterygidae, were lungfishes. Günther regarded the persistence of this group, virtually unaltered since the Devonian, as an excellent argument against Darwin. But others took a different view. Haeckel (e.g. 1889) used the Palaeozoic dipnoans to carry tetrapod ancestry back into the Devonian, while Huxley (1876) discussed the paired fins of *Neoceratodus* as forerunners of tetrapod limbs, and wrote of lungfishes as "the nearest allies of the Amphibia".

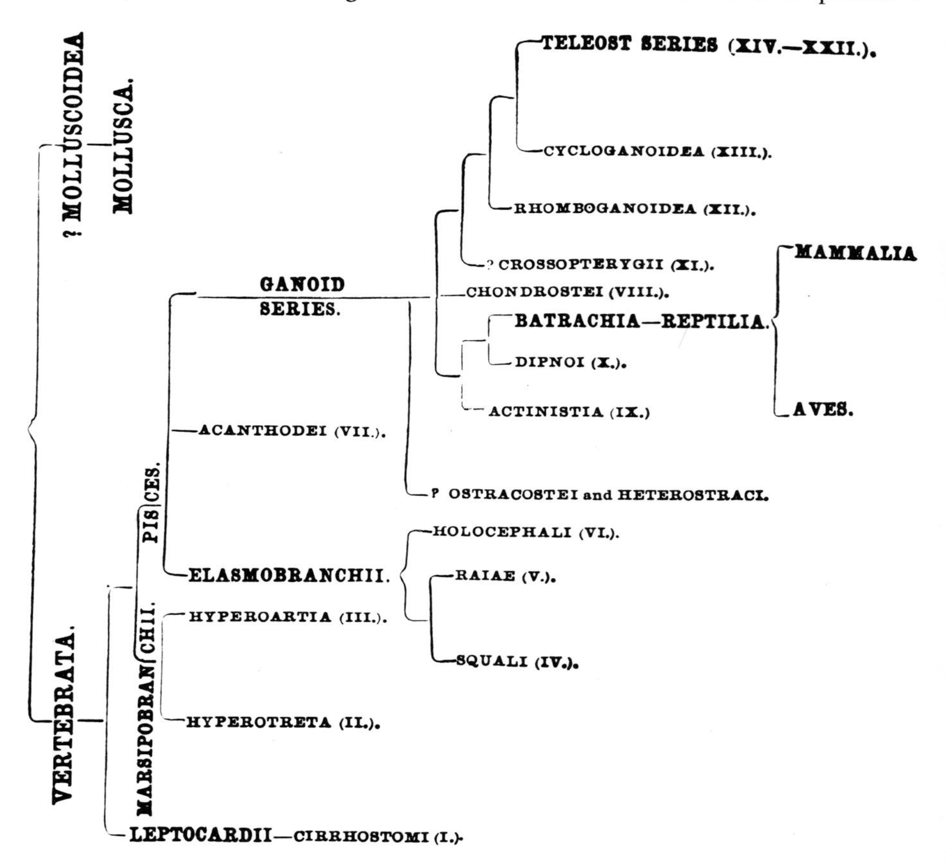

Fig. 1. A genealogical tree of the relations of the major groups of vertebrates (from Gill, 1872, p.xliii).

Gill (1872) also wrote of lungfishes as "most nearly related" to tetrapods, and expressed his opinions on the interrelationships of vertebrates in a diagram (Fig. 1), adapted from human genealogical trees, which is easily interpreted as a cladogram. So far as living fishes go, Gill's arrangement, with lungfishes the sister-group of tetrapods, coelacanths the sister of those two combined, and crossopterygians (*Polypterus*) included amongst the actinopterygians, with a query, because of fin structure, is still about the best available.

Thus during the latter part of the nineteenth century, evolutionists accepted lungfishes as either the ancestors or the nearest relatives of tetrapods. The query against the crossopterygians in Fig. 1 is symptomatic of fluctuating opinions on the content and position of that group (for a summary see Schaeffer, 1965). Crossopterygians were regarded as descendants of dipnoans (e.g. Huxley, 1880), ancestors of dipnoans (e.g. Haeckel, 1889) and relatives of actinopterygians (Fig. 1). But following Cope (1892), Kingsley (1892) and Pollard (1892), late nineteenth century opinion was crystallized by Baur's (1896) paper on labyrinthodonts, and Dollo's (1896) on lungfishes. These authors agreed that tetrapods could not be descended from lungfishes, but that crossopterygians, now containing the living polypterids, the coelacanths, and the Palaeozoic rhipidistians, were ancestral to both tetrapods and lungfishes. The reasons why dipnoans were replaced by crossopterygians as the ancestors of tetrapods are tangled, because some authors argued from the structure of *Polypterus*, and others from what was then known of the structure of Palaeozoic crossopterygians and dipnoans. But in general one may summarize these reasons in two statements: (i) lungfishes were too specialized in the jaws, dentition and other characters to have been ancestral to tetrapods; (ii) crossopterygians, as represented by *Polypterus*, *Eusthenopteron* etc., were sufficiently primitive to have been ancestral to tetrapods, and shared certain special similarities with them (e.g. folded teeth of rhipidistians and labyrinthodonts; occiput of *Polypterus* and labyrinthodonts–Pollard, 1892, p.411).

Dollo's tree (Fig. 2) is a visual summary of late nineteenth and early twentieth century opinion: the crossopterygians were ancestral to both lungfishes and tetrapods. In cladistic terms, this scheme implies only that crossopterygians are a paraphyletic group, with no synapomorphies, which may contain relatives of lungfishes, of tetrapods, and of both groups combined. As for lungfishes, Haeckel (1905, p.572) discussed the pre-Darwinian controversy over whether they

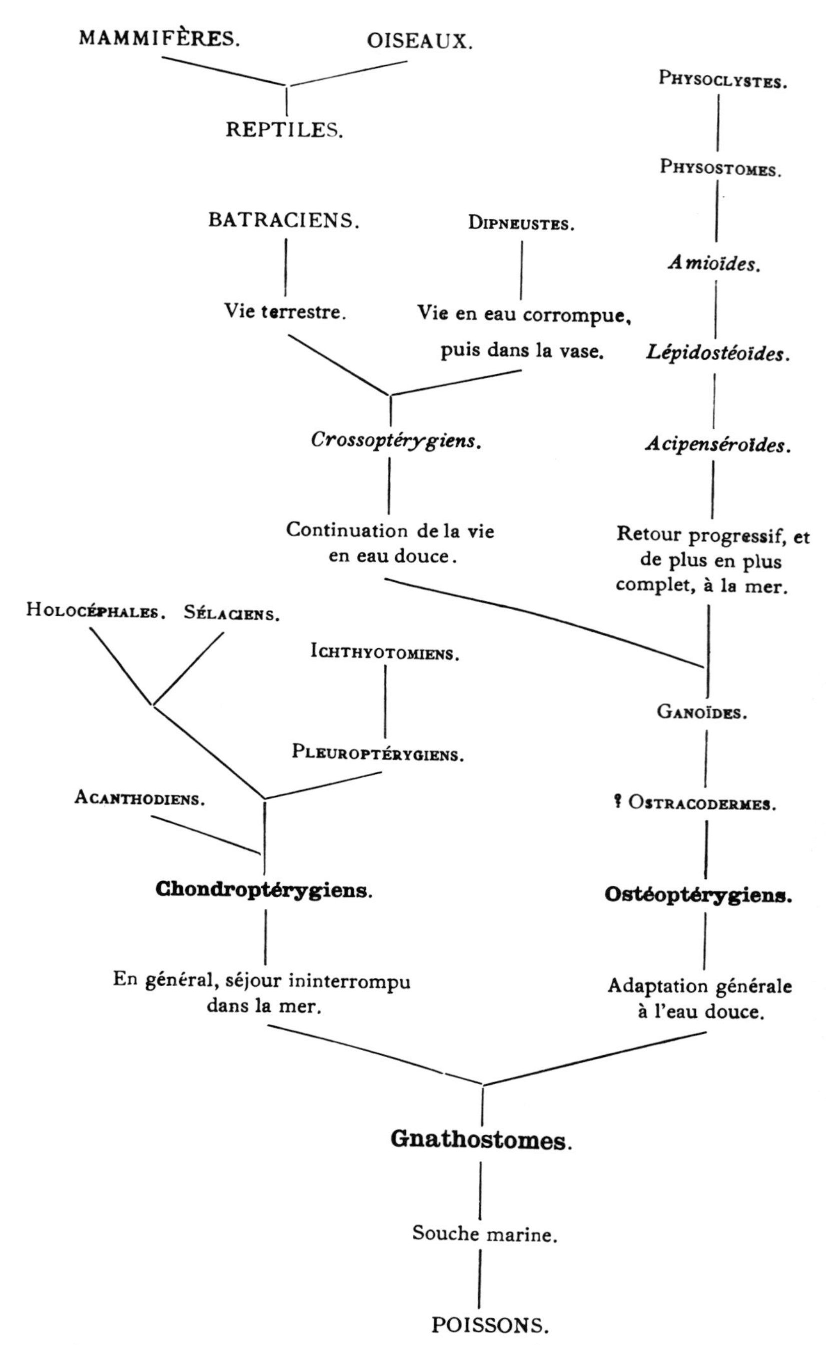

Fig. 2. The phylogeny of gnathostome vertebrates (from Dollo, 1896, p.113).

are fishes or amphibians in the text accompanying a tree similar to Fig. 2. He wrote: "the characters of the two classes are so far united in the dipneusts that the answer to the question depends entirely on the definition we give of 'fish' and 'amphibium' ". Just so, but Haeckel does not go on to define, or characterize fishes and amphibians, because that problem was no longer of interest. If evolution is true, tetrapods must have evolved from fishes, so it would be futile to continue to argue about how the groups were distinguished. And the knowledge that tetrapods must have originated during the Palaeozoic directed attention away from diagrams like Gill's (Fig. 1) and towards the tree format that Haeckel had invented, with an implicit (e.g. Fig. 2) or explicit (e.g. Fig. 3) (Haeckel, 1866) time-scale. Such diagrams lead naturally to interest in ancestral (paraphyletic) groups, to a vague concept of phylogenetic relationship and to the view that the problem of tetrapod origins is the province of the palaeontologist, attempting to close the gap between known Palaeozoic amphibians and crossopterygians.

One more step was necessary before the field was left to palaeontology—elimination of *Polypterus*, which in the words of Schaeffer (1965) had acted like a "well-intentioned impostor" in discussions of tetrapod origins. The view that *Polypterus* was wrongly associated with fossil crossopterygians strengthened gradually, due mainly to Goodrich's work (1901, 1909, 1924, 1928), and equally gradually it dropped out of such discussions. We should not forget, however, that some (e.g. Pollard, 1891, 1892) found *Polypterus* just as effective a tetrapod ancestor as others have found rhipidistians.

In the first half of this century, the major reviews of tetrapod origins are those of Gregory (1915), Watson (1926), Jarvik (1942) and Westoll (1943). Gregory concentrated on the forelimb, and Watson on establishing trends in Palaeozoic amphibians which could be extrapolated back into the rhipidistians. Westoll supplemented Watson's work with new information on rhipidistians, integrated with information on the Devonian ichthyostegids, described by Säve-Söderbergh (1932). Watson and Westoll concluded that Palaeozoic tetrapods converged on a single type, so that tetrapods are monophyletic, and that osteolepid (*sensu lato*) rhipidistians are the closest known approach to that type. Watson believed that the intracranial joint excluded osteolepids from the direct ancestry of tetrapods, but Westoll disagreed. As for lungfishes, Watson placed them as equally close to tetrapods and rhipidistians, in a trichotomy, whereas Westoll (also Lehmann and Westoll, 1952) derived lungfishes, like tetrapods, from osteolepids (cf. Fig. 2).

Jarvik (1942), in a review based on the snout, took a very different view from Watson and Westoll. Following Holmgren (1933), Jarvik argued that tetrapods are not monophyletic, and he derived urodeles from porolepiform rhipidistians, and the remaining tetrapods from osteolepiforms. Holmgren's (1933) idea, that urodeles are descended from dipnoans, can be seen as an attempt to reconcile an increasing conflict between neontology and palaeontology: neontologists (e.g. Kerr, 1932) repeatedly cited detailed resemblances between living dipnoans and amphibians, urodeles in particular, whereas palaeontologists found equally striking resemblances between Palaeozoic amphibians and rhipidistians. Jarvik resolved this conflict by arguing that dipnoans do not have choanae, so excluding them from relationship with rhipidistians and tetrapods. Jarvik's argument has since been generally accepted (see Panchen, 1967), though some (e.g. Jollie, Romer, Westoll) remain unconvinced.

Since Jarvik's (1942) monograph, discussions of tetrapod origins have shifted away from the general question of relationships with fishes, origin from rhipidistians being treated as established beyond doubt, and have concentrated instead on the "biology" of the transition, on particular aspects of it such as the homologies of dermal bones and on the issue of polyphyly versus monophyly (e.g. Szarski, 1962, 1977; Schmalhausen, 1968; Thomson, 1969). The long argument between Jarvik and advocates of tetrapod monophyly (e.g. Gross, Romer, Schaeffer, Szarski, Thomson, Vorob'eva, Westoll) will not be reviewed here. Note, however, that during that argument lungfishes tended to move further and further from tetrapods, being placed as the sister-group of all osteichthyans (e.g. White, 1965; Wahlert, 1968), or even as possible chondrichthyans (Jarvik, 1968).

Jarvik's idea that tetrapods are not monophyletic is typical of a line of thought which was fashionable during the middle third of this century, when many other vertebrate groups—agnathans, chondrichthyans, osteichthyans, holosteans, teleosts, reptiles, mammals and even birds—were held to be non-monophyletic by various authorities. Thus the argument about tetrapod origins was one amongst many. There seem to be only three ways of resolving such arguments. The first, and in my view the least profitable, is to attempt to stick to facts, and to argue about the interpretation of fossils (e.g. Thomson, 1964; Jarvik, 1966). In the end, such arguments are likely to founder in disputes over whether a particular fossil shows a suture or a crack, or over what structure passed through a particular foramen. A second approach is to argue on more general grounds, about the likelihood of parallelism in the evolution of particular structures (e.g. the mammalian ear, the tetrapod limb). But such arguments are only disguised or unanalysed versions (Gaffney, 1979b, p.99) of the third approach, the appeal to parsimony. In my view (Patterson, 1978), the claim that a group is non-monophyletic is defensible only as a claim for one particular theory of relationship between three (or more) groups rather than another. And such a claim can be countered only on grounds of parsimony; by arguing that all members of the group in question share characters which are most parsimoniously explained by common origin. As Gaffney (1979b) points out, such arguments need imply no assumptions about or knowledge of the evolutionary process: they concern character distribution alone, and so are equally appropriate in a non-evolutionary context. In the example of the case for an independent origin of urodeles and the remaining tetrapods, it would be necessary to show that the characters uniquely shared by urodeles and other tetrapods outnumber or outweigh the characters uniquely shared by urodeles and dipnoans (Holmgren's argument) or by urodeles and porolepids (Jarvik's argument). To do this, one would need a list of tetrapod characters. In other words, one would need to tackle the problem that Bischoff and Owen wrestled with 140 years ago—how are tetrapods characterized?

The reasons why the overriding criterion is parsimony, rather than the structure or stratigraphic succession of fossils, or appeals to evolutionary mechanisms (selection pressures etc.), have become clear over the last few years as the theory of cladism has been worked out, following Hennig (1966). Hennig provided, for the first time, a definition of relationship, a concept which had previously remained ambiguous, and showed that there was no necessary correlation between relationship and similarity, or characters in common. Instead, patterns of relationship are recognized by the distribution of synapomorphies (homologous character states). As these concepts have been discussed and developed, a new, more precise way of tackling problems like the origin of tetrapods has emerged. The traditional type of phylogenetic diagram (e.g. Fig. 3A,B), with its

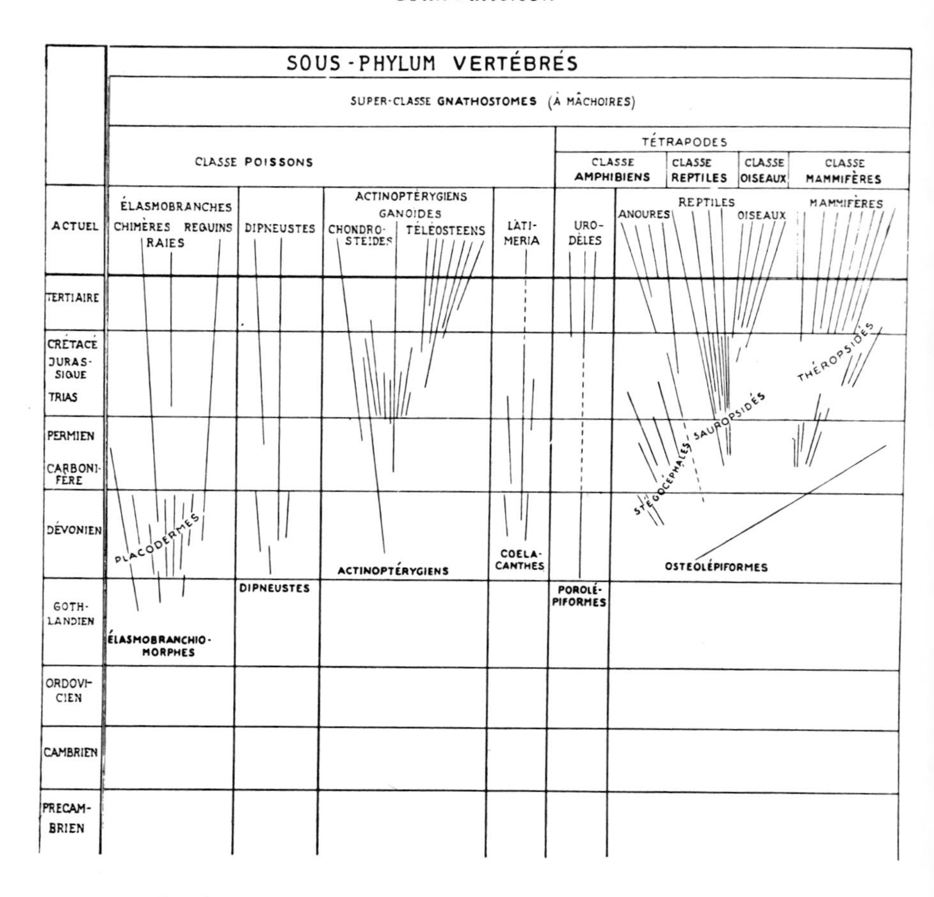

Fig. 3A. This figure and Fig. 3B show two representations of vertebrate phylogeny, (A) modified from Jarvik (1960, Fig. 28), (B) from Gross (1964, Fig. 3). In (A) the groups are left separate in the absence of concrete links between them, and the classes, above, are said to be more and more indefensible. In (B) the groups are linked by broken lines, and dotted lines separate classes and subclasses.

time-scale and its emphasis on fossils, illustrates the traditional way of tackling such problems—put crudely, by searching for fossils of the appropriate age, which will provide concrete links between groups, and so reveal the phylogeny (for an exposition of this method as applied to tetrapod origins see Romer, 1956). The fact that none of the groups in Fig. 3A,B are so linked might be taken as an indication that this method has not yet succeeded.

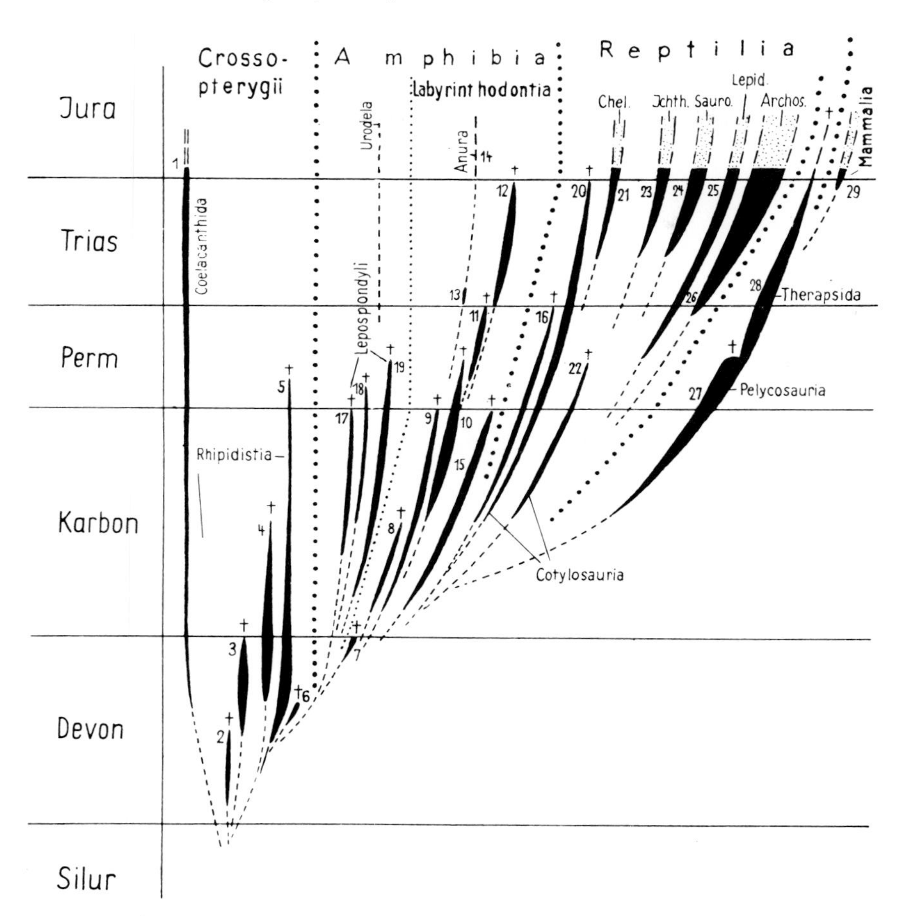

Fig. 3B. For legend see Fig. 3A.

In a cladogram (e.g. Figs. 1, 4), the type of diagram that Hennig advocated, all that is presented is a hypothesis about groups and their relationships. The advantages of such diagrams are their clarity and the ease with which they may be criticized. In evaluating or criticizing such a diagram, or hypothesis, we ask first what is the evidence that each group is monophyletic? In other words, what synapomorphies characterize each terminal taxon, and each junction in the diagram. The text accompanying the originals of Fig. 4 does not provide such information, and so far as I know no serious or comprehensive attempts have yet been made to produce a synapomorphy

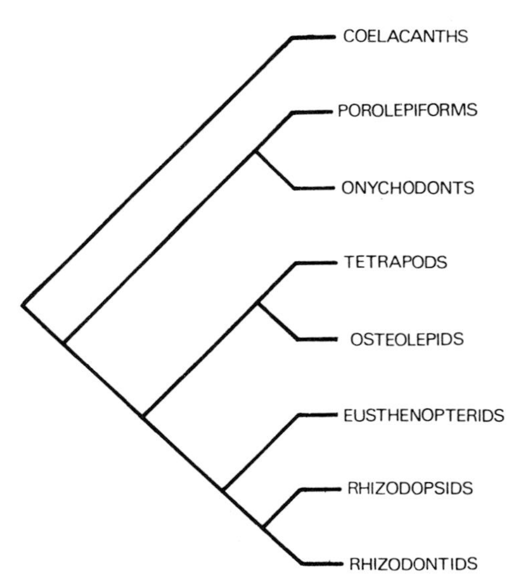

Fig. 4. A cladogram of crossopterygian and tetrapod vertebrates (adapted from Vorob'eva, 1977a, Fig. 22; 1977b, Fig. 8).

scheme for tetrapods and their relatives (but see Løvtrup, 1977; Gaffney, 1979a,b): in my view, such an attempt would be the best possible starting point for a discussion on tetrapod origins.

However, the method of synapomorphy schemes has been extensively applied in fishes. Largely by that method, another "well-intentioned impostor" has been detected, *Latimeria*. Since Huxley's 1861 paper, coelacanths have been regarded as crossopterygians, close relatives of rhipidistians and hence, during this century, of tetrapods. Characters apparently confirming this, such as an intracranial joint (Stensiö, 1932) and a choana (Nielsen, 1936), were found in fossil coelacanths. The discovery and description of *Latimeria* removed some of these characters (e.g. the choana), and revealed many autapomorphies (Bjerring, 1973, p.195) and little or no sign of tetrapod affinities. Nevertheless, as lungfishes were moved away from tetrapods, *Latimeria* took their place as the closest living relative of tetrapods (e.g. Figs. 3, 4; Romer, 1966, Fig. 67; Miles, 1975; Schultze, 1977). But recently, cladists have argued that coelacanths are the sister-group of lungfishes, rhipidistians and tetrapods (Miles, 1977), or of all other osteichthyans (tetrapods included—Wiley, 1979), or of chondrichthyans (Løvtrup, 1977).

These authors also list synapomorphies which reinstate lungfishes as the closest living relatives of tetrapods. A natural consequence of this is that the posterior nostrils of dipnoans may, after all, be choanae, as they were thought to be for so long (Gardiner, Chapter 8 of this volume). Thus, by a circuitous route, we are back to the position taken by Bischoff and other pre-Darwinians, or by Gill (Fig. 1) and Dollo (Fig. 2). So far as I know, the synapomorphies between lungfishes and tetrapods listed by Bischoff (1840b) remain uncontradicted, and to them may be added about 34 of the 57 characters listed by Kesteven (1950), the gill-arch muscles discussed by Wiley (1979), the ciliation of the larvae described by Whiting and Bone (1980), and so on.

As for the rhipidistians, we now have a great deal of anatomical information on various forms, *Eusthenopteron* in particular. But to the cladist, the rhipidistians remain where they were in Dollo's time (Fig. 2)—a paraphyletic group, some of which may be stem-group dipnoans, some stem-group tetrapods and some stem-group choanates. I believe (cf. Patterson, 1977, pp. 625–633) that individual rhipidistians cannot be allocated to those three groups, and ordered within them, until we have adequately characterized, by means of synapomorphies, the morphotypes of the three groups—tetrapods, dipnoans and choanates. To do this, we need a character phylogeny, or synapomorphy scheme of each group, based initially on the living forms. Miles (1977) has made a good start with the lungfishes, but we still lack any credible phylogeny of tetrapods, as Gaffney's recent papers (1979a,b) make plain. Gaffney was able to list 11 (1979a) or eight (1979b) synapomorphies of tetrapods as a whole, but several of these (Nos 1, 5 (in part), 8 (in part), 10, 11 in the 1979a list; a (in part), d, e, h (in part) in the 1979b list) are present in lungfishes, and these lists are specifically formulated to exclude osteolepiform rhipidistians, so evading the problem by assuming that it is already solved. As for Choanata, the taxon is characterized by the choana alone (Miles, 1977), or by the choana and a divided braincase (Gaffney, 1979b), the latter a character present in rhipidistians alone. The task of deciding whether certain rhipidistians are more closely related to tetrapods than to anything else demands a more extensive, and less partisan, characterization of these groups, which in turn demands a more convincing attack on the problem of tetrapod interrelationships, an attack which is likely to be led by neontology, not palaeontology (cf. Patterson, 1977, p. 621).

To sum up, since Darwin the problem of the origin of land vertebrates has been largely the province of palaeontologists, and their method has been to look for ancestors, or ancestral groups, in the fossil record. As in so many other cases, this method has led only to wrangles about polyphyly and the interpretation of fossils, and to unjustifiable statements about the supposed attributes of extinct paraphyletic groups. Dealing with another hotly disputed area of vertebrate phylogeny, the hominids, Tattersall and Eldredge (1977) (see also Eldredge, 1979, p. 168) argue that the accepted method has been to write scenarios—evolutionary trees "with an overlay of adaptational narrative"—and that the construction of trees, and of scenarios, is necessarily secondary to a more basic activity, constructing cladograms—seeking a precise pattern of relationship through character distribution. I suggest that we might make progress with the problem of tetrapods by taking that to heart, and going back to the problem that Owen and Bischoff argued about—what are the characters of tetrapods? With a comprehensive answer to that question, we could make better use of the fossil record, looking in it not for ancestors, but for the sequence in which those tetrapod characters arose.

REFERENCES

Baur, G. (1896). The Stegocephali. A phylogenetic study. *Anat. Anz.* **11**, 657–675.

Bischoff, T. L. W. von (1840a). "*Lepidosiren paradoxa.* Anatomisch utersucht und Beschrieben." Leipzig.

Bischoff, T. L. W. von (1840b). Description anatomique du *Lepidosiren paradoxa. Annls Sci. nat.* (2) **14**, 116–159.

Bjerring, H. C. (1973). Relationships of coelacanthiforms. *In* "Interrelationships of Fishes" (P. H. Greenwood, R. S. Miles and C. Patterson, eds), pp. 179–205. Academic Press, London and New York.

Cope, E. D. (1892). On the phylogeny of the Vertebrata. *Proc. Am. phil. Soc.* **30**, 278–281.

Dollo, L. (1896). Sur la phylogénie des dipneustes. *Bul. Soc. belge Géol. Paléont. Hydrol.* (9) **2**, 79–128.

Duméril, A. (1870). Le Lépidosiren et le Protoptére apartiennent á la classe des poissons ou ils sont les types de la sous-classe des Dipnés. *Annls Soc. linn. Angers* **12**, 140–149.

Eldredge, N. (1979). Cladism and common sense. *In* "Phylogenetic Analysis and Paleontology" (J. Cracraft and N. Eldredge, eds), pp. 165–198. Columbia University Press, New York.

Fitzinger, L. J. F. J. (1837). Vorläufiger Bericht über eine höchst interessante Entdeckung Dr Natterers in Brasil. *Isis, Jena* 1837, 379–380.

Gaffney, E. S. (1979a). Tetrapod monophyly: a phylogenetic analysis. *Bull. Carnegie Mus. nat. Hist.* **13**, 92–105.

Gaffney, E. S. (1979b). An introduction to the logic of phylogeny reconstruction. *In* "Phylogenetic Analysis and Paleontology" (J. Cracraft and N. Eldredge, eds), pp. 79–111. Columbia University Press, New York.

Gill, T. (1872). Arrangement of the families of fishes. *Smithson misc. Collns* **247**, 49 pp.

Goodrich, E. S. (1901). On the pelvic girdle and fin of *Eusthenopteron. Q. Jl microsc. Sci.* **45**, 311–324.

Goodrich, E. S. (1909). Vertebrata Craniata. First fascicle: cyclostomes and fishes. *In* "A Treatise on Zoology" (E. R. Lankester, ed.), Vol. 9. A. and C. Black, London.

Goodrich, E. S. (1924). The origin of land vertebrates. *Nature, Lond.* **114**, 935–936.

Goodrich, E. S. (1928). *Polypterus* a palaeoniscid ? *Palaeobiologica* **1**, 87–92.

Gregory, W. K. (1915). Present status of the problem of the origin of the Tetrapoda, with special reference to the skull and paired limbs. *Ann. N.Y. Acad. Sci.* **26**, 317–383.

Gross, W. (1964). Polyphyletische Stämme im System der Wirbeltiere ? *Zool. Anz.* **173**, 1–22.

Günther, A. C. L. G. (1872). Description of *Ceratodus*, a genus of ganoid fishes, recently discovered in rivers of Queensland, Australia. *Phil. Trans. R. Soc.* **161**, 511–571.

Haeckel, E. (1866). "Generelle Morphologie der Organismen." G. Reimer, Berlin.

Haeckel, E. (1889). "Natürliche Schöpfungs-Geschichte", 8th edn. G. Reimer, Berlin.

Haeckel, E. (1905). "The Evolution of Man." Watts, London.

Hennig, W. (1966). "Phylogenetic Systematics." University of Illinois Press, Urbana.

Holmgren, N. (1933). On the origin of the tetrapod limb. *Acta zool., Stockh.* **14**, 185–295.

Huxley, T. H. (1861). Preliminary essay upon the systematic arrangement of the fishes of the Devonian epoch. *Mem. geol. Surv. U.K.* Decade 10, 1–40.

Huxley, T. H. (1876). Contributions to morphology. Ichthyopsida No. 1. On *Ceratodus forsteri*, with observations on the classification of fishes. *Proc. zool. Soc. Lond.* 1876, 24–59.

Huxley, T. H. (1880). On the application of the laws of evolution to the arrangement of the Vertebrata and more particularly of the Mammalia. *Proc. zool. Soc. Lond.* 1880, 649–662.

Jarvik, E. (1942). On the structure of the snout of crossopterygians and lower gnathostomes in general. *Zool. Bidr. Upps.* **21**, 235–675.

Jarvik, E. (1960). "Théories de l'Évolution des Vertébrés Reconsidérées à la Lumière des Récentes Découvertes sur les Vertébrés Inférieurs." Masson, Paris.

Jarvik, E. (1966). Remarks on the structure of the snout in *Megalichthys* and certain other rhipidistid crossopterygians. *Ark. Zool. (2)* **19**, 41–98.

Jarvik, E. (1968). The systematic position of the Dipnoi. *Nobel Symposium* **4**, 221–245.

Kerr, J. G. (1932). Archaic fishes – *Lepidosiren, Protopterus, Polypterus* – and their bearing upon problems of vertebrate morphology. *Jena. Z. Naturw.* **67**, 419–433.

Kesteven, H. L. (1950). The origin of the tetrapods. *Proc. R. Soc. Vict.* **59**, 93–138.

Kingsley, J. S. (1892). The head of an embryo *Amphiuma*. *Am. Nat.* **26**, 671–680.

Krefft, G. (1870). Description of a gigantic amphibian allied to the genus *Lepidosiren*, from the Wide Bay District, Queensland. *Proc. zool. Soc. Lond.* 1870, 221–224.

Lehmann, W. M. and Westoll, T. S. (1952). A primitive dipnoan fish from the Lower Devonian of Germany. *Proc. R. Soc.* **B140**, 403–421.

Løvtrup, S. (1977). "The Phylogeny of Vertebrata." Wiley, London.

M'Donnell, R. (1860). Observations on the habits and anatomy of the *Lepidosiren annectens*. *Jl R. Dublin Soc.* **2**, 3–20.

Miles, R. S. (1975). The relationships of the Dipnoi. *Colloques int. Cent. natn. Rech. Scient.* **218**, 133–148.

Miles, R. S. (1977). Dipnoan (lungfish) skulls and the relationships of the group: a study based on new species from the Devonian of Australia. *Zool. J. Linn. Soc.* **61**, 1–328.

Nielsen, E. (1936). Some few preliminary remarks on Triassic fishes from East Greenland. *Meddr Grønland* **112** (3), 1–55.

Owen, R. (1841). Description of the *Lepidosiren annectens*. *Trans. Linn. Soc. Lond.* **18**, 327–361.

Owen, R. (1859). Palaeontology. *In* "Encyclopaedia Britannica", Vol. 17, pp. 91–176.

Panchen, A. L. (1967). The nostrils of choanate fishes and early tetrapods. *Biol. Rev.* **42**, 374–420.

Patterson, C. (1977). The contribution of paleontology to teleostean phylogeny. *In* "Major Patterns in Vertebrate Evolution" (M. K. Hecht, P. C. Goody and B. M. Hecht, eds), pp. 579–643. Plenum, New York.

Patterson, C. (1978). Arthropods and ancestors. *Antenna* **2**, 99–103.

Pollard, H. B. (1891). On the anatomy and phylogenetic position of *Polypterus*. *Anat. Anz.* **6**, 338–344.

Pollard, H. B. (1892). On the anatomy and phylogenetic position of *Polypterus*. *Zool. Jb.* **5**, 387–428.

Romer, A. S. (1956). The early evolution of land vertebrates. *Proc. Am. phil. Soc.* **100**, 157–167.

Romer, A. S. (1966). "Vertebrate Paleontology", 3rd edn. University of Chicago Press, Chicago.

Rosen, D. E., Forey, P. L., Gardiner, B. G. and Patterson, C. (in press). Lungfishes, tetrapods, paleontology and plesiomorphy. *Bull. Am. Mus. nat. Hist.*, in press.

Säve-Söderbergh, G. (1932). Preliminary note on Devonian stegocephalians from East Greenland. *Meddr Grønland* **94** (7), 1–107.
Schaeffer, B. (1965). The evolution of concepts related to the origin of the Amphibia. *Syst. Zool.* **14**, 115–118.
Schmalhausen, I. I. (1968). "The Origin of Terrestrial Vertebrates." Academic Press, New York and London.
Schultze, H-P. (1977). The origin of the tetrapod limb within the rhipidistian fishes. *In* "Major Patterns in Vertebrate Evolution" (M. K. Hecht, P. C. Goody and B. M. Hecht, eds), pp. 541–544. Plenum, New York.
Stensiö, E. A. (1932). Triassic fishes from East Greenland collected by the Danish expeditions in 1929–1931. *Meddr Grønland* 83 (3), 1–305.
Szarski, H. (1962). The origin of the Amphibia. *Q. Rev. Biol.* **37**, 189–241.
Szarski, H. (1977). Sarcopterygii and the origin of tetrapods. *In* "Major Patterns in Vertebrate Evolution" (M. K. Hecht, P. C. Goody and B. M. Hecht, eds), pp. 517–540. Plenum, New York.
Tattersall, I. and Eldredge, N. (1977). Fact, theory and fantasy in human paleontology. *Am. Scient.* **65**, 204–211.
Thomson, K. S. (1964). The comparative anatomy of the snout in rhipidistian fishes. *Bull. Mus. comp. Zool. Harv.* **131**, 313–357.
Thomson, K. S. (1969). The biology of the lobe-finned fishes. *Biol. Rev.* **44**, 91–154.
Vorob'eva, E. I. (1977a). Morfologiya i osobennosti evolyutsii kisteperykh ryb. *Trudy paleont. Inst.* **163**, 1–240.
Vorob'eva, E. I. (1977b). Filogeneticheskie svazi osteolepiformnykh Crossopterygii i ikh polozhenie v sisteme. *In* "Ocherki po filogenii i sistematike iskopaemykh ryb i beschelyustnykh", pp. 71–88. Paleont. Inst. Akademiya Nauk SSSR Moscow.
Wahlert, G. von (1968). "*Latimeria* und die Geschichte der Wirbelthiere. Eine evolutionsbiologische Untersuchung." Gustav Fischer, Stuttgart.
Watson, D. M. S. (1926). The evolution and origin of the Amphibia. *Phil. Trans. R. Soc.* **B214**, 189–257.
Westoll, T. S. (1943). The origin of the tetrapods. *Biol. Rev.* **18**, 78–98.
White, E. I. (1965). The head of *Dipterus valenciennesi* Sedgwick & Murchison. *Bull. Br. Mus. nat. Hist.* (Geol.) **11**, 1–45.
Whiting, H. P. and Bone, Q. (1980). Ciliary cells in the epidermis of the larval Australian dipnoan *Neoceratodus*. *Zool. J. Linn. Soc.* **68**, 125–137.
Wiley, E. O. (1979). Ventral gill arch muscles and the interrelationships of gnathostomes, with a new classification of the Vertebrata. *Zool. J. Linn. Soc.* **67**, 149–179.

8 | Tetrapod Ancestry: A Reappraisal

BRIAN G. GARDINER

Biology Department, Queen Elizabeth College (University of London), Campden Hill Road, London W8, England

Abstract: The dipnoans are the closest relatives of tetrapods. The Osteolepiformes are a paraphyletic or grade group with the Panderichthyidae the sister-group of dipnoans + tetrapods. The Porolepiformes are a non-choanate, monophyletic group considerably removed from tetrapod ancestry.

INTRODUCTION

For over a half-century following Watson's (1926) Croonian Lecture it has been customary to regard the crossopterygians and not the dipnoans as the closest relatives of tetrapods. The few dissenting voices have included Kesteven (1931, 1951) and Kerr (1932), who demonstrated the close relationship of the Dipnoi and the Amphibia, and Holmgren (1933) and Säve-Söderbergh (1934) who considered dipnoans to be more closely related to the urodeles. Despite the overwhelming support for Watson's proposed crossopterygian relationship (see for example Panchen, 1967) I shall argue that the dipnoans are the closest living relatives of tetrapods (Gardiner *et al.*, 1979), and further that they may be closer to tetrapod ancestry than any other group of fish, fossil or living. The conclusions summarized here represent a progress report in the preparation of a more detailed study by the author together with Drs Forey, Patterson and Rosen.

RELATIONSHIP BETWEEN LIVING DIPNOANS AND TETRAPODS

As recently as 1951 Kesteven (p.128) drew up a table in which he listed some 40 similarities between the Dipnoi and Amphibia includ-

Systematics Association Special Volume No. 15, "The Terrestrial Environment and the Origin of Land Vertebrates", edited by A. L. Panchen, 1980, pp. 177–185, Academic Press, London and New York.

ing, in my opinion, 34 derived features. It will suffice to emphasize just a few of these derived features which are unique to dipnoans and amphibians/tetrapods and to refer the reader to Kesteven (1951, p.128) and Løvtrup (1977, p.128) for a more complete list.

In both groups the pulmonary veins unite before emptying into an atrial cavity which is divided by an atrial septum. The epibranchial arteries of each side unite to form the aorta while the ventral aorta is shortened to form a partly divided conus arteriosus with three paired channels. Other shared derived characters include:

the structure and position of the glottis,
the possession of an epiglottis,
the development of an outgrowth from the ductus endolymphaticus,
the possession of the lens proteins D1, 2 and 5,
the possession of the bile salt bufol,
the similar mating calls of the males,
and last but by no means least the possession of an internal excurrent naris or choana.

Although from the outset the majority of authors interpreted the excurrent (internal) nostril in dipnoans as a choana homologous to that of tetrapods, Allis (1919, 1932a,b) considered it to be merely an osteichthyan posterior external nostril, not homologous with the tetrapod choana, which had migrated downwards and so came to lie inside the roof of the mouth. Allis's (1932b, p.666) view was subsequently championed by Jarvik (1942, p.274) who concluded (p.641) that the dipnoan posterior external nostril "has migrated downwards to the roof of the mouth and has become modified so as to resemble a choana". Most later authors including Panchen (1967), Bertmar (1969), Løvtrup (1977) and Miles (1977) have accepted Jarvik's (1942) interpretation. Nevertheless, additional evidence that the dipnoan posterior (internal) nostril is a true choana homologous with that of tetrapods has been provided by Miles (1977), albeit unwittingly, in his recent account of the Devonian lungfishes of Australia. In this paper Miles (Fig 57, 80a) described specimens of *Griphognathus* in which there is a marginal dentition (ectopterygoid of Miles) lateral to a completely bone-enclosed posterior naris. This set of lateral, tooth-bearing bones bites outside the lower jaw and must therefore be considered maxillary in position, and consequently Miles's (Figs 57,

80a, "p.n." posterior external nasal opening must be a choana in addition to being homologous with the excurrent naris of other osteichthyans and chondrichthyans.

RELATIONSHIPS OF FOSSIL GROUPS

1. *Porolepiformes*

The Porolepiformes are a monophyletic group with characteristic dendrodont teeth in which the pulp cavity is filled with osteodentine and the orthodentine has a complicated and regular folding, aptly described as "firelike branching" by Schultze (1970).

Porolepiformes share with coelacanths and osteolepids a braincase with an intracranial joint; while they share with coelacanths, onychodonts, osteolepids, tetrapods and dipnoans a sclerotic ring consisting of numerous small plates. They share with osteolepids, onychodonts and dipnoans both a submandibular series (modified branchiostegal rays) and "cosmoid" scales. All of these shared characters are here considered to be derived within the teleostomes and as such indicative of relationships.

Porolepiformes are also supposed to possess one other derived feature in common with osteolepids, dipnoans and tetrapods, namely a choana. The presence of a choana was first postulated by Jarvik (1942, Figs 36, 44) when he decided that the ventro-lateral notch in the postnasal wall of *Porolepis* and *Holoptychius* was in part for a fenestra endochoanalis. He developed this idea further in subsequent papers (1962, 1965, 1966, 1972) and has published several transverse sections of the snout of *Glyptolepis* in support of his theory. However, examination of these published sections, and in particular Fig. 8A–D in the 1972 paper, reveals only the smallest of openings in the roof of the palate. This, and a careful examination of specimens in the Institute of Geological Sciences Museum and The British Museum (Natural History), especially a neurocranium of *Glyptolepis* (B.M.N.H. P47838), has convinced me that the opening attributed by Jarvik to a choana is no more than a notch in the margin of the palate at the junction of four dermal bones (cf. Jarvik, 1972, pl.22, Fig. 1.). Moreover in *Porolepis* and *Glyptolepis* (Jarvik, 1942, Fig. 41; 1972, Fig. 7) there are in addition to this notch two external narial openings, whereas in dipnoans and tetrapods there is only a single external naris (the other external naris is presumed to have become the choana). These porolepiforms are the only fish in which it has been accepted that a choana and an excurrent nostril co-exist and are separate from one another.

From this evidence we may conclude that porolepids, like actinopterygians, coelacanths and onychodonts, are non-choanate and furthermore that they are the sister-group of the osteolepids + dipnoans + tetrapods combined, sharing with these latter groups a cosmoid scale and submandibular series.

2. Osteolepiformes

The Osteolepiformes are a paraphyletic or grade group ranging from the Middle Devonian to the early Permian. Unfortunately only a few members are at all well known. They are all assumed to have a true choana leading from the nasal cavity to the roof of the mouth and a single external nostril as in dipnoans and tetrapods. *Panderichthys* has been described as having two external nares (Vorobjeva, 1960, 1973): however, the smaller posterior opening between the lateral rostral and lachrymal does not appear to connect with the nasal capsule and is best considered as the opening of a gland such as the lachrymal organ. Since the choana is assumed to have evolved but once, then the Osteolepiformes could contain taxa which may be the sister-group of the lungfishes or the tetrapods, or the lungfishes + tetrapods.

3. Osteolepididae Cope 1889

The family includes such forms as *Osteolepis, Megalichthys, Thursius* and *Gyroptychius* and was the longest lived of the osteolepiform families. Its members all have slender bodies, thick rhombic scales, obtusely lobate paired fins and several of them developed ring-like centra (*Thursius, Megalichthys*). The family is doubtfully monophyletic.

Most of the genera have large replacement teeth (fangs) as in porolepids, but here the folding is described as polyplocodont (the orthodentine being simply and irregularly folded) with the bone of attachment extending between the folds while the pulp cavity is free. In the Middle Devonian *Osteolepis* the teeth are unfolded (Schultze, 1970).

The skull-table consists of elongate "parietals" (= postparietals of Westoll, 1943) flanked by paired "intertemporals" (= tetrapod supratemporal) and "supratemporals". Anteriorly the "frontals" (tetrapod parietals, *fide* Westoll) enclose the pineal foramen. There are therefore only two pairs of large bones that meet in the mid-line, the "parietals" and the "frontals".

Osteolepids share with dipnoans a biserial archipterygium, but as in *Eusthenopteron* the axis is shorter with fewer segments and reduced postaxial radials.

Osteolepids, however, share several derived features with lungfishes, including the presence of cosmine (Miles, 1977, p.311). However, cosmine is also said to be present in the head-bones of *Onychodus* (Jessen, 1966). The scales and dermal bones of the head of osteolepids and primitive dipnoans on the other hand frequently show Westoll lines. From this we may deduce that the osteolepidids share with fossil dipnoans a uniquely derived feature which is the ability to resorb and redeposit cosmine, in other words to rework the surface layers of the dermal bones.

Although on this evidence it is possible to regard the Osteolepididae as the sister-group of the lungfishes, the fact that lungfishes share a number of derived features in common with the tetrapods, including the loss of kinesis in the skull and palate, the presence of true post-parietals (see below) and pterygoids that meet anteriorly, means that it is preferable to consider the osteolepidids as the sister-group of the lungfishes + tetrapods.

4. Eusthenopteron

Eusthenopteron is probably the best known of all fossil fishes. The tail is almost symmetrical (diphycercal) with a prominent axial lobe, and the scales are cycloid and composed entirely of bone without a superficial covering of cosmine. Some authors (Andrews, 1973, p.143) consider that the absence of cosmine in *Eusthenopteron* is secondary, but it is also possible that the ability to rework cosmine had never been developed. Moreover the vertebral column of *Eusthenopteron* is primitive and similar to that in palaeoniscoids, embryos of *Amia* (Schaeffer, 1967, p.193) and coelacanths (Andrews,

1977, Fig. 5), whereas that of *Megalichthys* is more closely similar to that of stereospondyls (Panchen, 1977). In most other respects *Eusthenopteron* resembles the osteolepidids. Thus it has a sub-mandibular series, folded polyplocodont teeth, a hinged neuro-cranium and a similar disposition of dermal skull-bones. Furthermore, it shares, with osteolepidids, dipnoans and tetrapods, a choana and a large supraotic cavity correlated with the hypertrophy of the endolymphatic sac (Miles, 1977, p.100). *Eusthenopteron* also shares with the dipnoan *Griphognathus* a sublingual type of basihyal which Miles (1977, p.286) considers the result of parallel evolution. It seems more likely, however, that this is a derived feature and as such indicative of a closer relationship between *Eusthenopteron* and lungfishes.

Eusthenopteron then is the sister-group of the osteolepidis + lungfishes + tetrapods, sharing with them a choana and a supraotic cavity, and with the lungfishes a prominent basihyal.

5. Panderichthyidae Vorobjeva 1968.

The Panderichthyidae contain three genera, *Panderichthys*, *Obruchevichthys* and *Elpistostege*, of which *Panderichthys* is the only reasonably known form. In most characters the family appears to be similar to the Osteolepididae and to *Eusthenopteron*. However in two characters it resembles tetrapods.

Firstly *Panderichthys* has teeth in which the dentine is folded into meandering branches and in which the bone of attachment extends only a short distance between the folds. This type of tooth appears to be more complex than the polyplocodont form seen in *Eusthenopteron* and in some osteolepidids, but is identical with that of *Ichthyostega* and many labyrinthodonts. It may justly be described as labyrinthodont.

Secondly both *Panderichthys* (Vorobjeva, 1977, Fig. 2) and *Elpistostege* (Westoll, 1938, Fig. 2) have postparietals present in the skull-table (co-existent with undoubted frontals) and thus resemble *Ichthyostega* and tetrapods. Thus irrespective of how the bones are labelled there can be no doubt that the panderichthyids and tetrapods have one extra pair of large bones (three instead of the typical osteolepiform two) that meet in the mid-line in the skull-roof. The three pairs must therefore be the homologues of the tetrapod frontals, parietals and postparietals.

Lungfishes also share with panderichthyids and tetrapods two pairs of parietals in the skull-table (bones J and I of Thomson and Campbell, 1971; lateral parietals of Lehman, 1966) but in addition they possess a median parietal (bone B, Thomson and Campbell, 1971; central parietal, Lehman, 1966). A median parietal (= postparietal) is only found elsewhere in *Ichthyostega* and this is taken as additional support in favour of the close relationship postulated between dipnoans and tetrapods.

Thus in the form of the teeth and skull-roofing bones panderichthyids resemble tetrapods. However, if we believe that the loss of the intracranial joint and of the mobility of the palate occurred but once and that the meeting of the pterygoids anteriorly is another derived feature found only in dipnoans and tetrapods, then the panderichthyids must be the sister-group of the dipnoans + tetrapods and the folded labyrinthodont tooth is not necessarily synapomorphous (cf. *Lepisosteus*).

CONCLUSIONS

Starting from the premise that the intracranial joint and sclerotic ring of numerous plates are shared specializations of coelacanths, porolepids and primitive choanates, further advanced features within these groups are indicated in Fig. 1 in approximate order of origin.

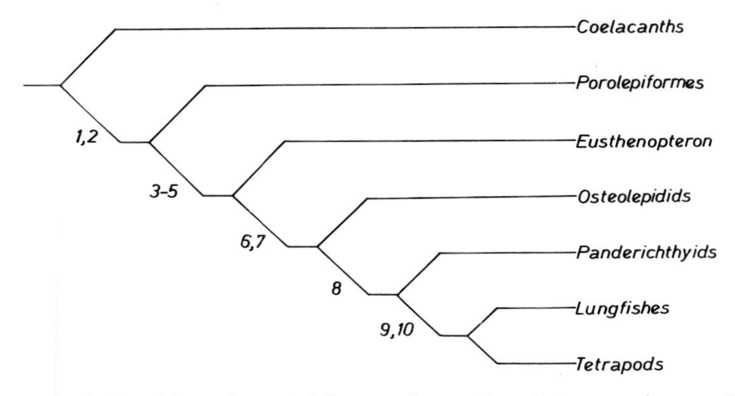

Fig. 1. Interrelationships of osteichthyans shown in relation to the acquisition of advanced characters. The numbered characters are explained in the text.

These advanced features of osteichthyans including the following:

(1) The possession of cosmoid scales.
(2) The presence of a submandibular series of bones.
(3) The development of a choana or internal excurrent naris.
(4) The development of a large supraotic cavity correlated with the hypertrophy of the endolymphatic sac.
(5) The possession of a sublingual type of basihyal.
(6) The development of Westoll lines.
(7) The formation of perichordal ring centra.
(8) The presence of true postparietals (or two pairs of parietals).
(9) Loss of the intracranial joint and mobility of the palate.
(10) Pterygoids meet anteriorly.

From this I conclude that the lungfishes are not only the closest living relatives of tetrapods but also that no known fossil group appears to lie between them.

REFERENCES

Allis, E. P. (1919). The lips and the nasal apertures in the gnathostome fishes. *J. Morph.* **32**, 145–206.

Allis, E. P. (1932a). Concerning the nasal apertures, the lachrymal canal and the bucco-pharyngeal upper lip. *J. Anat.* **66**, 650–658.

Allis, E. P. (1932b). The pre-oral gut, the buccal cavity and the buccopharyngeal opening in *Ceratodus. J. Anat.* **66**, 659–668.

Andrews, S. M. (1973). Interrelationships of crossopterygians. *In* "Interrelationships of Fishes" (P. H. Greenwood, R. S. Miles and C. Patterson, Eds), pp. 137–177. Academic Press, London and New York.

Andrews, S. M. (1977). The axial skeleton of the coelacanth, *Latimeria. In* "Problems in Vertebrate Evolution" (S. M. Andrews, R. S. Miles and A. D. Walker, eds), pp. 271–288. Academic Press, London and New York.

Bertmar, G. (1969). The vertebrate nose, remarks on its structural and functional adaptation and evolution. *Evolution* **23**, 131–152.

Gardiner, B. G., Janvier, Ph., Patterson, C., Forey, P. L., Greenwood, P. H., Miles, R. S. and Jefferies, R. P. S. (1979). The salmon, the lungfish and the cow: a reply. *Nature, Lond.* **277**, 175–176.

Holmgren, N. (1933). On the origin of the tetrapod limb. *Acta zool., Stockh.* **14**, 185–295.

Jarvik, E. (1942). On the structure of the snout of crossopterygians and lower gnathostomes in general. *Zool. Bidr. Upps.* **21**, 235–675.

Jarvik, E. (1962). Les Porolèpiformes et l'origine des urodèles. *Colloques int. Cent. natn. Rech. Scient.* No. 104, 87–101.

Jarvik, E. (1965). Specialisations in early vertebrates. *Annls Soc. r. zool. Belg.* **94**, 11–95.

Jarvik, E. (1966). Remarks on the structure of the snout in *Megalichthys* and certain other rhipidistid crossopterygians. *Ark. Zool.* **19**, 41–98.

Jarvik, E. (1972). Middle and Upper Devonian Porolepiformes from East Greenland with special reference to *Glyptolepis groenlandica* n.sp. *Meddr Grønland* **187** (2), 1–307.

Jessen, H. (1966). Die Crossopterygier des Oberon Plattenkalkes (Devon) der Bergisch–Gladbach–Paffrather Mulde (Rheinisches Schiefergebirge) unter Berücksichtigung von amerikanischem und europäischem *Onychodus*-Material. *Ark. Zool.* (2) **18**, 305–389.

Kerr, J. G. (1932). Archaic fishes – *Lepidosiren, Protopterus, Polypterus* – and their bearing upon problems of vertebrate morphology. *Jena. Z. Naturw.* **67**, 419–433.

Kesteven, H. L. (1931). Contributions to the cranial osteology of the fishes. The skull of *Neoceratodus fosteri*: a study in phylogeny. *Rec. Aust. Mus.* **18**, 236–265.

Kesteven, H. L. (1951). The origin of the tetrapods. *Proc. R. Soc. Vict.* **59**, 93–139.

Lehman, J-P. (1966). Les Actinoptérygiens, crossoptérygiens, dipneustes. *In* "Traité de Paléontologie" (J. Piveteau, ed.), Vol. 4, (3), 442 pp. Masson, Paris.

Løvtrop, S. (1977). "The Phylogeny of Vertebrata." Wiley, London.

Miles, R. S. (1977). Dipnoan (lungfish) skulls and the relationships of the group: a study based on new species from the Devonian of Australia. *Zool. J. Linn. Soc.* **61**, 1–328.

Panchen, A. L. (1967). The nostrils of choanate fishes and early tetrapods. *Biol. Rev.* **42**, 374–420.

Panchen, A. L. (1977). The origin and early evolution of tetrapod vertebrae. *In* "Problems in Vertebrate Evolution" (S. M. Andrews, R. S. Miles and A. D. Walker, eds), pp. 289–318. Academic Press, London and New York.

Säve-Söderbergh, G. (1934). Some points of view concerning the evolution of the vertebrates and the classification of this group. *Ark. Zool.* **26** (17), 1–20.

Schaeffer, B. (1967). Osteichthyan vertebrae. *Zool. J. Linn. Soc.* **47**, 185–195.

Schultze, H-P. (1970). Folded teeth and the monophyletic origin of tetrapods. *Am. Mus. Novit.* No. 2408, 1–10.

Thomson, K. S. and Campbell, K. S. W. (1971). The structure and relationships of the primitive Devonian lungfish – *Dipnorhynchus sussmilchi* (Etheridge). *Bull. Peabody Mus. nat. Hist.* **38**, 1–109.

Vorobjeva, E. I. (1960). On the crossopterygian fish *Panderichthys* from the U.S.S.R. *Paleont. Zh.* 1960, No. 1, 87–96. (In Russian).

Vorobjeva, E. I. (1973). Einige Besonderheiten im Schädelblau von *Panderichthys rhombolepis* (Gross) (Pisces, Crossopterygii). *Palaeontographica* **143A**, 221–229.

Vorobjeva, E. I. (1977). Morphology and features of evolution of crossopterygian fishes. *Trudy̅ Paleont. Inst.* **163**, 1–239. (In Russian.)

Watson, D. M. S. (1926). Croonian Lecture: The evolution and origin of the Amphibia. *Phil. Trans. R. Soc.* **B214**, 189–257.

Westoll, T. S. (1938). Ancestry of the tetrapods. *Nature, Lond.* **141**, 127–128.

Westoll, T. S. (1943). The origin of the tetrapods. *Biol. Rev.* **18**, 78–98.

9 | The Ecology of Devonian Lobe-finned Fishes

KEITH STEWART THOMSON

Department of Biology and Peabody Museum of Natural History, Yale University, New Haven, Connecticut, U.S.A.

Abstract: Contrary to the commonly held view, the evidence of environments of deposition and the patterns of distribution of Devonian lobe-finned fishes do not indicate that any of the major groups were primary freshwater fishes. Possibly the Eusthenopteridae show the most continental of habitats, but all groups of lungfishes, coelacanths and rhipidistians demonstrate a very strong and consistent connection with near-shore marine conditions. This view is entirely consistent with revised considerations of the evidence from patterns of ionic and osmotic regulation in fishes. Urea synthesis via the ornithine–urea cycle and osmotic regulation via urea retention are to be seen as ancient gnathostome adaptations and the presence of such mechanisms in all lobe-finned fishes makes possible the wide marine as well as freshwater distribution of the group. The sparse available evidence concerning Devonian amphibians shows the same pattern in which migration across salt water barriers was possible.

The fundamental adaptations of Rhipidistia *sensu lato* are all those of purely aquatic animals and terrestrial locomotion in these forms was negligible. The basic pattern of the fore and hind limbs of tetrapods is foreshadowed in the osteolepid rhipidistians (the closest lobe-finned family to the tetrapods) by virtue of functional adaptations of the paired fins for life in shallow water.

The first tetrapods evolved in continuously moist, well-vegetated environments, probably in coastal lowland estuaries where abundant fish food was available in the water. Feeding on land was a post-Devonian development. The first amphibians were highly dependent upon permanent waters and had a moist skin.

Systematics Association Special Volume No. 15, "The Terrestrial Environment and the Origin of Land Vertebrates", edited by A. L. Panchen, 1980, pp. 187–222, Academic Press, London and New York.

On the basis of present knowledge and particularly the absence of knowledge concerning Devonian Gondwanaland, it is not possible to determine the geographical regions in which any lobe-finned group or the first tetrapods evolved.

INTRODUCTION

"The early amphibians . . . [were] adaptations from fluviatile fishes . . . The nature of the geologic record of amphibians indicates that they evolved under conditions marked by seasonal dryness and inhabited river plains far from the sea." This statement, which could probably stand unmodified in any modern textbook of zoology, in fact is taken (by combining two separate sentences) from the classic paper of Barrel (1916). Probably more than any other, this work has set the pattern for all modern treatments of the ecology of the fish–amphibian transition. Barrel analysed the distribution of fossil fish remains of the "Siluro-Devonian" and concluded that all vertebrates had arisen in fresh waters rather than the seas, and that amphibians arose in warm, seasonally arid climates in which both locomotion on land and lung-breathing are necessary adaptations for survival.

Since 1916 the question of the origin of the tetrapods has been refined in the phylogenetic sense and virtually all authorities now agree that tetrapods arose from crossopterygian fishes usually called Rhipidistia, and that the transition occurred no earlier than the beginning of the Middle Devonian. The most recent analyses place the family Osteolepidae as the closest known relatives of the tetrapods. Questions of evolutionary morphology have been attacked boldly since 1916, although many real problems remain. Despite the many structural similarities between osteolepid crossopterygians and tetrapods, there are many fundamental differences that cannot easily be explained in terms of the gradual "unfolding" of a rhipidistian type.

Perhaps the field of enquiry that had advanced least since Barrel's time is, however, that of the "biology" of the lobe-finned fishes and first tetrapods. What sorts of animals were these? How and where did they live? What were their special physiological adaptations, ecological requirements, behavioural specializations; even, what were their most simple feeding and locomotor patterns? In order to come to explain the origin of tetrapods we must have detailed analyses of the fossil record in all these areas and more. And yet, of course, the fossil record is notoriously incomplete and we are often forced or seduced into the dangerous ground of extrapolation and argument by analogy. The four genera of living lobe-finned fishes play an extremely prominent part in these discussions. But here again, the more firmly we try to grasp the problem, the more it slips away from us. Our knowledge of the biology of the living lungfishes is disgracefully incomplete, that of the coelacanth *Latimeria* more understandably so. Probably the most neglected and most interesting living freshwater fish is the Australian lungfish *Neoceratodus*. The reader searches in vain for a comprehensive account of the diet of this fish, or of the oxygen balance in the waters in which it lives. In short, there is a very great amount of work remaining to be tackled before we can understand the biology of living and fossil lobe-finned fishes.

In 1969 I wrote a lengthy review "The Biology of Lobe-finned Fishes" in which I surveyed the ecology, physiology and functional morphology of all lobe-fins, from the early Devonian to the present. I will not attempt here to repeat the previous work. Since 1969 there has been a major increase in our knowledge of *Latimeria*, but relatively little advance with respect to fossil lobe-fins except in two areas that will be discussed in more detail below. The aim of the present study is to bring the general biological picture up to date and to concentrate for the first time upon the questions of the environments in which Devonian lobe-finned fishes lived, their biogeography and distribution, and to arrive at a new understanding of the environmental/physiological context in which lobe-finned fishes and the first tetrapods evolved.

As previously, I will use the familiar terms: Dipnoi, Crossopterygii, Coelacanthini and Rhipidistia for the major groups considered, without pre-judging the monophyletic reality of the last-named and without assuming any particular pattern of interrelationships.

THE BIOLOGY OF LOBE-FINNED FISHES

1. General Biology

The first thing that strikes one after looking at any reconstructions of lobe-finned fishes, is that they are typically and incontrovertibly fishes. Whereas it was possible for the early discoverers and students of modern Australian and South African lungfishes to mistake them for amphibians, when one looks at *Osteolepis, Eusthenopteron* or *Dipterus* one knows that one is dealing with a fish principally adapted for life in the water. Although the paired fin structure of some Rhipidistia is morphologically antecedent to that of tetrapods, the paired fins are not limbs. The fin rays are not reduced in number or length from what one might expect of a "good" fish. The anterior part of the palate in Rhipidistia (and possibly even lungfishes) shows a good choana, but the nasal passageway shows no indication that it was used for the conduction of an air current. There is no functional nasolachrymal duct, although its immediate anatomical precursor has perhaps been identified (Jarvik, 1942). The squamation forms a heavy exoskeleton and the vertebral column is weakly ossified both in the sense of there being a small volume of bone in the notochordal sheath and in that the ossifications present are not arranged so that the backbone could act in the tetrapod girder fashion (see, *inter alia*, Thomson and Bossy, 1971). The caudal fin is always large and, although it may change from a heterocercal to a symmetrical pattern, is always consistent with a fully aquatic locomotion. The lateral line and superficial latero-sensory receptor organs are well developed and there is evidence that electro-receptors are present in the superficial tissues (Thomson, 1976b). In short, these are not fishes that spent any time away from permanently wet conditions. That is not to say that they were always in standing or running water, but they were not adapted to stay out of it for any significant length of time.

On the other hand, it is only fair to point out that much of the above also applies to the first amphibians. A fundamental point, often missed in discussions of tetrapod origins, is that the transition from fish to amphibian was not a full transfer from water to life on land. *Ichthyostega* obviously spent a great deal of time in the water, fed on fish, swam extremely well and reproduced in water. Technically, the principal points of difference are the absence of adult gills and the development of tetrapod paired limbs. But I do not believe, from evidence available, that *Ichthyostega* could live on land anywhere except in conditions of high humidity – in swamps and dense moist vegetation around bodies of water. The real conquering of the land came later. It is an interesting question to what extent the first amphibians were even independent of the water for respiration. My own feeling is that they were dependent upon a moist skin at all times and that possibly the ability of the lungs to eliminate CO_2 was a very late development in amphibians (although certain of the late Palaeozoic and Triassic forms may have been further evolved in this direction than many modern amphibians (cf. Thomson, 1969b; Packard, 1976).

Although the living coelacanth *Latimeria* has recently been shown to be ovoviviparous (Smith *et al.*, 1975; cf. Griffith and Thomson, 1973), conditions in living lungfishes and amphibians suggest that the rhipidistian ancestors of tetrapods were merely oviparous.

2. Feeding and Swimming

The feeding mechanisms of rhipidistian fishes continue to be very unclear. It is now generally agreed that the intracranial joint of the crossopterygian skull was the focus of a major set of intracranial movements and the bare elements of the mechanics of the joint system have now been outlined elsewhere (Thomson, 1967). The intracranial joint of the living coelacanth *Latimeria* is of course a possible model here (see progressively refined accounts in Thomson, 1967; Alexander, 1973; Robineau and Anthony, 1973; Lauder, 1980). The consensus seems to be that in *Latimeria* and probably all coelacanths the intracranial joint was used to produce a type of suction feeding. It is equally clear that, although many of the ligamentous and muscle connections may be identical, the oblique angle of the rhipidistian gape, except possibly in some holotoptychoids, makes a suction mechanism impossible. We are then left with the interpretation that the joint in rhipidistians increases the range of possible orientation of the gape, that some mechanical advantage may be imparted to the bite and that both of these might be significant for fishes living in shallow waters, where the whole body cannot be oriented to allow an approach to the prey head on or from slightly below (Thomson, 1967, 1969a). I have also sought to show that the intracranial biomechanics change with relative elongation of the snout and that such an elongation is probably associated with increased absolute size (Thomson, 1967, 1968, 1969a). A series of morphological stages (not yet a phylogenetic series) can be assembled between Middle Devonian osteolepids and the first Amphibia in terms of elongation of the anterior parts of the skull and progressive reduction of the angle through which the joint is moved, ending in the amphibian condition in which the joint itself is apparently fused over. This mechanical interpretation has important bearing on the old but now essentially settled question of the homology of parietals and postparietals in the fish–tetrapod transition (e.g. Westoll, 1943).

The diet of rhipidistians must have varied considerably according to their absolute size. But all were fishes that either pursued their prey more or less actively or that lay in wait and then took their prey in a short lunge. The models of modern salmon and pike seem appropriate. Studies of *Latimeria* (Griffith and Thomson, 1972; Locket and Griffith, 1972) now suggest that in all lobe-finned fishes the unique structure of the second dorsal and the anal fins is associated with a pattern of slow swimming in which forward progression is the result of lateral sculling of these fins while the fish holds station or creeps up on its prey. Other details of crossopterygian body shape, position of fin insertion and swimming patterns have been discussed elsewhere (Thomson and Hahn, 1967; Thomson, 1969a). An interesting fact that has emerged from study of *Latimeria* (partial review in Millot *et al.*, 1978) is that this fish lacked the basic metabolic mechanisms for continued active swimming. This is most probably a particular specialization of certain coelacanths, rather than a common feature of all crossopterygians.

A great deal of information has been made available recently concerning both the detailed structure and the function of the paired fins in lobe-finned fishes. The structure of the paired fins and girdles of rhipidistians has been well discussed by Andrews and Westoll (1970a, 1970b) and Rackoff (1976) (see also Thomson and Rackoff, 1974; Rackoff, Chapter 11 of this volume). However, it must be noted that we still lack important details concerning the internal structure of the paired fins of "porolepiform" rhipidistians – a serious gap in our knowledge of early fishes as a whole. The work of Rackoff shows that in internal structure the Upper Devonian osteolepid *Sterropterygion brandei* (Thomson, 1972) is closer to that of tetrapods even than the previously well described *Eusthenopteron foordi*. The principal points of close agreement between this osteolepid and the tetrapods are in the orientation of the scapulocoracoid (see also Thomson and Rackoff, 1974; Janvier, 1978), the shape and torsion of the humerus and details of muscular arrangements.

Perhaps more interesting than the strictly anatomical comparisons in Rackoff's work on the osteolepid paired limbs, however, are his studies of the mechanics of the limbs. Rackoff is able to show clearly that the preaxial edge of the pectoral fin in *Sterropterygion* was dorsal (cf. Romer and Byrne, 1931) but that of the pelvic fin was ventral. This fish had the tetrapod pattern of a rotatory "elbow" joint in the fore limb and a hinge joint at the "wrist", while in the hind limb the hinge joint is at the "knee" and the rotatory joint is at the "ankle". These important differences between the fore and hind limbs of all tetrapods are explainable in terms of the functions of their fin-precursors in the crossopterygian fishes, and are confirmed even by motion picture analysis of a living *Latimeria* (Locket and Griffith, 1972; Rackoff, 1976). The fore and hind paired fins in rhipidistian fishes had quite different functions. The pectoral fins were predominantly used in weight support and in contributing to the posture of the head and anterior trunk. When the fish was in relatively deep water, potential prey could be attacked from behind, above, or below, by means of the fish taking a suitable line of approach, plus the unique flexibility of the intracranial joint system. However, when the fish was lying upon a solid substrate under water and more especially when it was lying only partially submerged in shallow water, the pectoral fins were essential to prop up the whole front part of the body so that the mouth could be opened fully and the gape directed more variably in the sagittal plane. Furthermore, when the fish was only partially submerged it became for the first time "heavy" and its weight tended to compress the trunk and make lung-breathing more difficult, just when it was most needed. Support of the weight of the trunk was thus essential to lung-breathing and it is not impossible that up-and-down movement of the trunk, produced by limb movements, could have aided in ventilation. Finally, once the fish came to feel the effects of gravity, propping up the front end was most useful in locomotion.

The major propulsive organ of the rhipidistian fish (and also the earliest tetrapods) was the tail. The effect of the tail pushing the body forward was facilitated if the front of the body were propped up, with the fore limbs not acting as agents for propulsion, but more as stilts. In developing a foreward thrust against the substrate, both in semi-terrestrial and underwater locomotion, the pectoral limbs were not as useful as the pelvic fins. Movements of the pelvic fins, in concert with the strong lateral flexure of the posterior part of the body, would have been important both in providing the maximum grip with which to push against the substrate, and in actively contributing to the thrust. The pelvic fins/limbs were not needed in weight support. At this stage in the development of terrestrial locomotion it was not necessary or even desirable to get the whole body up off the substrate, because the tail still provided the major force.

Thus it is possible to reconstruct the early stages of terrestrial locomotion in a manner that explains the different positions of hinge and rotatory joints in the fore and hind limbs in both tetrapods and fishes, and also to explain the differential development of the fore and hind limbs in the fishes (the fore limbs being far more "advanced" towards the tetrapod condition), in terms of quite different initial functions for the two sets of paired limbs.

The evidence from *Sterropterygion*, from the shoulder girdles of other osteolepids and also from *Eusthenopteron*, shows that these crossopterygian fishes could readily move through shallow waters and across wet swamps and marshes as, indeed, can many living fishes. In terms of the locomotor apparatus, therefore, these fishes are clearly prototetrapods. However, the fact that the fin rays are unreduced in size and number, and that the distal parts of the pentadactyl limb are still undeveloped, forces us to conclude that these were largely the adaptations of fishes, rather than land animals. They must be viewed, therefore, in terms of having specific advantage for life in and around the water, rather than for life on land. For an interesting discussion of the subsequent evolution of the tetrapod condition in terms of limb structure and gait, see Edwards (1977).

I would like to suggest an additional selective factor acting upon the prototetrapod limb. If it is possible (indeed likely) that many rhipidistian fishes bred in shallow freshwater pools somewhat away from the main bodies of permanent water, the ability to move and breathe in foul water not deep enough to cover the fish becomes of special importance. One can readily reconstruct that many aspects of specialized behaviour associated with reproduction, from finding a mate to defending a nest site and young, required particular patterns of posture and movement that added to the more general adaptive context.

3. Sensory Biology

Very little has been added recently to our knowledge of this subject. The suggestion that the pore-canal system of early vertebrates was a system for electroreception (Thomson, 1976b) while of great interest in general vertebrate sensory biology, says very little about the specifics of rhipidistian biology, except that it would have added to their abilities as predators, particularly in murky waters. It is worth noting, however, that basic principles would lead us to expect a significant difference in the length of the pore-canal itself (leading from the surface to the receptor cells) between the exclusively salt-water and freshwater fishes. That this has not yet been observed in lobe-finned fishes may indicate that the fishes which have been looked at were adapted for a broad range of environmental conditions. It would be extremely interesting if it should turn out that the peculiar rostral organ of coelacanths is an electrosensory organ (see McAllister, 1980).

4. Osmoregulation

Evidence is developed below that many fossil lobe-finned fishes lived in salt water and/or are distributed without apparent bar across the saltwater barriers separating Devonian land masses. Such a reconstruction obviously requires a physiological explanation. In a recent study (Thomson, 1979) I attempted to summarize the available data on ionic composition and patterns of osmotic and ionic regulation in early fishes, with the aim of shedding some new light upon the old question of the original environment of the craniates. In this study, somewhat to my initial surprise, I concluded that there was no firm geological or physiological evidence for a long enough period of purely freshwater environments for early craniates, early gnathostomes or early osteichthyans, that would account for the reduced ionic concentration of the blood of these fishes in comparison with that of sea water or the blood of the hagfish *Myxine*. Therefore the traditional explanation that such lower ionic concentrations are an adaptation to freshwater existence is thrown into question and, even more significantly, the point of origin of the two major patterns of osmotic regulation under stress among fishes – urea synthesis and retention (chondrichthyans, coelacanths, lungfishes) – and swallowing of salt water with preferential ion excretion (actinopterygians) requires re-thinking. My conclusions are that ionic concentrations were reduced in all lines of early fishes as a result of an adaptation of the earliest vertebrate stock (excluding the craniate hagfishes and their allies as more primitive in this respect) having to do fundamentally with cell physiological processes, and that this occurred when this stock was still living in the sea. The adaptation of urea synthesis via the ornithine–urea cycle and its retention for osmotic balance was concomitantly added to selective ion excretion as a fundamental adaptation of this early marine vertebrate stock. It is retained as a primitive condition in chondrichthyan fishes (Pang *et al.*, 1977), in lungfishes (Goldstein and Forster, 1970), in the coelacanths (Pickford and Grant, 1967) and in tetrapods. Thus the tetrapod pattern of urea production, which is essential for life on land, represents a direct continuation of an adaptation that must have existed in all lobe-finned fishes (Thomson, 1971). The loss of the urea synthesis and retention mechanisms in actinopterygians is, however, probably an adaptive response to freshwater conditions, but it probably occurred well after the first radiations of the Chondrostei had evolved.

This study of osmoregulatory patterns was completed before I was invited to prepare the present synthesis and it is a matter of some gratification to see that not only do the described patterns of marine habit and distributions of lobe-finned fishes (see below) require such an explanation, but also that the emerging conclusion that early lobe-finned fish groups were marine in habit (see also Thomson, 1969b; Vorobjeva, 1975) provides confirmatory evidence for the general thesis of an absence of major early freshwater radiations of the bony fishes. For our present purposes the essential point to be made is that the physiological adaptation of urea synthesis and retention, which is common to lobe-finned fishes and tetrapods, is primitively a marine adaptation, nicely constituting a pre-adaptation for life on land (both sea and land being for bony fishes a physiological desert) but one that, while it allows lobe-finned fishes to exploit both salt and freshwater conditions, equally and somewhat counter-intuitively, would have allowed the first tetrapods to live in the sea. Whether they did or not is an entirely different question.

5. Air-breathing

A closely related question in the environmental physiology of lobe-finned fishes is the origin of air breathing. An interesting debate has arisen in the recent literature concerning the environments in which lung-breathing arose. The traditionally held view is Barrel's (1916), that air-breathing arose in tropical fresh waters. It is buttressed by observations of the distribution of air-breathing in modern fishes (heavily in favour of tropical freshwater forms; Carter, 1957) and the physiology of modern lungfishes (review in Thomson, 1969a). But if modern interpretations of the palaeontological evidence and analysis of osmotic patterns among all fishes tend to show that gnathostomes and osteichthyans evolved in marine conditions (see above), this then leaves the proposition of a freshwater context for the evolution of the lung in some doubt.

Packard (1974, 1976) has proposed a counter view of the origin of lungs and air-breathing, according to which these adaptations may have arisen in marine conditions, specifically shallow marine lagoons. The argument, in brief, states that under all conditions the solubility of oxygen in sea water is less than in fresh water (up to 25–30% less) and that air-breathing would greatly increase the property known as "metabolic scope". Graham *et al.* (1978) have shown that Packard's argument on metabolic scope is incorrect and they attempt to show that conditions in Silurian seas did not fit Packard's interpretation. Graham *et al.* also note that under similar conditions of temperature the partial presure of oxygen in salt and fresh water may only differ by a small amount and the total deficiency in oxygen uptake might only be 3–11%. They state that "probably this deficiency is too small to establish a selective advantage sufficient to lead to the evolution of air-breathing" (1978, p. 459). However, such a difference is well within the range allowed for (rightly or wrongly) in most microevolutionary schemes of natural selection.

My own researches on this problem lead me to disagree with some of Graham *et al.*'s other arguments. One problem is that it is inappropriate to lean too heavily on analogy from modern fishes. The fact that salmon and eels can "migrate to and from salt water apparently without encountering major limitations in their gill ventilatory capacity, although compensatory changes in hemoglobin–O_2 affinity do take place" is beside the point. The special adaptations of modern anadromous fish do not tell us much about the origin of lungs in the Silurian. The lungfish data are more important and must be taken seriously. But I am inclined to question whether the absence of air-breathing from the majority of modern marine fishes (even the near-shore ones) can be given much weight. Fundamentally, there is no conclusive evidence to show that air-breathing could *not* have evolved in marine conditions, or that air-breathing could *only* have arisen in fresh water. Packard, on the other hand, while perhaps over-stating his case, has done a valuable service in pointing out the fact that if an argument is to be made on the availability of oxygen then the seas are no less a candidate and, fractionally more a candidate, than fresh waters. Useful new information on this question must await an increase in our knowledge of Silurian osteichthyans and careful analysis of the environments of Devonian forms, and at this point it is necessary to turn to the lobe-fins.

ENVIRONMENTS OF DEPOSITION AND DISTRIBUTIONS OF LOBE-FINNED FISHES

The first requirement for a basic ecology of lobe-finned fishes is a firm set of data on the environments in which these fishes and their tetrapod descendants evolved. As noted in the Introduction, all modern views on the subject stem from the work of Barrel (1916) and, following later authorities such as Romer (1955), the notion that amphibians arose in fresh water from freshwater ancestors and that these rhipidistian ancestors also arose in fresh water has become embedded in our thinking. The fact that many fossil lungfishes, the majority of fossil coelacanths, the living coelacanth *Latimeria*, many groups of fossil amphibians and a large proportion of Rhipidistia were (are) wholly marine in their distribution and physiology, has been largely ignored until recently (but see Thomson, 1969b; Vorobjeva, 1975). The consequences of such data for the ecology and evolutionary biology of lobe-finned fishes and amphibians have never been properly explored.

Evidence is available from two sources. First and foremost, one has the evidence of the rocks themselves; the particular environmental context which has been preserved along with the fossil; the evidence as to associated fauna and flora; the ecological and biological evidence concerning the physical environment – salt or freshwater; energy conditions of the water; oxygenation of the water; and so on. Secondly, there is the less direct evidence that can be gained from a consideration of the distribution of the fossil in space and time. To take an obvious example, when the osteolepid genus *Latvius* is found in purely marine back-reef conditions in the Frasnian of Germany, even though it is also found in an intermontane lake basin in the Fammenian of Nova Scotia, we can be fairly sure that *Latvius* is a genus of fish that is physiologically capable of surviving in and migrating through saltwater conditions. But when we find the lungfish *Soederberghia* in two purely freshwater deposits, one in Australia and one in North America, it is necessary to have a very clear notion of the relationship of continental masses during Devonian time in order to discuss the biology of the fish. In practice the data are rarely completely clear cut. As a prelude to treatment of the data, it is necessary to discuss some of the serious problems involved with the data available.

In Barrel's analysis (1916), all occurrences of early vertebrates in near-shore marine conditions had to represent washing of remains *post mortem* into the ocean from the coastal-plain rivers in which they really lived. Such *a priori* reasoning remains a problem, particularly when it comes to recording single occurrences of vertebrate fossils. As an example of the tension between geological evidence and interpretation we might take Denison's (1951) study of late Devonian "freshwater fishes" of the western United States. In this work, Denison, who along with the late Walter Gross has done more than anyone in sorting out the environmental context of early fossil fishes, states that the majority of the fossils he described "are found in lower near-shore transgressive facies . . . as far as the geological evidence goes, all could be littoral marine fishes . . . " Denison, (1951, p. 255). But the conclusion remains that of these antiarchs, lungfishes and crossopterygians "most . . . were inhabitants of freshwater streams and were washed after death into the edges of the advancing Chemung sea" (Denison, 1951, p. 266). One point of evidence that is usually used in corroboration of such interpretations is the abraded and fragmentary nature of the fossils, the mechanical trauma of which is used as evidence of transport from the coastal rivers. However, all pelagic and littoral marine fish remains (as opposed to the remains of sessile, burrowing or otherwise protected invertebrates) would be likely to show such mechanical comminution, as may be determined by any walk along any beach. Further, freshwater fish remains, whole or fragmented, are extraordinarily rare on modern ocean beaches.

The point of view will be taken in this paper that the conditions of the environment of deposition *must* be taken first of all at their face value. If a fossil is found in marine conditions then we should consider very seriously the possibility that it represents a group that could live in salt water, before proceeding to the alternative hypothesis that it really came from somewhere else. In practice there are indeed some "intermediate" sort of deposits where it is impossible to make an objective "face value" determination of the environment of deposition. But these are relatively rare.

In this study I propose to discriminate between three simple categories of environment of deposition for lobe-finned fishes. Following Allen *et al.* (1967), among others, we can divide continental environments into two classes, *internal* and *external. Intermontane basins* are essentially landlocked with a pattern of internal drainage. These large or small basins, of which the best examples might be the Escuminac Bay deposits of Quebec (Frasnian) and the Old Red Sandstone deposits of Scotland (Middle Devonian), never show marine invertebrate fossils or any physical evidence of marine conditions. However, they do (particularly the Scottish Old Red Sandstone) occasionally show evidence of hypersalinity. For discussion of the geological context of such intermontane basinal deposition, see Dineley and Williams (1967), Friend (1969) and Donovan (1973), among many others. The *extramontane* deposits of the Devonian are exemplified by the coastal-plain deposits of the Catskill of North America (see Allen and Friend, 1968; Burtner, 1963). The occurrences of fish fossils are less abundant and much more localized than in the case of the lacustrine internal deposits. But still there is an absence of wholly marine invertebrates, and the pattern of physical evidence suggests large rivers, braided streams and a variety of flood-plain conditions. Judging from the great size of some of the rhipidistian fishes found in these coastal-plain deposits, the rivers must in certain cases have been very large.

There is room for a major difference of opinion as to whether the general continental environments in which lobe-finned fishes lived in the Devonian were arid, semi-arid (seasonal) or continuously humid. In older views (again starting with Barrel – see discussion in

Inger, 1957 and Thomson, 1969a) an unnecessary concentration was given to arid conditions. In any case, I think it is now more generally agreed that, in terms of the fishes and amphibians with which we are concerned here, the permanent aquatic ecosystems and surrounding land vegetation provided a series of habitats in which conditions were permanently humid, although seasonal aridity changed the overall context and even the spatial distribution of such humid vegetated zones. The first land vertebrates did not crawl out of the water and live on sun-baked mudflats. They must have lived in swamps and marshes, near water, and in the shade provided by the newly evolving canopy of later Devonian land plants.

The third category of environmental depositional conditions in which lobe-finned fishes are found is purely *marine*. In fact, this means mostly near-shore deposits, within the continental shelf. Fossils have been found in a variety of marine shallow-water conditions, ranging at the extreme to complex reef and lagoonal conditions where relatively undisturbed conditions have led in certain cases (Bergish-Gladbach, Germany; Gogo, Western Australia) to well-preserved intact materials in which fine details of skeletal structure have been retained. The more common marine deposit is from a higher energy situation and here we find only fragmented and abraded specimens.

The raw data on the environments of deposition offer relatively crude evidence for a reconstruction of lobe-finned fish ecology and physiology. The data can be refined if we interpret them in their biogeographic context. For the purposes of this paper, I will use the convenient maps of continental positions constructed by Smith *et al.* (1973) with small modifications based on the reconstructions of Zeigler *et al.* (1977), House (1973) and Cocks and McKerrow (1973) (Figs 1 and 2) despite their evident inadequacy (see Tarling, Chapter 2 of this volume). However, not only is the general relationship of the major crustal blocks important to our study, the particularly relevant points are the extent of the continental shelves and the extent of the emergent landmasses. Fundamentally, we need to know both the areas in which our three categories of environment are distributed and also their potential interconnections. All current models allow considerable movement of landmasses and shallow epicontinental marine zones during the Devonian, but certain basic features persisted. Western Gondwanaland remained separate from

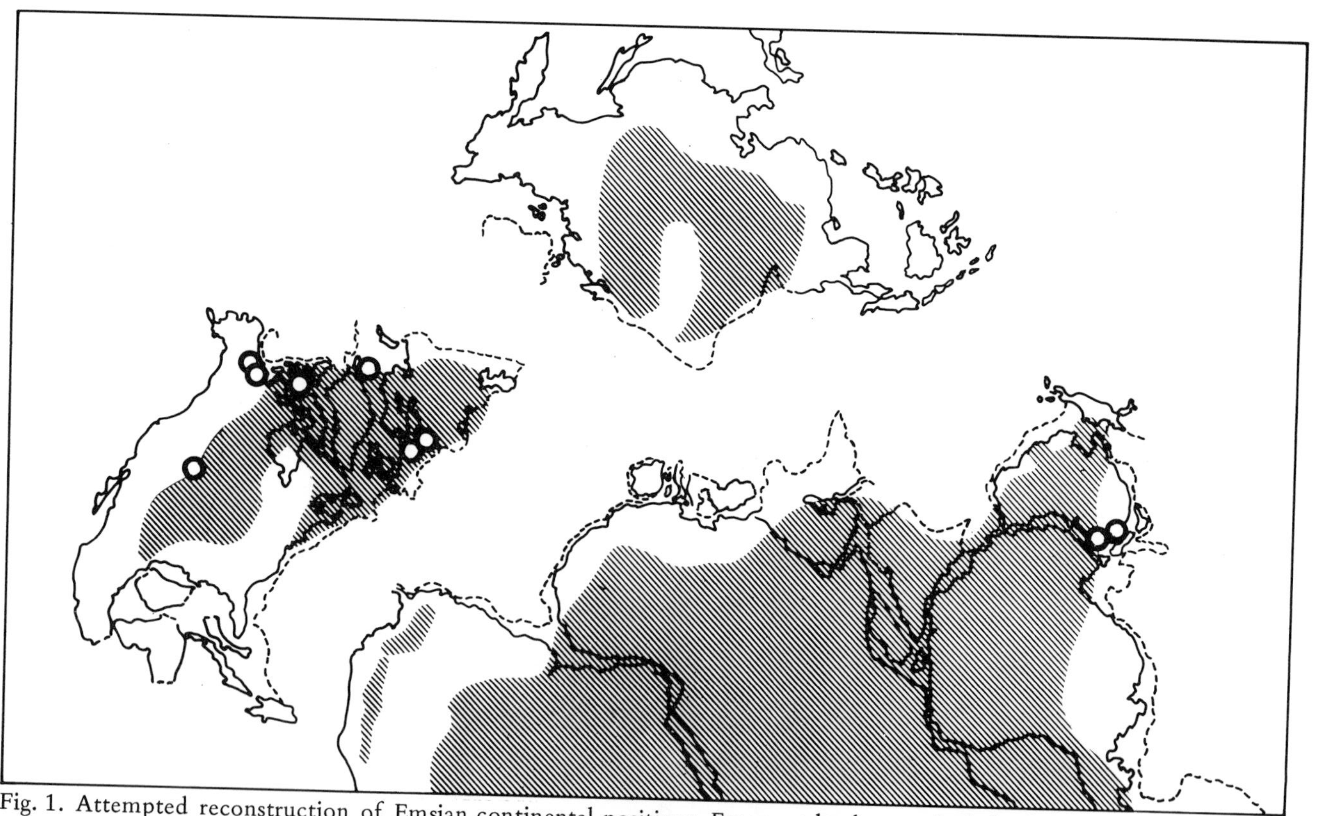

Fig. 1. Attempted reconstruction of Emsian continental positions. Emergent land areas shaded. The circles mark the ocurrences of Lower Devonian to Emsian lungfishes and porolepiforms. Only the Spitzbergen record is definitely freshwater in environment.

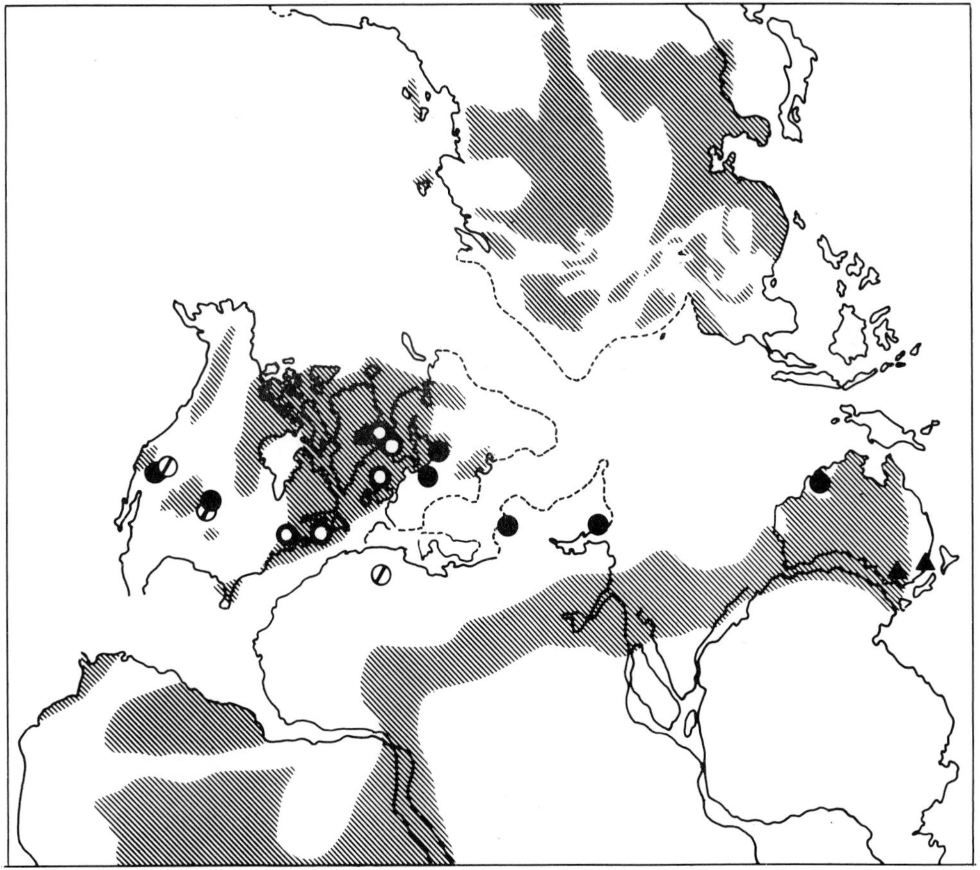

Fig. 2. Attempted reconstruction of Frasnian continental positions. Emergent land areas shaded. The open circles mark sites from which Middle Devonian freshwater osteolepids have been recovered. The closed circles mark all Devonian marine occurrences. The open circles with a bar mark Frasnian marine eusthenopterid occurrences. The closed triangles mark Upper Devonian tetrapod occurrences.

Europe–North America in the "Atlantic" region until the early Carboniferous, at the very least, and there was no land connection here between this Eur-American unit and South America or Africa by which continental vertebrates physiologically incapable of surviving in salt water could have migrated. Such an "African" land connection did not, in the model used here, develop until the later Carboniferous. Eur-America also remained separate from Asia during the Devonian, again with no dry land connections. On the other hand, there was dry land connection (the Old Red Landmass) of North America, Greenland and western Europe through all this period.

While these palaeobiogeographical models seem to be becoming well established and accepted, there remain some serious lacunae in our knowledge, particularly of the nature of the relationship between Eur-America and Eastern Gondwanaland (Australia and Antarctica). It is possible (see Tarling, Chapter 2 of this volume) that Australia and Antarctica were much closer to eastern Eur-America than is shown in Figs 1 or 2.

It is my conviction that biogeographical evidence can never be used to establish a case in palaeogeography, merely to frame the question. But it has been suggested that the patterns of distribution of the Devonian fishes *Bothriolepis, Remigolepis, Soederberghia* and *Groenlandaspis* together with the Devonian tetrapods (in Australia, Antarctica, Greenland and eastern Canada) pose a biogeographical problem for the sort of geography depicted in Figs 1 and 2. It must be noted that these distributions are only problematical if the fishes concerned were *exclusively* adapted for freshwater life and incapable of living in salt water. Further, no movement of continental blocks can ensure a continuity of *land* connections, and the distribution of seas and emergent land areas in the Devonian seems to preclude such land-bridges between Eur-America and Australia, however closely the blocks were associated. Nevertheless, as discussed below, the geographical context is important for understanding the biology of Devonian fishes and amphibians, and therefore all possibilities must be kept in mind.

1. *Analysis of Depositional and Biogeographical Data*

In order to analyse the data on the environments and distribution of fossil lobe-finned fishes of the Devonian, it is convenient to set up a series of models against which particular patterns of evidence can be weighed. The models specify certain requirements of evidence and can be applied at any taxonomic level. It will be seen that full interpretation requires phylogenetic data, and that ingredient is added in the discussion that follows. The models are set up to be as discrete as possible but, again, where conflicting results might obtain, phylogenetic discussion is necessary to resolve the question.

Marine model: variant I. The group under consideration arose in marine conditions and radiated only in shallow marine conditions. Requirement: fossils only be found in marine conditions. Prediction: widely distributed through shallow seas.

Marine model: variant II. The group arose in wholly marine conditions but part of the diversification of the group includes species that invaded the extramontane freshwater environment. Requirement: fossils found only in marine and extramontane environments. Prediction: widely distributed again via shallow seas.

Marine model: variant III. The group arose in marine conditions but proceeded through marine model II to the successful invasion by certain members of the intermontane environments. Requirement: fossils in all three environments. Prediction: the freshwater forms narrowly and the marine forms widely distributed.

Freshwater model: variant I. The group arose in fresh water and is restricted to fresh water. Requirements: no marine fossils. Prediction: fossils will be found only in the two continental environments, each group is restricted to a given landmass. Coast-wise changes in distributional pattern are possible through stream capture.

Freshwater model: variant II. From a freshwater origin, tolerance of marine conditions evolved in certain taxa and distribution is possible through marine conditions. All fishes nevertheless lived the vast majority of their lives in freshwater conditions. In this case the fossil data would be the same as for freshwater model I, but a few marine occurrences might be found. Prediction: distribution around a given landmass would be wide, distribution between widely separated landmasses would be unlikely, but not impossible.

Putting the matter very simply, if the lobe-finned fishes were principally marine groups that later invaded fresh water, we would expect marine forms to be numerically more abundant and more widely dispersed than extramontane forms, and extramontane forms more so than intermontane forms. On the other hand, given the limits of intracontinental distributions, if the lobe-finned fishes were principally freshwater forms, we would expect roughly the reverse pattern. There will, however, be a stage at which, if the data are sufficiently incomplete, it will not be possible to distinguish the two models, I leave the reader to decide whether the data are better than this or not.

The picture is made the more complicated if we consider the possibility of anadromy.

Anadromous model I. From a marine origin, the group consists of taxa, or includes some taxa, that are anadromous, living most of their life at sea in the open ocean where preservation is very unlikely but entering coastal and fresh waters to breed and here becoming preserved as fossils. In this case, a very wide distribution of fossils is possible, with preservation in extramontane deposits being most likely and in marine deposits less likely. I would contend that it is not possible to establish the anadromous model with certainty on the basis of environmental and distributional data alone. Some further evidence relating more directly to physiology would be needed and such evidence is possibly available for certain taxa (see note below).

For the purposes of applying the above models to the data on lobe-finned fishes in the Devonian, it is most convenient to work with the genus as the lowest taxonomic level. Species-level systematics is very inadequate for most groups and in any case the generic level allows a convenient level of generality. The family is the most convenient higher taxonomic grouping, although, for the lungfishes, the family-level taxonomy of the Devonian forms is still in a state of flux. We will discuss each major group in chronological order, starting with the lungfish.

(a) Dipnoi. The sole early Devonian lungfish *Uranolophus* (western U.S.A.), is probably marine, although Denison (1968) leaves the matter open, tending towards the freshwater context. The early Middle Devonian *Dipnorhynchus* clearly fits marine model I: it is found in purely marine deposits at two very distant points, Germany (Lehman and Westoll, 1952) and Australia (Hills, 1933; Campbell and Thomson, 1971). The environment of the late early Devonian *Melanognathus* (District of McKenzie, Canada) is unclear, but it is definitely not intermontane. The early Eifelian *Stomiakykus*, described by Bernacsek (1977) from the Yukon, Canada, is also marine. Of the other Middle Devonian lungfishes, only *Dipterus*

(world-wide) and *Pentlandia* (Scotland) have been found in purely intermontane conditions. *Dipterus* (although very poorly defined – the genus probably includes about five unrecognized genera) is also found in purely marine and extramontane conditions as well, and this distribution continues into the late Devonian. The other Middle Devonian lungfishes fit marine models I or II, as do *Sunwapta, Griphognathus, Holodipterus, Chirodipterus* and *Grossipterus* from the late Devonian. Gregory *et al.* (1977 and personal communication) have discovered a Givetian/Frasnian fauna in central Nevada which is probably from a near-shore marine environment. Two lungfishes are found, a form close to *Griphognathus* and a "dipterid". In the late Devonian, however, there is also a diversity of genera found only in the intermontane basins of North America and Greenland. Most are (apparently) endemic to single basins. One exception is *Rhynchodipterus*, found in both Europe and Greenland. Only one late Devonian continental lungfish is found more widely distributed than *Rhychodipterus. Soederberghia* is found at Escuminac, in Canada, and also in eastern Australia (Campbell and Bell, 1977); for discussion see below.

The lungfishes as a whole clearly fit marine models I and II and must have arisen from marine ancestors. At some time in the Middle Devonian, lungfishes started invading fresh waters along the lines of marine models II and III. From this, while many genera remained purely marine, intermontane basins were penetrated and here a number of genera with highly restricted distributions evolved. Beyond the Devonian, lungfishes seem to be more completely restricted to continental conditions and thereafter the freshwater model might be thought to apply.

(b) Onychodontiformes. The onychodonts are found predominantly in marine conditions throughout the Devonian and there seems no reason to suppose that there was any stage in their history during which they were restricted to living in fresh water. This entire group therefore fits only marine model I or II.

(c) Rhipidistia. The Rhipidistia present an equally complex picture as the lungfishes. Perhaps the earliest record of a rhipidistian is *Powichthys,* from the early Lower Devonian of Arctic Canada (Jessen, 1975). This curious form, which Jessen includes with the porolepi-

forms although it also has certain features reminiscent of lungfishes or brachopterygians, occurs in definitely marine conditions. Another marine, early Devonian porolepiform from Arctic Canada is known but not yet described (Dineley and Bernacsek, personal communication). The only other definitely early Devonian porolepiform is *Porolepis* itself, which occurs in both fully marine (Poland) and continental (Spitzbergen) conditions. *Heimenia* comes from the early/Middle Devonian boundary in Spitzbergen (?freshwater) and the *Melanognathus* locality in Arctic Canada (uncertain). The genus *Holoptychius* (Middle and late Devonian) is found principally in extramontane or, less commonly, marine conditions and its distribution especially in the late Devonian is essentially world-wide. It seems to me that *Holoptychius* fits an anadromous model. Most species of *Holoptychius* have the large size and body shape of oceanic fishes and the distribution seems to be the most telling point, together with its very rare occurrence in any intermontane basins. *Glyptolepis* on the other hand, presents a more difficult situation because no incontrovertibly marine species are known. There are several possibilities. *Glyptolepis* could have arisen from a single anadromous or marine stock and then have become distributed through marine or coastal conditions (marine model III), or the genus could have arisen from a freshwater isolate (derived perhaps from *Holoptychius*) as part of marine model III and then become distributed through freshwater conditions only. These possibilities are so close that one cannot distinguish between them. However, there is a third possibility. Given the extreme closeness of *Holoptychius* and *Glyptolepis,* the latter might in fact have arisen several times in parallel from *Holoptychius* stocks being confined to continental distributions. All of these take us beyond the limits of our analysis. The principal point remains that all porolepiforms have a very close connection with marine conditions and their distributions are probably best explained by variations of the marine model, including origin from marine ancestors.

"*Osteolepiforms*". The Osteolepidae are apparently first represented by an undescribed early Devonian record from China, the environment for which is not available. In the Middle Devonian, the earliest record may be a new discovery in eastern Australia of osteolepid scales in a (probably extramontane) continental deposit of early Eifelian age (G. C. Young, personal communication). After this the family is represented by several genera all of which occur in intermontane or extramontane basins (*Osteolepis, Thursius,* in Scotland; *Canningius* in Greenland; *Gyroptychius* in Scotland, Greenland and Norway). However, definite marine records of Middle Devonian age also occur – for example the material from Iran described by Janvier (1978) – and fragmentary osteolepid material is in fact available world-wide in a variety of environments. In the late Devonian, the picture is just as complicated. In the Frasnian, osteolepids are known from completely marine deposits, the inter-reef Gogo deposit of Western Australia (Brunton *et al.*, 1967; Gardiner and Miles, 1975) and the back-reef calcareous flagstones of Bergisch-Gladbach (see Ørvig, 1960; Jessen, 1966; Schultze, 1969; Jux, 1964; and see above for lungfishes), and from central Nevada (Gregory *et al.*, 1977). The genus *Latvius*, which is found in the Bergish-Gladbach locality, is also found in the Baltic Frasnian (Kokenhusen: marine) but is, however, also the only rhipidistian present in the intermontane Albert Formation (Fammenian) of New Brunswick, Canada (Greiner, 1977). Other late Devonian osteolepid records of *Glyptopomus, Gyroptychius, Panderichthys, Megistolepis, Thaumatolepis* and *Sterropterygion* seem all to be from extramontane conditions. But fragmentary Upper Devonian osteolepid remains are found in a variety of shallow marine environments from Turkey (Janvier, 1978) to the Soviet Union and western North America. The enigmatic form *Elpistostege* Westoll from Escuminac Bay in Canada seems to be definitely a rhipidistian rather than a tetrapod as is shown by the new discoveries by Vorobjeva in the Soviet Union (1975, 1977). It is closely related to if not identical with *Panderichthys* (Osteolepidae *sensu lato*).

The family Osteolepidae extends also into the Carboniferous and Permian in the form of *Megalichthys, Ectosteorhachis* and *Lohsania*: these occurrences are largely, but not exclusively, continental. (It will be noted that Janvier *et al.* (1978) have postulated the derivation of the *Megalichthys*-type of osteolepid from Devonian shallow-water marine environments of Northern Gondwanaland.)

If one is not overly influenced by the excellent preservation of osteolepid fishes from the Middle Devonian intermontane basins of Scotland and Greenland, one can appreciate that the Osteolepidae are in fact well represented in all types of environment and are extremely widely distributed geographically. The Frasnian records are particularly interesting because it is clear that a quite consistent fauna including marine osteolepids and lungfishes, best represented at Gogo and Bergisch-Gladbach, was extremely widely distributed in shallow marine conditions (for discussion see Gardiner and Miles, 1975). There can be little possibility that this fauna became distributed so widely merely by stream capture across extramontane coastal margins. Distribution must have occurred through the shallow-water marine environment. In addition, the special evidence of cosmine distribution in osteolepid fishes (Thomson, 1975, 1976a; see below) suggests that many of the osteolepid genera were in fact anadromous. I am inclined to think that the family Osteolepidae as a whole fits marine models II and III, better than it fits freshwater model II.

The family Eusthenopteridae appears first in the Scottish Middle Devonian in the form of *Tristicopterus alatus* from Achanarras and the Orkneys (intermontane). In the Givetian/Frasnian fauna from central Nevada mentioned above, Gregory *et al* (1977 and personal communication) I have discovered a very large marine eusthenopterid. In the late Devonian the genus *Eusthenopteron* is widely distributed in the Escuminac Bay locality (intermontane) where *E. foordi* is numerically the most abundant osteichthyan. Outside of Canada, *Eusthenopteron* species are found in the extramontane deposits of Europe. The related genus *Eusthenodon* is found in the intermontane basin of East Greenland and in the extramontane desposits of the Soviet Union (this last transferred to the new genus *Jarvikina* by Vorobjeva, 1977). *Litoptychius bryanti* (described by Denison, 1951) is found in more or less marine conditions, but Denison believes that the remains were transported from an estuarine environment or from nearby freshwater streams. Lehman (1977) has recently described eusthenopterid remains from a marine environment in Morocco. The other eusthenopterid records (for example *Hyneria*, Thomson, 1968; *Platycephalichthys*, e.g. Vorobjeva, 1977) are from extramontane conditions. The genus *Sauripterus* is placed by Andrews and Westoll (1970) in the Rhizodontidae. It is from the extramontane Catskill of Pennsylvania, U.S.A.

In summary, the Eusthenopteridae and the one Devonian rhizodontid are continental in distribution, except for the Moroccan form and possibly *Litoptychius. Eusthenopteron* and *Eusthenodon* are widely distributed within the Old Red Landmass, but no further. No eusthenopterid has been described from Australia, although the enigmatic genus *Canowindra* (described by Thomson, 1973) might be either related to the Holoptychidae or the Rhizondontidae. It is found in a definitely continental deposit. The evidence is equivocal: either the Eusthenopteridae and their close relatives the Rhizodontidae fit a freshwater model, principally variant I, or they fit (from the Nevadan and the Moroccan record) also marine model II.

I greatly regret that at the present time it is not possible to include in the full analysis the exciting discoveries of new rhipidistians from the Soviet Union (Vorobjeva, 1977) but I would hazard the guess that the evidence they offer, while of the greatest importance in study of the adaptive radiations and evolution of the Rhipidistia and of tetrapod origins, will not significantly change the pattern of environment and distribution that has been discussed in the present paper.

(d) Coelacanthini. The Devonian coelacanths are all marine forms; the group as a whole fits marine models I and II.

(e) Tetrapoda. The distribution and environment of deposition of Devonian tetrapods are most interesting. So far there are three definite records of tetrapods. The most famous is the East Greenland occurrence of an uppermost Devonian fauna including at least three taxa of ichthyostegal Amphibia (Säve-Söderbergh, 1932; Jarvik, 1952, 1961). These are preserved in an intermontane basinal setting. The two other records are in Australia. Warren and Wakefield (1972) described an amphibian trackway from Victoria, proposing a (late) Frasnian Age. The setting is definitely continental. Campbell and Bell (1977) described an amphibian lower jaw from the late Frasnian or early Fammenian of New South Wales, also from a continental (possible extramontane) setting. Both Australian records are distinctly older than the Greenland record.

Following the Australian discoveries, Panchen (1977) and Janvier (1978) have suggested that the first Amphibia may have arisen in Gondwanaland. My view is that it is less important to attempt to find the exact centre of origin of the first tetrapods (interesting as that might be) than to analyse the overall pattern of their distribution. And in fact with only two data points (Greenland and Australia), both late Devonian in age, there simply are insufficient data to propose a centre of origin. The most important point about the amphibian *distribution* is that it points up once again the physiological and biogeographical problems that have already been encountered with the lobe-finned fishes. If Australia were separate from Eur-America in the Middle and late Devonian, even though all known records are from continental deposits, we must assume that the first Amphibia, like so many of their lobe-finned cousins, could migrate freely across shallow saltwater barriers. The only alternative is a dry land connection between Australia and Greenland.

2. Note on the Population of Intermontane Basins and the Old Red Landmass

One result of the present study is to show that the wide distribution of many if not all lobe-finned fishes can be explained in terms of migration across salt water barriers. It is not possible currently to analyse the full composition of all faunas in which lobe-finned fishes are found, although the simple analysis of Gardiner and Miles (1975) of the widespread marine Frasnian faunas including osteolepids and lungfishes (see also Gregory *et al.*, 1977) is an indicator of what may eventually be possible. However, we are in a position to analyse the varying composition of intermontane basinal faunas. Under the old assumption that lobe-finned fishes were primarily freshwater forms, it was difficult to explain the fact that Middle Devonian intermontane basins contain largely osteolepids and lungfishes bearing a full cosmine cover, while Frasnian basins largely lack such forms, except by involving a phylogenetic shift (see Thomson, 1976a). With the evidence that the lobe-finned fishes are primarily marine, we can see the intermontane basinal faunas as having developed, not from one another but independently, by multiple invasions from marine and extramontane faunas. Thus the broad difference among basins, as well as certain points in common, can be explained in terms of the changes of colonization of basins from without. This is the explanation of the presence of *Latvius* (if correctly identified) in the intermontane Albert Formation of New Brunswick while elsewhere it is found only in strictly marine lagoonal conditions.

3. Note on the Problem of Anadromy

It is of course a caricature merely to distinguish marine and freshwater patterns of overall biology for fishes. We know from living fishes that a very large number of groups show a tolerance of both salt and freshwater conditions. The anadromous fishes such as salmoniforms are a good example of the way in which the categories can merge. However, even here there are some important distinctions.

(1) It is always easier for a marine fish to enter fresh water than for a freshwater group long adapted to fresh water to enter salt water. The reason lies in the overall osmolarity of the blood in all gnathostomes, the osmotic pressure of which is very far below that of sea water. The constant evolutionary tendency is therefore for saltwater forms to colonize fresh water rather than the reverse.

(2) Given this, it is probable that some form of anadromy is a common intermediate stage in this process, in which fishes come to enter fresh water to breed (because of a more protected ecology, because of lower stress, in order to take advantage of seasonal food abundances for the young fishes) while living the majority of their lives at sea (in order to take advantage of more equable environments and a greater food abundance).

(3) For any anadromous fish, fossil remains are more likely to be found in extramontane continental conditions than in marine conditions, for these are just the sort of fishes that, when in the sea, would have been widely dispersed in open waters rather than localized in shallow water, near-shore environments.

(4) This last statement (3) must however, be qualified, on the basis of our knowledge of living fishes, for it is the large anadromous fishes that live in the open oceans. There are many smaller sized fishes (like cyprinodonts) that move equally between shallow marine conditions and coastal streams.

We have seen above the difficulty of distinguishing an anadromous condition in the available environmental and distributional data. But we must be prepared to accept that it may have been reasonably common. In particular, forms that are widely distributed in extramontane conditions might well be taxa that were anadromous.

A special point that has been raised on several occasions (see Westoll, 1936; Thomson, 1976a; Greiner, 1978) concerns the pattern of distribution of fishes (principally lungfishes and osteolepids) showing extensive resorption of cosmine. I have argued, from consideration of the Permian and Carboniferous osteolepids, that the special patterns of resorption and regrowth of the cosmine cover of osteolepids might be an indication that such fishes occupied a wide variety of habitats from marine to fresh water, specifically that they might have been euryhaline and that cosmine resorption of the sort shown by *Ectosteorhachis* (Lower Permian, U.S.A.) might be an indication of anadromous life-history. If this is so, then it should follow that such fishes would be found principally in large coastal river systems, rather than intermontane lake basins, and this ought to explain the distribution of osteolepids in the Upper Devonian and later where, with the exception of *Latvius* from New Brunswick, they are absent from the intermontane basinal deposits. Contrary to Greiner (1978) I do not believe that the presence of *Latvius* in New Brunswick negates this argument, especially as *Latvius* is elsewhere known only from marine sediments (a point Greiner seems to have missed).

CONCLUDING DISCUSSION

The principal result of this study is the demonstration from both inferential physiological data and from the geological evidence for environments and distributions, that all major groups of Devonian lobe-finned fishes were capable of living in saltwater conditions. None could be classified as a primary freshwater group in the usual biogeographical sense. Only a very few genera, principally in the late Devonian, were confined to freshwater environments and we have no solid grounds upon which to conclude that even these genera were incapable of surviving saltwater conditions.

The rhipidistian ancestors of the Amphibia (of which our closest known representatives so far are members of the family Osteolepidae) were fully adapted aquatic organisms incapable of surviving indefinitely out of the water, but with several interesting so-called pre-adaptations that allowed them to utilize the intermediate habitats between the water and land – the marshes, swamps and emergent vegetation of Devonian salt and freshwater environments. Lungs allowed them to survive foul water and temporary migrations out of the water. The particular structures of the paired limbs must be seen as adaptations for locomotion both against a solid substrate (submerged or emergent) and as a special adaptation improving feeding, respiration *and* locomotion in shallow waters where the body was not fully supported, and amidst wet vegetation completely out of the water. Possibly there was a major role in reproductive behaviour also. The characteristic differences between tetrapod fore and hind limbs are explainable in terms of the special conditions applying in these intermediate habitats.

Occupation of these semi-terrestrial habitats, whether on a seasonal or purely sporadic basis, conferred a significant advantage on the rhipidistian lobe-finned fishes. There may have been a greater abundance of prey, security from large predators, and protection of eggs, larvae and young.

The lobe-finned fishes as a whole used their special adaptations to occupy a very wide range of habitats in which they were the top predators, a role for which they vied, no doubt, with placoderms. Whereas the lungfishes concentrated upon a mostly durophagous diet and the coelacanths became specialized for suction feeding and a more sedentary, stealthy approach to food, the Rhipidistia remained active cruising predators dependent upon free-swimming prey. They found this prey in every environment from shallow epicontinental seas to intermontane lakes and streams. Many achieved a very large size, although the more continental ones were smaller. In terms of physiology of the lobe-finned fishes, fresh water or salt water offered no serious bar to migration following the best food supplies, and it is not really necessary to try to assort different fishes to single environmental patterns. The lobe-finned fishes colonized the whole tropical and subtropical world. They migrated principally through the shallow seas and entered all coastal rivers. Only the fishes found as fossils in the intermontane basins seem to have been restricted in their geographical distribution and even here we cannot tell whether this was due to geography or physiology.

The large coastal rivers of the margins of Devonian continents meandered through broad alluvial plains and the low-lying marshes where the land joined the sea must have been ideal breeding grounds for invertebrates and fishes alike. We do not need to assume that the breeding and principal feeding grounds for these fishes were in the same place. It is possible, even necessary in some cases, to reconstruct that the fishes fed in the seas and entered fresh waters to breed. However, if we concentrate upon the particular sorts of fishes that were the immediate ancestors of the tetrapods we must conclude that they spent most of their time in shallow waters and found their prey in the most productive of environments – swamps and marshes, and shallow marine lagoons. The entry of large numbers of lobe-finned fishes into freshwater environments seems not to have occurred until the Middle and late Devonian and thus coincides with the development of the first continental ecosystems including vascular plants (see papers by Edwards (Chapter 4), Rolfe (Chapter 6) and Scott (Chapter 5) in this volume). The first ventures of bony fishes into continental waters are likely to have been seasonal. The whole sequence of lobe-finned migrations into fresh water and the evolution of tetrapods was accomplished in a very short time.

However, this picture of the lobe-finned fishes does not fully explain the origin of tetrapods. The problem is that we cannot characterize the biology of the first amphibians as readily as we can their rhipidistian ancestors. A major new conclusion of the present study is that the first amphibians in all probability were physiologically equipped to survive salt water, having had the same set of physiological adaptations as their lobe-finned ancestors. Corroboration of this is afforded by the fact that several derived groups of extinct amphibians were actually restricted to marine environments. This then leads us to propose a new and disturbing question: can we any longer unquestioningly assume that the first Amphibia were in any sense fully freshwater animals?

The question is important for several reasons. First the answer would help us to explain the unusual pattern of distribution of Devonian tetrapod fossils – in the apparently widely separate continental environments of eastern Australia and western Greenland. No doubt, other amphibian fossils will be found from other geographical regions as well. Will they also be only from continental deposits? The question is also important in attempting to understand the overall biological ecological setting of the fish–tetrapod transition. The following possibilities may be proposed.

(1) The Amphibia arose from a rhipidistian stock that had not lost its direct connection with salt water. One can imagine such animals entering fresh waters to take advantage of seasonal food abundances (particularly invertebrates) and to breed in safety. Certain stages (young adults?) may have returned to sea to feed, either in coastal lagoons and swamps or in more open water. This marine phase produced wide distribution. Full restriction to freshwater environments for the whole life-history will have come later as both the environmental contexts changed and the biology of the organisms became more specialized.

(2) The opposite view is that Amphibia arose in a freshwater con-

text from fishes that had already renounced marine conditions. Thereupon, all distribution had to be through continental conditions. As all the early Amphibia were clearly bound to habitats in and around water, the world-wide distribution of freshwater Amphibia requires migrations in conjunction with stream drainages and passage from basin to basin becomes more difficult (but not impossible) to explain. The present understanding of Devonian biogeography makes this model an impossibility as it does not allow for any continental dry land connections between Australia and the Old Red Landmass.

Further, evidence from invertebrate distributions (Ziegler *et al.*, 1977; Boucot, 1974) suggests that the faunas of eastern Australia were quite distinct from those of Eur-America for all of the Devonian. They all form part of an Old World Province and this is reflected in the potential for marine distribution of the lobe-finned fishes. But the differences seem to imply a considerable separation between the continental masses.

There is only one other lobe-finned fish that has the same pattern of continental distribution as the first tetrapods, the lungfish *Soederberghia*, which can be safely assumed to have had close marine ancestors. There are no other examples of fishes found in intermontane basins being so widely distributed without direct fossil evidence of marine occurrences as well. The other fishes that have an Arctic–Antarctic distribution (*Remigolepis, Groelandaspis* and *Bothriolepis*) have sufficient direct evidence of marine or extramontane occurrences that we do not need to hope for drastic revision of Devonian geography to account for their migrations through fresh waters.

Here the matter must rest for the moment. The situation can be clarified (but probably never conclusively proved or disproved) by the discovery of other patterns of lobe-finned distributions and by confirmation of one or other geographical model. Our purpose here is well served if we become aware of the first of the two alternatives given above, and perhaps I may state an intuitive preference for this "saltwater connection".

ACKNOWLEDGEMENTS

I am grateful to G. C. Young (Australia), G. M. Bernacsek (Dar es Salaam), J. S. Rackoff (Pennsylvania) and J. T. Gregory (California) for sharing with me results of work in progress, and D. Schindel for valuable discussions. This study has been supported in part by grant DEB-77-08412 of the National Science Foundation. Maps prepared by Linda Price Thomson.

REFERENCES

Alexander, R. McN. (1973). Jaw mechanisms of the coelacanth *Latimeria*. *Copeia* 1973, 156–158.

Allen, J. R. L. and Friend, P. F. (1968). Deposition of the Catskill facies, Appalachian region, with notes on some other Old Red Sandstone basins. *Geol. Soc. Am. spec. Pap.* No. 106, 21–74.

Allen, J. R. L., Dineley, D. L. and Friend, P. F. (1967). Old Red Sandstone basins of North America and northwest Europe. *In* "International Symposium on the Devonian System" (D. H. Oswald, ed.), pp. 69–98. American Association of Petroleum Geologists, Calgary, Alberta.

Andrews, S. M. and Westoll, T. S. (1970a). The postcranial skeleton of *Eusthenopteron foordi* Whiteaves. *Trans. R. Soc. Edinb.* **68**, 207–329.

Andrews, S. M. and Westoll, T. S. (1970b). The postcranial skeleton of rhipidistian fishes excluding *Eusthenopteron*. *Trans. R. Soc. Edinb.* **68**, 391–489.

Barrel, J. (1916). Influence of Silurian–Devonian climates on the rise of air-breathing vertebrates. *Bull geol. Soc. Am.* **27**, 387–436.

Bernacsek, G. M. (1977). A lungfish cranium from the Middle Devonian of the Yukon Territory, Canada. *Palaeontographica* **157A**, 175–200.

Boucot, R. J. (1974). Silurian and Devonian biogeography. *In* "Paleogeographic Provinces and Provinciality" (A. A. Ross, ed.), pp. 165–176, Spec. Pub. Society of Economic Paleontologists and Mineralogists, Tulsa.

Brunton, C. H. C., Miles, R. S. and Rolfe, W. D. I. (1967). Gogo Expedition 1967. *Proc. geol. Soc. Lond.* **165**, 79–83.

Burtner, R. L. (1963). Sediment dispersal patterns within the Catskill facies of southeastern New York and northeastern Pennsylvania. *In* "Stratigraphy of Pennsylvania and Adjacent States" (V. C. Shepps, ed.), *Penn. geol. Surv. Bull.* **939**, 7–23.

Campbell, K. S. W. and Bell, M. W. (1977). A primitive amphibian from the late Devonian of New South Wales. *Alcheringa* **1**, 369–381.

Carter, G. S. (1957). Air breathing. *In* "The Physiology of Fishes" (M. E. Brown, ed.), Vol. 1, pp. 59–79. Academic Press, New York and London.

Cocks, L. R. M. and McKerrow, W. S. (1973). Brachiopod distributions and faunal provinces in the Silurian and Lower Devonian. *In* "Organisms and Continents through Time" (N. F. Hughes, ed.), *Paleont. Ass. spec. Pap.* No. 24, 291–364.

Cowles, R. B. (1958). Additional notes on the origin of tetrapods. *Evolution* **12**, 419–421.

Denison, R. H. (1951). Late Devonian fresh-water fishes from the western United States. *Fieldiana: Geology* **11**, 221–261.

Denison, R. H. (1968). Early Devonian lungfishes from Wyoming, Utah and Idaho. *Fieldiana: Geology* **17**, 353–413.

Dineley, D. L. and Williams, B. J. P. (1968). Sedimentation and paleoecology of the Devonian Escuminac Formation and related strata, Escuminac Bay, Quebec. *Geol. Soc. Am. spec. Pap.* No. 106, 241–264.

Donovan, R. M. (1973). Basin margin deposits of the Middle Old Red Sandstone at Dirlot, Caithness. *Scott. J. Geol.* 9, 203–211.

Edwards, J. L. (1977). The evolution of terrestrial locomotion. *In* "Major Patterns in Vertebrate Evolution" (M. K. Hecht *et al.*, eds), pp. 553–578. Plenum, New York.

Friend, P. F. (1969). Tectonic features of Old Red sedimentation in North Atlantic borders. *In* "North Atlantic Geology and Continental Drift" (G. M. Kay, ed.), *Pet. Geol. Mem.* No. 12, 703–710.

Gardiner, J. B. and Miles, R. S. (1975). Devonian fishes from the Gogo Formation, Western Australia. *In* "Problèmes Actuels de Paléontologie" (J-P. Lehman, ed.), *Colloques int. C.n.R.S.* **218**, 73–79.

Goldstein, L. and Forster, R. P. (1970). Nitrogen metabolism in fishes. *In* "Comparative Biochemistry of Nitrogen Metabolism" (J. W. Campbell, ed.), pp. 496–518. Academic Press, New York and London.

Graham, J. B., Rosenblatt, R. H. and Gans, C. (1978). Vertebrate air breathing arose in fresh waters and not in the ocean. *Evolution* **32**, 459–463.

Gregory, J. T., Murphy, T. G. and Reed, J. W. (1977). Devonian fishes in Central Nevada. *In* "Western North America: Devonian" (M. S. Murphy *et al.*, eds), *Univ. Cal. Riverside Mus. Contr.* 4, 112–120.

Greiner, H. R. (1977). Crossopterygian fauna from the Albert Formation, New Brunswick, Canada, and its stratigraphic-paleoecological significance. *J. Paleont.* **50**, 44–56.

Greiner, H. (1978). Late Devonian facies inter-relationships in bordering areas of the North Atlantic and their palaeogeographic implications. *Palaeogeogr., Palaeoclimat., Palaeoecol.* **25**, 241–263.

Griffith, R. W. and Thomson, K. S. (1972). Observations on a dying coelacanth. *Am. Zool.* **12**, 730.

Griffith, R. W. and Thomson, K. S. (1973). *Latimeria chalumnae:* reproduction and conservation. *Nature, Lond.* **242**, 617–618.

Hills, E. S. (1933). On a primitive dipnoan from the Middle Devonian rocks of New South Wales. *Ann. Mag. nat. Hist.* **11**, 634–643.

House, M. R. (1973). An analysis of Devonian goniatite distributions. *In* "Organisms and Continents in Space and Time" (N. F. Hughes, ed.), *Paleont. Ass. spec. Pap.* No. 12, 305–318.

Inger, R. F. (1957). Ecological aspects of the origins of the tetrapods. *Evolution* **11**, 373–376.

Janvier, P. (1978). Vertébrés dévoniens de nouveaux gisements du Moyen-Orient. *Ann. Soc. géol. N.* **97**, 373–382.

Janvier, P., Termier, G. and Termier, H. (1979). The osteolepiform rhipidistian fish *Megalichthys* in the Lower Carboniferous of Morocco, with remarks on the palaeobiogeography of the Upper Devonian and Permo-Carboniferous osteolepids. *Neues Jb. Geol. Paläont.* 1979, 7–14.

Jarvik, E. (1942). On the structure of the snout in crossopterygians and lower gnathostomes in general. *Zool. Bidrag. Upps.* **21**, 237–675.

Jarvik, E. (1952). On the fish-like tail in the ichthyostegid stegocephalians. *Meddr Grønland* **14**(12), 1–90.

Jarvik, E. (1961). Devonian vertebrates *In* "Geology of the Arctic" (G. O. Raasch, ed.), pp. 197–204. University of Toronto Press.

Jarvik, E. (1967). On the structure of the lower jaw in dipnoans: with a description of an early Devonian dipnoan from Canada, *Melanognathus canadensis* gen. et sp. nov. *J. Linn. Soc. (Zool.)* **47**, 155–184.

Jessen, H. (1973). Weitere Fischreste aus dem oberen Plattenkalk der Bergisch-Pfaffrather Mulde (Oberdevon. Rheinisches Schiefergebirge). *Palaeontographica* **143A**, 159–187.

Jessen, H. L. (1975). A new choanate fish, *Powichthys thorsteinssoni* n.g., n.sp., from the early Lower Devonian of the Canadian Arctic Archipelago. *In* "Problèmes Actuels de Paléontologie" (J-P. Lehman, ed.), *Colloques int. C.n.R.S.* **218**, 213–222.

Jux, U. (1964). Zur stratigraphischer Gliederung des Devon profils von Bergisch-Gladbach (Rheinisches Schiefergebirge). *Decheniana* **117**, 159–174.

Lauder, G. V. (1980). The role of the hyoid apparatus in the feeding mechanism of the coelacanth *Latimeria chalumnae, Copeia* (in press).

Lehman, J-P. (1977). Sur la présence d'un ostéolépiforme dans le Dévonien supêrieur du Tafilaler. *C.r. Acad. Sci., Paris* **2850**, 151–153.

Lehman, W. M. and Westoll, T. S. (1952). A primitive dipnoan fish from the Lower Devonian of Germany. *Proc. R. Soc. Lond.* **B140**, 403–421.

Locket, N. A. and Griffith, R. W. (1972). Observations on a living coelacanth. *Nature, Lond.* **237**, 175.

McAllister, D. E. (1980). Review: Anatomie de *Latimeria chalumnae,* Volume III. *J. Fish. Res. Bed Can.* (1980).

Millot, J., Anthony, J. and Robineau, J. (1978). "Anatomie de *Latimeria chalumnae*", Vol. 3. Centre Nationale de la Recherche Scientifique, Paris.

Ørvig, T. (1960). New finds of acanthodians, arthrodires, crossopterygians, ganoids and dipnoans in the upper Middle Devonian calcareous flags (Oberer Plattenkalk) of the Bergisch Gladbach-Pfaffrath Trough. *Paläont. Z.* **34**, 295–355.

Ørvig, T. (1969). Vertebrates from the Wood Bay Group and the position of the Emsian–Eifelian boundary in the Devonian of Vestspitsbergen. *Lethaia* **2**, 273–328.

Packard, G. C. (1974). The evolution of air-breathing in Paleozoic gnathostome fishes. *Evolution* 28, 320–325.

Packard, G. C. (1976). Devonian amphibians: did they excrete carbon dioxide via skin, gills or lungs? *Evolution* **30**, 270–280.

Panchen, A. L. (1976). Geographical and ecological distribution of the earliest tetrapods. *In* "Major Patterns in Vertebrate Evolution" (M. K. Hecht. *et al*, eds), pp. 723–738. Plenum, New York.

Pang, P. K. T., Griffith, R. W. and Atz, J. W. (1977). Osmoregulation in elasmobranchs. *Am. Zool.* **17**, 365–378.

Pickford, G. E. and Grant, F. B. (1967). Serum osmolarity in the coelacanth, *Latimeria chalumnae*: urea retention and ion regulation. *Science* 155, 568–570.

Rackoff, J. S. (1976). "The Osteology of *Sterropterygion* (Crossopterygii: Osteolepidae) and the Origin of Tetrapod Locomotion." Ph.D. Thesis, Yale University.

Rayner, D. H. (1963). The Achanarras limestone of the Middle Old Red Sandstone, Caithness, Scotland. *Proc. Yorks. geol. Soc.* **34**, 117–138.

Ritchie, A. (1974). From Greenland's icy mountains. *Aust. nat. Hist.* **18**, 28–35.

Robineau, D. and Anthony, J. (1973). Biomécanique du crâne de *Latimeria chalumnae* (poisson crossopterygien coelacanthide). *C.r. Acad. Sci., Paris* **276**, 1305–1308.

Romer, A. S. (1955). Fish origins – fresh or salt water? *Deep Sea Res.* **3** (Suppl.), 261–280.

Romer, A. S. (1958). Tetrapod limbs and early tetrapod life. *Evolution* **12**, 365–369.

Romer, A. S. and Byrne, F. (1931). The pes of *Diadectes*: notes on the primitive limb. *Palaeobiologica* **4**, 25–48.

Säve-Söderbergh, G. (1932). Preliminary note on Devonian stegocephalians from East Greenland. *Meddr Grønland* **94**(7), 1–105.

Schultze, H-P. (1969). *Griphognathus* Gross, ein langschnauziger Dipnoer aus dem Overdevon von Bergisch-Gladbach (Rheinisches Schiefergebirge) und von Lettland. *Geologica Palaeontologica* **3**, 21–79.

Smith, A. G., Bryden, J. C. and Drewry, G. E. (1973). Phanerozoic world maps. *In* "Organisms and Continents through Time" (N. F. Hughes, ed.), *Paleont. Ass. spec. Pap.* No. 12, 1–42.

Smith, C. L., Rand, C. S. and Atz, J. W. (1975). *Latimeria*, the living coelacanth, is ovoviviparous. *Science* **190**, 1105–1106.

Szarski, H. (1976). Sarcopterygii and the origin of tetrapods. *In* "Major Patterns in Vertebrate Evolution" (M. K. Hecht *et al.*, eds), pp. 517–540. Plenum, New York.

Thomson, K. S. (1967). Mechanisms of intercranial kinetics in fossil rhipidistian fishes (Crossopterygii) and their relatives. *J. Linn. Soc. (Zool.)* **178**, 223–253.

Thomson, K. S. (1968). A new Devonian fish (Crossopterygii: Rhipidistia) considered in relation to the origin of the Amphibia. *Postilla* **124**, 1–13.

Thomson, K. S. (1969a). The biology of lobe-finned fishes. *Biol. Rev.* **44**, 91–154.

Thomson, K. S. (1969b). The environment and distribution of Paleozoic sarcopterygian fishes. *Am. J. Sci.* **267**, 457–464.

Thomson, K. S. (1971). Adaptation and evolution of early fishes. *Q. Rev. Biol.* **46**, 139–166.

Thomson, K. S. (1972). New evidence on the evolution of paired fins of Rhipidistia and the origin of the tetrapod limb, with description of a new genus of Osteolepidae. *Postilla* **157**, 1–7.

Thomson, K. S. (1973). Observations of a new rhipidistian from the Upper Devonian of Australia. *Palaeontographica* **143A**, 209–220.

Thomson, K. S. (1975). The biology of cosmine. *Bull. Peabody Mus. nat. Hist.* **40**, 1–59.

Thomson, K. S. (1976a). The faunal relationships of rhipidistian fishes (Crossopterygii) from the Catskill (Upper Devonian) of Pennsylvania. *J. Paleont.* **50**, 1203–1208.

Thomson, K. S. (1976b). On the individual history of cosmine and a possible electro-receptive function of the pore-canal system in fossil fishes. *In* "Problems in Vertebrate Evolution" (S. M. Andrews *et al.*, eds), pp. 247–270. Academic Press, London and New York.

Thomson, K. S. (1980). Environmental factors in the evolution of vertebrate endocrine systems. *In* "Evolution of Vertebrate Endocrine Systems" (P. K. T. Pang and A. Epple, eds). Texas Tech. University Press, Lubbock (in press).

Thomson, K. S. and Bossy, K. H. (1971). Adaptive trends and relationships in early Amphibia. *Forma Functio* 3, 7–31.

Thomson, K. S. and Campbell, K. S. W. (1971). The structure and relationships of the primitive dipnoan fish *Dipnorhynchus sussmilchi* Etheridge. *Bull. Peabody Mus. nat. Hist.* **39**, 1–109.

Thomson, K. S. and Hahn, K. V. (1968). Growth and form in fossil rhipidistian fishes (Crossopterygii). *J. Zool. Lond.* **156**, 199–223.

Thomson, K. S. and Rackoff, J. S. (1974). The shoulder girdle of the Permian rhipidistian fish *Ectosteorhachis nitidus* Cope: structure and possible function. *J. Paleont.* **156**, 199–223.

Vorobjeva, E. (1975). Some peculiarities in evolution of the rhipidistian fishes. *In* "Problèmes Actuels de Paléontologie" (J-P. Lehman, ed.), *Colloques int. C.n.R.S.* **218**, 223–230.

Vorobjeva, E. (1977). Morphology and evolution of rhipidistian fishes. *Akad. Nauk. U.S.S.R. Trudy paleont. Izvestia* **163**, 1–239.

Warburton, F. E. and Denman, N. S. (1961). Larval competition and the origin of tetrapods. *Evolution* **15**, 566.

Warren, J. W. and Wakefield, N. A. (1972). Trackways of tetrapod vertebrates from the Upper Devonian of Victoria Australia. *Nature, Lond.* **238**, 469–470.

Westoll, T. S. (1936). On the structures of the dermal ethmoid shield of *Osteolepis. Geol. Mag.* **73**, 157–171.

Westoll, T. S. (1943). The origin of the tetrapods. *Biol. Rev.* **18**, 78–98.

Ziegler, A. M., Scotese, C. R., Johnson, M. E., McKerrow, W. S. and Bombach, R.K. (1977). Paleozoic biogeography of continents bordering the Iapetus (Pre-Caledonian) and Rheic (Pre-Hercynian) oceans. *In* "Paleontology and Plate Tectonics" (R. W. West, ed.), *Milwaukee Public Mus. spec. Publ. Biol. Geol.* No. 2, 1–22.

10 Osteolepid Remains from the Devonian of the Middle East, with Particular Reference to the Endoskeletal Shoulder Girdle

PHILIPPE JANVIER

Department of Vertebrate Palaeontology, University of Paris VI, 4, place Jussieu, 75230 Paris, France

Abstract: Remains of cosmine-covered Osteolepiformes are recorded from several Devonian vertebrate-bearing localities of the Middle East, namely in north-eastern Iran and in south-western Turkey. In these localities, the accompanying vertebrate and invertebrate fauna, as well as the sedimentology, indicate that the osteolepids lived in a marine, shallow-water, near-shore environment. In the Middle Devonian of Iran and in the Upper Devonian of Turkey, relatively well preserved endoskeletal shoulder girdles of osteolepids have been discovered and are described in this paper. They emphasize the remarkable diversity of that part of the skeleton in the Osteolepiformes. However, they all show a typical triradiate structure only met with in the Osteolepiformes. The question of whether the Osteolepiformes are monophyletic or paraphyletic is discussed on the basis of the anatomy of the endoskeletal shoulder girdle. Although cranial anatomy indicates that the tetrapods are more closely related to the Osteolepididae than to any other group of the Osteolepiformes, it seems that the triradiate endoskeletal shoulder girdle of the Osteolepiformes represents a condition from which that of the tetrapods cannot be derived. Therefore, it is suggested that the Osteolepiformes may be monophyletic and represent the sister-group of the tetrapods.

INTRODUCTION

Abundant vertebrate remains have been recorded from Middle and Upper Devonian exposures of various countries of the Middle East,

Systematics Association Special Volume No. 15, "The Terrestrial Environment and the Origin of Land Vertebrates", edited by A. L. Panchen, 1980, pp. 223-254, Academic Press, London and New York.

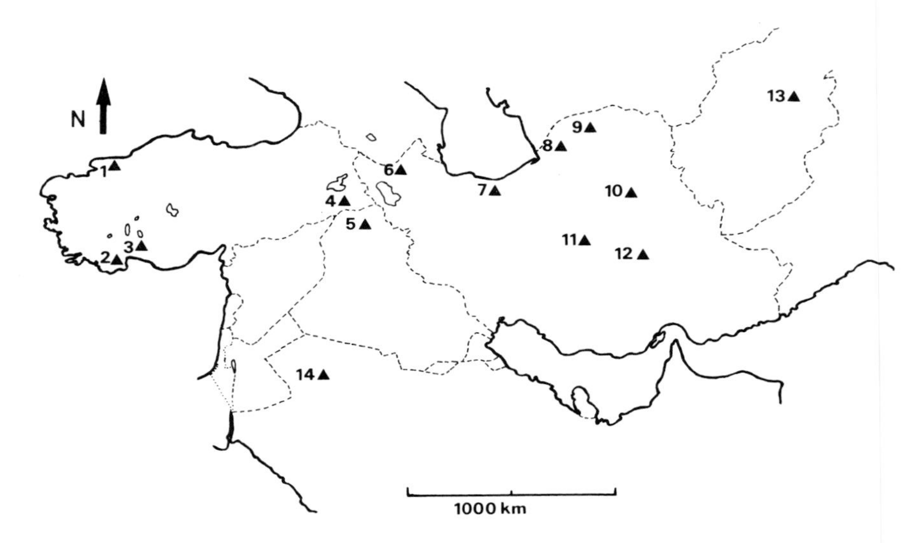

Fig. 1. The Devonian vertebrate-bearing localities of the Middle East. (1) Istambul region; (2) Armutgözlek Tepe, near Antalya; (3) Bademli; (4) Great Zap Valley; (5) Pirispiki red beds; (6) Zonuz; (7) Djeiroud; (8) Khush-Yeilagh; (9) Robat-e-Gharabil; (10) Tabas; (11) Yazd; (12) Kerman region; (13) Dasht-e-Nawar; (14) Al Jawf. Osteolepid remains are recorded from 2, 4, 8 and 12.

since Rieben (1935) mentioned the occurrence of fragmentary fish remains in the Upper Devonian of Azerbaidjan. Most of the vertebrate-bearing localities of the Middle East are situated in Iran (Huckriede *et al.*, 1962; Schultze, 1973; Janvier, 1974, 1976; Janvier and Martin, 1978), but new localities have recently been recorded from Turkey (Janvier and Marcoux, 1976; Janvier and Ritchie, 1977), Iraq (Seilacher, 1963), Saudi Arabia (Powers, 1968), Afghanistan (Boutière and Brice, 1966; Janvier, 1976) and the Jammu and Kashmir district of India (Gupta and Denison, 1966; Gupta and Turner, 1973; Gupta and Janvier, 1979) (Fig. 1).

The Devonian vertebrate material from the Middle East consists predominantly of placoderm remains, which are associated with fragmentary remains of actinopterygians, dipnoans, coelacanthiforms, struniiforms and rhipidistians. Some elasmobranch, acanthodian and agnathan (thelodonts) remains may also occur among this material. As to the choanate fishes, osteolepiforms have been discovered in four localities of the Middle East: Khush-Yeilagh (Eifelian) and Ravar (Frasnian) in Iran, and Ermutgözlek Tepe (? Frasnian) and the Great Zap Valley (Famennian) in Turkey (Fig. 1).

GEOLOGICAL AND FAUNAL ENVIRONMENT OF THE DEVONIAN OSTEOLEPIFORMS FROM THE MIDDLE EAST

The four osteolepiform-bearing localities just mentioned represent four different lithofacies and faunal assemblages. Therefore, a brief account of the fauna found in association with these osteolepiform remains may throw some light on the problem of the environment of these choanate fishes.

1. Khush-Yeilagh

This locality is situated near the Khush-Yeilagh Pass, between the towns of Shahrud and Shahpasand, in Eastern Alborz (Fig. 1,*8*). The fish remains come from a centimetre-thick bone-bed situated within a thick arkosic layer which forms the basal part of the Khush-Yeilagh Formation (Stampfli, 1978). This layer is overlain conformably by marine carbonaceous beds which are referred to the Lower or Middle Givetian (Stampfli, 1978). In some places, the top part of this arkosic layer grades into a carbonaceous sandstone containing *Amaltheolepis*-like thelodont scales as well as some conodonts suggestive of Eifelian or Upper Emsian age (B. Hamdi, personal communication). Consequently, the Khush-Yeilagh bone-bed is older than the Givetian and is possibly of Emsian or Eifelian age.

The rich fish assemblage collected from this bone-bed is still under study, but a provisional fish faunal list may be given here:

Elasmobranchii: "*Ctenacanthus*" sp.
Placodermi : Ptyctodontidae indet., ?*Gerdalepis* sp.,
Antiarcha indet., *Holonema* sp., Coccosteidae indet., Groenlandaspididae indet., Phlyctaeniidae indet.
Acanthodii: Ischnacanthiformes indet., *Gyracanthus* sp., ?Climatiidae indet.
Dipnoi: Dipnoi indet. (cf. *Stomiahykus*)
"Crossopterygii": *Onychodus* cf. *sigmoides* Newberry, Holoptychiidae indet., Osteolepididae indet.

These fishes are associated with large but indeterminable plant remains. The matrix of this bone-bed is extremely hard and the specimens can only be prepared in negative, by etching away the bone, and making casts with silicone rubber.

2. Ravar

This locality is situated about 8 km west of Ravar, in central Iran. The fish remains are found in a flaggy limestone exposed in the Tangil-e-ab-Garm gorge crossing the Band-e-Anâr Range. This limestone may easily be dated as Lower Frasnian by the invertebrate fauna (mainly brachiopods and conodonts).

The following fish faunal list may be given for this locality:

Elasmobranchii: *Ctenacanthus* sp.
Placodermi: ?*Eastmanosteus* sp., *Holonema* cf. *radiatum* Obrouchev, Ptyctodontida indet., *Byssacanthus* sp.
Acanthodii: *Persacanthus* cf. *kermanensis* Janvier
Actinopterygii: *Moythomasia* sp.
Dipnoi: *Rhinodipterus* sp.
"Crossopterygii": *Diplocercides* sp., *Strunius rolandi* (Gross), ?*Onychodus* sp., Osteolepididae indet.

This limestone layer also contains abundant remains of brachiopods, bryozoans and corals which indicate a marine environment.

3. Armutgözlek Tepe

The red sandstone of the Armutgözlek Tepe is exposed in the Sarcinar Dag, 5 km west of Kemer, in the Antalya Bay, Turkey. It belongs to a thick Palaeozoic series included in the Upper Antalya Nappes (Brunn *et al.*, 1970; Janvier and Marcoux, 1977). This sandstone layer contains relatively abundant fish remains suggestive of an Upper Devonian (possibly Frasnian) age. The provisional fish faunal list for this locality is:

Elasmobranchii: *"Ctenacanthus"* sp.
Placodermi: Ptyctodontida indet., *Bothriolepis* cf. *canadensis* Whiteaves, *Groenlandaspis seni* Janvier and Ritchie, *Holonema* sp.
Acanthodii: Acanthodii indet.
Dipnoi: *Rhinodipterus* sp., ?*Oervigia* sp.
"Crossopterygii": Osteolepididae indet.

The fish remains are associated with abundant plant remains and had to be prepared in negative. This thick and massive sandstone layer contains, in some places, several discontinuous *Tigillites*-bearing beds. These trace fossils consist of closely set vertical tubules which are supposed to be the natural casts of plant stems buried *in situ*. The basal part of each of these *Tigillites*-bearing beds shows a horizontal network of traces which may correspond to the rhizomes of these plants.

The presence of *Tigillites* indicates that the environment of deposition of this sandstone was very near to the shore, but still under marine influence, since undoubtedly marine fish remains, like plates of *Holonema* and *Ctenacanthus* spines, are found in the same layer and probably were brought in from the open sea by the tide.

4. Great Zap Valley

A new Devonian fish locality has recently been discovered by Dr O. Monod (Orsay, France) in a thick Palaeozoic series exposed in the Great Zap Valley in south-eastern Turkey. The fish remains are associated with abundant ostracods which are suggestive of an uppermost Devonian age (possibly Famennian). Some samples from this locality yielded scales and teeth of undetermined elasmobranchs, actinopterygians and osteolepids.

According to the ostracod assemblage and to the lithology, the fish-bearing layer may have been deposited in a shallow bay or a lagoon.

5. Palaeoecological Remarks

During Middle and Upper Devonian times, the Middle East region was occupied by an epicontinental sea surrounding small islands (Rzhonsnitskaya, 1968; Stampfli, 1978). These islands lined the northern edge of the Arab–African Platform, which probably was a part of the Gondwanian landmass. Stampfli (1978, p.47) has pointed out that the Middle Devonian sedimentation in Iran is characterized by a great diversity of facies, probably due to the presence of several depositional basins. During this period, reliefs of Silurian age were abraded and covered, later in the Devonian, by very shallow epicontinental seas. This shallow-water environment, with vast tidal flats surrounding the merging land became fairly common from Turkey to

Kashmir as early as the Eifelian. From that time until the very end of the Devonian there is in the Middle East, as well as in other parts of the world, a clear distinction between the vertebrate faunas associated with a typically marine environment (with abundant brachiopod, echinoderms and corals) on one hand, and that associated with an arenaceous ("Old Red Sandstone") environment. Although intermediate facies containing mixed faunas may occur, the former facies generally yields large arthrodires (dinichthyids), elasmobranchs, dipnoans, actinopterygians, struniiforms and coelacanthiforms, whereas the latter yields abundant remains of bothriolepidid antiarchs, groenlandaspids and osteolepiforms. In the case of the Middle East localities, the marine carbonaceous limestone of the Kerman region (Iran) is particularly rich in large dinichthyiid and holonematid remains and poor in antiarch and osteolepid remains (only found in the Ravar locality). On the contrary, the Armutgözlek Tepe red sandstone yielded mainly bothriolepid, groenlandaspid and osteolepid remains, but only a few plates of *Holonema* and spines of ctenacanthid sharks. In some cases, like the Khush-Yeilagh bone-bed, remains of fishes belonging to both of these ecological assemblages are present. This also seems to be the case in the inter-reef facies of the Gogo locality in Australia (Gardiner and Miles, 1975), where osteolepids and bothriolepids are found in associations with holonematids and abundant actinopterygians.

These two major tendencies of Devonian vertebrate palaeoecology are generally much better marked in the Upper Devonian than in the Lower or Middle Devonian. This is probably due to the fact that the surface occupied by tidal flats increased considerably in the late Middle Devonian and early Upper Devonian times.

Among the osteolepiforms, members of the Osteolepididae undoubtedly lived in a marine environment, but were probably more or less bound to the tidal flat and could survive in pools during the low tide. This may also have been the mode of life of the bothriolepid antiarchs and groenlandaspid arthrodires. It seems reasonable to suggest that life on the tidal flat has been for the osteolepids one of the steps leading to the terrestrial environment.

THE ENDOSKELETAL SHOULDER GIRDLE OF THE OSTEOLEPIFORMS

The endoskeletal shoulder girdle of osteolepiforms consists of a single bone, often referred to as a "scapulocoracoid", and this is attached to the cleithrum by the means of three buttresses. It has been described in detail in *Eusthenopteron foordi* Whiteaves (Andrews and Westoll, 1970a) and in *Ectosteorhachis nitidus* Cope (Thomson and Rackoff, 1974). It is also known in some other forms, *Megalichthys hibberti* Ag. (Andrews and Westoll, 1970b), *Rhizodus hibberti* (Ag. and Hibbert) and *Strepsodus sauroides* Binney (Andrews and Westoll, 1970b), but in these cases it is often crushed or distorted and does not allow detailed study.

The osteolepid remains described here, with particularly well preserved endoskeletal shoulder girdles, have been discovered in two of the above-mentioned localities: Khush-Yeilagh in Iran and the Armutgözlek Tepe in Turkey. Since these remains cannot be determined at specific or even generic level, I refer to them as "osteolepid A" and "osteolepid B", respectively, in the following description.

1. *Osteolepid A*

Osteolepid A, from the Emsian or Eifelian Khush-Yeilagh bone-bed, is represented by some isolated bones, namely a shoulder girdle, a lower jaw and a fronto-ethmoidal shield. Since the material collected during the 1976 and 1977 expeditions has not yet been fully investigated, more osteolepid remains may be expected from this locality. This osteolepid is probably one of the stratigraphically oldest known osteolepiforms, together with *Thursius macrolepidotus* Sedg. and Murch. from Scotland and *Thursius talsiensis* Vorobjeva (1971) from Latvia.

(a) Cleithrum. The outline of the cleithrum is basically the same as in other osteolepids (Fig. 2A) and more closely resembles that of *Osteolepis macrolepidotus* (Jarvik, 1948, Fig. 25D) than that of *Thursius pholidotus* (Jarvik, 1948, Fig. 68R). As in the former, the posterior part of its outer surface is smooth and cosmine-covered, whereas its anterior part is ornamented with parallel and vertical ridges. In distinction to the cleithrum of *Eusthenopteron foordi* (Andrews and Westoll, 1970a, Fig. 1), that of osteolepid A is devoid of an anterior process passing laterally to the dorsal part of the clavicle. The ventral division of the cleithrum is gently curved anteromedially, (Fig. $2B_1$), but its curvature is less marked than in *E. foordi.* The internal surface of the ventral division shows an anteromedially directed ridge (*r.*) which may correspond to that described at the

same place in *E. foordi* (Andrews and Westoll, 1970a, Fig. 1C, *"ri.Clm.v."*) and in some other osteolepiforms (Janvier and Marcoux, 1976, Fig. 1A). The anterior edge of the dorsal division is slightly curved in an anteromedial direction and must have lined the posterior part of the branchial cavity. The posterior and ventral edges of the ventral division are lined by a narrow depressed area which is also found in some other osteolepids (Janvier and Marcoux, 1976, Fig. 1A, "*pect.*") and probably represents the part of the cleithrum overlapping the foremost row of trunk scales.

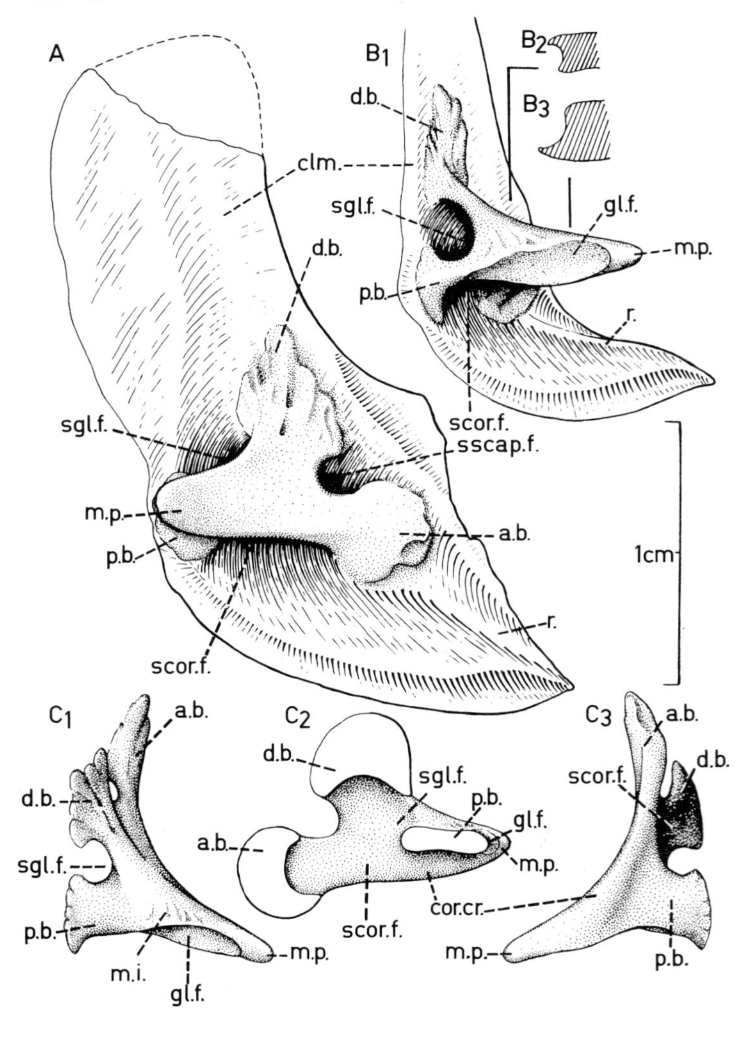

The cleithrum of *E. foordi* is remarkable in having a prominent anterior process (Andrews and Westoll, 1970a, Fig. 1C, "*pr.Clm.*"), whereas all the other known osteolepiform cleithra only show a slight anterior angle or even no indication of such an angle. In the Porolepiformes, such an anterior process or angle is totally lacking (Jarvik, 1972, Figs 51,52). Although the cleithrum of the latter is completely different from that of the osteolepiforms, it may be suggested that the lack of a prominent anterior process represents a primitive condition for the choanates in general.

(b) Endoskeletal shoulder girdle. The endoskeletal shoulder girdle consists of a relatively small central part, triangular in shape in medial aspect, attached to the internal surface of the cleithrum by means of three endoskeletal processes or buttresses (Fig. 2). The posterior buttress (*p.b.*) extends medially to the glenoid fossa (*gl.f.*), which is transversally elongate and faces posterolaterally. The dorsal buttress (*d.b.*) extends dorsally from the midpart of the bone. The anterior buttress (*a.b.*) extends almost to the anterior margin of the cleithrum, in an anterior or slightly anteroventral direction. The two latter buttresses have been referred to the "supraglenoid" and "infraglenoid"

Fig. 2. Osteolepididae gen. et. sp. indet. ("osteolepid A"), uppermost Emsian or Lower Eifelian, Khush-Yeilagh, Eastern Alborz, Iran. (A) Cleithrum and endoskeletal shoulder girdle in medial view (Iran Museum of Natural History, No. D.F. 0088); (B) same specimen in posterior view (B_1) and longitudinal sections through the lateral (B_2) and medial (B_3) parts of the glenoid fossa respectively; (C) endoskeletal shoulder girdle of the same specimen in dorsal (C_1), lateral (C_2) and vental (C_3) view.

Key to abbreviations. *a.b*, anterior buttress; *a.scl.*, subclavian artery; *clm.*, cleithrum; *cor.*, coracoid plate; *cor.cr.*, coracoid crest; *cor.pr.*, coracoid process; *d.b.*, dorsal buttress; *d.f.*, dental fossae; *d.m.*, dorsal muscular unit; *e.n.*, external naris; *f.*, foramen piercing the midpart of the endoskeletal shoulder girdle in the Actinopterygii; *f.m.d.*, fossa for the dorsal muscular unit; *f.m.v.*, fossa for the ventral muscular unit; *gl.f.*, glenoid fossa; *hu.*, humerus; *m.i.*, possible area of insertion for a dorsal muscle; *m.p.*, pointed medial process; *m.t.*, main teeth; *nas.c.*, nasal cavities; *olf.c.*, canal for the olfactory nerve; *orb.*, orbital notch; *p.b.*, posterior buttress; *pi.*, pineal foramen; *p.l.sc.*, processus lateralis of the scapular division; *p.m.sc.*, processus medialis of the scapular division; *prn.f.*, prenasal fossa; *scor.f.*, supracoracoid foramen; *sgl.f.*, supraglenoid foramen; *sscap.f.*, anterior opening of the subscapular fossa; *v.g.*, vascular groove.

buttresses respectively by Andrews and Westoll (1970a), whereas Thomson and Rackoff (1974) referred to the three buttresses as the "lateral, dorsal and anterior" buttresses respectively.

The areas for the attachment of the dorsal and anterior buttresses onto the cleithrum are roughly semilunar in shape (Fig. $2C_2$). This is due to the fact that the lateral face of these buttresses is slightly concave or flattened. The section of the posterior buttress is elongate in shape (Fig. $2C_2$) and extends in anteroposterior direction. Contrary to the condition in *E. foordi*, the limits of the distal part of the buttresses may be easily determined in this specimen, thanks to the small irregular endoskeletal outgrowths that occur near the periphery of their distal expansion.

In dorsal and ventral aspects, the endoskeletal shoulder girdle is roughly triangular in shape (Fig. $2C_1$,C_3). Dorsally, and near the dorsal edge of the glenoid fossa, it shows a roughened area (*m.i.*) which may correspond to some muscular insertion. The most peculiar characteristic of this specimen, when compared to endoskeletal shoulder girdles of other osteolepiforms, is the posteromedial pointed process (*m.p.*) which extends medially to the glenoid fossa. The anteromedial surface of this process is continuous with the medial surface of the central part of the endoskeletal shoulder girdle.

The shape of the glenoid fossa, also, is unique among the few osteolepiforms where it has been recorded. As mentioned above, it is transversally elongated, extending from the medial part of the posterior buttress to the lateral part of the pointed medial process referred to above. As in other osteolepiforms, it is slightly pear-shaped (Fig. 2B) and "screw-shaped"; that is, its medial part faces slightly upwards (Fig. $2B_3$), whereas its lateral part faces slightly downwards (Fig. $2B_2$). There is no indication of a glenoid foramen of the kind described in *E. foordi* (Andrews and Westoll, 1970a, Fig. 4) and in *Ectosteorhachis nitidus* (Thomson and Rackoff, 1974, Fig. 1).

The central part of the endoskeletal shoulder girdle shows a ventromedial crest (*cor.cr.*) uniting the anterior buttress to the pointed medial process, and which may represent the primary coracoid division of this bone. Therefore, it will be referred to here as the "coracoid crest".

The large foramina opening medially between the three

buttresses have been homologized with the three main openings found in the endoskeletal shoulder girdle of some primitive tetrapods (Romer, 1922, 1924; Miner, 1925; Andrews and Westoll, 1970a). Therefore, they have received the names of "supraglenoid foramen", "supracoracoid foramen" and "opening of the subscapular fossa", respectively (Andrews and Westoll, 1970a, pp.226–227, Fig. 5). In osteolepid A, the supraglenoid foramen (*sgl.f.*) is comparatively small, whereas the supracoracoid foramen (*scor.f.*) is remarkably large. The anterior opening of the subscapular fossa (*sscap.f.*) is very small (even smaller than the supraglenoid foramen) and is almost completely surrounded by the distal expansions of the dorsal and anterior buttresses (Fig. 2A). The subscapular fossa itself is a relatively large space limited laterally by the mesial face of the cleithrum and mesially by the lateral face of the central part of the endoskeletal shoulder girdle.

The orientation of this shoulder girdle is difficult to determine since it has only been found by itself, but it may have been positioned as shown on Fig. 2A, by comparison with articulated osteolepids from Europe (Jarvik, 1948) and North America.

(c) Other skeletal elements referred to osteolepid A. Some other osteolepid remains have been found in the same bone-bed as the shoulder girdle just described. Since their size fits that of the latter relatively well, they are provisionally referred to the same taxon and will be briefly described here.

(*i*) *Fronto-ethmoidal shield.* An imperfect snout (Fig. 3), or fronto-ethmoidal shield, shows the same general outline as that of typical osteolepids from Scotland. The natural cast of its internal cavities is partly preserved, namely that of the canals for the olfactory nerves (*olf.c.*) and of the nasal cavities (*nas.c.*).

In dorsal view, the impression of the ventral surface of the dermal bones shows a posterior median roughened area corresponding to the attachment of the underlying endocranium. This roughened surface is not visible in front of the pineal foramen, where the endoskeleton was still present before the bone was etched away. The pineal foramen (*pi.*) is relatively small and situated on a level with the posterior part of the orbital notch (*orb.*), as in some other osteolepids such as *Thursius* (Jarvik, 1948, Fig. 66; Vorobjeva, 1977, Fig. 25). The orbital notch is preserved only on the right side

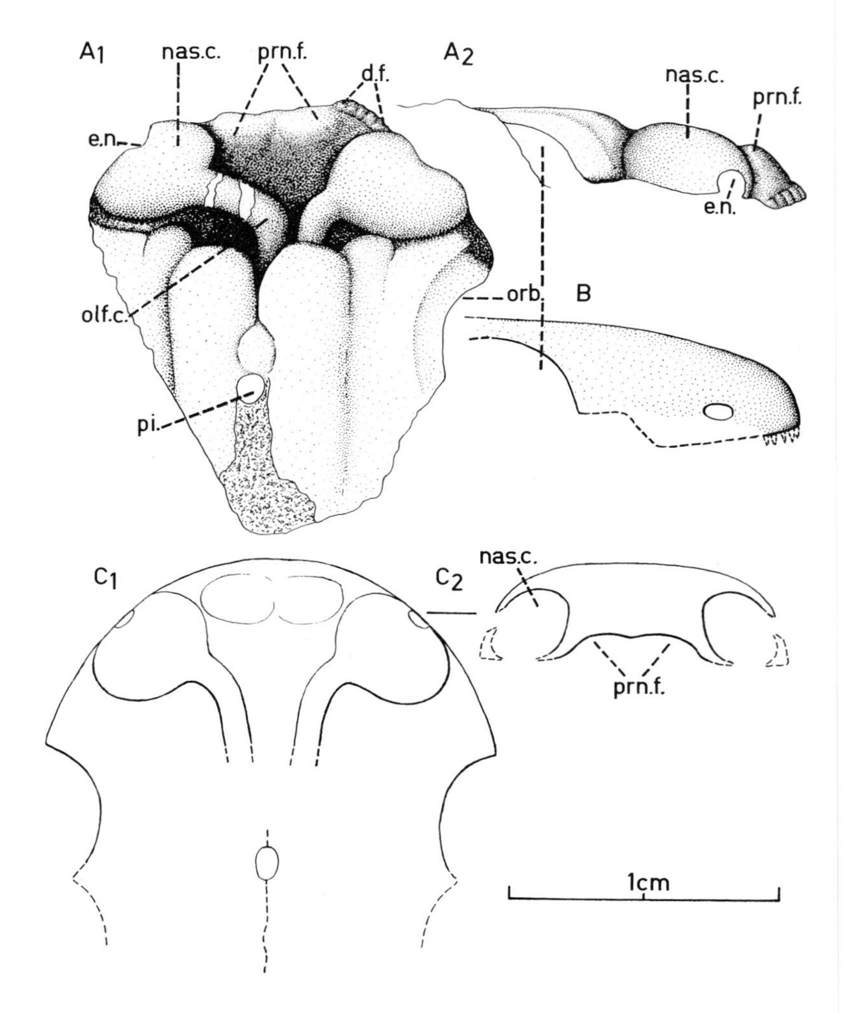

Fig. 3. Osteolepididae gen. et sp. indet. ("osteolepid A"), uppermost Emsian of Lower Eifelian, Khush-Yeilagh, Eastern Alborz, Iran. (A) Imperfect fronto-ethmodial part of the head, preserved as a natural cast in dorsal (A_1) and lateral (A_2) view (Iran Museum of Natural History, No. D.F. 0086); (B) reconstruction of the external aspect of the same specimen, obtained from a cast of its counterpart, lateral view; (C) attempted reconstruction of the internal cavities of this specimen, showing the nasal cavities in their probable natural position (C_1) and transverse section through the anterior part of the specimen (C_2). See key to abbreviations in legend to Fig. 2.

and is relatively large. In front of the pineal foramen there is a small, rounded impression corresponding to a shallow pit on the ventral surface of the dermal bone. It may represent the location of the parapineal organ.

The canals for the olfactory nerves were slightly displaced to the right during fossilization. The actual course of these canals is reconstructed in Fig. $3C_1$. The oral margin of this snout is almost completely broken off and the position of the external nares is difficult to infer from this specimen (*e.n.*). However, a small portion of the oral margin is visible anteriorly to the right nasal cavity and shows small dental fossae (*d.f.*). The cast of the prenasal fossa (*prn.f.*) is transversally elongate in shape and relatively high (Fig. $3A_2$). It clearly shows a median groove corresponding to the median ridge which separates the two adjacent parts of the prenasal fossa in other osteolepiforms (Jarvik, 1948, Fig. 37, "*ri.in.*", 1966, Fig. 17B, "*d.pnc*"; Vorobjeva, 1977, Fig. 3, "*cpet*"). The prenasal fossa of osteolepid A had basically the same shape as that of *Thursius* (Vorobjeva, 1975, Fig. 3, 1977, Fig. 3B) or, even of *Megalichthys* (Jarvik, 1966, Fig. 17), and did not extend backwards as far as the internasal cavity of porolepiforms.

In all, this badly preserved snout referred to ostoepid A fits the general osteolepid pattern relatively well: this pattern may be a primitive condition for the Osteolepiformes.

(*ii*) *Lower Jaw.* An incomplete lower jaw of the right side showing typical osteolepid characteristics is also referred to osteolepid A (Fig. 4). Its general outline and features recall the internal face of the jaw of Osteolepididae gen. et sp. indet. drawn by Jessen (1966, Pl.10,2), although its anterior end seems to be somewhat lower and the adsymphysial dental plates are missing.

2. *Osteolepid B*

Osteolepid B has been found in the red sandstone of the Armutgözlek Tepe, in Turkey. Some remains of this form have already been briefly described (Janvier and Marcoux, 1976; Janvier, 1977b), but new material permits a complementary description of the shoulder girdle.

(a) Cleithrum. Contrary to that of the osteolepid A, the cleithrum of the osteolepid B does not show any pronounced mesial curvature.

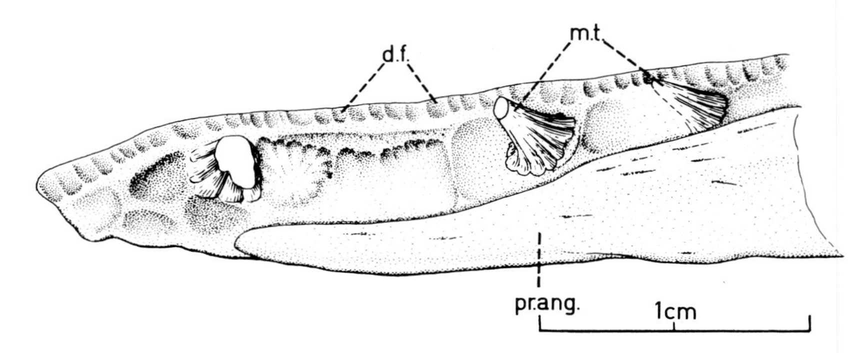

Fig. 4. Osteolepididae gen. et sp. indet. ("osteolepid A"), uppermost Emsian of Lower Eifelian, Khush-Yeilagh, Eastern Alborz, Iran. Lower jaw of the right side in medial view (Iran Museum of Natural History, No. D.F. 0090). See key to abbreviations in legend to Fig. 2.

Its ventral division is only slightly bent in a mesial or anteromesial direction. As pointed out by K. S. Thomson (personal communication, 1976), the orientation given to the cleithrum of this form in its preliminary description (Janvier and Marcoux, 1976, Fig. 1) is wrong. The dorsal tip should be rotated about 25° backwards and probably had the orientation shown here in Fig. 5A. It is approximately the orientation of the cleithrum in *Ectosteorhachis nitidus* (Thomson and Rackoff, 1974, Fig 1, Pl. 2,2). Furthermore, the cranial material referred to osteolepid B recently discovered in the Armutgözlek Tepe locality shows some possibly derived characteristics of the Megalichthyinae, namely the large anterior folded teeth borne by the posterior process of the premaxillary (Janvier, 1977b; Janvier *et al.*, 1979).

The medial face of the cleithrum shows the overlap area for the supracleithrum and, posteroventrally, an overlap area for the trunk scales.

The external surface of the cleithrum is only known for the ventral division. It shows an ornamentation of confluent ridges which fades away near the posterior margin of the bone, where a smooth cosmine covering persists. There is an overlap area for the clavicle, but no pronounced anterior process of the kind encountered in *E. foordi* (Janvier and Marcoux, 1976, Fig. 1C).

(b) Endoskeletal shoulder girdle. The endoskeletal shoulder girdle, or scapulocoracoid, is well preserved as a natural mould in the

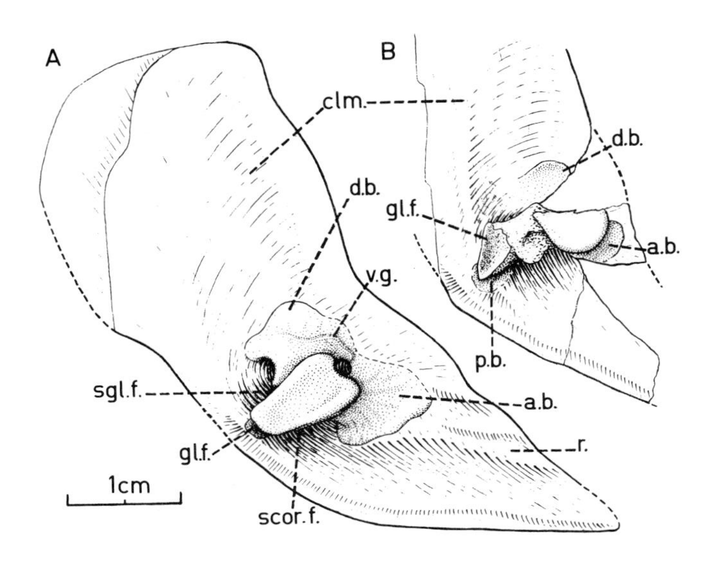

Fig. 5. Osteolepididae gen. et sp. indet. ("osteolepid B"), Upper Devonian (?Frasnian), Armutgözlek Tepe, Antalya region, Turkey. (A) Cleithrum and endoskeletal shoulder girdle of the left side, medial view (M.T.A. enstitüsü, Ankara, No. A.T. 01. Same specimen as in Janvier and Marcoux (1976) but redrawn after chemical preparation); (B) imperfect cleithrum and endoskeletal shoulder girdle, medial view. (M.T.A. enstitüsü, Ankara, No. A. T. 26.) See key to abbreviations in legend to Fig. 2.

specimen A.T.01a (Figs 5 and 6). Its overall morphology does not differ markedly from that of other osteolepiforms, that is, it possesses three buttresses and a pear-shaped glenoid fossa. When compared to the endoskeletal shoulder girdle of osteolepid A, its central part is more massive and the dorsal and anterior buttresses are much broader. The latter buttress is also more ventrally expanded. The glenoid fossa (Fig. 6B) is more rounded in shape and comparatively larger. One of the most peculiar features of this specimen is that the central part is produced dorsomedially and ventromedially into prominent ridges which are prolonged anteriorly by rounded processes (Fig. 6A). The mesial surface of the bone is also slightly concave and is prolonged posteriorly by a short and rounded process, corresponding to the pointed mesial process described in osteolepid A. Finally, a shallow sinuous groove is visible on the distal expansion of the dorsal buttress (*v.g.*, Figs 5A and 6A) and may represent the passage of the subclavian artery.

Some of the characteristics observed on specimen A.T.01a, namely

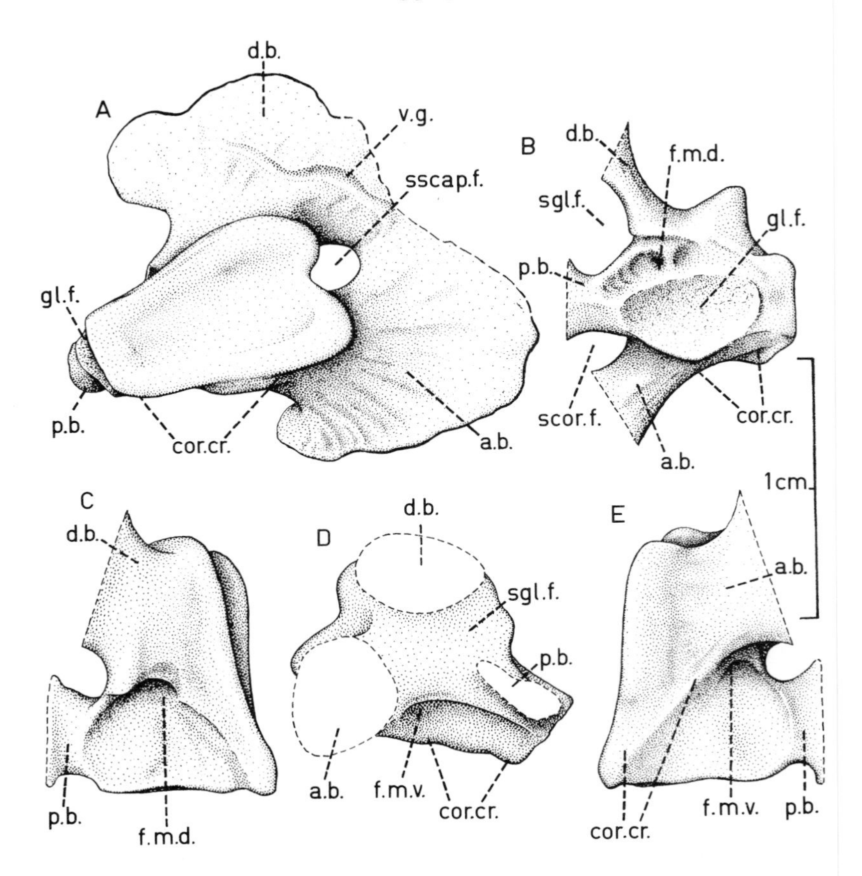

Fig. 6. Osteolepididae gen. et sp. indet. ("osteolepid B"), (Upper Devonian – ?Frasnian), Armutgözlek Tepe, Antalya region, Turkey. Endoskeletal shoulder girdle of the specimen A.T.01. (A) Medial view; (B) posterior, and slightly ventral view; (C) dorsal view; (D) lateral view; (E) ventral view. See key to abbreviation in legend to Fig. 2.

the ventromesial crest and the large size of the glenoid fossa, are confirmed by specimen A.T.26 (Fig. 5B) in which, however, the endoskeletal shoulder girdle is incomplete.

The details of the surface of the endoskeletal shoulder girdle of osteolepid B have been interpreted (Janvier and Marcoux, 1976, Fig. 2) in the light of the reconstruction of the pectoral fin musculature in *E. foordi* given by Andrews and Westoll (1970a, Figs 31 and 32). It is obvious that at least two main muscle masses, a dorsal one and a ventral one, were inserted on the endoskeletal shoulder girdle. They may have been inserted in the fossae observed on the dorsal and ventral faces of the central part respectively (*f.m.d.*, *f.m.v.*).

DISCUSSION

Leaving aside the much debated problem of the origin of the urodeles (Jarvik, 1942, 1962, 1966, 1972, 1975) the majority of the tetrapods, if not all of them, are generally considered as "descending" from the Osteolepiformes and as forming with the latter a systematic unit sometimes referred to as the "osteolepiform–tetrapod stock" (Jarvik, 1965, 1966, 1972). Actually, it is beyond doubt that the primitive tetrapods share with the known Osteolepiformes a certain number of characteristics (e.g. the processus dermintermedius) that are not found in any other group of Gnathostomata. Further, the similarities described by Jarvik (1942) in the snout region between Osteolepiformes and some primitive tetrapods are convincing enough. However, the main problem which remains to be solved in this field is to decide whether the tetrapods only share an immediate common ancestor with the Osteolepiformes as a group (Fig. 7B), or whether they are more closely related to one particular group of the Osteolepiformes than to the others (Fig. 7A). In other words, the question is whether the Osteolepiformes are monophyletic or paraphyletic. In the latter case, of course, the name "Osteolepiformes" may still be tolerated as a stem group.

Jarvik expressed no definitive opinion on which group of osteolepiforms may have given rise to the tetrapods (the Eutetrapoda, in the sense of this author), although he suggested that the "Anurans have developed from primitive Osteolepiformes" (Jarvik, 1942, p. 651). Thomson (1968) suggested that the ancestry of the tetrapods should be searched for among the Rhizodontidae (at that time rhizodonts included *Eusthenopteron*). He suggested that *Hyneria* shows some features approaching the stegocephalian skull-roof pattern. A similar opinion is put forward by Andrews and Westoll (1970b, p.482) concerning the relationships between the tetrapods and the "group containing *Eusthenopteron*". Gross (1964) suggested that the tetrapods were rooted near *Panderichthys*. More recently, Vorobjeva (1975, 1977) suggested that the tetrapods are more closely related to the Osteolepididae, and possibly even to one particular sub-family of that group (Fig. 7A), than to any other group of the Osteolepiformes.

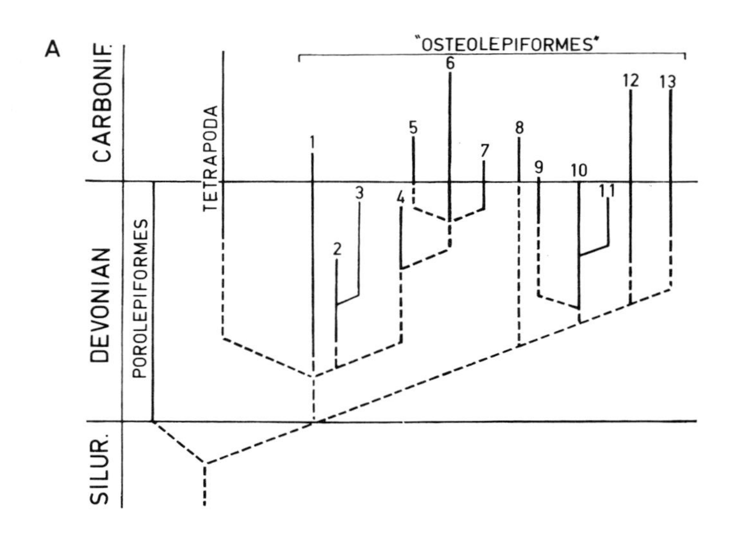

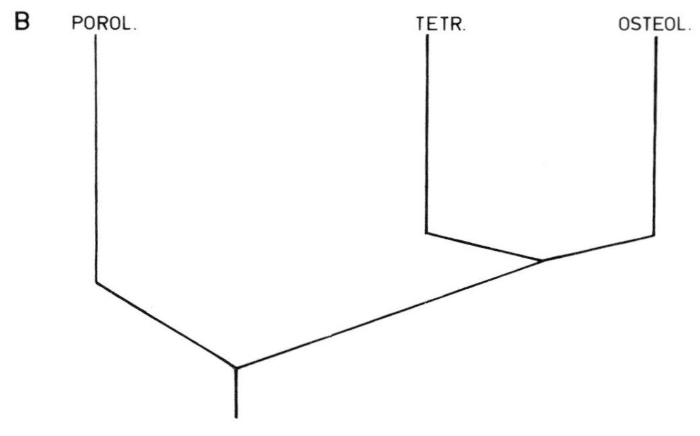

Fig. 7. (A) Interrelationships of the choanates, according to Vorobjeva (1977). (1 - 7) Osteolepididae; (8) Lamprotopeltidae; (9) Panderichthyidae; (10) Eusthenopterinae; (11) Platycephalichthinae; (12) Rhisodopsidae; (13) Rhizodontidae. (B) Interrelationships of the choanates, in the case when the Osteolepiformes are considered as monophyletic. Osteol., Osteolepiformes; Porol., Porolepiformes; Tetr., main tetrapod stock.

She pointed out some peculiar features in the cranial anatomy of *Gyroptychius* which recall certain stegocephalian characteristics.

Another possibility could be that none of the known Osteolepiformes shows more derived tetrapod characteristics than the others, and that all of them represent a monophyletic group which only shares with the tetrapods an immediate common ancestor. In that case, the tetrapod features found in some Osteolepididae would probably represent primitive characteristics of the osteolepiform–tetrapod ensemble. According to this view, the tetrapods and the Osteolepiformes would be sister-groups (Fig. 7B). The hypothesis of this type of relationship has not been popular, since it makes the "ancestors" of the tetrapods unknown as fossils and, consequently, is less satisfactory for eclectic palaeontologists.

Notwithstanding the fact that the cranial anatomy probably provides more significant characteristics, these two possible tetrapod–osteolepiform relationships will be discussed here on the basis of the structure of the endoskeletal shoulder girdle.

1. Possible Homologies in the Endoskeletal Shoulder Girdle of the Teleostomi

Although minor differences in the shape of the endoskeletal shoulder girdle of the various osteolepiform groups may occur, it is obvious that the triradiate shape of this bone is a common feature which does not occur in any other known teleostome group. It is almost certain that the endoskeletal shoulder girdle of the Porolepiformes (Andrews and Westoll, 1970b, Fig. 18; Jarvik, 1972, Fig. 52B) differs greatly from that of the Osteolepiformes in the way it is attached to the cleithrum. It also differs from that of the Coelacanthiformes (Millot and Anthony, 1958) in which the endoskeletal shoulder girdle is a massive knob of bone attached to the cleithrum. On the contrary, an endoskeletal shoulder girdle of the fossil and extant primitive Actinopterygii (namely the Acipenseridae and the Palaeonisciformes) may show some similarities to that of the Osteolepiformes (Fig. 8A). On the basis of these similarities, homologies have been proposed by Romer (1924): (see also Andrews and Westoll, 1970a, Fig. 5). The thorough description of the shoulder girdle in several fossil actinopterygian genera by Jessen (1972) gives further material for this comparison. It appears that the processus medialis of the scapular division (*prm.sc.*, Fig. 8A) in actinopterygians (= "mesocoracoid" arch) may be homologous to the dorsal buttress of that of the Osteolepiformes. The medium-sized foramen piercing the midpart of the endoskeletal shoulder girdle in the Palaeonisciformes ("*kmp*", Jessen, 1972, Figs 9 and 10) and in the Acipenseridae (Fig. 8A) may be homologous to the supracoracoid foramen of the Osteolepiformes (and primitive tetrapods). Consequently, the division of the midpart of the actinopterygian shoulder

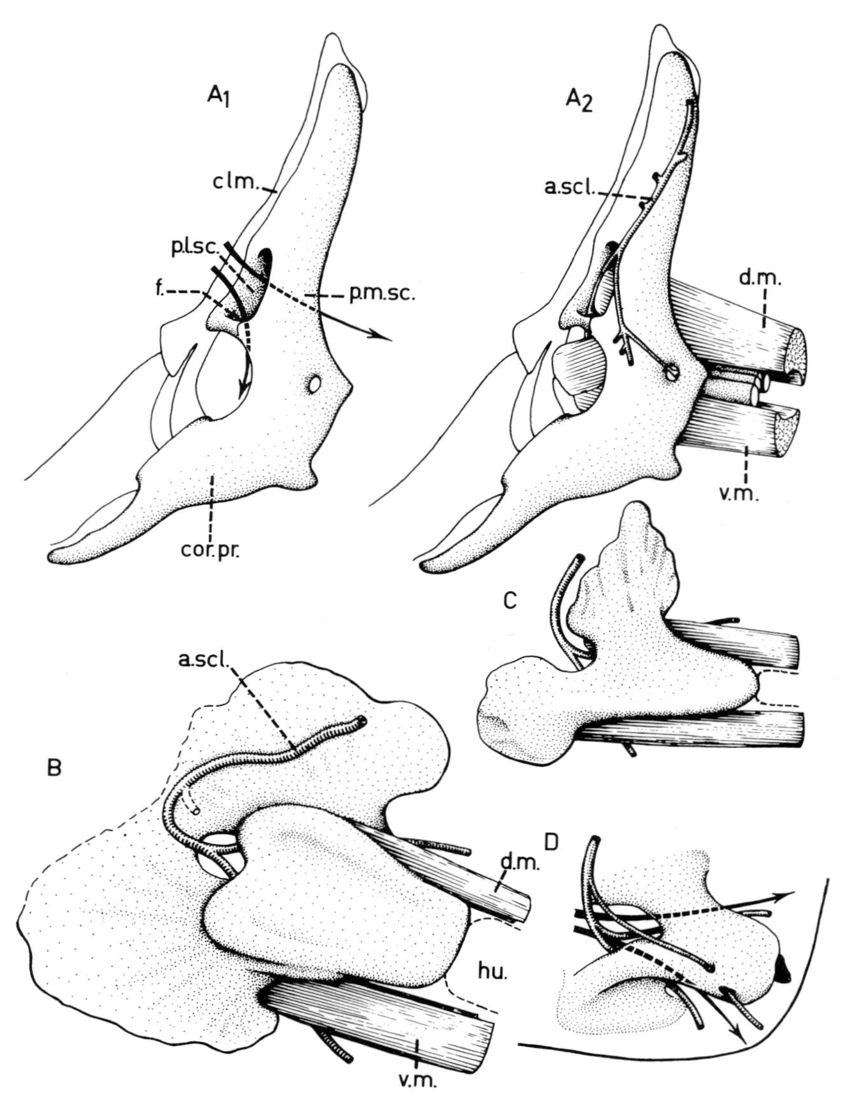

girdle extending anteriorly to the foramen in question may correspond to the anterior buttress in the Osteolepiformes. Finally, the coracoid process (*cor.pr.*) of primitive actinopterygians ("*cor*", Jessen, 1972, Figs 2, 9 and 10) occupies the same position as the coracoid crest of the Osteolepiformes and the coracoid plate of the tetrapods. Besides the shape of the glenoid fossa, the main difference between the endoskeletal shoulder girdle of the primitive Actinopterygii and that of the Osteolepiformes rests on the fact that in the former the girdle is attached to the cleithrum by the whole surface of a large endoskeletal plate, the processus lateralis of the scapular division (*p.l.sc.*, Fig. 8A), whereas in the latter it is attached only by means of the separate buttresses. Consequently, the scapular fossa of the Osteolepiformes is closed laterally by the cleithrum, whereas the "upper muscular canal" of the Actinopterygii (Jessen, 1972, Pl. 13,3, "*o.m.*"), which may correspond to the subscapular fossa, is closed laterally by the *processus lateralis* of the scapular division.

As to the soft anatomy, very little can be said of the shoulder girdle region of the Osteolepiformes. However, it seems reasonable to assume that the musculature from girdle to fin consisted at least of a dorsal and a ventral main muscle mass (*d.m.*, *v.m.*, Fig. 8B,C), as in primitive actinopterygians. These muscles allowed vertical movements whereas rotations were effected by smaller muscles, the insertion of which cannot be located with certainty on the shoulder girdle. The two main muscle masses were probably inserted in the dorsal and

---

Fig. 8. The musculature and the main arteries of the shoulder girdle in the Actinopterygii and the Osteolepiformes. (A) *Acipenser ruthenus* L., endoskeletal and exoskeletal shoulder girdle of a 24 mm embryo, medial view (redrawn from Jessen, 1972, Pl.13, and simplified). In A_1, the arrows indicate the position of the foramina which are supposed here to be homologous to the supraglenoid and supracoracoid foramina of the Osteolepiformes respectively. In A_2, the dorsal (adductor) and ventral (abductor) main muscular units are restored, as well as the subclavian artery. (B–D) Endoskeletal shoulder girdle of three Osteolepiformes, medial view. The subclavian artery and the brachial arteries are reconstructed on the basis of the groove described in osteolepid B. The dorsal and ventral main muscular units are reconstructed only in (B) and (C). A small medial artery passing through the glenoid foramen is reconstructed in (D). (B) Osteolepid B; (C) osteolepid A; (D) *Eusthenopteron foordi* (from Andrews and Westoll, 1970a). See key to abbreviations in legend to Fig. 2.

ventral fossae described in osteolepid B (*f.m.d.*, *f.m.v.*, Fig. 6), but it is not impossible that some lateral components of these masses passed through the supraglenoid and supracoracoid foramina respectively, and were inserted on some part of the walls of the subscapular fossa, as suggested by Andrews and Westoll (1970a, Fig. 31, "*m.sc.hum.*", "*m.s.cor.*"). This may be true in particular for the dorsal muscle mass in osteolepid A (Fig. 8C). The coracoid crest of the Osteolepiformes enclosed the ventral muscle mass medially and thus occupied the same position as the coracoid division of the Actinopterygii (*cor.pr.*, Fig 8A).

The vascularization and innervation of the shoulder girdle and of the proximal part of the fin in the Osteolepiformes may not have been fundamentally different from that of the primitive Actinopterygii. If the groove observed in osteolepid B (*v.g.*, Fig. 7) represents the trace of the a. subclavia (Fig. 8B), this vessel occupied the same position as that of the Acipenseridae in relation to the scapular division. This artery entered the subscapular fossa through its anterior opening (between the dorsal and anterior buttresses) and may have sent off branches to the dorsal and ventral muscle masses respectively. These arterial branches passed through the supraglenoid and supracoracoid foramina respectively (Fig. 8B,C,D). In osteolepiforms having a glenoid foramen (*Eusthenopteron foordi, Ectosteorhachis nitidus*) a medial branch, issuing from the subclavian artery, may have passed through this foramen and irrigated the ventromedial part of the pectoral fin (Fig. 8D), in the same way as the metapterygial main artery in the Acipenseridae (Jessen, 1972, Pl.13,3, "*mam*").

2. *Phylogenetic Remarks*

The endoskeletal shoulder girdle of the Teleostomi is represented, in the Devonian, as two major patterns: a massive knob like that of the Coelacanthiformes and of the Dipnoi on one hand (see however the Addendum at the end of this paper), and the largely perforated or triradiate one of the Actinopterygii, the Osteolepiformes and the tetrapods on the other. It is difficult to determine which of these two types of endoskeletal shoulder girdle is plesiomorphous for the Teleostomi, since outgroup comparison with that of the Elasmobranchiomorphi or that of the Acanthodii does not provide much information. However, the endoskeletal shoulder girdle of *Acanthodes*

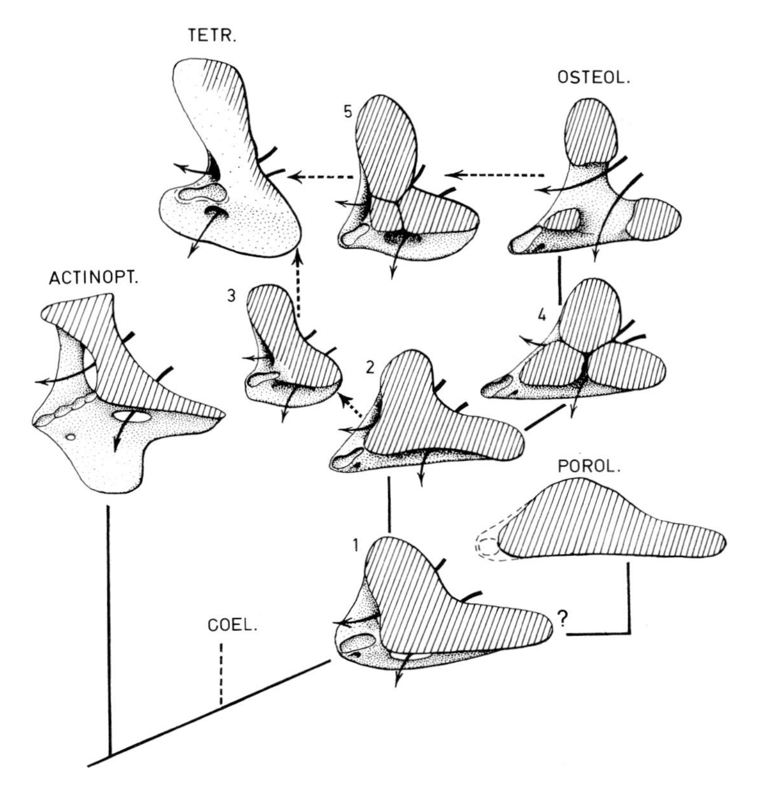

Fig. 9. Evolution of the endoskeletal shoulder girdle in the Teleostomi. In this diagram, the presence of passageways through the endoskeletal shoulder girdle for the brachial nerves and vessels is supposed to be plesiomorphous for the Teleostomi. Endoskeletal shoulder girdle in lateral view. Surface in contact with the cleithrum is cross-hatched. (1) Hypothetical choanate ancestor; (2) hypothetical common ancestor of the tetrapods and the Osteolepiformes; (3) hypothetical pretetrapod (the Osteolepiformes being monophyletic); (4) hypothetical pre-osteolepiform; (5) hypothetical pretetrapod (the Osteolepiformes being paraphyletic and the tetrapods "descending" from one particular group of the Osteolepiformes). Actinopt., Actinopterygii; Coel., Coelacanthiformes; Tetr., tetrapods; Osteol., Osteolepiformes; Porol., Porolepiformes.

(Moy-Thomas and Miles, 1971, Fig.4.12) resembles in some respects that of some primitive actinopterygians, and shows a foramen piercing its midpart ventrally. If the Acanthodii are considered as the sister-group of the Teleostomi (Miles, 1973, 1975), this may indicate that the type of endoskeletal shoulder girdle found in the Actin-

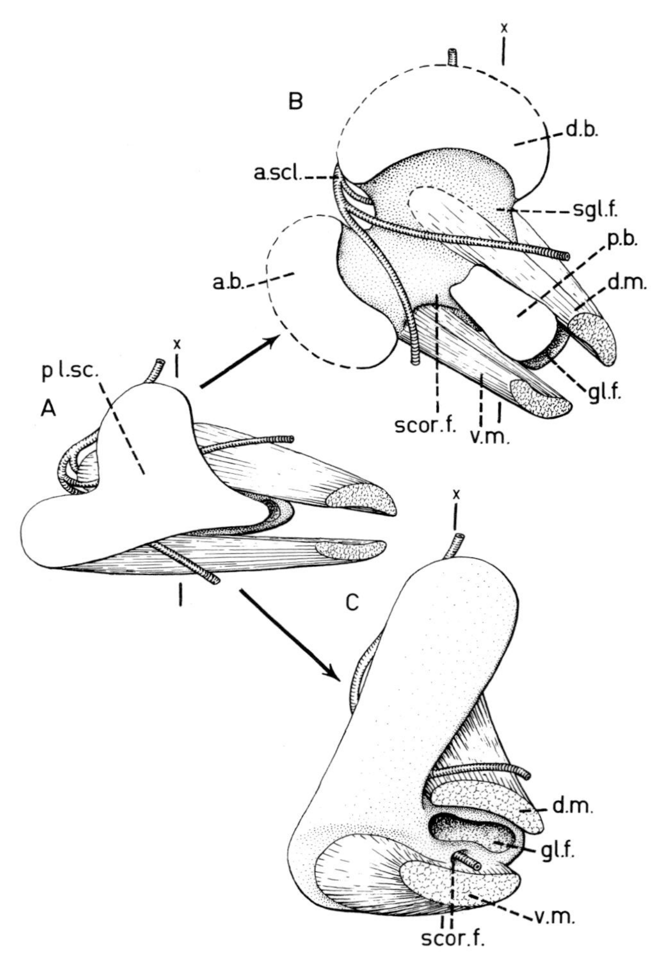

Fig. 10. The endoskeletal shoulder girdle and some associated muscles and arteries in the tetrapods, the Osteolepiformes and their hypothetical common ancestor, lateral view. (A) Hypothetical common ancestor; (B) generalized osteolepiform pattern; (C) generalized primitive tetrapod pattern. See key to abbreviations in legend to Fig. 2.

opterygii and the Osteolepiformes is plesiomorphous for the Teleostomi. The small, massive and knob-like endoskeletal shoulder girdle of the Coelacanthiformes and post-Devonian Dipnoi would, in this case, represent convergently derived conditions.

Considering this working hypothesis, we may first note that the endoskeletal shoulder girdle of the Osteolepiformes differs from that

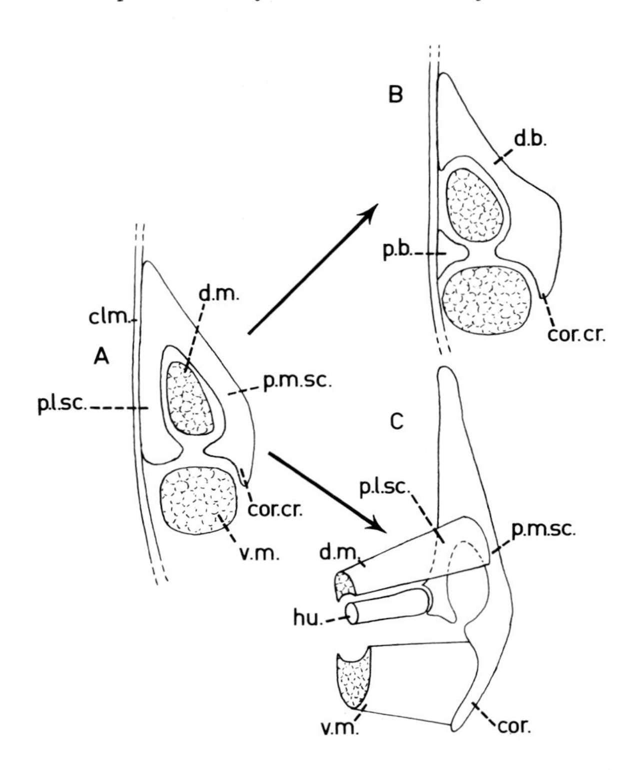

Fig. 11. Transverse section through the midpart of the endoskeletal shoulder girdle of the Osteolepiformes, the tetrapods and their hypothetical common ancestor. The position of these sections is indicated on Fig. 10 (X). (A) Hypothetical common ancestor; (B) generalized osteolepiform pattern; (C) generalized primitive tetrapod pattern. See key to abbreviations in legend to Fig. 2.

of the Actinopterygii, e.g. by the absence of any equivalent of the processus lateralis of the scapular division, and consequently the subscapular fossa is closed laterally by the cleithrum. This latter state undoubtedly represents a derived condition. In the Porolepiformes, there is no evidence that the endoskeletal shoulder girdle was attached to the cleithrum by three separate buttresses. On the contrary, it seems to be attached by the means of a single large plate, possibly corresponding to the processus lateralis of the primitive teleostome endoskeletal shoulder girdle (Fig. 9). Thus, the lack of processus lateralis and the presence of three buttresses separately attached on the cleithrum may reasonably be considered as an autapomorphy of the Osteolepiformes (Figs 9, 10 and 11).

Turning now to the question of osteolepiform–tetrapod relationships, we may note that the endoskeletal shoulder girdle of most of the early stegocephalians shows the same cavities and foramina as that of the Osteolepiformes, that is, the subscapular fossa, the supraglenoid and the supracoracoid foramina. The coracoid plate is most probably derived from the part of the bone extending ventromedially to the supracoracoid foramen, and thus is at least topographically homologous to the ridge referred to here as the "coracoid crest" in the Osteolepiformes. The supraglenoid foramen occupied the same position as that in the Osteolepiformes. The main difference between the endoskeletal shoulder girdle of the tetrapods and that of the Osteolepiformes is the fact that the subscapular fossa of the former is closed laterally by a thick wall of endoskeletal bone (Fig. 9). Consequently, the generally accepted ancestor–descendant relationship between the Osteolepiformes and the tetrapods implies that this endoskeletal wall of the subscapular fossa is a neoformation. It may have appeared by fusion of the distal expansions of the buttresses (Fig. 9,*5*), as is also partly the case in osteolepid B (Fig. 6A). The other possibility advocated here is that the tetrapods and their fish "ancestors" never possessed three completely separate buttresses and were derived from a form still having an endoskeletal lateral wall of the subscapular fossa (Figs. 8,*3* and 10A). This latter possibility is one of the most parsimonious interpretations. The phylogenetic consequences have, however, to be tested by similar study of other anatomical features.

In sum, the absence of any triradiate endoskeletal shoulder girdle in the early tetrapods indicates that this characteristic represents an autapomorphy of the Osteolepiformes and consequently that the tetrapods are the sister-group of the latter. Certain resemblances between the stegocephalians and the Osteolepididae may be due to the fact that the latter still possess a great number of plesiomorphous characteristics and, consequently, are morphologically closer to the hypothetical common ancestor of the tetrapods and the Osteolepiformes than is, for instance, *Eusthenopteron*.

3. Remarks on the functional anatomy

One of the major morphological changes that has occurred during the fish–tetrapod transition concerns the orientation of the glenoid fossa.

In the choanate fishes, as in the majority of other bony fishes, the glenoid fossa faces backwards and slightly laterally. In the early tetrapods, it faces laterally or slightly posterolaterally. During this migration, the dorsal muscle mass still remained inserted dorsally into the glenoid fossa or near the supraglenoid foramen (Romer and Byrne, 1931, Fig. 8A), and probably did not increase considerably in size. By contrast, the ventral muscle mass became much enlarged and acquired a transverse orientation, as did the whole limb. Consequently, its proximal end had to be inserted onto a laterally facing surface. This functional problem has been solved by the development of the "coracoid crest" which limited medially the space for the ventral muscle mass in the choanate fishes (Fig. 11A,B). This crest therefore developed into a coracoid plate onto which the proximal end of the ventral muscle mass could easily expand. The development of the coracoid plate and of the ventral muscle mass probably represent major events in the evolution of the tetrapod fore limb, since the ability of the limb to support the weight of the body is in great part due to the development of its ventral musculature.

CONCLUSIONS

(1) The endoskeletal shoulder girdle in two osteolepids from the Devonian of the Middle East is described. In the osteolepid A, from the Emsian or Eifelian of north-eastern Iran, the glenoid fossa is transversely elongated, the supracoracoid foramen is very large and there is a medial pointed process extending medially to the glenoid fossa. In the osteolepid B, from the Upper Devonian of south-western Turkey, the central part of the endoskeletal shoulder girdle is produced into ventromedial and dorsomedial crests. The ventromedial crest prolongs the coracoid crest anteriorly. The distal part of the dorsal and ventral buttresses is considerably expanded and the former bears a faint groove, possibly for the subclavian artery.

(2) The endoskeletal shoulder girdle of the Teleostomi is briefly discussed. It is suggested that the presence of passageways for the blood vessels and the nerves through the endoskeletal shoulder girdle may be a synapomorphy of the Acanthodii and the Teleostomi (in that case a plesiomorphous characteristic for the Teleostomi) or, at

least, a synapomorphy of the Teleostomi. However, it would be retained only by the Actinopterygii, the Osteolepiformes, the tetrapods and possibly also the Porolepiformes (for the Dipnoi, see Addendum).

(3) The triradiate endoskeletal shoulder girdle of the Osteolepiformes, with distally separate buttresses and with a subscapular fossa closed laterally by the cleithrum, may represent a derived characteristic. In the tetrapods, the subscapular fossa of the endoskeletal shoulder girdle is closed laterally by an endoskeletal wall, as in the actinopterygians and in the hypothetical teleostome ancestor. Consequently, the derivation of the tetrapod from any osteolepiform in which the endoskeletal shoulder girdle is known implies the neoformation of this lateral wall. On the other hand, derivation of the tetrapods from hypothetical choanate fish still possessing this wall and already having the synapomorphies of the tetrapods + Osteolepiformes, is more parsimonious. According to this view, the tetrapods would be the sister-group of the Osteolepiformes.

(4) In contradistinction to this latter view, some features of the cranial anatomy of early representatives of the Osteolepididae and of primitive tetrapods may be considered as synapomorphies and, consequently, indicate that the tetrapods are more closely related to the Osteolepididae (or even to the Gyroptychiinae, according to Vorobjeva, 1975, 1977) than to any other group of the Osteolepiformes. In this case, the Osteolepiformes would represent a stem group and would be paraphyletic. The choanate fishes probably underwent a mosaic evolution, similar to that of the therapsids, for example, and the possibility cannot be excluded that some supposed "stegocephalian characteristics" appeared independently in several families of the Osteolepiformes.

ADDENDUM

During the present symposium, Dr B. Gardiner (London) presented entirely new views on the phylogeny of the choanates (see Chapter 8 of this volume). According to him, the Osteolepiformes would be a paraphyletic stem group from which were derived both the Dipnoi and the Tetrapoda. These views are quite different to those expressed in the present paper and seem to be inconsistent with the autapo-

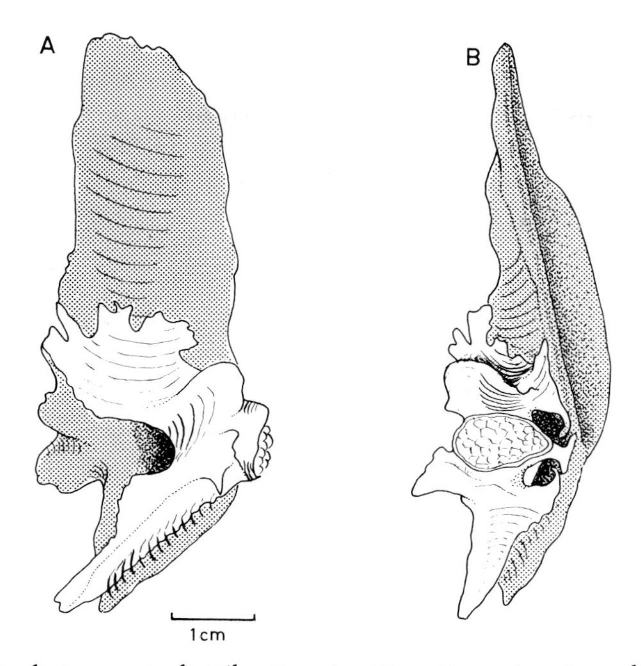

Fig. 12. *Chirodipterus australis* Miles, Frasnian, Gogo Formation, Australia, BMNH P.52570. Cleithrum and endoskeletal shoulder girdle of right side in medial (A) and posterior (B) view. Cleithrum dotted. Camera lucida drawing, with permission of Dr P. Forey (British Museum) and Dr S. M. Andrews (Royal Scottish Museum). Note the remarkable triradiate structure of the endoskeletal shoulder girdle, which resembles very much that of osteolepid B (Figs 5 and 6) in having the same ventromedial and dorsomedial crests.

morphous state of the triradiate endoskeletal shoulder girdle of the Osteolepiformes. However, Dr Gardiner kindly pointed out to me the fact that the endoskeletal shoulder girdle of the primitive lungfish *Chirodipterus*, from the Frasnian of Gogo, Australia, is triradiate in very much the same manner as that of the Osteolepiformes (unpublished material under study by Dr Mahala Andrews). In the meantime, Dr Peter Forey confirmed this observation by sending me a freehand sketch of the endoskeletal shoulder girdle of *Chirodipterus* (Fig. 12). The latter is triradiate, with *separate* buttresses and medial crests which recall those described here in osteolepid B from Turkey. If the lateral endoskeletal wall of the subscapular fossa has already disappeared in the primitive lungfishes, it would suggest that the Osteolepiformes and the Dipnoi share an immediate common ancestor. As to the affinities of the tetrapods, the problem remains the same as discussed here, as long as the endoskeletal lateral wall of the subscapular fossa is considered as a plesiomorphous characteristic. Considering these new facts, the tetrapods would be the sister-group of the ensemble Dipnoi + Osteolepiformes.

ACKNOWLEDGEMENTS

I would like to express my gratitude to the staff of the Iran National Museum of Natural History, in particular to His Excellency Eskandar Firouz and Dr Hind Sadek-Kooros, for providing me with field assistance. I thank also Dr Jean Marcoux, who discovered the Armutgözlek Tepe locality, as well as Alain Blieck and Michel Martin for their collaboration in collecting the material described here. I am very grateful to Dr Peter Forey for having improved and discussed this paper.

REFERENCES

Andrews, M. S. and Westoll, T. S. (1970a). The postcranial skeleton of *Eusthenopteron foordi* Whiteaves. *Trans. R. Soc. Edin.* **67**, 207–329.

Andrews, M. S. and Westoll, T. S. (1970b). The postcranial skeleton of rhipidistian fishes excluding *Eusthenopteron. Trans. R. Soc. Edin.* **68**, 391–489.

Boutière, A. and Brice, D. (1966). La série dévonienne de Chaghana-Oudjerak (province de Ghazni, Afghanistan). *C. r. Acad. Sci., Paris* **263**, 1940–1942.

Brunn, J. H., de Graciansky, P. C., Gutnic, M., Juteau, T., Lefèvre, R., Marcoux, J., Monod, O. and Poisson, A. (1970). Structures majeures et corrélations stratigraphiques en les Taurides occidentales. *Bull.Soc.géol.France* **12**, 515–556.

Gardiner, B. and Miles, R. S. (1975). Devonian fishes of the Gogo Formation, Western Australia. *Colloques int. C.n.R.S.* No. 218, 73–78.

Gross, W. (1964). Polyphyletische Stämme im System der Wirbeltiere. *Zool. Anz.* 1973, 1–22.

Gupta, V. J. and Denison, R. H. (1966). Devonian fishes from Kashmir, India. *Nature, Lond.* **211**, 177–178.

Gupta, V. J. and Janvier, P. (1979). A review of the Devonian vertebrate localities of the Indian Himalayas (Kasmir, Ladakh and Kumaun), with remarks on their stratigraphical and palaeobiogeographical significance. *In* "Himalayan Geology", pp. 78–83. Hindustan Press, Delhi.

Gupta, V. J. and Turner, S. (1973). Oldest Indian fish. *Geol. Mag.* **110**, 483–484.

Huckriede, R., Kursten, M. and Wenzlaff, H. (1962). Zur Geologie des Gebietes zwischen Kerman und Zagand (Iran). *Beih. Geol. Jb.* **51**, 1–197.

Janvier, P. (1974). Preliminary report on Late Devonian fishes from Central and Eastern Iran. *Geol. Surv. Iran Rept.* **31**, 1–48.

Janvier, P. (1977a). Les poissons dévoniens de l'Iran central et de l'Afghanistan. *Mem. h.sér. Soc. géol. France* **8**, 277–289.

Janvier, P. (1977b). Vertébrés dévoniens de deux nouveaux gisements du Moyen-Orient. Le probléme des relations intercontinentale au Paléozoïque moyen vu à la lumière de la paléobiogéographie des Rhipidistiens ostéolépiformes et des premiers tétrapodes. *Ann. Soc. géol. N.* **97**, 373–382.

Janvier, P. and Marcoux, J. (1976). Remarques sur la ceinture scapulaire d'un poisson choanate Ostéolépiforme des grés rouges dévoniens de l'Armutgözlek Tepe (Taurus Lycien occidental, Turquie). *C. r. Acad. Sci., Paris* **283**, 619–622.

Janvier, P. and Marcoux, J. (1977). Les grès rouges de l'Armutgözlek Tepe: leur faune de poissons (Antiarches, Arthrodires et Crossopterygiens) d'âge Dévonien supérieur (Nappes d'Antalya, Taurides occidentales, Turquie). *Rev. Geol. méd.* **4**, 183–188.

Janvier, P. and Martin, M. (1978). Les vertébrés dévoniens de l'Iran central. I: Dipneustes. *Géobios* **11**, 819–833.

Janvier, P. and Ritchie, A. (1977). Le genre *Groenlandaspis* Heintz (Pisces, Placodermi, Arthrodira) dans le Dévonien d'Asie. *C. r. Acad. Sci., Paris* **284**, 1385–1388.

Janvier, P., Termier, G. and Termier, H. (1979). The osteolepidiform rhipidistian fish *Megalichthys* in the Lower Carboniferous of Morocco, with remarks on the palaeobiogeography of the Upper Devonian and Permo-Carboniferous osteolepidids. *Neu. Jb. Geol. Paläont. Mh.* **1**, 7–14.

Jarvik, E. (1942). On the structure of the snout of crossopterygians and lower gnathostomes in general. *Zool. Bidrag. Upps.* **21**, 235–675.

Jarvik, E. (1948). On the morphology and taxonomy of the Middle Devonian osteolepid fishes of Scotland. *K. svenska Vetensk Akad. Handl.* (3)**25**, No. 1, 1–301.

Jarvik, E. (1962). Les Porolepiformes et l'origine des Urodèles. *Colloques int. C.n.R.S.* No. 104, 87–101.

Jarvik, E. (1965). Die Raspelzunge der Cyclostomen und die pentadactyle Extremität der Tetrapoden als Beweise für monophyletische Herkunft. *Zool. Anz.* **175**, 101–143.

Jarvik, E. (1966). Remarks on the structure of the snout in *Megalichthys* and certain other rhipidistid crossopterygians. *Ark. Zool.* **19**, 41–98.

Jarvik, E. (1972). Middle and Upper Devonian Porolepiformes from East Greenland with special reference to *Glyptolepis groenlandica* n.sp. *Meddr Grønland* **187**(2), 1–307.

Jarvik, E. (1975). On the saccus endolymphaticus and adjacent structures in osteolepiforms anurans and urodeles. *Colloques int. C.n.R.S.* No. 218, 192–211.

Jessen, H. (1972). Schultergürtel und Pectoralflosse bei Actinopterygiern. *Foss. Strata* **1**, 1–101.

Miles, R. S. (1973). Relationships of acanthodians. *In* "The Interrelationships of Fishes" (P. H. Greenwood, R. S. Miles and C. Patterson, eds), pp. 63–103. Academic Press, London and New York.

Miles, R. S. (1975). The relationships of the Dipnoi. *Colloques int. C.n.R.S.* No. 218, 133–148.

Millot, J. and Anthony, J. (1958). "Anatomie de *Latimeria chalumnae*", Vol. 1. Centre National de la Recherche Scientifique, Paris.

Miner, R. W. (1925). The pectoral limb of *Eryrops* and other primitive tetrapods. *Bull. Am. Mus. nat. Hist.* **51**, 145–312.

Moy-Thomas, J. A. and Miles, R. S. (1971). "Palaeozoic Fishes." Chapman and Hall, London.

Powers, R. W. (1968). Saudi Arabia. *In* "Lexique Stratigraphique Internationnal" (L. Dubertret, ed.), Vol. III, "Asia", pp. 1–177. Centre National de la Recherche Scientifique, Paris.

Rieben, H. (1935). Contribution à la géologie de l'Azerbaidjan Persan. *Bull. Soc. Neuchât. Sci. nat.* **59**, 19–144.

Romer, A. S. (1922). The locomotor apparatus in certain primitive mammal-like reptiles. *Bull. Am. Mus. nat. Hist.* **46**, 117–606.

Romer, A. S. (1924). Pectoral limb musculature and shoulder-girdle structure in fish and tetrapods. *Anat. Rec.* **27**, 119–143.

Romer, A. S. and Byrne, F. (1931). The pes of *Diadectes*: notes on the primitive limb. *Palaeobiologica* **4**, 24–48.

Rzhonsnitskaya, M. A. (1967). Devonian of U.S.S.R. *Calgary Alberta Soc. Pet. Geol.* 1, 331–347.

Schmidt, W. (1963). Ein Rhenanide aus der unteren Mittel Devon der Turkei. *Neu. Jb. Geol. Paläont.* **118**, 217–230.

Schultze, H. P. (1973). Large Upper Devonian arthrodires from Iran. *Fieldiana: Geology.* **23**, 53–78.

Seilacher, A. (1963). Kaledonischer Unterbau der Irakiden. *Neu. Jb. Geol. Paläont.* **118**, 527–542.

Stampfli, G. M. (1978). "Etude Géologique générale de l'Elburz Oriental au Sud de Gonbad-e-Qabus. Iran N–E." Thesis No. 1868, Université de Genève.

Thomson, K. S. (1968). A new Devonian fish (Crossopterygii, Rhipidistia), considered in relation to the origin of the Amphibia. *Postilla* **124**, 1–13.

Thomson, K. S. and Rackoff, J. S. (1974). The shoulder girdle of the Permian rhipidistian fish *Ectosteorhachis nitidus* Cope: structure and possible function. *J. Paleont.* **48**, 170–179.

Vorobjeva, E. (1975). Some peculiarities in evolution of the rhipidistian fishes. *Colloques int. C.n.R.S.* No. 218, 224–230.

Vorobjeva, E. (1977). Morfologia y ossobennosty evoliutii kisteperik ryby. (Morphology and evolutionary features of the crossopterygian fishes.) *Acad. Nauk. SSSR Trudy paleont.* Izvestia **163**, 1–239 (in Russian.)

11 | The Origin of the Tetrapod Limb and the Ancestry of Tetrapods

JEROME S. RACKOFF

Department of Geology and Geophysics, Yale University, New Haven, Connecticut, U.S.A. (Present Address: Department of Biology, Bucknell University, Lewisburg, Pennsylvania, U.S.A.)

Abstract: The paired fins of *Sterropterygion brandei*, an osteolepidid rhipidistian from the Upper Devonian of Pennsylvania, resemble the limbs of tetrapods in a suite of derived character states. Pectoral resemblances include high humeral torsion, humeral shaft and the detailed arrangement of muscular processes. A rotary elbow joint is recognized, but appears to be primitive for the Osteolepidida. The robust fin endoskeleton and fin musculature, and the fusion of the scapulocoracoid with the dermal girdle also suggest terrestrial abilities. The pelvic fin reveals the first rhipidistian tarsal joint of tetrapod design. Elbow and tarsal joints employ differential joint mobility to create oblique axes of maximum flexure by which distal parts of the fins are twisted forward, maximizing the conversion of muscular effort into forward thrust. Stout post-axial processes bordering the rotary joints provide the origin for the controlling musculature. The pattern of rotary joints presages that seen in the limbs of tetrapods. Functional explanations are provided for the differences between pectoral and pelvic fins, and for the changes in the fin-to-limb transition.

The tetrapod-like limb morphology is associated with a *dorsal* orientation of the preaxial edge of the pectoral fin, as postulated by Romer and Byrne (1931). A mobile shoulder joint, free of the constraints of basal scutes (which are independent of the fin lobe) allows terrestrial contact. In this derived fin orientation, functional integration of proximal and distal pectoral muscles was maintained by a different relative orientation of the postaxial process (entepicondyle). The tetrapod resemblances of the humerus are functionally interrelated, and dependent upon this entepicondylar position.

Systematics Association Special Volume No. 15, "The Terrestrial Environment and the Origin of Land Vertebrates", edited by A. L. Panchen, 1980, pp. 255–292, Academic Press, London and New York.

Based on weighted shared derived character states, the tetrapods (considered to be monophyletic) share a more recent common ancestry with *Sterropterygion* than with *Eusthenopteron*. The minimum age of origin of tetrapods is lower Famennian.

INTRODUCTION

One of the striking aspects of tetrapod limb design is the contrasting pattern of joint mobility in the fore and hind limbs. The elbow is capable of rotation as well as flexion, whereas the wrist is a simple hinge. Exactly the opposite is true of the hind limb, where the ankle is the rotary joint and the knee the hinge joint. The rotary joints allow the feet to be twisted into a forward position, so that the propulsive force generated by the limb musculature can be effectively

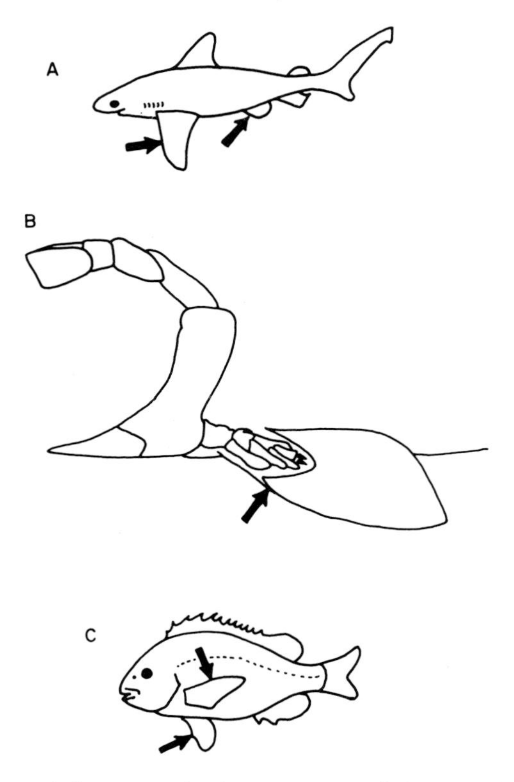

Fig. 1. Orientation of the preaxial edges (arrows) of the paired fins. (A) Shark; (B) *Eusthenopteron* pectoral fin (redrawn from Andrews and Westoll, 1970a); (C) teleost.

translated into forward motion. But why should this rotation occur in the proximal joint (elbow) in one case, and the distal one (ankle) in the other, and how did this contrasting pattern originate?

Romer and Byrne (1931), who first called attention to these problems, realized that the answers must be sought among the Devonian rhipidistian fishes that were ancestral to the tetrapods. They noted that in many primitive fishes the preaxial edges of the paired fins were directed ventrolaterally (Fig. 1A,B). In most modern bony fishes, however, the preaxial edges of the pectoral fins have been rotated to a dorsal position (Fig. 1C). Romer and Byrne suggested that this radical difference in the orientation of the pectoral and pelvic fins could account for the differences between tetrapod fore and hind limbs, and thus they predicted a preaxial-edge-up orientation for the pectoral fins of the rhipidistian ancestors of the tetrapods. Not accepting that a simple reorientation of known fin structure was possible, they chose to reverse accepted homologies for radius and ulna, thus effecting a reorientation by definition alone. This error forced them to postulate highly unnatural twists of the fin in order to place it in an appropriate position for terrestrial propulsion. The altered homologies, strange torsion and reversed functions of flexor and extensor musculature could not be accepted by other workers (Gregory, 1935a; Gregory and Raven, 1941; Westoll, 1943), and ultimately caused rejection of the entire theory, including the seminal suggestion of a dorsal preaxial edge. Subsequent research has confirmed the ventrolateral orientation of the preaxial edge of the pectoral fin in all previously known rhipidistians (Andrews and Westoll, 1970a,b). Most authors have thus tended to emphasize the fundamental similarities in structure (Westoll, 1943), function (Szarski, 1962), and evolutionary history (Eaton, 1951) of the rhipidistian pectoral and pelvic fins.

When closely compared, however, the pectoral and pelvic fins of lobe-finned fishes display a number of functionally significant differences. The more limited mobility of the pelvic fins has been described for the dipnoan *Neoceratodus* (Dean, 1906), and the coelacanth *Latimeria* (Thomson, 1966). The pelvic fins of rhipidistians are always much smaller than the pectorals, and Westoll (1943) considered the presence of a stout postaxial process (entepicondyle) on the humerus, but not the femur, to be a major problem of locomotor mechanics. The pelvic fin endoskeleton of the

Upper Devonian rhipidistian *Eusthenopteron* thus resembles the pectoral fin minus the humerus and radius (Andrews and Westoll, 1970a). Thomson (1969, p. 139, 1972) has provided the only explanation for the size difference, noting that the pectoral fins must continuously support the anterior part of the body during terrestrial progression, lest the weight of the body compress the lungs and prevent pulmonary respiration. The pelvics, which function only intermittently, need not be developed so strongly, nor in the same manner.

Any comprehensive theory of tetrapod limb origins must thus provide (a) a functional explanation for the differences in the pectoral and pelvic fins of rhipidistians (including the contrasting patterns of joints and postaxial processes), and (b) a relationship between these fin differences and the contrasting patterns of joint mobility in the fore and hind limbs of tetrapods. Due to the generally undifferentiated nature of the pelvic fin endoskeleton, the origin of the hind limb has been neglected by all but a few workers (Romer and Byrne, 1931; Gregory and Raven, 1941; Jarvik, 1965a), and no satisfactory comprehensive solution to these problems has hitherto been offered.

NEW EVIDENCE ON THE PAIRED FINS OF THE RHIPIDISTIA

Andrews and Westoll (1970a,b) have summarized a great deal of information about the structure of the paired fins in the Rhipidistia, including principally the well-known genera *Eusthenopteron* and *Sauripterus*. No attempt will be made here to repeat their excellent review of the literature. The present study arises from a detailed description of the paired fins and girdles of a new rhipidistian fish, *Sterropterygion brandei* Thomson 1972 (Family Osteolepididae) from the Upper Devonian of Pennsylvania (Rackoff, 1976). This fish provides the first detailed evidence of fin structure in the Osteolepididae, and reveals many important structural and functional attributes that clarify the problem of tetrapod origins.

In the holotype (Yale Peabody Museum 6721; Fig. 2), the fins are preserved in natural position with internal structures three-dimensionally intact. It is possible to see the correct orientation of the pre- and postaxial margins of the fins, and the nature of the elbow and ankle joints. This fish gives the first evidence of the critical functional

Fig. 2. *Sterropterygion brandei* Thomson. Right lateral view of whole specimen prior to preparation.

differences between fore and hind limbs to which Romer and Byrne (1931) referred.

The pectoral fins are ventrolaterally located and possess large fin lobes accounting for one-half of the total fin length. Fulcral scales and stout preaxial lepidotrichia clearly indicate the *dorsal* orientation of the preaxial fin margins. One of the major barriers to accepting such a fin orientation in a tetrapod ancestor has been the problem of establishing fin–substrate contact: a fin with a dorsally oriented preaxial margin would require considerably greater mobility than a ventrolaterally extended one. Such mobility would appear to be impossible because of the enlarged dorsal and ventral scutes which characteristically flank the bases of osteolepid pectoral fins. Jarvik (1944, 1948, Fig. 30, p. 111) has described these scutes as integral parts of the fin lobe which serve to attach the fins to the body wall, thus creating a long-based appendage of severely limited mobility. In *Sterropterygion*, however, the well-preserved basal scutes are clearly independent of the fin base, and thus provide no barrier to pectoral fin mobility (Fig. 3; refigured from Thomson, 1972, p. 6). The scutes, which become thinner away from the fin lobe, serve instead as streamlining devices which smooth the contour of the body when the pectoral fins are fully reflected against the flank (Rackoff, 1976; see also Breder, 1926, Fig. 72, p. 247, on "axillary scale" function in *Elops saurus*). This is surely the case in all osteolepidids.

Pectoral fin mobility is confirmed by the structure of the shoulder joint. The glenoid is an essentially oval, posteroventrally facing surface with a slight twist in the dorsolateral margin, which thus faces posteroventromedially. The humeral head is of the same size as the glenoid, and is capable of rocking movements (both dorso-ventral and side-to-side) as well as considerably rotary movement. The role of shoulder mobility in effecting fin–substrate contact was not considered by Romer and Byrne (1931), but it is supported by evidence of fin use in a live captured specimen of the coelacanth *Latimeria* (Locket and Griffith, 1972). Analysis of a motion picture taken by R. W. Griffith shows that the pectoral fin, the orientation of which is analogous to that of *Sterropterygion*, is lowered to a ventrolaterally extended position by means of the extensive shoulder mobility. Other indications of the capacity for fin–substrate contact in *Sterropterygion* are the great relative length of the fin lobe and robustness of its endoskeletal elements (Thomson, 1972), as well as

Fig. 3. *Sterropterygion brandei* Thomson. Dorsal basal scute of right pectoral fin. The specimen was preserved with the preaxial edge lifted slightly away from the body. Body scales clearly surround the base of the scute, isolating it from the fin lobe.

Fig. 4. *Sterropterygion brandei* Thomson. Right pectoral fin in lateral view, with scales of the fin

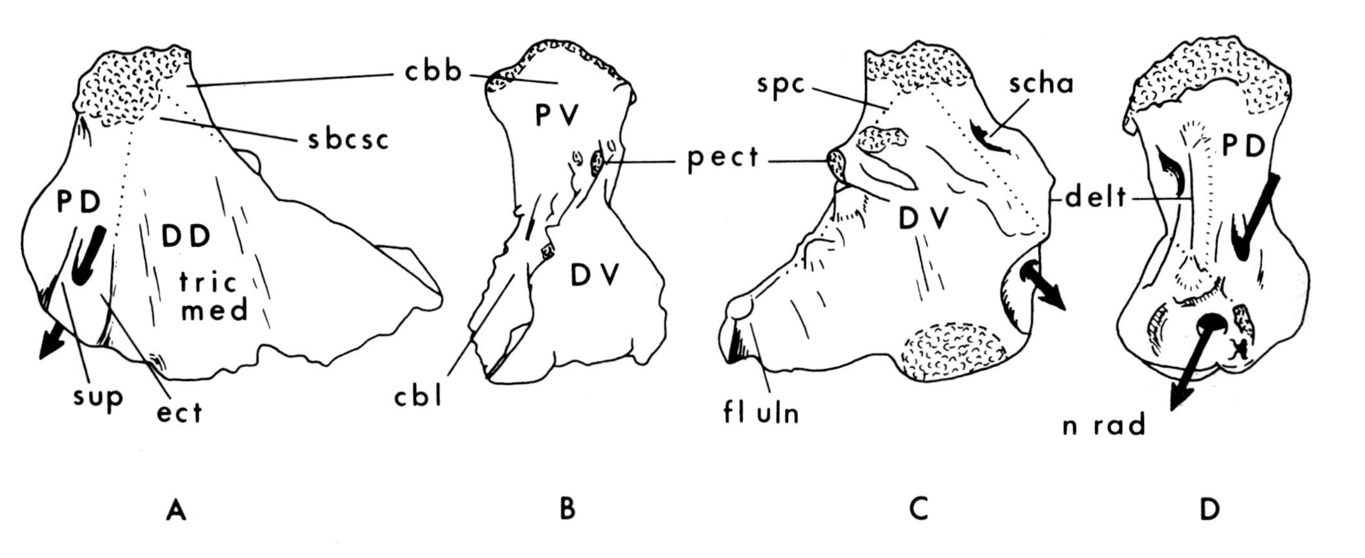

Fig 5. Left humerus of *Sterropterygion*. (A) Dorsal; (B) postaxial; (C) ventral; (D) preaxial. cbb, coracobrachialis brevis; cbl, coracobrachialis longus; delt, deltoideus; DD, distal dorsal triangular area; DV, Distal ventral triangular area; ect, ectepicondyle; fl uln, ulnar flexors; n rad, radial nerve; pect, pectoralis; PD, proximal dorsal triangular area; PV, proximal ventral triangular area; scha, scapulohumeralis anterior; sbcsc, sub-coracoscapularis; spc, supracoracoideus; sup, supinator; tric med, medial head of triceps.

the fusion of the scapulocoracoid with the dermal shoulder girdle, a strengthening feature hitherto seen among the osteolepids only in *Ectosteorhachis* (Thomson and Rackoff, 1974).

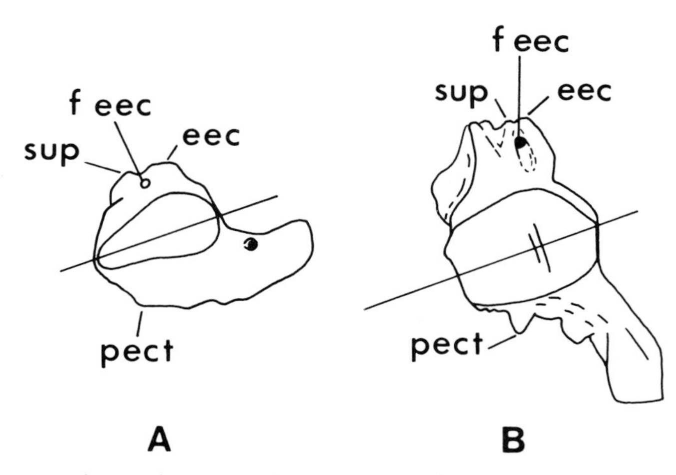

Fig. 6. Proximal articular views of right humeri of (A) *Eusthenopteron* (modified from Andrews and Westoll, 1970a) and (B) *Sterropterygion*, with long axes of humeral heads aligned. eec, ectepicondyle; f eec, foramen ectepicondyle; pect, pectoralis; sup, supinator.

The pectoral fin endoskeleton (Fig. 4) is of the biseriate type, with preaxial radial elements diverging from successive postaxial axial mesomeres. The robust humerus (first axial mesomere) has a series of well-defined muscular processes, including a large, distally-tapering postaxial process (the tetrapod entepicondyle). The arrangement of these processes (Fig. 5) differs significantly from that of *Eusthenopteron* (Andrews and Westoll, 1970a), but homologies can be confidently established by comparing proximal articular views of the two humeri with the long axes of the humeral heads aligned (Fig. 6). It is evident that a functionally constant relationship exists in both genera between the long axis of the humeral head and the location of the muscular processes on the ventral (pectoralis) and dorsal (deltoid-supinator-ectepicondyle) surfaces. (Further detailed research on other osteolepiform rhipidistians will of course be necessary to establish the generality of this relationship.) Effective aquatic locomotion apparently necessitated such constancy of the mechanical system controlling fin motion about the principal glenoid axis. These design constraints were abandoned during the rhipidistian–amphibian transition only after the development of the permanently flexed elbow and horizontal humerus—permanent commitments to terrestrial existence.

The very different appearances of the humeri of *Sterropterygion* and *Eusthenopteron* are thus primarily the result of the differences in the relative orientation of the postaxial process. Since this structure provides the origin for the distal flexor musculature, the morphological differences between the humeri reflect differing patterns of functional integration of distal and proximal musculature. Andrews and Westoll (1970b, p. 404) have similarly noted that bicondylar torsion of the humerus (discussed below) reflects the orientation of the elbow musculature. The patterns of muscular integration in *Sterropterygion* and *Eusthenopteron* are ultimately related to the different orientations of the pectoral fins. A functionally integrated system of fin muscles is thus maintained in genera with contrasting fin positions by adjustments in the relative location of the postaxial process.

The conspicuous humeral processes are convenient landmarks for reconstructing the four triangular areas that characterize the tetrahedral humerus of early tetrapods (Romer, 1922), and the humerus of *Eusthenopteron* (Andrews and Westoll, 1970a). Within this geometric framework, the attachments of other muscles, not represented by bony processes, are readily interpolated (Fig. 5). Thus restored, the humerus of *Sterropterygion* displays a remarkable suite of tetrapod resemblances (Figs 5 and 7). Viewed preaxially or postaxially, the humerus has a constricted "waist", the first indication of a humeral shaft among the rhipidistians. The deltoid process and the ectepicondylar foramen (for the radial nerve) are, as in tetrapods, more preaxially located than in *Eusthenopteron*. The medial triceps head, ineffective in *Eusthenopteron* (Andrews and Westoll, 1970a), is a true elbow extensor. A discrete pectoralis process exists, in contrast to the continuous ventral diagonal ridge of *Eusthenopteron*. There is also an increased differentiation of protractors (supracoracoideus) and retractors (coracobrachialis brevis et longus). The functions of protraction and retraction were later assumed, in tetrapods, by other limb muscles, but the capacity for these functions in *Sterropterygion* suggests an essentially tetrapod pattern of limb movements.

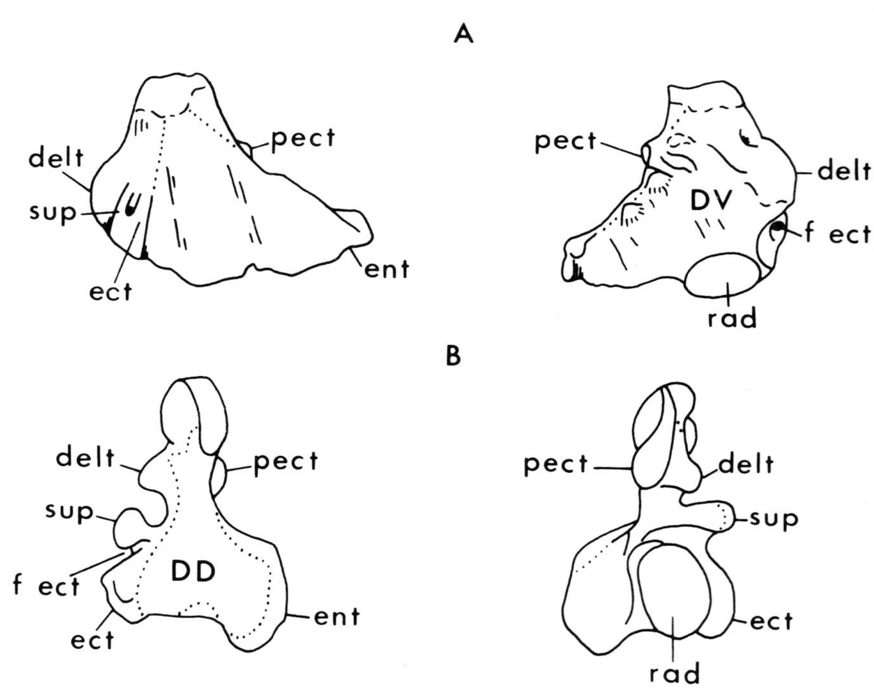

Fig. 7. Left humeri of (A) *Sterropterygion* and (B) *Eryops*. Views of (A) are dorsal (left) and ventral (right); those of (B) are postaxial (left) and preaxial (right). DD, Distal dorsal triangular area; delt, deltoideus; DV, distal ventral triangular area; ect, ectepicondyle; ent, entepicondyle; f ect, foramen ectepicondyle; pect, pectoralis; rad, articular facet for radius; sup, supinator.

Sterropterygion also resembles early tetrapods in the high degree of "twist" or torsion of the humerus (Fig. 8). Various methods of measuring this torsion have been employed (see Andrews and Westoll, 1970a, for a review), but the planes-of-distal-flattening method is the most functionally significant for the present comparisons, since it employs the long axis of the humeral head and the plane of the entepicondyle as referents. The latter, as indicated previously, is the functionally relevant variable in osteolepiform humeral morphology. By this method, the mild torsion of *Eusthenopteron* (21°) contrasts sharply with the high values for *Sterropterygion* (74°) and tetrapods (e.g. *Eryops* = 88°). Bicondylar torsion measurements (using entepicondyle–ectepicondyle axis as a distal referent) produce comparable relative values: *Eusthenopteron*, 50°; *Sterropterygion*, 84°; *Eryops*, 98°.

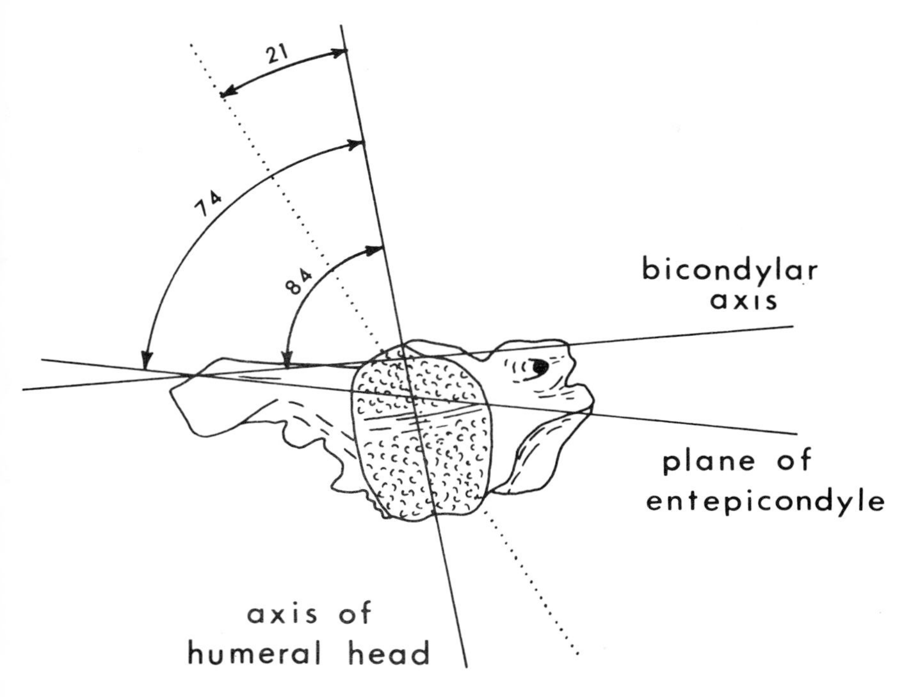

Fig. 8. Left humerus of *Sterropterygion* viewed perpendicular to articular head, showing torsion measurements. Dotted line represents relative position of plane of entepicondyle in *Eusthenopteron* when the long axes of the humeral heads of the two genera are superimposed.

Andrews and Westoll (1970a) speculated that torsion increased rapidly during the rhipidistian-amphibian transition. The present evidence suggests, however, that high torsion values were established early in the pre-tetrapod lineage in association with a unique dorsal orientation of the preaxial edge of the pectoral fin. During the most accelerated phase of the rhipidistian-amphibian transition, only a modest further increase in torsion was required to achieve tetrapod values. High humeral torsion was independently developed in *Rhizodopsis* (Andrews and Westoll, 1970b), but it is not known to have been associated with a dorsal preaxial edge in this genus.

The distal humeral joint surfaces are distinctive in their contrasting degrees of mobility. The radial facet curves strongly onto the ventral surface and allows the radius, at full flexion, to make an angle of at least 45° with the humerus. The facet for the ulna (second axial mesomere) is flatter, more fully distal, and permits more limited

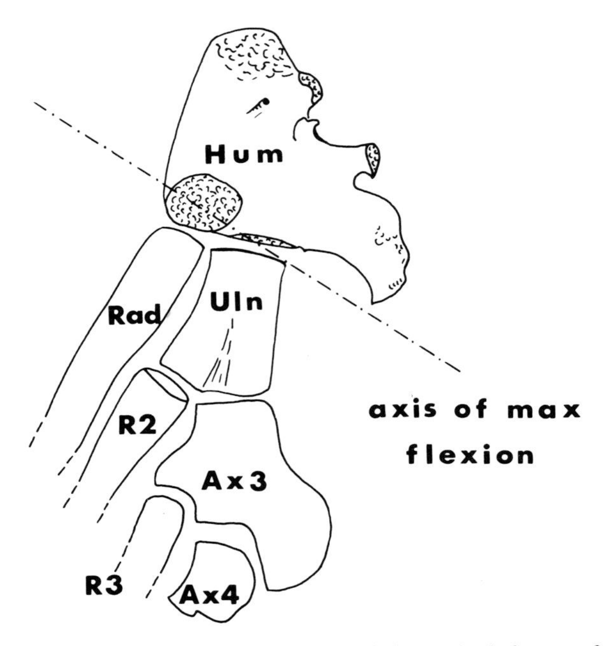

Fig. 9. Ventral view of the right pectoral fin endoskeleton of *Sterropterygion*, showing oblique axis of maximum elbow flexion. Ax3, third axial element; Ax4, fourth axial element; Hum, humerus; R2, second preaxial radial; R3, third preaxial radial; Rad, radius, Uln, ulna.

flexion. At full flexure of the elbow, the radius moves proximal and ventral to the ulna, forming an oblique axis of flexion (Figs 9 and 10) that tends to twist the distal part of the fin into a more forward position; preaxial rotation of radius and ulna increases this effect (Andrews and Westoll, 1970a, Text-fig. 30 and p. 298). The forwardly directed foot thus developed by specialization of joint surfaces, rather than a radical inward torsion of the fin axis causing interference and suppression of the second preaxial radial (Westoll, 1943; cf. Holmgren, 1949; Eaton, 1951).

Both the complex of proximal pectoral muscles with tetrapod homologues and the rotary elbow joint are also present in *Eusthenopteron* (Andrews and Westoll, 1970a), and are apparently primitive osteolepiform adaptations to an aquatic environment. The precise tetrapod-like *arrangement* of the muscles in *Sterropterygion* was the direct result of a unique (among rhipidistians) preaxial-edge-up fin orientation.

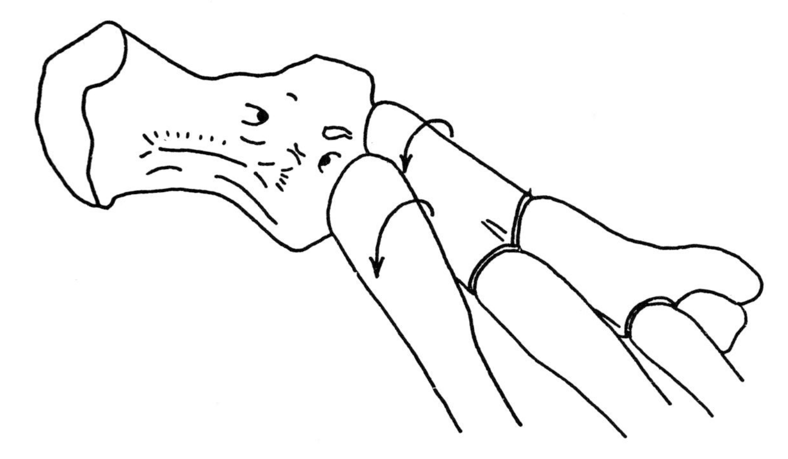

Fig.10. Preaxial view of extended left pectoral fin of *Sterropterygion.* At maximum flexure, the radius moves more ventral and proximal than the ulna, twisting the plane of the distal fin lobe forward; preaxial rotation of radius and ulna (arrows) increases this effect (see Andrews and Westoll, 1970a, Text-fig. 30 and p. 297). The same mechanism occurs in the fibulo-tarsal joint of the pelvic fin of *Sterropterygion.*

The distal elements of the pectoral fin are quite broad, suggesting considerable mechanical strength, but the joints separating these elements allow only a modest degree of flexure, with no indication of an incipient carpus.

The pelvic fins (Fig. 2), with *ventrally* directed preaxial margins, are smaller than the pectorals, but have the same robust proportions (fin lobe length = ½ fin length). The proximal endoskeleton is little differentiated. The featureless, rod-like femur, lacking a postaxial process, articulates proximally with an oval acetabulum which faces backwards and slightly outwards. The distal articular facets of the femur are not well exposed, but appear to allow only limited hinge-like flexion along a single axis.

Distally, the pelvic endoskeleton displays a degree of differentiation greater than that of any other known rhipidistian. The fibula (second axial mesomere) of other osteolepiform genera (e.g. *Eusthenopteron, Osteolepis*; Andrews and Westoll, 1970a,b) typically displays a thin, distally-expanding postaxial process which serves to support the overlapping bases of the lepidotrichia; it clearly lacks the mechanical strength to serve as a site for muscle attachment (cf. Eaton, 1951). The fibula of *Sterropterygion* (Fig. 11), in contrast, is a stout,

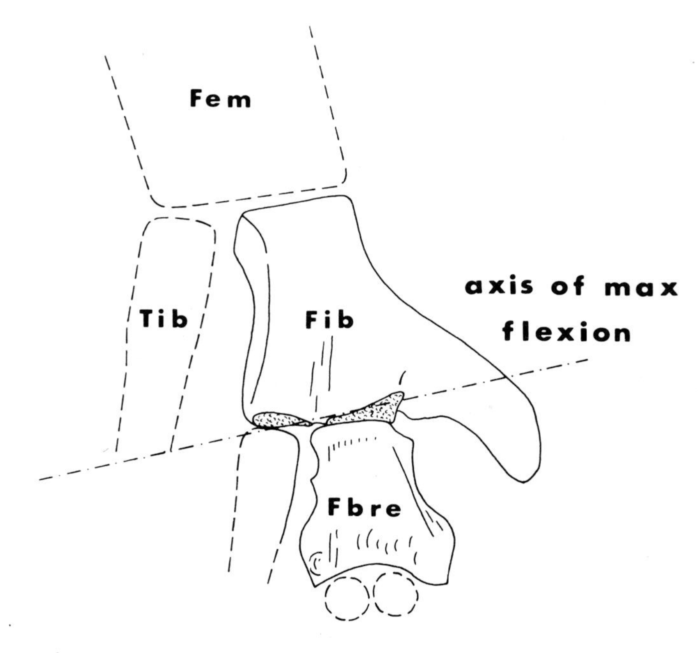

Fig.11. Dorsal view of the left pelvic fin endoskeleton of *Sterropterygion*, showing oblique axis of maximum tarsal flexion. Fbre, fibulare; Fem, femur; Fib. fibula; Tib, tibia.

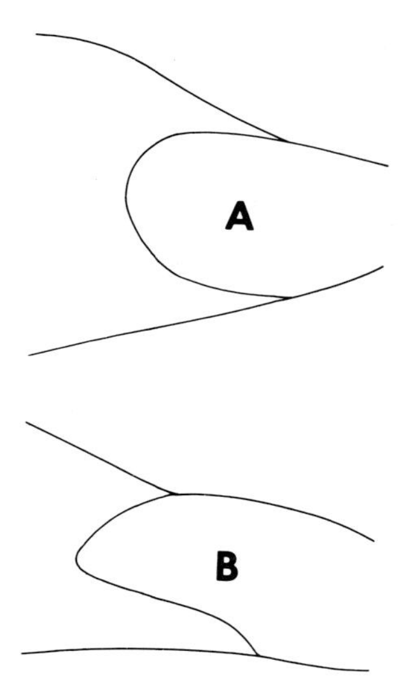

Fig.12. A comparison of (A) pectoral and (B) pelvic fin lobe shapes in *Sterropterygion*.

hook-shaped, distally-tapering structure (with consequent reduction of surface area) extending dorsal to the plane of the fibular shaft. The latter two features make it most unsuitable for lepidotrichial support, but the stoutness of the process provides a suitable base for muscular origin. The fibular postaxial process is, in fact, closely similar to that of the humerus (the entepicondyle) from which the elbow flexors originate. Associated with this muscular process is a marked asymmetry in the external shape of the pelvic fin lobe (Fig. 12), in which the preaxial lepidotrichia extend much further proximally than do the postaxial ones. This suggests the transformation of a lepidotrichial process into one for muscular support, forcing the postaxial lepidotrichia to recede distally to new points of attachment. Progressive muscularization of the fin endoskeleton was ultimately accompanied by complete loss of the lepidotrichia during the rhipidistian–amphibian transition. The morphology of the fibula suggests a way in which the process may have been initiated.

The most striking feature of the fibula is the nature of the distal articular surfaces (Fig. 11). The facet for the second preaxial radial is principally distal, but extends a short distance onto the dorsal surface, indicating the capacity for some dorsiflexion (probably accompanied by rotation). The neighbouring articular area for the third axial element has a tapering extension much further onto the dorsal surface, providing greater powers of dorsiflexion. This narrow extension served as a "track" to guide the movement of the broad axial element, which automatically rotated to allow dorsal sliding of its postaxial edge.

At full dorsiflexion, the third axial is thus displaced significantly proximal to the radial, creating an oblique (proximo-postaxial to distal-preaxial) axis of flexure (Fig. 11), which, exactly analogous to the elbow joint, serves to twist the distal part of the fin into a more forward position. This, then, is the first evidence of the development of a functional tarsal joint of tetrapod design within the Rhipidistia.

The elbow and tarsal joints of *Sterropterygion* are remarkably similar in structure and function. Both have oblique axes of flexion, bordered proximally by stout, distally-tapering postaxial processes. The strong musculature originating on these processes controls the bending *plus* forward turning of the relatively inflexible fin lobe. The deviation of the postaxial processes from the planes of their axial mesomeres affords a mechanical advantage to the musculature by increasing the angle of application (cf. Andrews and Westoll, 1970b,

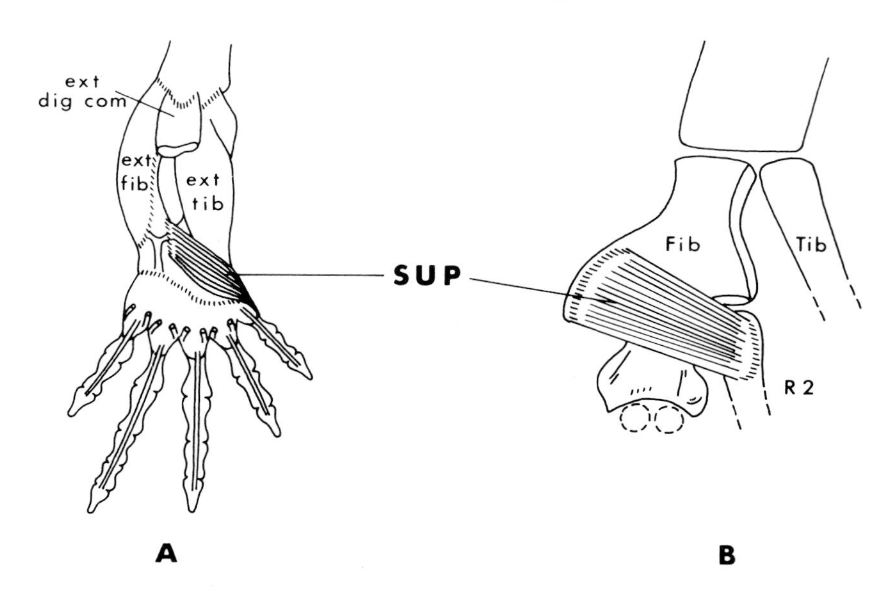

Fig.13. Comparison of the supinator muscle in the hind limb of (A) a urodele and (B) ***Sterropterygion***. ext dig com, extensor digitorum communis; ext fib, extensor fibularis; ext tib, extensor tibialis; Fib, fibula; R2, second pelvic preaxial radial; SUP, supinator; Tib, tibia. Urodele figure modified from Schaeffer, 1941.

p. 404). Forward rotation at the elbow was accomplished by the obliquely directed elbow flexors. In the tarsus, a comparable oblique muscle, here termed the "supinator" (Fig. 13A), extended from the postaxial process to various ridges and processes on the dorsal (extensor) side of the distal fin elements (especially the third axial mesomere, Fig. 11). This muscle is believed to be homologous, at least in part, to the diagonally oriented urodele "supinator" (Fig. 13B; supinator pedis of Wilder, 1912; abductor and extensor digit I of Schaeffer, 1941), which raises the medial border of the foot at the end of the propulsive phase of the locomotor cycle, and helps to maintain the forward orientation of the foot; true supination of the mammalian type does not, however, occur in amphibians Schaeffer, 1941). Together, the elbow and tarsal joints of *Sterropterygion* provide a pattern that accurately presages the problematical joint pattern of tetrapods: proximal rotation of the fore limb, and distal rotation of the hind limb. Furthermore, the enigmatic differences in the pattern of postaxial processes of pectoral and pelvic fins (Westoll, 1943) is readily explained when one considers that these processes provide the origin for the musculature effecting joint rotation.

The consistent association of robust postaxial processes and rotary joints in *Sterropterygion* and *Eusthenopteron* suggests a means of evaluating joint mobility in other forms with biseriate fins, even if the actual joint surfaces are not preserved. Thus the lack of a postaxial process on the first pectoral mesomere of the enigmatic Upper Devonian rhipidistian *Canowindra* (Thomson, 1973) suggests a nonrotary elbow joint. The curious proximo-distal arrangement of the usually oblique series of dorsal processes (ectepicondyle-supinator-deltoid) of the humerus seems to confirm this prediction. Similarly, the widespread occurrence of stout humeral postaxial processes in the Osteolepidida, and in *Sauropterus* (Andrews and Westoll, 1970a,b), suggests that the rotary elbow joint was a primitive aquatic adaptation in these groups. Stout postaxial processes have also been reported for the limb of *Hesperoherpeton* (Eaton and Stewart, 1960), but here they are uniquely arranged on four successive mesomeres. Although without precedent among known Rhipidistia, the logical possibility of such a successive series is supported by Westoll's (1943) presumption that each mesomere had the capacity to produce a postaxial process. This potential, however, appears to be utilized in only two functional situations: (a) where a broad, flat surface is required for lepidotrichial support, and thus develops on distal mesomeres, and (b) where a stout surface is required for the origin of muscles controlling a (proximal or distal) rotary joint. Thus at least some, and probably all, of the processes of *Hesperoherpeton* bore musculature. Assuming the extreme case of four successive rotary joints, the functional significance of such an arrangement would depend entirely upon the degree of twist possible at *each* joint. Thus successive small increments of forward twist could produce a foot that was fully forward-facing. If, however, the elongate and slender limb bones allowed greater flexibility and increased torsion at each joint, the limb would be a biomechanical impossibility, for it would tend to spiral inward! The intention of the foregoing discussion has not been to resolve the problem of *Hesperoherpeton*, be it rhipidistian or immature amphibian; such a resolution is clearly impossible in the absence of preserved joint surfaces. These comments serve, instead, to indicate potentially fruitful functional approaches to such problems, utilizing the new information derived from the morphology of *Sterropterygion*.

THE DIFFERENT FUNCTIONS OF FORE AND HIND APPENDAGES

The pectoral fins of fishes primitively had an attachment with the axial skeleton via the dorsal elements of the exoskeletal shoulder girdle. The latter evolved prior to the paired fins (Jarvik, 1965b) in order to brace the structurally weak region of the gills, and only later became associated with the endoskeletal girdle, to which the developing pectoral fins attached. Thus dermal elements are never associated with the pelvic girdle (although such an association has been described for the arthrodire *Sigaspis* by Goujet, 1973), and this girdle primitively lacks an axial connection. Since the propulsive thrust of tetrapod limbs is most effective when transmitted to the axial skeleton, one might wonder why the tetrapods did not, in effect, become "front-wheel-drive vehicles", rather than the reverse.

There were certainly significant advantages to be realized by losing the axial connection and developing a mobile neck. The latter would compensate for the inability of a clumsy-bodied primitive tetrapod to change direction easily, and would allow more effective use of the cranial sense organs by avoiding the wide yawing of the head that is typical of terrestrial locomotion in a fish out of water (see Miner, 1925).

The retention of this nexus during at least the early phases of the rhipidistian–amphibian transition was related, however, not to its role in transmitting propulsive thrust, but rather to its overriding importance in permitting pulmonary respiration. As Thomson (1972) has indicated, the pectoral limbs were preoccupied with the functional role of supporting the anterior trunk above the ground, so that the weight of the body would not compress the lungs. But even in the elevated body position, the lungs could be compressed against the relatively broad surface of the ventral shoulder girdle unless this was prevented by the axial connection, which maintained a fairly constant distance between the vertebral column (from which the visceral organs are suspended) and the ventral girdle. The loss of this connection in *Ichthyostega* followed the development of a more effective mechanism: a cuirass of stout dorsal ribs upon which the animal could rest without obstructing respiration. It remains to be seen whether the latter development was common to all early tetrapods, or merely one of a variety of solutions to the respiratory problem.

Loss of the axial connection had two important consequences. By freeing the shoulder girdle, it allowed the development of the characteristic "under-slung" suspension of the body on the tetrapod fore limb, and the acquisition of a new shock-absorbing function. Additionally, it relieved the pectoral fin of its incessant supportive role, and the size-dominance of this appendage was gradually lost. Fore and hind limbs thus appear to be nearly equal in size in *Ichthyostega* (Jarvik, 1955, Fig. 11, p. 152), and in Upper Devonian tetrapod trackways, the impression of the pes is already larger than that of the manus (Warren and Wakefield, 1972).

The relative increase in size of the pelvic appendage is associated with the establishment of a connection with the vertebral column via a sacral rib. This nexus made possible the efficient transmission of propulsive forces to the axial skeleton, and is most probably associated with increasing use of the hind limb to provide propulsive thrust. Earlier in the rhipidistian–amphibian transition, the primary source of thrust for terrestrial locomotion was undoubtedly the same as that for aquatic locomotion: posteriorly-directed sine wave contractions of the lateral body musculature. In this locomotor pattern (termed a "trot" in salamanders; Edwards, 1977), the fins were at first used only as stationary pivots, but it is likely that the pelvic fins gradually provided a small but increasing component of the propulsive thrust (cf. Edwards, 1977).

Two contrasting trends were thus manifested in the pectoral and pelvic fins of the rhipidistian ancestors of tetrapods. The pectoral fins began as stout supportive structures, important both at rest and during locomotion, and underwent a gradual reduction in functional importance. The pelvic fins at first provided intermittent terrestrial support (only during active locomotion), but were soon modified to provide propulsive thrust. Efficient translation of muscular force into forward movement required forward turning of the feet. The reasons for this rotation *proximally* in the fore limb can be effectively explained by examining *aquatic* functions of the pectoral fin.

The rotary elbow joints of *Sterropterygion* and *Eusthenopteron* probably originated to turn the ventral fin surface *backward* for slow paddling, rather than to turn the foot *forward*. The proximal site of this rotation, at the elbow, allowed a considerable part of the distal fin lobe to make propulsive contact with the water. The pattern of movements of the fin during slow paddling was similar to that in occasional terrestrial excursions, and the elbow joint was thus preadapted to twist the "foot" into a more advantageous anterolateral (and ultimately forward) position. For paddling, as for terrestrial support, distal differentiation was unnecessary. The former required a relatively rigid flat plate, while the latter needed a rigid strut (although passive sagging under the weight of the body could initiate a hinge-like carpus; Andrews and Westoll, 1970a). In addition, both functions required considerable proximal strength and mobility.

Specialization of the pelvic fins for a locomotor role on a solid substrate also required a means of turning the distal fin segment forward. It is not evident, however, why this rotation developed at the ankle, rather than the knee. If there was no selective pressure favouring a particular joint, perhaps its development in a distal (tarsal) location occurred by chance alone; it was simply one of several possible solutions to a functional problem that happened to evolve.

THE TRANSITION TO TETRAPOD LIMB STRUCTURE

Despite striking similarities between the limbs of *Sterropterygion* and tetrapods, major structural reorganization was still required during the rhipidistian-amphibian locomotor transition. It is now possible, however, to specify more precisely the nature of some of these changes.

Expansion of the scapulocoracoid during the transition was accompanied by reduction of the dermal girdle, with important changes in musculature (Rackoff, 1976). Loss, however, of the connection of the dermal girdle to the back of the head, which was required for unimpeded pulmonary respiration, occurred only after the development of the rib cage.

Greater proximal mobility of the pectoral appendage was gradually achieved by migration of the glenoid from a posterolateral to a lateral position. The screw-shaped glenoid appears to be primitive for the Order Osteolepidida of Andrews and Westoll (1970b), since it is possessed by all members of this taxon for which evidence is available. The detailed structure of the screw-shaped surface is quite variable (Andrews and Westoll, 1970a,b; Thomson and Rackoff, 1974; Rackoff, 1976), and doubtlessly served to structurally define different patterns of fore limb movement. The functional significance of particular variations is not, however, adequately understood. Andrews and Westoll (1970a) suggested that the original function of the screw-shaped glenoid was to provide lift without braking action by raising the leading edge of the fin. With the evolution in *Sterropterygion* of a pectoral fin which lies flat against the flank, such lift could be achieved by *lowering* the preaxial edge.

The basic pattern of tetrapod humeral morphology was already established in *Sterropterygion*, but several important changes occurred at the fish – tetrapod transition. The humeral head, subequal in size to the glenoid in known rhipidistians, increased rapidly in size late in the transition. Shaft elongation continued, and torsion increased modestly. The oblique axis of elbow flexion, developed for effective slow paddling, was pre-adaptive for terrestrial existence, and radial and ulnar facets migrated further ventrally to form the permanently flexed elbow joint. Less strength was required to maintain the flexed elbow than to bend the inflexible rhipidistian fin lobe. There was thus a reduction of the very large distal ventral triangular area of the humerus (Figs 5 and 7), from which the massive rhipidistian elbow flexors originated. Formerly separated by this large flexor surface, the pectoralis muscle was then free to approach the deltoid.

Preaxial migration of the pectoralis to form a delto-pectoral crest was the result of changing muscle functions associated with the new limb posture. Formerly inserting on either side of the long axis of the humeral head, the pectoralis and deltoid of *Sterropterygion* were opposing muscles. Due to the unique orientation of the trailing pectoral fin (preaxial-edge-up and appressed to the flank), the pectoralis process served as a lever arm for lowering the fin and drawing it away from the body. This latter function is "abduction" in the strict sense, but it is exactly the opposite of standard usage of the term in higher tetrapods, where "abduction" from the mid-plane of the body is accompanied by elevation of the limb. Similarly, the deltoid process served as a lever arm for elevation and "adduction" of the lowered fin. The advent of the permanently flexed elbow and horizontal humerus served as a commitment to terrestrial existence, and made inoperative those design constraints for aquatic locomotion which had maintained the conservative arrangement of the proximal humeral muscles. With the tetrapod humerus permanently "abducted" from the flank, the old pectoralis role was lost. Preaxial migration of the pectoralis process allowed this muscle to assume new roles in terrestrial support and locomotion.

The relatively undifferentiated distal part of the fin required more extensive changes. Andrews and Westoll (1970a) noted that terrestial contact would result in passive bending of the distal fin in the right direction, thus initiating the formation of a carpus. Incorporation of the free distal end of the radius into the carpus was a necessary sequel. In both the pectoral and the pelvic fins, digits had to be formed out of the distal mesenchyme, resulting in a "dramatic change in a canalized morphogenetic system" (Schaeffer, 1965). There is as yet no adequate explanation for the basic pattern of five digits (or seven, including prehallux and postminimus) in both manus and pes.

Finally, loss of the lepidotrichia and the fin web had to occur to allow the muscular lobe of the fin more effective propulsive contact with the ground. Gunter (1956) considered the loss of the fin web in itself as definitive of the origin of the tetrapods. Gregory (1915, p. 364), however, noted that various living fishes have reduced their dermal rays without any impairment of their locomotor abilities in the water. Eaton (1951) thus considered the loss of the lepidotrichia to be a rhipidistian innovation for manoeuvrability among dense weeds, and hence a pre-adaptation for terrestrial experiments. But the very tetrapod-like limb morphology of *Sterropterygion*, suggestive of effective terrestrial locomotor abilities, was achieved without any significant diminution of the lepidotrichia. Rather than being a rhipidistian innovation that *allowed* terrestrial conquest, loss of fin rays thus seems to have been a late occurrence which improved the effectiveness of limb-substrate contact.

In the pelvic fin of known rhipidistians, limited proximal differentiation obscures the details of the transformation to a functional hind limb. Increased mobility was achieved by the lateral migration of the acetabulum and greater flexion at the knee. Permanent knee flexure allowed femoral retractors to provide thrust, while femoral abductors and ventral flexors supported the body (see Schaeffer, 1941). There was a migration of musculature onto the developing ischium and, dorsally, onto the iliac blade. The greater division of labour among the proximal muscles allowed increasingly precise movements, and led to standardization of the locomotor pattern.

Distal changes in the hind limb, and particularly the origin of the tetrapod tarsus, can be reconstructed with greater precision. An oblique joint axis (extending further distally on the tibial side) is repeatedly encountered in the hind limbs of primitive tetrapods (von Meyer, 1858; Romer and Byrne, 1931; Schaeffer, 1941), but it is more distally located than its analogue in *Sterropterygion*. The proximal joint axes of early tetrapods did not extend across the entire width of the tarsus, and are thus of minimal importance (Schaeffer, 1941). In *Trematops* (Schaeffer, 1941), the greatest flexion is thus tarso-metatarsal, but the joint plane distal to the fibulare, fourth centrale and tibiale is also important. Motion along these tetrapod axes was clearly impossible for *Sterropterygion*, for the differentiation of the distal elements had not yet occurred, and, of the elements bordering the latter joint axis, only the fibulare homologue can be confidently identified. The functional demands of terrestrial locomotion thus resulted in the initial development of an oblique rhipidistian joint axis in a proximal, fibulo-tarsal location.

The tibia of *Sterropterygion*, however, is longer than the fibula, and would interfere with fibulo-tarsal flexure were it not for the capacity of the free distal end of the tibia to move independently to accommodate itself to tarsal motion. The forward turning of the rhipidistian "foot" undoubtedly facilitated the ultimate incorporation of the tibia into the tarsus by causing this element to lie closer to the ankle joint. Similar arguments for the incorporation of the radius into the carpus have been proposed by Westoll (1943). There is currently no fossil evidence for tibial incorporation prior to the tetrapod stage but, once it occurred, the length of the tibia must have disrupted the fibulo-tarsal joint axis on its preaxial side. For uninterrupted adaptive functioning of the tarsus during the locomotor transition, new joint axes must already have been developing distally, among the newly differentiated elements of the tarsus and pes.

It is thus apparent that the essential pattern of the tetrapod ankle joint was established before the primary weight-transmitting function of the tetrapod tibia, with its more extensive proximal articulation, had evolved. Although the free distal end of the rhipidistian tibia may have acted as an auxiliary strut during contact with the substrate, the main weight-bearing axis of the rhipidistian paddle was clearly the series of axial mesomeres, for the distal elements of this series converge on the fibula. The major transfer of weight from fibula to tibia via the pronator profundus and interosseous cruris (Schaeffer, 1941), had to await the incorporation of the tibia into the tarsus. The presence in the Rhipidistia of homologous muscles spanning tibia and fibula would, however, have facilitated this incorporation.

The origin of the more distal limb elements remains unclarified. There is clearly an inadequate number of distal ossifications in *Sterropterygion* to allow precise homologies with the tetrapod tarsus and pes. Since the proliferation of ossifications out of the distal mesenchyme required a reorganization of the developmental system of the limbs, it is unlikely that fossil intermediates will ever be discovered to clarify this aspect of the transition; clearly, morphogenetic studies can provide the only new evidence. Differences in gross limb morphology must be attributable to variant systems of genetic regulation and developmental control. Lewis and Holder (1977) have suggested that a unitary mechanism of developmental controls exists in the limbs of all tetrapods, with quantitative variations in parameters producing variants on a basic topological scheme. A simple set of developmental signals provides proximo-distal and antero-posterior positional information to the differentiating cells, which thus become non-equivalent to each other. Particularly intriguing is the evidence (Lewis, 1975, 1976; Lewis and Holder, 1977) that each segment of the limb is formed by one generation of cells emerging from the undifferentiated region of the distal mesenchyme. One might presume, therefore, that a relatively minor genetic change increasing the potency of the mesenchymal cells (i.e. the maximum number of cell divisions that were possible) could cause a proliferation of distal elements and initiate digit formation. This cellular mechanism may thus unify and explain a variety of older terminology: the "budding" from two primitive centralia identified by many workers on urodele development (Zwick, 1898; Rabl, 1910; Holmgren, 1933, 1939), the "polyisomerism" of Gregory (1935b, 1949) and the "neomorphs" of Westoll (1943). Interestingly, the morphogenetic processes involved in limb modelling in insects are fundamentally similar to those involved in digit formation in vertebrates (Whitten, 1969).

By whatever means it occurred, the proliferation of distal elements increased the flexibility of the tarsus, and less force was then required to re-position the foot. The rhipidistian "supinator" muscle was thus reduced, and with it the postaxial process; the latter is absent in *Ichthyostega*.

The degree of forward turning of the fins of *Sterropterygion* depends upon the degree of protraction that was possible, as well as the magnitude of lateral body undulations and the inherent rotary capacities of the elbow and tarsal joints. It is unlikely that the "feet" could accomplish more than a moderate departure from the lateral position. Further improvements in the orientation of the feet would be expected during the rhipidistian–amphibian transition and, by the time the digits had formed, the feet should have been almost fully forward, with no rotational slippage during the stride. This had indeed been the case for all known early tetrapod footprints (Romer and Byrne, 1931; Romer, 1940; Schaeffer, 1941; Westoll, 1943), until the discovery in the Upper Devonian rocks of eastern Victoria, Australia, of the earliest tetrapod trackways (Warren and Wakefield, 1972). The best preserved set of these possibly ichthyostegid footprints reveals well-developed digits, and indicates that the feet were rotated only 4° anterior to the lateral axis (86° from the midline). *Sterropterygion* could certainly exceed this degree of forward rotation, particularly with the assistance of lateral body undulations. The Australian trackway thus suggests a rather specialized locomotor mechanism, perhaps one in which the usual sinuous body motion is absent. Among living salamanders, Edwards (1976) indicates that the girdle rotation that accompanies sinuous locomotion is minimal in "elongate" forms with small legs and feet. But the robust limbs of *Ichthyostega* accord better with Edwards' "robust" salamander pattern, in which girdle rotation contributes relatively more propulsive thrust. A second Australian trackway, produced perhaps by another species of ichthyostegid, does show evidence of body undulation, but limb dragging has obscured the foot orientation.

If the laterally facing tracks are indeed representative of the basal tetrapod stock, then further improvements in foot angulation may not have occurred until after the digits were fully developed (cf. Morton, 1926). Alternatively, the evidence may suggest that the ichthyostegids were an aberrantly specialized side-branch of the prototetrapod stock, and are an inappropriate model for locomotor evolution in the remaining tetrapods. Further clarification of this matter must await a more complete analysis of the trackways, and a detailed postcranial study of *Ichthyostega*.

The essential mechanics of tetrapod locomotion were thus present in the limbs of at least one rhipidistian. But was *Sterropterygion* actually capable of terrestrial locomotion? The powerful pectoral fin was certainly capable of raising the anterior part of the body to allow pulmonary respiration. The pelvic fin, however, is considerably smaller in size. Nonetheless, the fin lobe is relatively large, and the axial elements are broad and robust. In these features, the limbs of *Sterropterygion* as well as of other rhipidistians demonstrate that terrestrial locomotion was at least *possible*. A tarsal joint of tetrapod design, however, occurs only in *Sterropterygion*, and it is difficult to conceive of any selective pressure in the water that could have led to its development. Only the stresses of *actual* terrestrial locomotion could have effected its evolution.

PHYLOGENETIC CONSIDERATIONS

Despite attempts to relate tetrapods, at least in part, to dipnoans (e.g. Holmgren, 1933 and later papers) or other groups (Kesteven, 1950), the rhipidistian fishes have been generally recognized as the ancestors of tetrapods since the late nineteenth century (Cope, 1892; Baur, 1896). The osteolepiform rhipidistians appear to be most suitable as tetrapod precursors (e.g. Schultze, 1977), but distinctive features of urodele anatomy and development have long suggested to Jarvik (1942 and later papers) and his supporters an independent derivation of this group from the Porolepiformes. The majority of workers have, however, endorsed a monophyletic origin of the tetrapods (see recent review by Szarski, 1977). The latter view is further supported by a unitary system of developmental controls in the limbs of all tetrapods (Lewis and Holder, 1977; cf. Nieuwkoop and Sutasurya, 1976, for developmental evidence supporting polyphyly), as well as the occurrence of a uniform but long neglected tetrapod pattern of rotary limb joints (Romer and Byrne, 1931).

It is thus appropriate to seek to identify the osteolepiform subtaxon with which the tetrapods share the most recent common ancestry. Because of distinctive features of the shoulder girdle, pectoral fin and axial skeleton, Andrews and Westoll (1970b) (see also Andrews, 1973) removed *Rhizodus*, *Strepsodus* and *Sauripterus* from this assemblage, and placed them in a separate Order Rhizodontida. *Sauripterus*, formerly popular as a basis for theories of limb origins (for review see Andrews and Westoll, 1970b), is thus effectively removed from considerations of tetrapod ancestry. The remaining osteolepiform genera (Order Osteolepidida of Andrews and Westoll) are grouped in three families: Eusthenopteridae, Osteolepididae and Rhizodopsidae. The last of these was considered by Andrews and Westoll (1970b) to be a late, divergently specialized osteolepidid side-branch which was convergent with the Eusthenopteridae.

If, however, the late stratigraphic occurrence of the Rhizodopsidae is not used as a basis for phylogenetic reconstruction (Schaeffer *et al.*, 1972), several distinctive features of the paired fins of *Rhizodopsis* merit serious consideration in any analysis of tetrapod relationships. *Rhizodopsis* has a high degree of humeral torsion (Andrews and Westoll, 1970b; actual measurements could not be made on the available material) as in *Sterropterygion* and tetrapods (although the phylogenetic significance of this feature has been questioned by Andrews and Westoll, 1970a). Of greater significance is the preaxial distribution of the muscular processes on the dorsal surface of the humerus; these correspond closely with the pattern seen in *Sterropterygion* and tetrapods (cf. Plate IIC of Andrews and Westoll, 1970b, with Figs 5 and 7 of this paper). There is thus reason to suspect that these pectoral fin features were also associated with a dorsal orientation of the preaxial edge in *Rhizodopsis*. An additional observation by Andrews and Westoll (1970b, p. 400) on the pelvic fin of this genus is noteworthy: "the fin rays on the preaxial side appear longer than those on the postaxial, although this may be due to preservation". It is not clear from this statement whether the length differential occurs in the proximal or distal part of the lepidotrichia. A proximal asymmetry like that of *Sterropterygion* would suggest a postaxial process from which musculature originated to control a rotary tarsal joint. The incompleteness of the material prevents more detailed comparisons at this time, but a close relationship between *Rhizodopsis* and *Sterropterygion* remains a strong possibility which should be subjected to further testing.

In each of the two remaining families of the Osteolepidida, differing combinations of tetrapod "resemblances" (*sensu* Hennig, 1965) are displayed in different genera. The multiple independent acquisition of character states appears to have been common. In an attempt to resolve the confusing mosaic of primitive and derived characters found in different genera (DeBeer, 1954), and to determine which taxon share the most recent common ancestry with tetrapods, one would clearly benefit from the comparison of a large number of characters (e.g. Hennig, 1965, 1966). Regrettably, however, the vagaries of preservation make many of the osteolepiform genera non-comparable. Thus *Hyneria*, with tetrapod-like cranial proportions (Thomson, 1968) is known only from the head region, whereas *Sterropterygion* is lacking most of the head. The present analysis is thus a more modest attempt to assess the significance of selected osteolepiform postcranial characters for tetrapod relationships. *Eusthenopteron foordi* is the best known member of the Eusthenopteridae, and has long been favoured in discussions of tetrapod limb origins (see Andrews and Westoll, 1970a), while *Sterropterygion brandei* is clearly the most tetrapod-like species in the Osteolepididae. The restricted question which will now be addressed is which of these two species shares a more recent common ancestry with the tetrapods.

Many of the postcranial character states which *Eusthenopteron* shares with tetrapods fall into one of the following categories: (a) they are primitive, and thus useless for assessing relationships; (b) they are part of a morphocline, the polarity of which cannot currently be established with certainty; (c) they are derived, but have been acquired independently in the Osteolepididae. Some examples of these are discussed briefly below.

The fusion of the scapulocoracoid with the dermal girdle, a requirement for effective terrestrial locomotion, was previously known in *Eusthenopteron*, but not in any osteolepidids (Andrews and Westoll, 1970a,b). It has since been discovered in two osteolepidids: *Ectosteorhachis* (Thomson and Rackoff, 1974), and now *Sterropterygion*. It occurs also in some porolepiforms (Gross, 1936; Andrews and Westoll, 1970b), and perhaps in the rhizodontiform *Sauripterus* (see Gregory, 1935a, Fig. 1), and may thus be primitive. It is judged likely, however, that fusion is the derived state, independently acquired in various forms, since these occurrences are regularly correlated with other indicators of mechanical strength in locomotion, or with large body size (if *Sauripterus* has fusion).

The screw-shaped glenoid surface of tetrapods is found in all of the Osteolepidida that have been examined, and nowhere else. It thus appears to be a primitive character state in this taxon, and can reveal nothing about relationships. But references to "the" screw-shaped glenoid mask a broad spectrum of variability in the shape, degree of twist and orientation of the twisted surfaces of the glenoid (Thomson and Rackoff, 1974; Rackoff, 1976). Potentially significant characters may well exist here, but the functional significance of these variations for shoulder mobility is inadequately understood.

The pectoral fin endoskeleton of *Eusthenopteron* and *Sterryopterygion* is a uniseriate archipterygium, which is a primitive feature of widespread occurrence among crossopterygians (Andrews, 1973). Variations on the basic pattern are seen in different crossopterygian groups, but, within the Order Osteolepidida, the pattern is quite consistent, and is a suitable starting point for tetrapod limb origins. Tetrapod resemblances in *Eusthenopteron* are more marked proximally; the humerus bears a complex of muscular processes with tetrapod homologues, and distally displays a rotary "elbow" joint of tetrapod design. These features, however, appear to be widespread in both families of the Osteolepidida, and hence must be regarded as primitive features in the group.

Eusthenopteron's principal claim to tetrapod relationship resides in its axial skeleton. But even the nature of the compound centra of this genus, which has served as a model for the origin of tetrapod vertebrae, cannot be used as a guide to relationship. Rhipidistian vertebral variability is extensive, and unpublished studies of *Sterropterygion*, *Ectosteorhachis* and *Rhizodopsis* further extend the range of known variation. With a limited knowledge of the functional significance of this variation, and with considerable debate over the homologies of the central elements, correct assessment of primitive and derived character states is currently impossible. After a recent review of this evidence, Panchen (1977a, p. 303) concluded that "vertebrae of *Eusthenopteron* need not, of necessity, be regarded as representing a condition ancestral to that of tetrapods and those of *Osteolepis* are more likely candidates".

The one feature of the axial skeleton that *Eusthenopteron* alone shares with tetrapods is the presence of bicipital dorsal ribs (Andrews and Westoll, 1970a). Andrews (1973, p. 160) suggested that unossified ribs may be the primitive crossopterygian condition. Ribs would thus have been formed as ossifications in the myosepta whenever mechanical demands of the locomotor pattern made such structures selectively advantageous. Dorsal rib facets have also been observed in the osteolepidid *Ectosteorhachis* (Thomson and Vaughn, 1968). Although these facets are unexpectedly high on the holospondylous centra of this genus, and thus not comparable in detail to the ribs of *Eusthenopteron*, the capacity for dorsal rib formation has been demonstrated in the Osteolepididae as well as the Eusthenopteridae. Outside of the Osteolepidida, crossopterygian ribs are found only among the coelacanthiforms, and these are pleural ribs. It is noteworthy to report, therefore, that pleural ribs have also been observed on the anterior trunk vertebrae of *Sterropterygion*. The shared possession in both of these groups of a preaxial-edge-up pectoral fin orientation in addition to pleural ribs is perhaps more than coincidence; there may be a significant functional association of these character states.

In summary, it must be emphasized that any hypothesis of a unique relationship between *Eusthenopteron* and tetrapods is based almost exclusively on one shared derived character state: bicipital dorsal ribs.

In contrast, a suite of close resemblances to tetrapods has been demonstrated in the paired fins of *Sterropterygion*: humeral shaft, high humeral torsion, preaxial location of deltoid and ectepicondylar foramen, functional medial triceps, differentiation of protractors and retractors, discrete pectoralis process, rotary tarsal joint. As demonstrated earlier, however, many of these features are not independent variables; they are functionally dependent upon the relative location of the postaxial process of the humerus (entepicondyle). Independent consideration of all of these characteristics would unduly bias the evidence in favour of a *Sterropterygion*–tetrapod relationship. Following Hecht and Edwards' (1977, p. 8) definition of a character as "a suite of correlated attributes whose homologous states vary concordantly within the taxa under consideration and are usually functionally and/or developmentally related", a single complex character can be recognized in the proximal pectoral fin. The principal variable, the relative location of the postaxial process, here determines the tetrapod-like *arrangement* of the dorsal and ventral muscle processes (in contrast to the mere *presence* of such processes in other Osteolepidida). Humeral torsion appears also to be dependent on the principal variable. Shaft-formation may or may not deserve independent character status; its presence is not evident from the incomplete material of *Rhizodopsis* (Andrews and Westoll, 1970b, Plate IIC) which otherwise displays such striking similarities to *Sterropterygion.* The presence of a distinct pectoralis process would seem to bear no necessary relationship to the location of the entepicondyle; the pectoralis section of a continuous ventral ridge could as easily occupy any position relative to the entepicondyle. Thus one complex, functionally interrelated character, and one or two additional characters, are employed in the pectoral fin, while the single significant pelvic fin character is the rotary tarsal joint (including joint surfaces and postaxial process modified for muscular origin). All of these are clearly derived character states that are shared with the tetrapods.

The contradictory tetrapod resemblances presented by *Eusthenopteron* and *Sterropterygion* can be resolved in various ways. The most parsimonious hypothesis (with due respect for the fallibility of parsimony) is that tetrapods share a more recent common ancestry with *Sterropterygion* than they do with *Eusthenopteron.* This hypothesis requires fewer independent acquisitions of character states than its alternative. The common ancestor of *Sterropterygion* and tetrapods would be required to lack ribs, which would subsequently develop in parallel in the tetrapod line. An alternative although somewhat less parsimonious sequence is possible if the common ancestor possessed pleural ribs in functional association with the preaxial-edge-up fin orientation; these would then be lost, and dorsal ribs acquired, as the tetrapods evolved.

Confidence in this most parsimonious hypothesis, based on admittedly few characters, can be increased by character weighting. Schaeffer *et al.* (1972) note that a "high level of adaptive specificity" may permit relationships to be assessed on few characters. Hecht and Edwards (1977) have argued that equal weighting of characters is illusory, and that maximum weight should be given to those characters bearing the most information. Schlee's (1969, p. 129) methodological definition for distinguishing synapomorphy from convergence also suggests maximal information: "Homologue formation, whose identity is ensured by rich and corresponding structures, which accomplish in a unique ("new") way a function that is fulfilled in all other groups by another formation (often less differentiated, often less rich in structures)".

It is suggested that the limited number of characters employed in assessing the sister-group status of *Sterropterygion* and tetrapods satisfy the above criteria, and should be weighted heavily. The proximal functional complex of the pectoral fin fulfills Hecht and Edwards' (1977) criteria for their weight categories IV (highly integrated functional complex) and V (innovative and unique; maximum weight), and the tarsal joint falls within category V. The transformational series leading to the tetrapod limbs offer numerous detailed homologies, and provide a coherent functional picture of locomotor changes during the rhipidistian–amphibian transition.

STRATIGRAPHICAL COMMENT

Amphibian fossils pre-dating the Greenland ichthyostegids have recently been discovered in Australia, pushing back the "age of origin" (Hennig, 1965) of the tetrapods, and posing interesting palaeogeographic problems (see Panchen, 1977b). The tetrapod trackways of Warren and Wakefield (1972) were reported to be probably Frasnian. Also Frasnian is the problematical *Elpistostege* of Canada (Westoll, 1938), but this has been considered a rhipidistian of the Family Panderichthyidae (Vorobjeva, 1973). A newly discovered mandible from the Middle Famennian of Australia has been described as an ichthyostegid and designated *Metaxygnathus denticulus* (Campbell and Bell, 1977). It is reported to be stratigraphically close to the trackway horizon of Warren and Wakefield (1972). The latter would thus appear to be younger than originally reported. There is, then, no unequivocal evidence of tetrapods from the Frasnian.

Sterropterygion was first reported (Thomson, 1972) from the Upper Devonian Susquehanna Group, probably Catskill Formation, of northern Pennsylvania. Field work by the author confirms the Catskill identity of the outcrop, and indicates the Duncannon Member (uppermost Catskill) of Faill and Wells (1974). Correlations in this area are imprecisely established, but the Duncannon Member appears to correspond roughly to the upper half of the Famennian (Oliver *et al.*, 1969, Fig. 4, p. 1006).

With the existence of definitive tetrapod characters in the Middle Famennian of Australia, the minimum age of origin of the tetrapods appears to have been lower Famennian.

ACKNOWLEDGEMENTS

This paper is derived in part from a dissertation submitted to Yale University in partial fulfilment of the requirements for the Ph.D. Degree. I am most grateful to K. S. Thomson for the opportunity to study the fossil material, and for continual advice, support and discussion. I am indebted to K. S. Thomson for reviewing this manuscript, to J. H. Ostrom, E. Simons and J. A. W. Kirsch for constructive criticism of earlier versions of this work, and to E. Cotter for stratigraphic discussions. Part of this work was supported by a faculty summer research grant from the Arthur Vining Davis Fund, Bucknell University, and by NSF Grant DEB76-21311.

REFERENCES

Andrews, S. M. (1973). Interrelationships of crossopterygians. *In* "Interrelationships of Fishes" (P. H. Greenwood, R. S. Miles and C. Patterson, eds), pp. 137–177. Academic Press, London and New York.

Andrews, S. M. and Westoll, T. S. (1970a). The postcranial skeleton of *Eusthenopteron foordi* Whiteaves. *Trans. R. Soc. Edinb.* **68**, 207–329.

Andrews, S. M. and Westoll, T. S. (1970b). The postcranial skeleton of rhipidistian fishes excluding *Eusthenopteron. Trans. R. Soc. Edinb.* **69**, 391–489.

Baur, G. (1896). The Stegocephali. A phylogenetic study. *Anat. Anz.* **11**(22), 657–673.

Breder, C. M., Jr (1926). The locomotion of fishes. *Zoologica* 4(5), 159–297.

Campbell, K. S. W. and Bell, M. W. (1977). A primitive amphibian from the Late Devonian of New South Wales. *Alcheringa* **1**, 369–381.

Cope, E. D. (1892). On the phylogeny of the Vertebrata. *Proc. Am. phil. Soc.* **30**, 278–285.

Dean, B. (1906). Notes on the living specimens of the Australian lung-fish, *Ceratodus forsteri*, in the Zoological Society's collection. *Proc. zool. Soc. Lond.* 1906, 168–178.

DeBeer, G. A. (1954). *Archaeopteryx* and evolution. *Adv. Sci.* **42**, 1–11.

Eaton, T. H. (1951). Origin of tetrapod limbs. *Am. Midl. Nat.* **46**, 245–251.

Eaton, T. H. and Stewart, P. L. (1960). A new order of fish-like Amphibia from the Pennsylvanian of Kansas. *Univ. Kans. Publs Mus. nat. Hist.* **12**, 217–240.

Edwards, J. L. (1976). "A Comparative Study of Locomotion in Terrestrial Salamanders." Ph.D. Dissertation, University of California, Berkeley.

Edwards, J. L. (1977). The evolution of terrestrial locomotion. *In* "Major Patterns in Vertebrate Evolution" (M. K. Hecht, P. C. Goody and B. M. Hecht, eds), pp. 553–577. Plenum, New York.

Faill, R. T. and Wells, R. B. (1974). Geology and mineral resources of the

Millerstown Quadrangle, Perry, Juniata, and Snyder Counties, Pennsylvania. *Pa. geol. Surv. Atlas* **136**, 1–276.

Goujet, D. (1973). *Sigaspis*, un nouvel Arthrodire du Dévonien Inférieur du Spitzberg. *Palaeontographica* **143**, 73–88.

Gregory, W. K. (1915). Present status of the problem of the origin of the tetrapods, with special reference to the skull and paired limbs. *Ann. N.Y. Acad. Sci.* **26**, 317–383.

Gregory, W. K. (1935a). Further observations on the pectoral girdle and fin of *Sauripterus taylori* Hall, a crossopterygian fish from the Upper Devonian of Pennsylvania, with special reference to the origin of the pentadactylate extremities of tetrapods. *Proc. Am. phil. Soc.* **75**, 673–690.

Gregory, W. K. (1935b). Reduplication in evolution. *Q. Rev. Biol.* **10**, 272–290.

Gregory, W. K. (1949). The humerus from fish to man. *Am. Mus. Novit.* No. 1400, 1–54.

Gregory, W. K. and Raven, H. C. (1941). Studies on the origin and early evolution of paired fins and limbs. *Ann. N.Y. Acad. Sci.* **42**, 273–360.

Gross, W. (1936). Beiträge zur Osteologie baltischer und rheinischer Devon-Crossopterygier. *Paläont. Z.* **18**, 129–155.

Gunter, G. (1956). Origin of the tetrapod limb. *Science* **123**, 495–496.

Hecht, M. K. and Edwards, J. L. (1977). The methodology of phylogenetic inference above the species level. *In* "Major Patterns in Vertebrate Evolution" (M. K. Hecht, P. C. Goody and B. M. Hecht, eds), pp. 3–51. Plenum, New York.

Hennig, W. (1965). Phylogenetic systematics. *Ann. Rev. Ent.* **10**, 97–116.

Hennig, W. (1966). "Phylogenetic Systematics", 251 pp. University of Illinois Press, Urbana.

Holmgren, N. (1933). On the origin of the tetrapod limb. *Acta zool., Stockh.* **14**, 185–295.

Holmgren, N. (1939). Contribution to the question of the origin of the tetrapod limb. *Acta zool., Stockh.* **20**, 89–124.

Holmgren, N. (1949). On the tetrapod limb problem—again. *Acta zool., Stockh.* **30**, 485–508.

Jarvik, E. (1942). On the structure of the snout of crossopterygians and lower gnathostomes in general. *Zool. Bidr. Upps.* **21**, 235–675.

Jarvik, E. (1944). The dermal bones, sensory canals and pit-lines of the skull in *Eusthenopteron foordi* Whiteaves, with some remarks on *E. säve-söderberghi* Jarvik. *K. svenska VetenskAkad. Handl.* **21**(7), 1–32.

Jarvik, E. (1948). On the morphology and taxonomy of the Middle Devonian osteolepid fishes of Scotland. *K. svenska VetenskAkad. Handl.* **25**(1), 1–301.

Jarvik, E. (1955). The oldest tetrapods and their forerunners. *Sci. Monthly, N.Y.* **80**, 141–154.

Jarvik, E. (1965a). Die Raspelzunge der Cyclostomen und die pentadactyle Extremität der Tetrapoden als Beweise fur monophyletische Herkunft. *Zool. Anz.* **175**, 101–143.

Jarvik, E. (1965b). On the origin of girdles and paired fins. *Israel J. Zool.* **14**, 141–172.

Kesteven, H. L. (1950). The origin of the tetrapods. *Proc. R. Soc. Vict.* **59**, 93–138.

Lewis, J. H. (1975). Fate maps and the pattern of cell division: a calculation for the chick wing-bud. *J. Embryol. exp. Morphol.* 33, 419–434.

Lewis, J. H. (1976). Growth and determination in the developing limb. *In* "Vertebrate Limb and Somite Morphogenesis" (D. A. M. Balls and J. Hinchcliffe, eds), pp. 215–228. Cambridge University Press.

Lewis, J. and Holder, N. (1977). The development of the tetrapod limb: embryological mechanisms and evolutionary possibilities. *In* "Major Patterns in Vertebrate Evolution" (M. K. Hecht, P. C. Goody and B. M. Hecht, eds), pp. 139–148. Plenum, New York.

Locket, N. A. and Griffith, R. W. (1972). Observations on a living coelacanth. *Nature, New Biol.* **237**, 175.

Meyer, H. von. (1858). "Reptilien aus der Steinkohlen-Formation in Deutschland", 126 pp. Cassel.,

Miner, R. W. (1925). The pectoral limb of *Eryops* and other primitive tetrapods. *Bull. Am. Mus. nat. Hist.* **51**, 145–312.

Morton, D. J. (1926). Notes on the footprint of *Thinopus antiquus*. *Am. J. Sci.* **12**, 409–414.

Nieuwkoop, P. D. and Sutasurya, L. A. (1976). Embryological evidence for a possible polyphyletic origin of the recent amphibians. *J. Embryol. exp. Morphol.* **35**, 159–167.

Oliver, W. A., Jr, De Witt, W., Jr., Dennison, J. M., Hoskins, D. M. and Huddle, J. W. (1969). Devonian of the Appalachian Basin, United States. *In* "International Symposium on the Devonian System" (D. H. Oswald, ed.), Vol. 2, pp. 1001–1040. Alberta Society of Petroleum Geologists, Calgary.

Panchen, A. L. (1977a). The origin and early evolution of tetrapod vertebrae. *In* "Problems in Vertebrate Evolution" (S. M. Andrews, R. S. Miles and A. D. Walker, eds), pp. 289–318. Academic Press, London and New York.

Panchen, A. L. (1977b). Geographical and ecological distribution of the earliest tetrapods. *In* "Major Patterns in Vertebrate Evolution" (M. K. Hecht, P. C. Goody and B. M. Hecht, eds), pp. 723-738. Plenum, New York.

Rabl, C. (1910). "Bausteine zu einer Theorie der Extremitaten der Wirbeltiere", 290 pp. Leipzig.

Rackoff, J. S. (1976). "The Osteology of *Sterropterygion* (Crossopterygii: Osteolepidae) and the Origin of Tetrapod Locomotion". Ph.D. Thesis, Yale University.

Romer, A. S. (1922). The locomotor apparatus of certain primitive and mammal-like reptiles. *Bull. Am. Mus. nat. Hist.* 46, 517–606.

Romer, A. S. (1940). Review of "Die Fossilen Amphibien", O. Kuhn, Berlin, 1939. *J. Paleont.* **14**, 389–391.

Romer, A. S. and Byrne, F. (1931). The pes of *Diadectes*: Notes on the primitive limb. *Palaeobiologica* 4, 24–48.

Schaeffer, B. (1941). The morphological and functional evolution of the tarsus in amphibians and reptiles. *Bull. Am. Mus. nat. Hist.* **78**(6), 395–472.

Schaeffer, B. (1965). The rhipidistian-amphibian transition. *Am. Zool.* **14**, 115–118.

Schaeffer, B., Hecht, M. K. and Eldredge, N. (1972). Phylogeny and paleontology. *Evol. Biol.* 6, 31–46.

Schlee, D. (1969). Hennig's principle of phylogenetic systematics, an "intuitive, statistico-phenetic taxonomy"? *Syst. Zool.* **18**, 127–134.

Schultze, H-P. (1977). The origin of the tetrapod limb within the rhipidistian fishes. *In* "Major Patterns in Vertebrate Evolution" (M. K. Hecht, P. C. Goody and B. M. Hecht, eds), pp. 541–544. Plenum, New York.

Szarski, H. (1962). The origin of the Amphibia. *Q. Rev. Biol.* **37**, 189–291.

Szarski, H. (1977). Sarcopterygii and the origin of the tetrapods. *In* "Major Patterns in Vertebrate Evolution" (M. K. Hecht, P. C. Goody and B. M. Hecht, eds), pp. 517–540. Plenum, New York.

Thomson, K. S. (1968). A new Devonian fish (Crossopterygii: Rhipidistia) considered in relation to the origin of the Amphibia. *Postilla* **124**, 1–13.

Thomson, K. S. (1969). The biology of the lobe-finned fishes. *Biol. Rev.* **44**, 91–154.

Thomson, K. S. (1972). New evidence on the evolution of the paired fins of Rhipidistia and the origin of the tetrapod limb, with a description of a new genus of Osteolepidae. *Postilla* **157**, 1–7.

Thomson, K. S. (1973). Observations on a new rhipidistian fish from the Upper Devonian of Australia. *Palaeontographica* **143**, 209–220.

Thomson, K. S. and Rackoff, J. S. (1974). The shoulder girdle of the Permian rhipidistian fish *Ectosteorhachis nitidus* Cope: structure and possible function. *J. Paleont.* **48**, 170–179.

Thomson, K. S. and Vaughn, P. P. (1968). Vertebral structures in Rhipidistia (Osteichthyes, Crossopterygii) with description of a new Permian genus. *Postilla* **127**, 1–19.

Vorobjeva, E. J. (1973). Einige besonderheiten in Schadelbau von *Panderichthys rhombolepis* (Gross), (Pisces, Crossopterygii). *Palaeontographica* **143**, 221–229.

Warren, J. W. and Wakefield, N. A. (1972). Trackways of tetrapod vertebrates from the Upper Devonian of Victoria, Australia. *Nature, Lond.* **238**, 469–470.

Westoll, T. S. (1938). Ancestry of the tetrapods. *Nature, Lond.* **141**, 127–128.

Westoll, T. S. (1943). The origin of the primitive tetrapod limb. *Proc. R. Soc. Lond.* **B131**, 373–393.

Whitten, J. (1969). Cell death during early morphogenesis: parallels between insect limb and vertebrate limb development. *Science* **163**, 1456–1457.

Wilder, H. H. (1912). The appendicular muscles of *Necturus maculosus*. *Zool. Jb.* Suppl. **15**(2), 383–424.

Zwick, W. (1868). Beitrage zur Kenntnis des Baues und der Entwicklung der Amphibien-Gliedmassen, besonders von Carpus und Tarsus. *Z. Wiss. Zool.* **63**, 62–114.

12 The Hyomandibular as a Supporting Element in the Skull of Primitive Tetrapods

ROBERT L. CARROLL

Redpath Museum, McGill University, Montreal, Canada

Abstract: The Lower Carboniferous labyrinthodont amphibian *Greererpeton* has a large hyomandibular supporting the braincase on the cheek. Functionally this is a continuation of the pattern of rhipidistian fish. In the transition to amphibians, the skull becomes more freely movable on the trunk and a more solid attachment between the dermal skull and the braincase evolves. The hyomandibular may have been retained as a connecting link in the ancestors of all amphibian orders. Integration of the occipital plate and the skull-table and/or cheek occurs in various ways and at various times in the different orders. In most, but not all, groups this is followed by the reduction of the size of the stapes and, in many orders, by its integration into the middle ear as part of an impedance matching system. This occurs early in the Carboniferous in the anthracosaurs and loxommatids but later in temnospondyls. Microsaurs and their descendants, the apodans and primitive salamanders, retain the mechanical connection between the stapes and the cheek. Early reptiles also retain the hyomandibular as a supporting element for the braincase.

It has long been universally accepted that the hyomandibular loses its function of supporting the jaws on the braincase in the transition from rhipidistian fish to amphibians, becomes much reduced in size and takes on the function of the stapes. In fact, the hyomandibular is retained as a supporting element in several, and perhaps all, lines of primitive tetrapods.

Systematics Association Special Volume No. 15, "The Terrestrial Environment and the Origin of Land Vertebrates", edited by A. L. Panchen, 1980, pp. 293-317, Academic Press, London and New York.

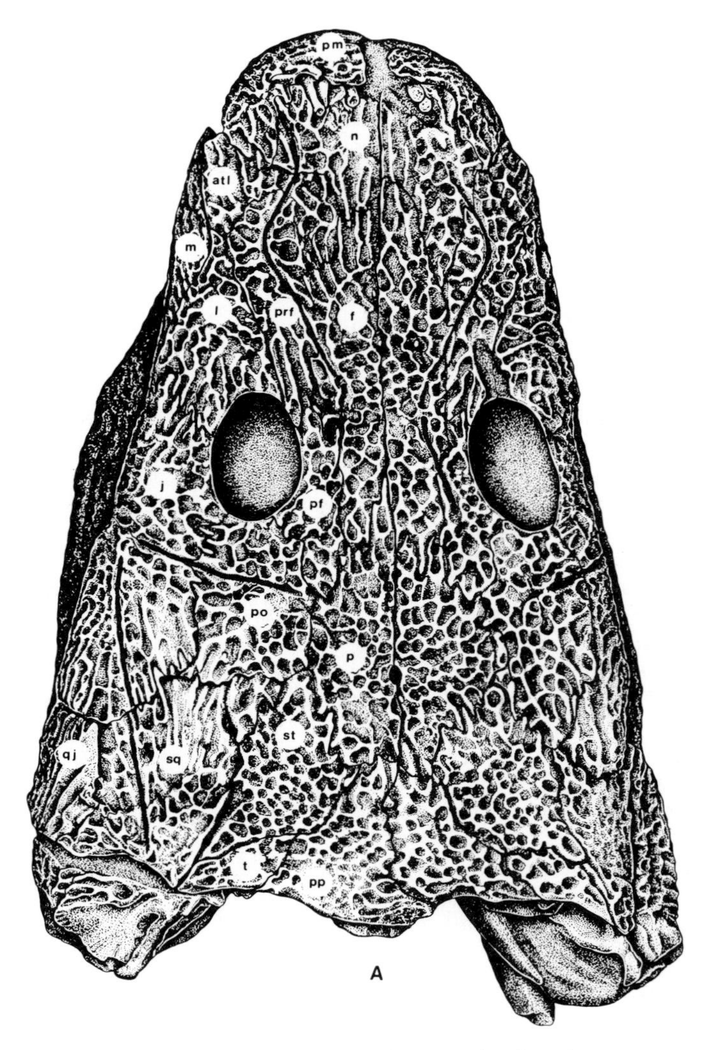

Fig. 1 Skull of *Greererpeton* in (A) dorsal, (B) palatal and (C) occipital views (Cleveland Museum of Natural History No. 11068), × ¾; (D) restoration of occiput, based on several specimens. The braincase is only very weakly attached to the skull-roof by the limited dorsal extremities of the exoccipitals. The otic capsule apparently has no ossified connection with the skull-roof, but is supported by the hyomandibular which has extensive contact with the palatoquadrate and pterygoid.

Key to abbreviations. a, angular; art, articular; art fa hyo, articular facet for hyomandibular; atl, anterior tectal; bo, basioccipital; eo, exoccipitial;

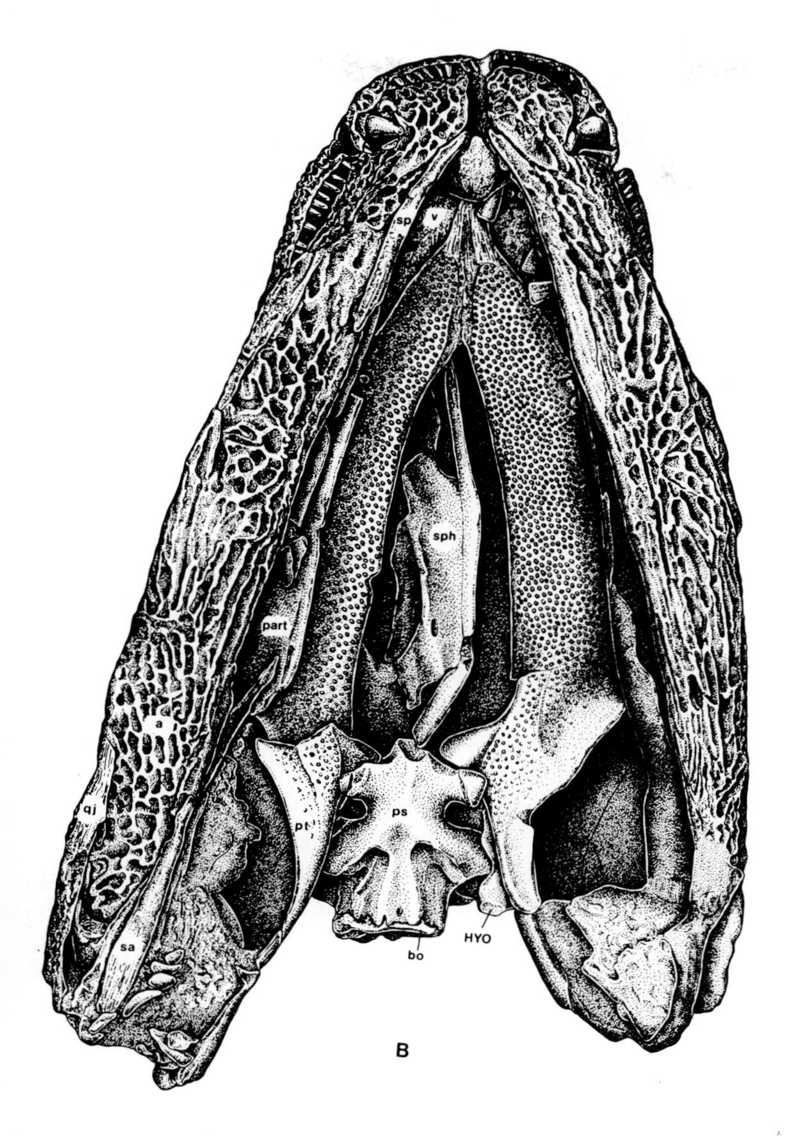

ept, epipterygoid; f, frontal; HYO, hyomandibular; j, jugal; l, lacrimal; llc, lateral line canal groove; m, maxilla; n, nasal; o, opisthotic; p, parietal; part, prearticular; pf, postfrontal; pm, premaxilla; po, postorbital, pp, postparietal; pre op, preopercular; prf, prefrontal; ps, parasphenoid; pt, pterygoid; PTF, posttemporal fossa; q, quadrate; qj, quadratojugal; S, stapes; sa, surangular; so, supraoccipital; sp, splenial; sph, sphenethmoid; sq, squamosal; st, supratemporal; t, tabular; v, vomer.

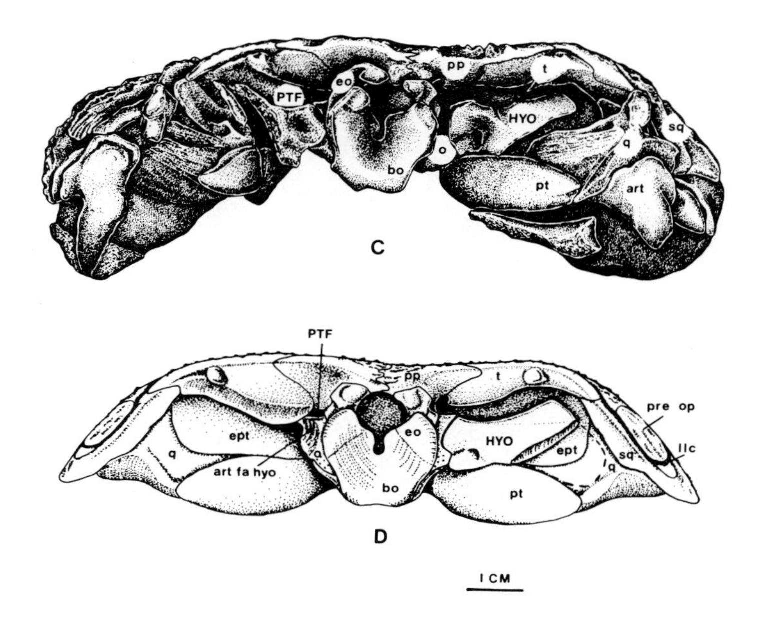

Fig. 1. *continued.*

A large hyomandibular unquestionably forms a link between the braincase and the cheek in several well-preserved skulls of the primitive labyrinthodont amphibian *Greererpeton* from the Lower Carboniferous (or Namurian; Holmes, Chapter 14 of this volume) of West Virginia (Fig. 1). As seen in occipital view, the hyomandibular has a widely expanded plate pressed against the palatoquadrate, above the rather narrow quadrate ramus of the pterygoid. Distally, the hyomandibular ends abruptly, but the configuration of the quadrate and pterygoid suggests that it was continued in cartilage to the level of a ridge just above the articulating surface of the quadrate. *Greererpeton* has no otic notch, or other evidence of tympanic support; nor is it likely that this animal, with well-developed lateral line canals and rather weak limbs (Fig. 2), would have had much use for receiving high frequency airborne vibrations.

The configuration of the dermal bones of the skull-roof, the pattern of sculpturing and the rhachitomous structure of the vertebrae all support the inclusion of *Greererpeton* among the primitive temnospondyl amphibians (Romer, 1969). It is probably close to

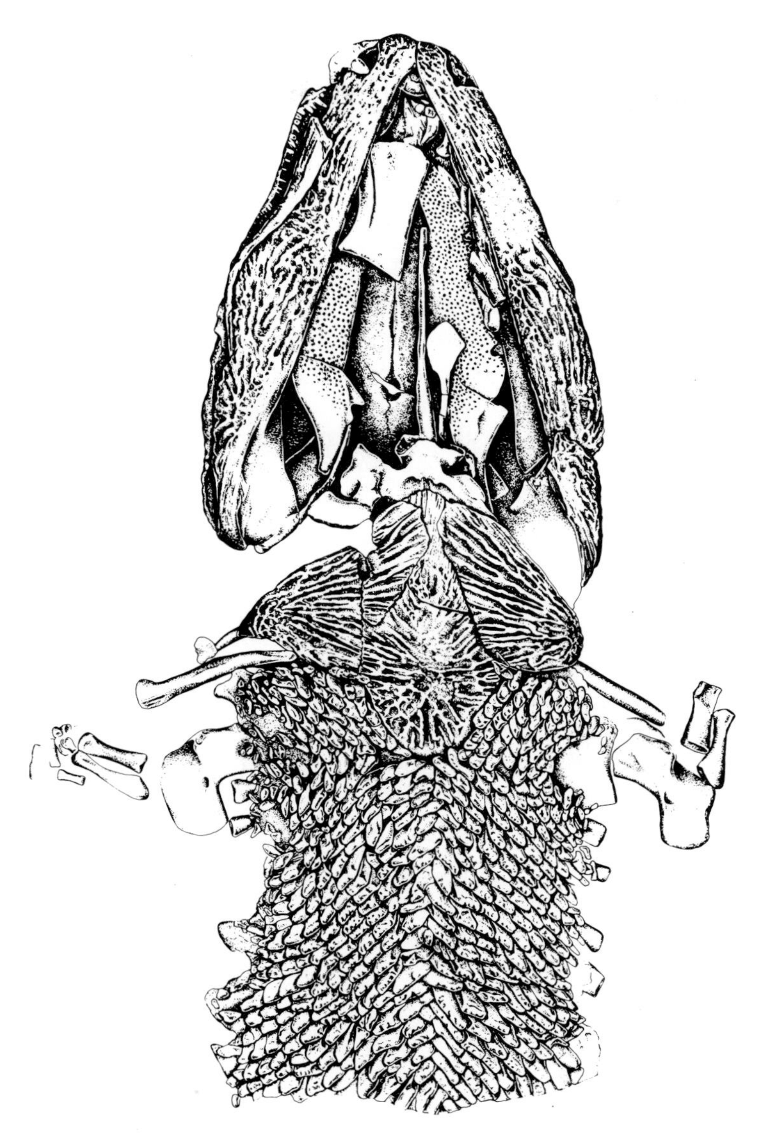

Fig. 2. Ventral view of the anterior portion of the skeleton of *Greererpeton* (Cleveland Museum of Natural History, No. 11090), × ½.

the ancestry of the Pennsylvanian genera *Colosteus* and *Erpetosaurus*. It is apparently most closely related to *Pholidogaster* (Panchen, 1975) from the British Lower Carboniferous. Apart from the ichthyostegids,

Greererpeton and *Pholidogaster* are among the oldest known labyrinthodonts.

One might think that the retention of the hyomandibular was simply a relict feature of a single lineage of rather archaic amphibians, until one evaluates the function of this element in the support of the braincase, not only in *Greererpeton*, but in primitive tetrapods in general.

If we consider the rhipidistian fish (Thomson, 1969), there are few well-defined areas of attachment between the braincase and the skull. This is only natural, considering that the skull of these fish is structurally an extension of the trunk (Fig. 3). The skull-table and cheek are essentially a continuation of the operculum and elements of the dermal shoulder girdle. The notochord extends through the base of the otic-occipital portion of the braincase so that it is firmly, but not rigidly, attached to the vertebral column. There is, however, no well-developed articulating surface between the occipital surface and the cervical vertebrae (Fig. 5). Some mobility of the head on the trunk was surely possible, but of limited extent.

In contrast, even the most primitive tetrapod, *Ichthyostega* (Jarvik, 1952), has lost nearly all of the opercular series, and there is no bony connection between the skull and the dermal shoulder girdle. It is probable that the primitive rhipidistian association between the otic-occipital portion of the braincase and the notochord is retained (as restored by Jarvik). This is, however, lost in all other known Palaeozoic amphibians. Some specialization of the skull-roof

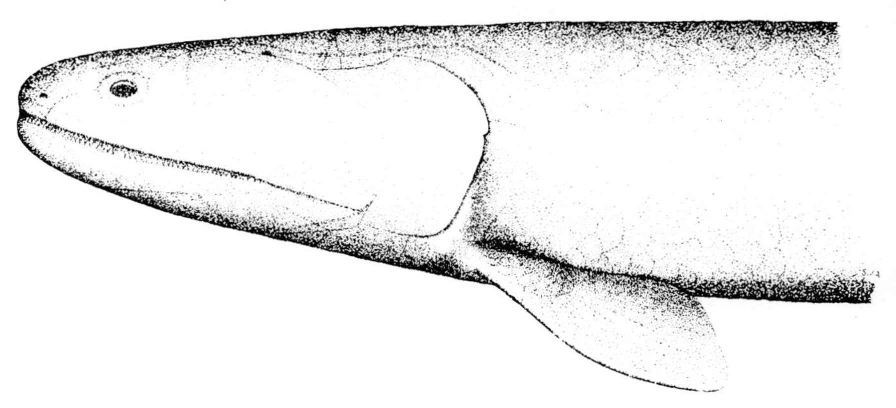

Fig. 3. Restoration of the anterior portion of the body of *Eusthenopteron* (from Andrews and Westoll, 1970).

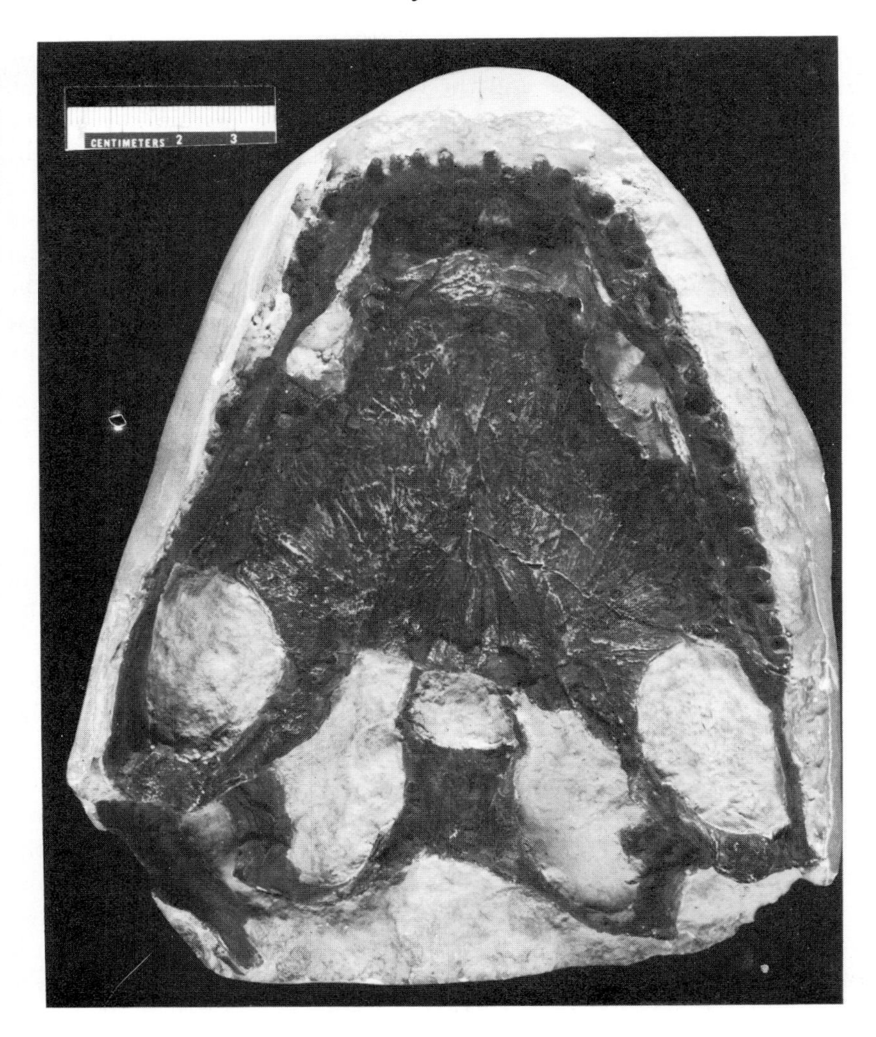

Fig. 4. Palate of *Ichthyostega*. The otic-occipital portion of the braincase has fallen out. The anterior extent of the opening for the notochord is seen at the level of the basicranial articulation.

occurred in *Ichthyostega* to provide for its attachment to the braincase, but this is evidently still weak, for in one of the skulls showing this area (Fig. 4) the otic-occipital portion of the braincase has fallen out. In two other specimens, however, it remains in place.

Obviously one of the various mechanical problems in the

rhipidistian–amphibian transition is that of evolving a close association between the dermal skull and the braincase. The highly kinetic skull of rhipidistians has a plethora of mobile units. Within the dermal skull, the skull-table (itself in two pieces), the operculum and the cheek are all movable relative to one another. The palate articulates with the braincase. Within the braincase, the otic-occipital and ethmoid units articulate with one another. The hyomandibular forms a movable link between the braincase, the operculum and the jaws (Fig. 7). The functional significance of the mobility of these elements in both feeding and respiration has been discussed by Thomson (1967).

I will be concentrating on the following areas of contact:

skull-table to cheek
braincase to skull-table and cheek
hyomandibular to braincase and cheek

It has not been possible to specify any particular genus or even family of rhipidistian fish as the most appropriate ancestors for amphibians, whether they are considered monophyletic or polyphyletic. This may mean either that we lack the fossils of the appropriate rhipidistian group, or that adaptive differences between rhipidistians and amphibians are so great that we would not recognize a specific ancestor at the rhipidistian level.

Picking any particular rhipidistian for comparison with amphibians may be hazardous if we consider details of the anatomy, but for the changes that I would like to consider, the structure of the skull-table, cheek and back of the braincase seem sufficiently similar in all rhipidistians that almost any well-known genus will serve. *Eusthenopteron* thus provides a good basis for comparison since it is well known and representative of the general anatomical pattern from which most, if not all, amphibians must have evolved.

Recent information on the Panderichthyidae (Gardiner, Chapter 8 of this volume), suggests that this family has a pattern of the dermal skull-roof closer to that of labyrinthodonts than any other rhipidistian family. The back of the braincase, and its association with the skull-roof, however, have not been described.

In rhipidistians, as exemplified by *Eusthenopteron*, and *Ectosteorhachis* there is no special area of attachment between the dorsal surface of the braincase and the skull-table, but a nearly flat abutment of the two elements (Fig. 5). This is, however, enough to

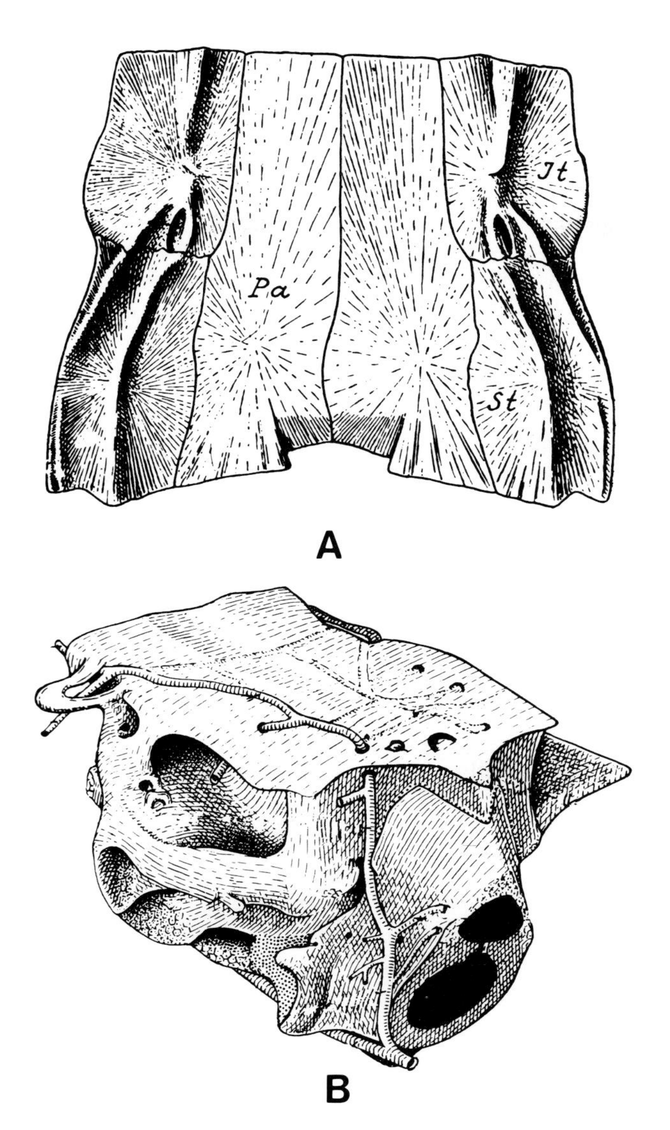

Fig. 5. (A) A ventral view of the posterior portion of the skull-roof of *Eusthenopteron* (from Jarvik, 1954). Bones labelled It, Pa and St are identified, respectively, as supratemporal, postparietal and tabular in tetrapods. The tabular is grooved for support of the crista parotica of the braincase. (B) Oblique posterior-dorsal view of the braincase of *Eusthenopteron* (from Jarvik, 1975). The dorsal surface of the braincase shows no specialized surface for connection to the skull-roof. An essentially similar structure is seen in the Lower Permian rhipidistian, *Ectosteorhachis* (Romer, 1937, Fig. 1).

keep them solidly attached in most fossils. The margin of the otic capsule fits into a narrow groove in a bone that Jarvik (1954) termed the supratemporal, but which other palaeontologists would term the tabular in analogy with this bone in tetrapods. Although this line of attachment appears quite weak, it is significant, since it is in the same position as a major contact in labyrinthodonts. The occipital plate has no well-defined area of attachment to the bones that will form the back of the skull-table in primitive tetrapods, the postparietals. The hyomandibular articulates at two points with the otic capsule proximally, and runs down along the margin of the palatoquadrate to end cartilaginously, just above the articulating surface of the quadrate. The dermal cheek and operculum are movable relative to both the underlying endochondral bones and the skull-table.

The only interdigitating connection between the back of the braincase and the rest of the skull is via the narrow paroccipital processes (Fig. 6). When the close association between the braincase and the notochord, and between the dermal skull and the shoulder girdle, are lost in the origin of amphibians, this would clearly be inadequate to resist relative movement of the elements.

Several alternative ways of strengthening the attachment of the braincase to the skull-roof would be possible. The contact between

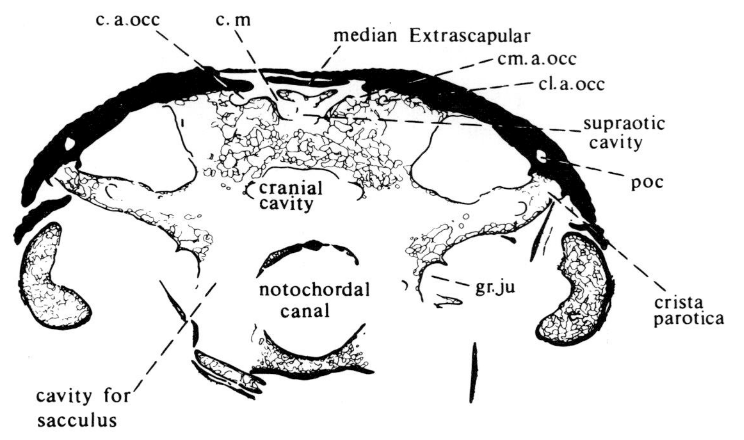

Fig. 6. Section through the skull of *Eusthenopteron* showing nature of contact between the braincase and the skull-roof (from Jarvik, 1975). c.a.occ, Canal for occipital artery; cl.a. occ and cm.a.occ, lateral and medial branches of canal for occipital artery; gr. ju, groove for jugular vein; poc, postotic sensory canal.

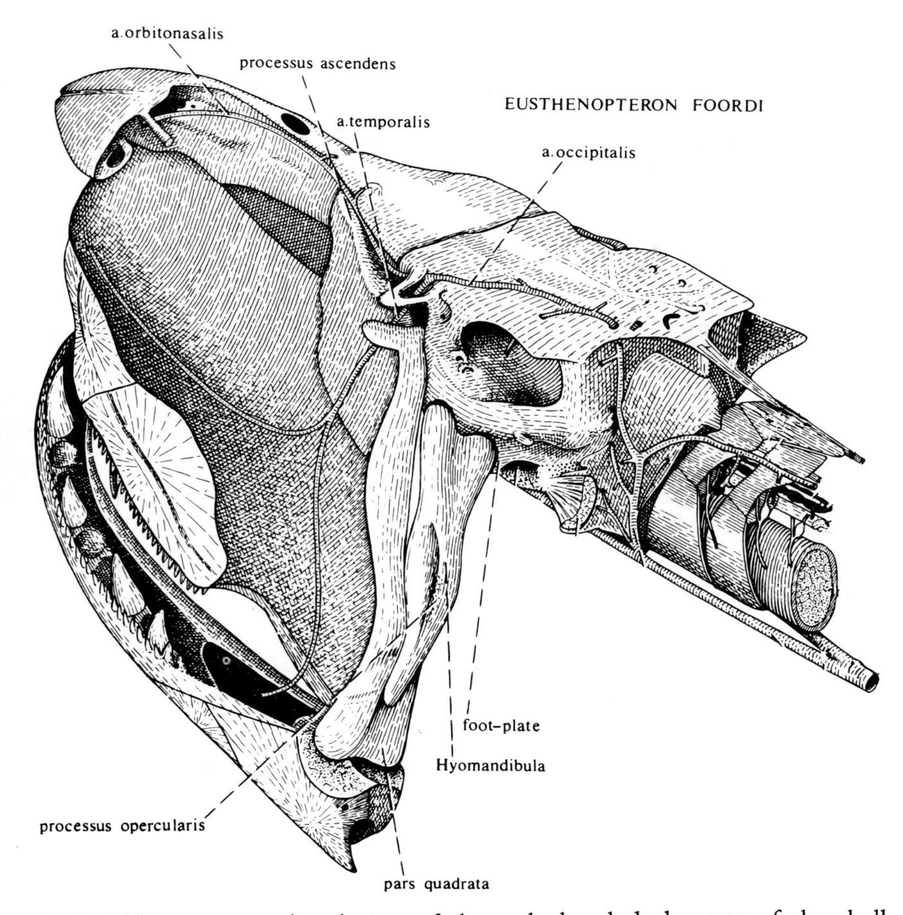

Fig. 7. Oblique posterodorsal view of the endochondral elements of the skull of *Eusthenopteron* (from Jarvik, 1975), showing the manner in which the back of the braincase is linked to the palatoquadrate via the hyomandibular.

the otic capsule and the tabular could be strengthened. Contact could be gained between the otic-occipital plate and the postparietals, or, if the cheek and the skull-table were to become suturally attached, support could be maintained by the hyomandibular. At least two different ways of attachment are clearly evidenced by the early labyrinthodonts. Anthracosaurs (Fig. 8) as early as the Viséan, show a well-ossified opisthotic which is broadly in contact with a ventral elaboration of the tabular, and the exoccipitals are in contact with either the postparietals directly, and/or via a supraoccipital (Panchen, 1975). The mobility of the skull-table on the cheek is retained. The hyomandibular or stapes is not described in any embolomere but at this stage it was presumably already small, for, without a strong attachment between the skull-table and the cheek, there would have been little functional reason for a large hyomandibular to be retained to link the cheek with the braincase.

Anthracosaurs are among the first tetrapods to develop an otic notch that can be reasonably considered to have supported a tympanum. Knowledge of the impedance matching system in living tetrapods (Wever and Werner, 1970) suggests that a tympanum would only have served to transmit airborne sound efficiently if the stapes were light and had a small foot plate. A stapes of this nature is known in *Seymouria* (White, 1939), but not in other anthracosaurs. *Eoherpeton* (Panchen, 1975), the oldest known anthracosaur, has little overlap of the skull-table beyond the cheek, and has only a short otic notch (but see Panchen, Chapter 13 of this volume). The configuration of this animal suggests that the notch may have evolved within the anthracosaurs.

In contrast to anthracosaurs and loxommatids, early temnospondyl

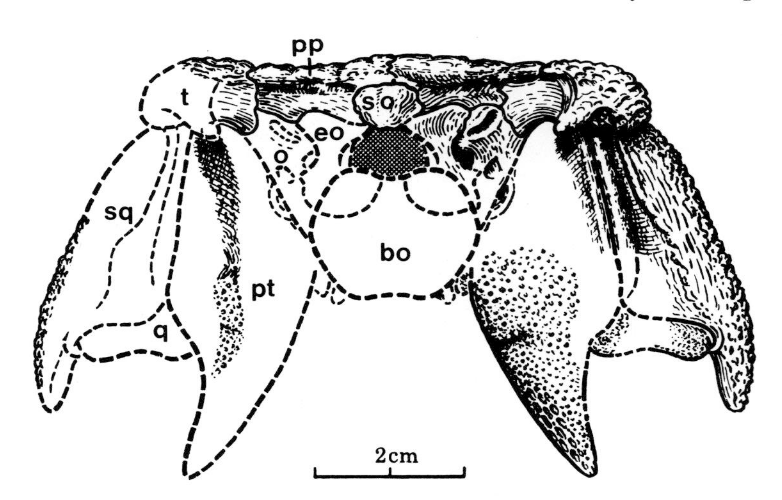

Fig. 8. Occipital view of the early anthracosaur *Eoherpeton* (from Panchen, 1975). The skull-table and cheek are not suturally attached, but the opisthotic and exoccipitial are firmly connected to the tabular and postparietal. See key to abbreviations in legend to Fig. 1.

amphibians quickly evolved a firm union between the skull-table and the cheek, but most were slow in establishing a strong contact with either the otic capsule and the tabular, or between the occipital plate and the postparietals. In *Greererpeton*, support of the braincase is achieved by the retention of a large hyomandibular linking the cheek to the braincase. Later temnospondyls, such as the Lower Permian *Edops* (Romer and Witter, 1942), which still closely resemble *Greererpeton* in the structure of the occiput, have established a solid connection between the exoccipitals and the postparietals, but the opisthotic is still incompletely ossified. The contact between the braincase and the skull-table is, nevertheless, strong enough that the hyomandibular has been reduced to the dimensions of a stapes (Fig. 9), and the otic notch structure presumably supports a tympanum. The pattern of the otic notch and stapes in such temnospondyls as *Edops* can be traced via the dissorophids and *Doleserpeton* (Bolt, 1977) to *Triadobatrachus* and frogs. Bolt suggested that the great taxonomic diversity of the dissorophids may be associated with the use of vocalization in species recognition, as in most groups of modern frogs.

The reduction of the hyomandibular to a stapes unquestionably

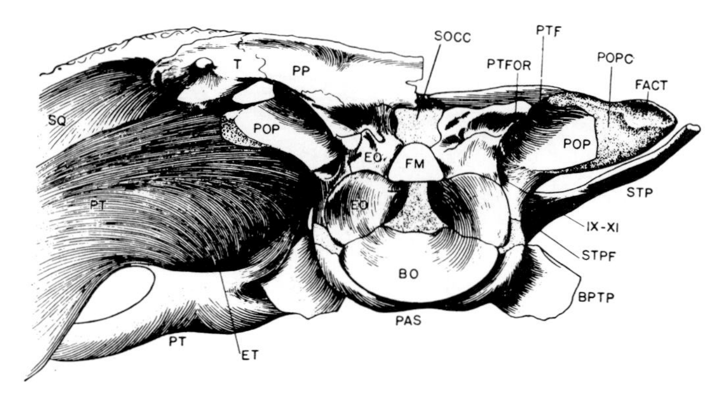

Fig. 9. Occiput of *Edops* (from Romer and Witter, 1942). BPTP, Basipterygoid process; FACT, restored facet of paroccipital process for tabular; POP, paroccipital process; POPC, cartilaginous region of paroccipital process; PTF, post-temporal fossa; PTFOR, post-temporal foramen; SOCC, supraoccipital cartilage; STP, stapes; STPF, stapedial foramen. As in *Greererpeton*, the only bony connection between the skull-roof and the braincase is provided by the exoccipitals. This is, however, much more extensive than in the earlier genus.

occurred separately in temnospondyls and anthracosaurs, as did the elaboration of an otic notch.

Another group, usually considered as temnospondyls, shows a markedly different pattern of the occiput.

Loxommatoids (Beaumont, 1977) are also known from the Viséan. They show an occipitial structure which combines the strong attachment of skull-table and cheek typical of temnospondyls with a firm union between the opisthotic (combining the supraoccipital) and the tabular, otherwise developed earlier in anthracosaurs (Fig. 10). They have an otic notch, but the stapes has not been described.

This pattern suggests that the loxommatids evolved their type of braincase support separately from that of genera such as *Greererpeton*. On quite different grounds, Panchen (Chapter 13 of this volume), has suggested that loxommatids are more closely related to the anthracosaurs than to the temnospondyls. This would accord with the pattern of the occiput.

Among labyrinthodonts, the initial divergence in the nature of the association between the braincase and the skull-roof may be traced to ancestors close to the rhipidistian–amphibian boundary that had slightly different proportions of the skull-roof. Primitive temnos-

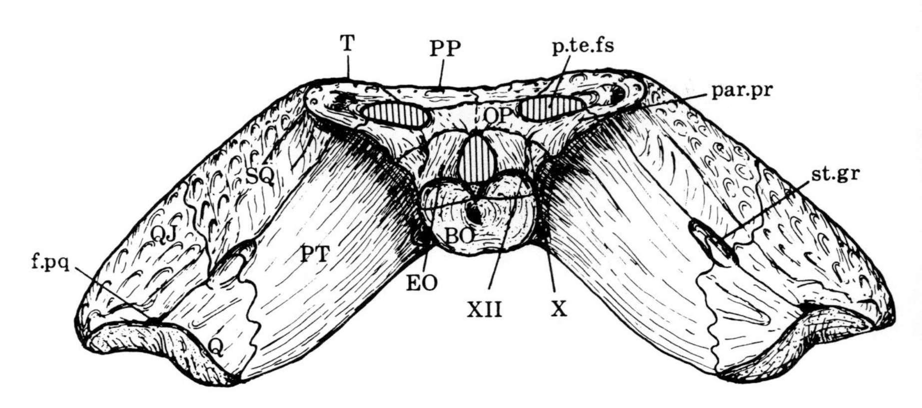

Fig. 10. Occipital view of the early loxommatid *Megalocephalus* (from Beaumont, 1977). f. pq, Paraquadrate foramen; p. te. fs, post-temporal fossa; st. gr, groove for stapedial process to quadrate (this groove bears the same relationship to surrounding structures as does the lateral line canal groove associated with the preopercular in *Greererpeton*). Bone labelled OP includes the area occupied by both the supraoccipital and the opisthotic in other early tetrapods.

pondyls (other than loxommatids) appear to have a wider skull-table than anthracosaurs. This may have made elaboration of the otic capsule beyond the limits of the braincase awkward, whereas anthracosaurs with a narrow skull-roof achieved this association early in their evolution. Most embolomeres retained a movable articulation between the skull-table and the cheek. A notable exception is *Anthracosaurus* (Panchen, 1977) which developed a strong connection, possibly in relationship with the use of extremely large marginal teeth.

Crassigyrinus (Panchen, Chapter 13 of this volume; Smithson, Chapter 16 of this volume), has been identified as an ancestral batrachosaur (a group used to include that genus, anthracosaurs and Loxommatoidea). *Crassigyrinus* is primitive in the nature of the pattern of the bones of the skull-table, but the long tabular horns are accepted as a shared batrachosaur synapomorphy. The elongation of the tabular horns may be associated with the presence of deep otic notches. The skull-table is narrow in this genus, and the cheeks are extremely broad. If the long tabular horns and deep notches are truly primitive for this genus, and for batrachosaurs in general, it may be assumed that the hyomandibular modified its supporting function very early in the evolution of this group. Despite the primitive nature of the pattern of the bones of the skull-table and apparently the vertebrae, it could be argued that the development of a deep otic notch was a specialization of this genus, which is known no earlier in the fossil record than *Eoherpeton* and *Loxomma*.

Aistopods, nectrideans, adelogyrinids, archerontiscids and lysorophids are not included in this consideration since their occipital structure is either not sufficiently known or the earliest adequately known genera are too highly specialized to be compared either with rhipidistians or with any other groups of primitive tetrapods.

The earliest adequately known reptiles are already specialized beyond the level of the early temnospondyls and anthracosaurs, as well as being much smaller in size. Both captorhinomorphs and pelycosaurs retain evidence of their ancestry in the pattern of their hyomandibular (Fig. 11). It clearly retains a supporting function in both groups. In neither of these groups is the opisthotic well-ossified in primitive members (Romer and Price, 1940; Carroll and Baird, 1972), and it faces toward the cheek, rather than toward the skull-table. There may have been some mobility between the skull-table

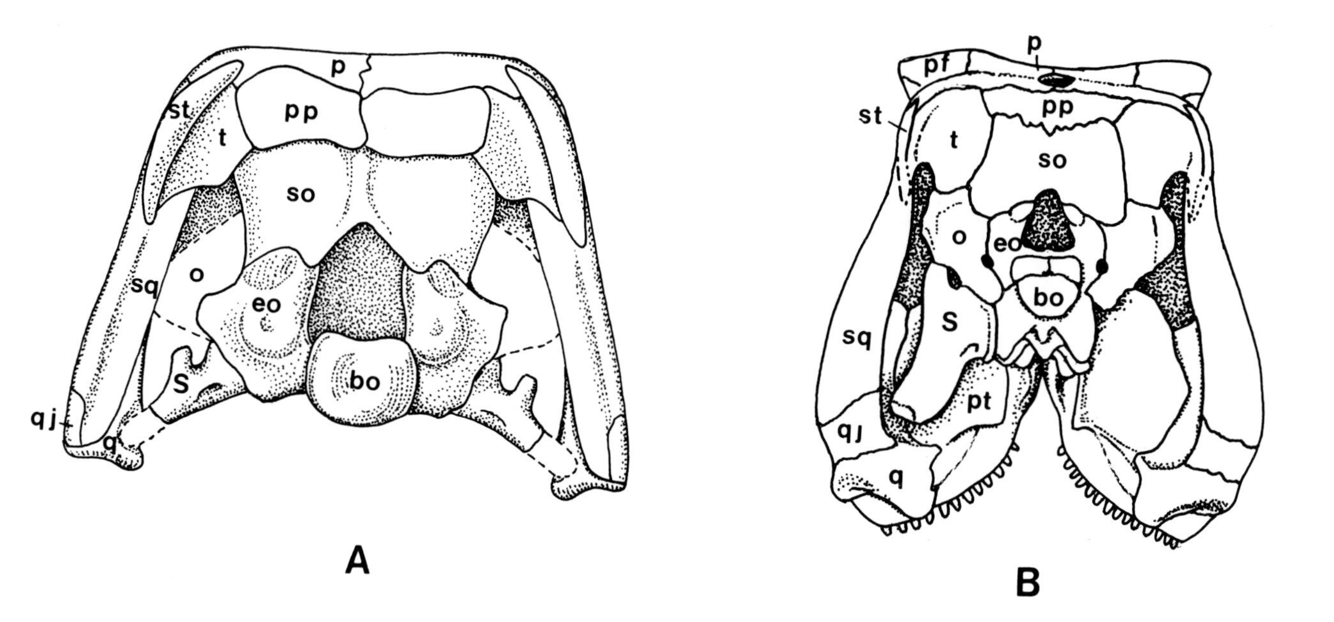

Fig. 11. Occiputs of (A) the primitive captorhinomorph *Paleothyris* and (B) the primitive pelycosaur *Opiacodon* (from Romer and Price, 1940). In both these genera, the braincase is connected to the skull-roof via a large supraoccipital. The opithotic is loosely, if at all, associated with the skull-roof, and does not reach the cheek. The stapes is a large element, linked by cartilage to the quadrate. It retains a supporting role. See key to abbreviations in legend to Fig. 1.

and the cheek in the most primitive of reptiles, but the magnitude of movement must have been small enough that the hyomandibular was held in position between the quadrate and the braincase.

In both pelycosaurs and romeriid captorhinomorphs, contact between the occipital plate and dermal bones of the skull-table is via a large supraoccipital. In the captorhinomorphs (as established by Heaton, 1978) and, as far as we know, primitively for all reptiles, this attachment remains loose rather than sutural, and the braincase could move relative to the skull-roof and palate.

The supraoccipital could rotate under the postparietals, and the basisphenoid moved against the basicranial articulation formed by the pterygoid and epipterygoid. The braincase uses the hyomandibular somewhat as an axis which rotates on the quadrate. It thus retains, in a modified form, its primitive function in supporting the braincase against the cheek.

As in temnospondyls, advanced reptiles elaborated the otic capsule laterally, freeing the hyomandibular from a supporting role. In most groups the otic capsule reaches the cheek rather than the skull-table. The procolophonids appear exceptional in that the paroccipital process reaches the tabular. A light-weight, columnar stapes was evolved independently in four or five lines of advanced reptiles (Romer, 1967). This happened last of all in the line leading to mammals.

Lizards, rhynchosaurs and archosaurs in the Lower Triassic have narrow, rod-like stapes. The tympanum presumably attached to the widely emarginated quadrate, as in modern lizards. It is not possible to say whether this structure evolved in a single group of ancestral diapsids, or arose separately in the ancestors of each of these groups. Turtles certainly evolved this pattern separately. Millerosaurs, prolacertids and procolophonids also appear to have separately evolved an impedance matching system in the middle ear, subsequent to the freeing of the stapes from a supporting role. In these groups the tympanum is supported primarily by the squamosal and/or quadratojugal rather than by the quadrate.

Pelycosaurs diverged from the main reptilian lineage at the base of the Pennsylvanian. Primitively, the opisthotic is not strongly integrated with the cheek and there may have been some mobility between the supraoccipital and the postparietal (Reisz, 1972). In the sphenacodonts, the occiput becomes strongly integrated with the

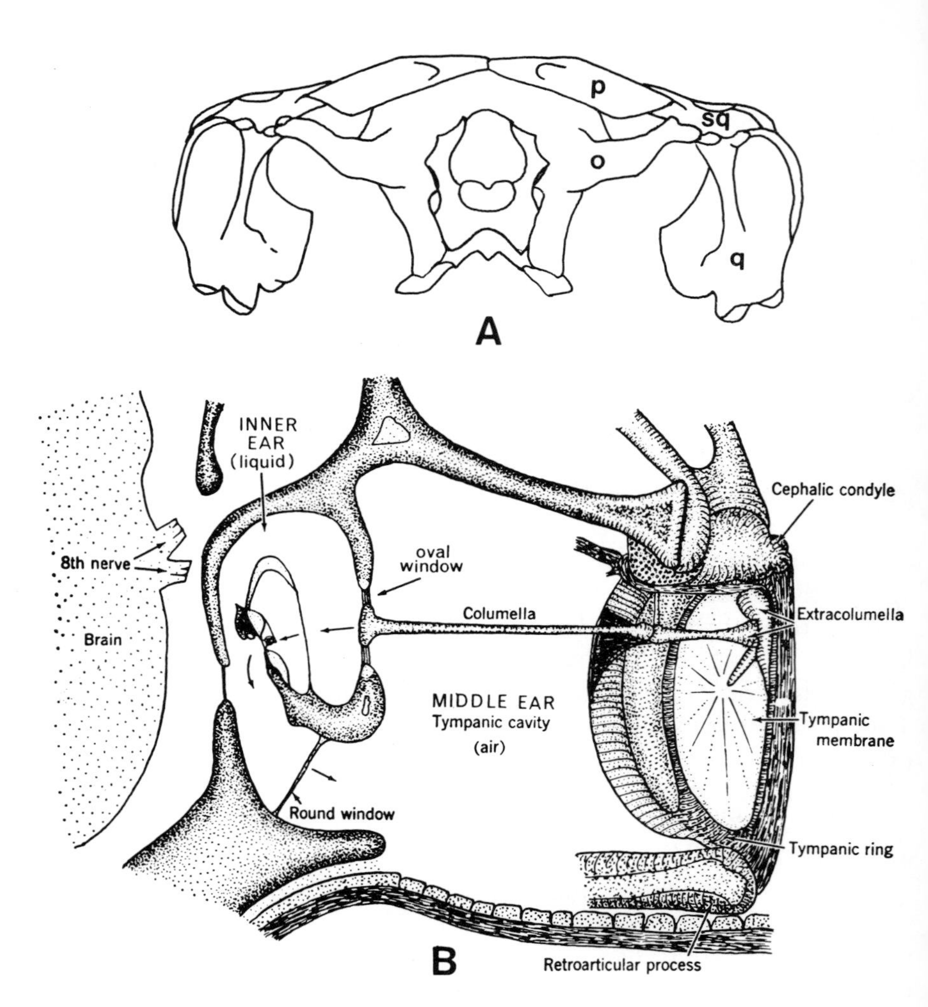

Fig.12. (A) Occiput of the Upper Triassic lizard *Kuehneosaurus* (from Robinson, 1962) showing the attachment of the paroccipital process to the dorsal end of the quadrate and the squamosal. This attachment frees the stapes from a supporting role. This pattern was initiated by the late Permian or early Triassic in several lines of diapsid reptiles. See key to abbreviations in legend to Fig. 1. (B) Diagrammatic representation of the ear of a lizard illustrating the basic components of an impedance matching system. The large tympanum transmits vibrations via the small, light columella (suspended in an air-filled chamber) to the oval window of the middle ear. The greater area of the tympanum relative to that of the foot plate of the stapes serves to amplify mechanically the force impinging on the ear drum in order to compensate for the differential in density between the fluid of the inner ear and the outside air (from Robinson, 1973).

cheek and the skull roof. Such a strong attachment leads to a freeing of the hyomandibular from a supporting role in other reptilian groups and a reduction in its size. This is not the case in sphenacodonts, which retain a massive stapes. This is presumably associated with the entirely different evolutionary pattern followed by the middle ear structures in the line leading from therapsids to mammals (Allin, 1975).

I shall have to disagree with Carroll, (1969, 1970a,b) who maintained that the reptilian hyomandibular had evolved from a reduced or at least dorsally oriented stapes in primitive anthracosaurs. Knowledge of the function of the middle ear in living frogs, reptiles and mammals, makes it very unlikely that there was a tympanum as part of an impedance matching system in primitive reptiles, or their immediate ancestors (Fig. 12A). It has been argued that if primitive reptiles retained a large, ventrally oriented stapes as an inheritance from fish ancestors, they had not evolved from labyrinthodonts (Panchen, 1972). We know that early temnospondyls, as represented by *Greererpeton*, retain a fish-like orientation of the hyomandibular. It is probable that ancestral anthracosaurs did as well. Early reptiles still appear to share more significant features of the postcranial skeleton with anthracosaurs than with other known Palaeozoic amphibians, but their relationship would have to date from a time prior to the reduction of the hyomandibular and the solidification of the braincase to the skull-roof. This takes us back to the early Lower Carboniferous, if not before.

Diadectes and *Tseajia* (Heaton, Chapter 18 of this volume) may be taken as exemplifying another alternative pattern of the occipital–hyomandibular complex. The late appearance of these genera (Lower Permian) makes it difficult to sort out which features of their anatomy are primitive and which are advanced, but it is probable that their retention of a very large hyomandibular–stapes, is primitive. They possess this structure in conjunction with a large quadrate the configuration of which suggests support of a tympanum in *Tseajia;* in *Diadectes* the area of the tympanum is preserved as a thin ossified sheet. This is a combination of characters different in principle from both the impedance matching ears of modern reptiles, mammals and frogs, and from the type of middle ear seen in salamanders and apodans, and assumed to be present in primitive captorhinomorphs and pelycosaurs. The extremely large size of the tympanum in these genera may partially compensate for the massive nature of the stapes, and the orientation of the stapes may make it effective as a lever. It is conceivable that these genera evolved from the base of the anthracosaur line represented by *Eoherpeton*, and so have a separate ancestry from that of true reptiles. It is possible that this lineage never underwent the size reduction that is postulated to explain the origin of the reptilian reproductive pattern (Carroll, 1970b).

The anatomy of *Eoherpeton* (Smithson, Chapter 16 of this volume) shows it to be a large, robust, terrestrial form, as are *Diadectes* and *Tseajia*. There is no evidence of this type of hyomandibular–stapes in seymouriamorphs, limnoscelids or the Nycteroleteridae, classified with diadectids and *Tseajia* by Heaton (Chapter 18 of this volume).

Microsaurs evolved a fundamentally different sort of articulation between the occipital condyle and the first cervical vertebra to the complex pattern seen in labyrinthodonts and reptiles (Carroll and Gaskill, 1978), indicating a distinct ancestry, but they also retained the hyomandibular as a relatively large, ventrolaterally directed structure that linked the braincase with the cheek. Like reptiles, microsaurs initially had a weakly ossified otic capsule and retained or elaborated a movable articulation between the cheek and the skull-table (Fig. 13). No microsaur has a stapes of the shape or relative position to function in the manner of this bone in early reptiles, nor is the relationship of the supraoccipital functionally similar.

The descendants of microsaurs have retained the hyomandibular as

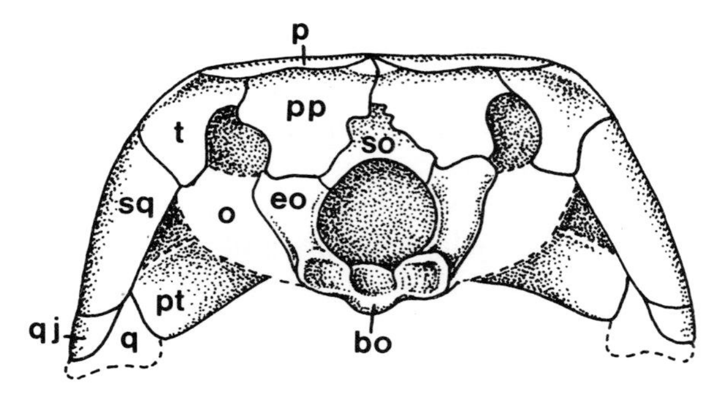

Fig. 13. Occiput of the early microsaur *Asaphestera*. The otic capsule is apparently not ossified, but the braincase is relatively firmly attached to the skull-table by the postparietals and the supraoccipital which broadly underlies them. The cheek is movable on the skull-table. The stapes is not known in this genus. See key to abbreviations in legend to Fig. 1.

a connecting structure in many lines to the present day. In apodans, this bone is capable of transmitting relatively low frequency vibrations between the cheek and the inner ear (Wever and Gans, 1976), but it retains the rhipidistian linkage between the braincase and the quadrate (Carroll and Currie, 1975). This is also the case, although in a slightly modified sense, in primitive salamanders (Carroll and Holmes, 1980). In hynobiids, primitive salamandrids and ambystomatids, the braincase is solidly integrated with the skull-roof,

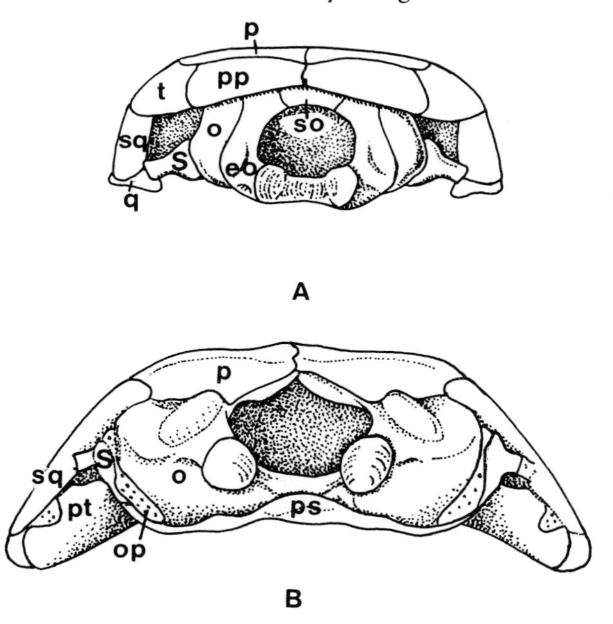

Fig.14. (A) Occiput of the Lower Permian microsaur *Hapsidopareion*. The otic capsule is relatively well ossified, but the cheek is not firmly attached to the skull-table. The stapes links the lower cheek and the braincase. (B) The primitive salamander *Hynobius* in occipital view showing the attachment of the stapes to the cheek region. In this genus and in primitive ambystomatids and salamandrids, the jaw suspension, consisting of the squamosal, quadrate and pterygoid, is somewhat movable on the skull-roof and the palate. Primitively, the squamosal has a hinge-like attachment to the parietal or to the dorsal surface of the otic capsule and the pterygoid slides across the basicranial articulation situated at the base of the otic capsule. The attachment of the stapes to the squamosal and/or quadrate limits the scope of this movement. This differs from the primitive tetrapod pattern in that the braincase is firmly anchored to the skull-roof. See key to abbreviations in legend to Fig. 1.

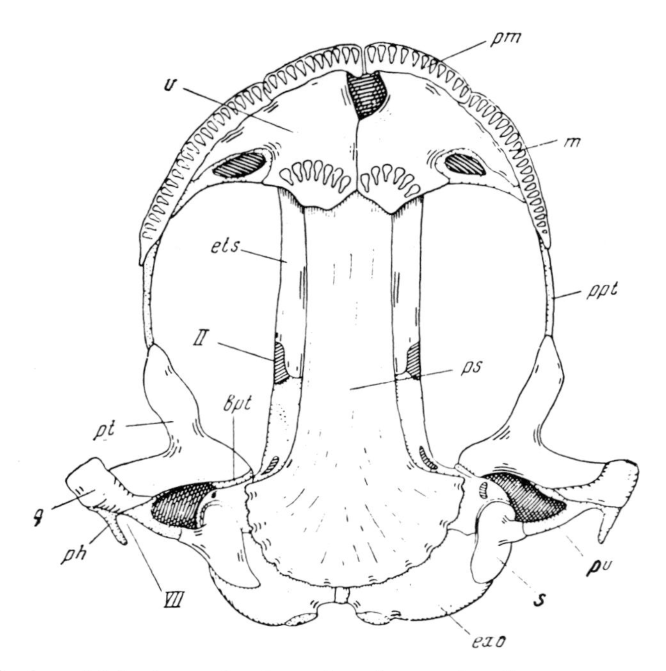

Fig. 15. The hynobiid salamander *Ranodon* (from Schmalhausen, 1968, Fig. 135) showing the nature of the connection between the stapes and squamosal. Bpt, Basicranial articulation on fused basisphenoid-prootic; ets, ethmosphenoid; exo, fused exoccipital and opisthotic; ph, hyoid process; ppt, pterygoid process of palatoquadrate cartilage; pv, processus ventralis stapedes; S, stapes. Other abbreviations as in the legend to Fig. 1.

but the cheek is movable (Figs 14 and 15). The hyomandibular, as in other tetrapods, serves to convey vibrations to the inner ear, but maintains the linkage between the braincase and the lower cheek. As shown by embryological studies (Fox, 1959), this continuity is maintained during development not only in the hynobiids, but also in cryptobranchids, proteiids and ambystomatids.

It may be a bit difficult to accept the primacy and basically primitive nature of the supporting role of the hyomandibular-stapes in modern salamanders and apodans as well as in primitive tetrapods, but I feel we have long over-extended the sound-conducting role that we see in modern mammals and reptiles to other animals with a basically different cranial anatomy.

ADDENDUM

Subsequent to the completion of this manuscript, Lombard and Bolt (1979) published a paper on the closely related topic of the evolution of the middle ear in tetrapods. Their approach was almost entirely different from that used here, being based primarily on the comparative anatomy of the structures of the middle ear in living tetrapods. Their conclusions were similar however; the tympanum was not present in primitive tetrapods, but evolved separately in several lines. The current paper might be considered as confirmation of what they term the "alternate" view of the evolution of the middle ear structures in tetrapods.

ACKNOWLEDGEMENTS

I would like to thank Drs David Dunkle and Michael Williams of the Cleveland Museum of Natural History for the loan of material of *Greererpeton* that formed the basis for this consideration. Collection and preparation of these specimens were performed by the staff of the Cleveland Musuem, supported by grants from the United States National Science Foundation. Drawings of these specimens were made by Pamela Gaskill, supported by a grant from the Canadian National Research Council. I thank Malcolm Heaton for permission to present unpublished ideas concerning the function of the hyomandibular and occiput in primitive reptiles.

REFERENCES

Allin, E. F. (1975). Evolution of the mammalian middle ear. *J. Morph.* **147**, 403–438.

Andrews, S. M. and Westoll, T. S. (1970). The postcranial skeleton of *Eusthenopteron foordi* Whiteaves. *Trans. R. Soc. Edinb.* **68**, 207–329.

Beaumont, E. H. (1977). Cranial morphology of the Loxommatidae (Amphibia; Labyrinthodontia). *Phil. Trans. R. Soc.* **B280**, 29–101.

Bolt, J. R. (1977). Dissorophoid relationships and ontogeny, and the origin of the Lissamphibia. *J. Paleont.* **51**, 235–249.

Carroll, R. L. (1969). Problems of the origin of reptiles. *Biol. Rev.* **44**, 151–170.

Carroll, R. L. (1970a). The ancestry of reptiles. *Phil. Trans. R. Soc.* **B257**, 267–308.

Carroll, R. L. (1970b). Quantitative aspects of the amphibian reptilian transition. *Forma Functio* **3**, 165–178.

Carroll, R. L. and Baird, D. (1972). Carboniferous stem-reptiles of the family Romeriidae. *Bull. Mus. comp. Zool. Harv.* **143**, 321–364.

Carroll, R. L. and Currie, P. J. (1975). Microsaurs as possible apodan ancestors. *Zool. J. Linn. Soc.* **57**, 229–247.

Carroll, R. L. and Gaskill, P. (1978). The Order Microsauria. *Am. phil. Soc. Mem.* **126**, 1–211.

Carroll, R. L. and Holmes, R. (1980). The skull and jaw musculature as guides to the ancestry of salamanders. *Zool. J. Linn. Soc.* **68**, 1–40.

Fox, H. (1959). A study of the development of the head and pharynx of the larval urodele *Hynobius* and its bearings on the evolution of the vertebrate head. *Phil. Trans. R. Soc.* **B242**, 151–204.

Heaton, M. J. (1978). "Cranial Soft Anatomy and Functional Morphology of a Primitive Captorhinid Reptil." Ph.D. Thesis, Department of Biology, McGill University.

Jarvik, E. (1952). On the fish-like tail in the ichthyostegid stegocephalians. *Meddr. Grønland* **114**(12), 1–90.

Jarvik, E. (1954). On the visceral skeleton in *Eusthenopteron* with a discussion of the parasphenoid and palatoquadrate in fishes. *K. svenska VentenskAkad. Handl.* (4)**5** No. 1, 1–104.

Jarvik, E. (1975). On the saccus endolymphaticus and adjacent structures in osteolepiforms, anurans and urodeles. *Colloques int. C.n.R.S.* No. 218, 191–211.

Lombard, R. E. and Bolt, J. R. (1979). Evolution of the tetrapod ear: an analysis and reinterpretation. *Biol. J. Linn. Soc.* **11**, 19–76.

Panchen, A. L. (1972). The interrelationships of the earliest tetrapods. *In* "Studies in Vertebrate Evolution – Essays Presented to Dr. F. R. Parrington, F.R.S." (K. A. Joysey and T. S. Kemp, eds), pp. 65–87. Oliver and Boyd, Edinburgh.

Panchen, A. L. (1975). A new genus and species of anthracosaur amphibian from the Lower Carboniferous of Scotland and the status of *Pholidogaster pisciformis* Huxley. *Phil. Trans. R. Soc.* **B269**, 581–640.

Panchen, A. L. (1977). On *Anthracosaurus russelli* Huxley (Amphibia: Labyrinthodontia) and the family Anthracosauridae. *Phil. Trans. R. Soc.* **B279**, 447–512.

Reisz, R. (1972). Pelycosaurian reptiles from the Middle Pennsylvanian of North America. *Bull. Mus. comp. Zool. Harv.* **144**, 27–62.

Robinson, P. L. (1962). Gliding lizards from the Upper Keuper of Great Britain. *Proc. geol. Soc. Lond.* No. 1601, 137–146.

Robinson, P. L. (1973). A problematic reptile from the British Upper Triassic. *J. geol. Soc. Lond.* **129**, 457–479.

Romer, A. S. (1937). The braincase of the Carboniferous crossopterygian *Megalichthys nitidus. Bull. Mus. comp. Zool. Harv.* **82**, 3–73.

Romer, A. S. (1967). Early reptilian evolution re-viewed. *Evolution* **21**, 821–833.

Romer, A. S. (1969). A temnospondylous labyrinthodont from the Lower Carboniferous. *Kirtlandia* No. 6, 1–20.

Romer, A. S. and Price, L. I. (1940). Review of the Pelycosauria. *Geol. Soc. Am. Spec. Pap.* No. 28, 1–538.

Romer, A. S. and Witter, R. V. (1942). *Edops*, a primitive rhachitomous amphibian from the Texas redbeds. *J. Geol.* 50, 925–960.
Schmalhausen, I. I. (1968). "The Origin of Terrestrial Vertebrates" (K. S. Thomson, ed.). Academic Press, New York and London.
Thomson, K. S. (1967). Mechanisms of intracranial kinesis in fossil rhipidistian fishes (Crossopterygii) and their relatives. *J. Linn. Soc. (Zool.).* **178**, 223–253.
Thomson, K. S. (1969). The biology of the lobe-finned fishes. *Biol. Rev.* **44**, 91–154.
Wever, E. G. and Gans, C. (1976). The caecilian ear: further observations. *Proc. natn. Acad. Sci. U.S.A.* 73, 3744–3746.
Wever, E. G. and Werner, Y. L. (1970). The function of the middle ear in lizards: *Crotaphytus collaris*, Iguanidae. *J. exp. Zool.* **175**, 327–342.
White, T. E. (1939). Osteology of *Seymouria baylorensis* Broili. *Bull. Mus. comp. Zool. Harv.* 85, 325–409.

13 The Origin and Relationships of the Anthracosaur Amphibia from the Late Palaeozoic

A. L. PANCHEN

Department of Zoology, The University, Newcastle upon Tyne, England

Abstract: The generally accepted classification of the Amphibia of the Palaeozoic and Triassic the "Stegocephalia") is into two major groups, the Lepospondyli and the Labyrinthodontia. It is now accepted that the former is an artificial grouping of at least three unrelated taxa. It is suggested here that the Labyrinthodontia are similarly polyphyletic, the diagnostic characters of the group simply being those that characterize primitive tetrapods. The "taxon" Labyrinthodontia is commonly divided into three orders. Of these the Ichthyostegalia includes exclusively Devonian forms; the Temnospondyli includes the vast majority of "labyrinthodonts" and may be polyphyletic; while the Batrachosauria (*sensu* Panchen: Anthracosauria auct.) appears to be a natural group. Apart from a series of primitive tetrapod features, the latter group is characterized by well-developed tabular horns forming the posterior corners of the skull-table (lost in seymouriamorphs), a rounded otic notch (paralleling some temnospondyls) and diplospondylous vertebrae in which the pleurocentrum is always the better developed central element. A phylogeny of the group is reconstructed. *Crassigyrinus scoticus* Watson from the late Viśean and Namurian of Scotland is a relict primitive batrachosaur from which the cranial anatomy of all known batrachosaurs can be derived.

INTRODUCTION

The generally accepted classification of the class Amphibia, as with so many other vertebrate groups, is that set out in the latest edition

Systematics Association Special Volume No. 15, "The Terrestrial Environment and the Origin of Land Vertebrates", edited by A. L. Panchen, 1980, pp. 319–350, Academic Press, London and New York.

of Romer's classic text (Romer, 1966). Amphibia are therein divided into three subclasses, the Lissamphibia, comprising the three extant orders and closely related fossil forms, the Lepospondyli, usually small Palaeozoic forms thought to have holospondylous vertebrae, and the Labyrinthodontia, frequently large and usually retaining the complex internal tooth structure of their presumed fish ancestors. However, more recent work has cast doubt on the integrity of all three of these groups. A more complete taxonomic history is given in Panchen (1977a).

Briefly, the term Lissamphibia was revived by Parsons and Williams (1962, 1963) to characterize the extant orders, using a series of characters, pre-eminent amongst which was the pedicellate character of the teeth. However, Carroll and Currie (1975) and Carroll and Gaskill (1978) have suggested the origin of the living Apoda, and conceivably the Urodela, from microsaur lepospondys, while the Anura (frogs and toads) seem closer to labyrinthodonts (Bolt, 1969). The Microsauria, the largest and most important group of the "Lepospondyli" are now known to have vertebrae in which the principal or only central element is the pleurocentrum (Panchen, 1977a; Carroll and Gaskill, 1978). Their whole centrum therefore appears to be of compound origin, like that of labyrinthodonts, and is not closely comparable to the undivided (holospondylous) centrum of Nectridea and Aistopoda, the other major lepospondyl groups (Panchen, 1977a). Proximity of relationship of the latter two is difficult to corroborate or refute (A. C. Milner, Chapter 15 of this volume).

The Labyrinthodontia is a much larger assemblage of extinct Amphibia and has been regarded as a natural group since the time of Nicholson and Lydekker (1889) ("Labyrinthodontia Vera"), a taxonomic concept strongly reinforced by the classic work of Watson (1917, 1919, 1926) on their origin and phylogeny. However, diagnostic characters unique to labyrinthodonts are difficult to find. Romer (1947) in his "Review of the Labyrinthodontia" diagnoses the group as follows: "Primitive apsidospondyls in which centra are formed by arch elements (pleurocentra or intercentra or both) and anuran specialisations are lacking". The first part of the diagnosis is tautologous, the subclass Apsidospondyli (since abandoned – Romer, 1966) being used to characterize amphibians with centra thought to be formed by "arch elements" (arcualia) in contrast to the Lepospondyli. The second part is non-diagnostic and merely represents Romer's opinion, since abandoned by him but revived by more recent workers, that the Labyrinthodontia and Anura together constitute a monophyletic group.

The eponymous tooth structure of the Labyrinthodontia cannot be used to characterize the group either. The complex infolding of the dentine to which the term refers is a character shared with crossopterygian fishes (Schultze, 1969) and the more specific structure (polyplocodont – Schultze) which characterizes most early labyrinthodonts is very close to that of the crossopterygian *Panderichthys.* Thus, if tetrapods are monophyletic, there is little doubt that the polyplocodont tooth was a primitive character of all tetrapods. Similar arguments may be put forward against other supposed diagnostic features of the Labyrinthodontia. The pattern of bones of the dermal skull-roof is easily derived from that of the Osteolepidida (Osteolepiformes of Jarvik) amongst the crossopterygians (Westoll, 1938, 1943; Parrington, 1967) and that which appears to be primitive for labyrinthodonts is probably primitive for all tetrapods. The evolution of tetrapod vertebrae has been reviewed recently (Panchen, 1977a) and there is little doubt that the "apsidospondyl" structure, with compound centra, is a direct inheritance from the fish condition, while one particular configuration, the rhachitomous vertebra, may be primitive for all tetrapods.

One feature of most labyrinthodonts does seem to be a shared derived character: that of an otic "notch" between the skull-table and cheek, which presumably accommodated a tympanum and is associated with a hyomandibular bone reduced in mass relative to the fish suspensory hyomandibula. The hyomandibula (columella auris or stapes) thus becomes an element in an impedance matching system coupling airborne sound waves to the fluid-filled receptor system of the inner ear (Carroll, Chapter 12 of this volume). However, there is evidence to suggest that this complex of characters arose polyphyletically within the labyrinthodonts.

SUBDIVISION OF THE LABYRINTHODONTIA

Early attempts to classify and subdivide the Labyrinthodontia were heavily dependent on the structure of the vertebrae and particularly on the nature of the (usually compound) centrum. Three principal

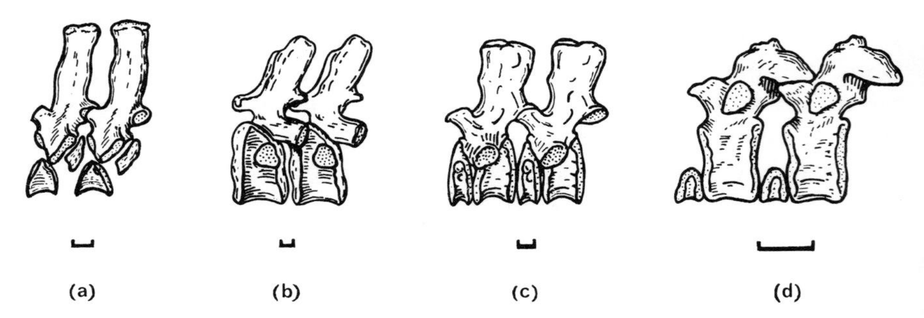

Fig. 1. Two successive vertebrae of "labyrinthodont" Amphibia in left ventral view (scale bars 1 cm). (a) Rhachitomous, *Eryops*; (b) stereospondylous, *Mastodonsaurus*; (c) embolomerous, *Eogyrinus*; (d) gastrocentrous, *Seymouria*. (a) After Moulton, (b) after Nilsson, (d) after White.

types of centrum were recognized: embolomerous (usually Carboniferous forms), rhachitomous (characteristic of the Permian but extending "downwards" and "upwards") and stereospondylous (exclusive to the Triassic) (Fig. 1). Watson (1919), by studies of apparent cranial trends, was able to corroborate the earlier suggestion that the stereospondylous vertebra, and the animals which it characterized, represented a culmination of trends seen in rhachitomous forms. Romer (1947, 1966) later revived Zittel's term Temnospondyli as an ordinal name for the whole group. In his review, Romer (1947) included the ichthyostegids, primitive but aberrant amphibia from the Devonian of East Greenland first described by Säve-Söderbergh (1932), within the Temnospondyli. Later, however, he removed them as a separate order of labyrinthodonts.

Before discovery of the ichthyostegids our knowledge of early Amphibia was dominated by Watson's studies (1926, 1929) of British Carboniferous species. Great emphasis was placed by Watson on the Anthracosauroidea, with primitive skulls and embolomerous vertebrae. They were contemporary with another group of large crocodile-like labyrinthodonts, the loxommatids, which outnumber anthracosaurs in most collections, notably in the Hancock Museum here in Newcastle where Watson's studies were largely conducted. Despite this the postcranial skeleton of loxommatids was and is virtually unknown and, bearing in mind the frequency of isolated embolomerous central elements in collections, Watson made the reasonable assumption that they, like the anthracosaurs, were embolomerous. Thus Watson was able to extrapolate the trends he showed in the skulls of rhachitomes and stereospondyls back to the presumed embolomeres.

Watson (1926, 1951, 1954) had assigned the anthracosaurs importance not only as a link between fish and amphibians, but also as one between amphibians and reptiles. Subsequently, Säve-Söderbergh (1934) produced a dichotomous classification of lower vertebrates in which he widely separated the Temnospondyli, assumed to be related to Anura, and the Anthracosauria [*sic*], assumed to be related to reptiles and thus to all amniotes. He also drew attention to a useful key feature of the skull separating the two groups (Fig. 2).

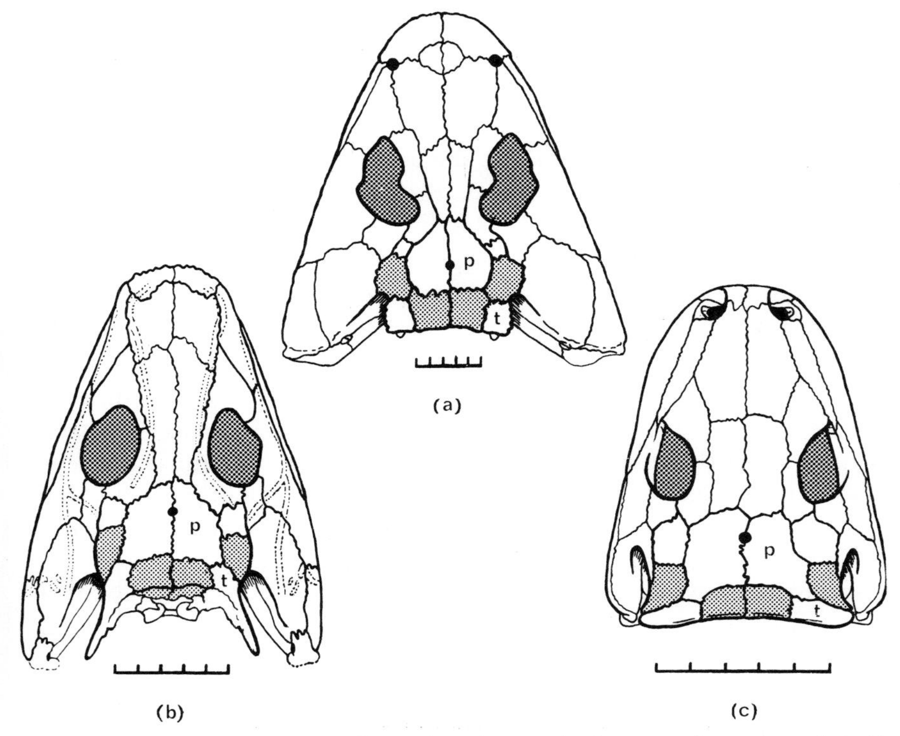

Fig. 2. "Labyrinthodont" skulls in dorsal view to show configuration of skull-table (scale bars 5 cm). (a) *Loxomma* (after Beaumont); (b) *Palaeoherpeton*; (c) *Seymouria* (after White). Supratemporal (antero-lateral) and postparietal stippled; p, parietal; t, tabular.

In temnospondyls the tabular bones, forming the posterior corners of the skull-table, are separated from the respective left and right parietals by a sutural junction between the postparietals and supratemporals, while in anthracosaurs the opposite arrangement prevails, with a tabular–parietal suture on each side separating the postparietal from the supratemporal. It is clear, however, (Panchen, 1964, 1970, 1975) that the ancestors of diagnostic anthracosaurs must have passed through a temnospondyl configuration in their ultimate derivation from osteolepiform fish. Thus the temnospondyl condition is primitive for tetrapods and no indication of monophyly within that group.

In Romer's (1947, 1966) classifications the Temnospondyli and the Anthracosauria are reunited within the Labyrinthodontia, maintaining Säve-Söderbergh's distinction at a lower taxonomic level and unequivocally uniting the loxommatids with the temnospondyls. A further complication ensued, however. The Permian terrestrial tetrapod *Seymouria*, long regarded as a primitive reptile, was accepted as an amphibian, largely because of the undoubted amphibian status of other seymouriamorphs, such as *Kotlassia* (Bystrow, 1944) and *Discosauriscus* (Špinar, 1952). Its links appeared to be with the anthracosaurs, as Watson had suggested, and it had the diagnostic configuration of the skull-table. Therefore Romer (1947) expanded the use of the term Anthracosauria to include the Seymouriamorpha. At about the same time, Efremov (1946) had proposed the taxon Batrachosauria to denote a new subclass of tetrapods including only the Seymouriamorpha and various forms *incertae sedis* (Heaton, Chapter 18 of this volume). Olson (1947) presented a similar concept "Parareptilia" to include seymouriamorphs, the disputed palaeozoic Diadectomorpha and the extant Chelonia, but subsequently (Olson, 1962) moved to Efremov's batrachosaur concept.

In my review of the Anthracosauria (s.s.) (Panchen, 1970) I "extended down" the term Batrachosauria in agreement with Kuhn (1965) and thus unwittingly produced a semantic controversy with North American colleagues. They retained Romer's usage and (with Romer) used Embolomeri in the sense in which Watson, Säve-Söderbergh and I had used the term Anthracosaur(oid)ia. The subsequent description of non-embolomerous anthracosaurs, however, necessitated some revision of the classification on both sides.

In 1973 I gave a revised account of *Crassigyrinus scoticus* Watson and suggested that it represented a monotypic taxon, the Palaeostegalia, within the Labyrinthodontia but outside and equal in status to the Ichthyostegalia, Temnospondyli and Batrachosauria. However, this suggestion based on only the cheek and lateral snout of a single skull met with little enthusiasm. Fortunately, however, the situation has now changed radically (see below).

Before turning to the Anthracosauria, one further problem must be aired. Carroll (Chapter 12 of this volume) has shown that the Lower Carboniferous temnospondyl *Greererpeton* (and presumably the closely related *Pholidogaster* – Panchen, 1975) retains a fish-like suspensory hyomandibular and thus lacks the sound-conducting stapes and associated otic notch of most, but not all, other temnospondyls. This exciting discovery seems to me to threaten the integrity of the Temnospondyli as a monophyletic taxonomic group. The problem parallels that of the "Labyrinthodontia" as a whole. The diagnostic characters of the Temnospondyli (especially the configuration of the skull-table and the, primitively, rhachitomous vertebrae) are almost all primitive for all tetrapods. Now that at least the family Colosteidae (*Greererpeton et al.*) are known to lack the characteristic labyrinthodont ear region a fragmentation of the temnospondyls is inevitable, unless some overriding shared derived character can be found. I will suggest below that perhaps it can, but at the expense of banishing the loxommatids from the Temnospondyli.

THE ANTHRACOSAURIA

Before the present study the Batrachosauria (*sensu* Panchen) were divided as follows (Panchen, 1975):

Order BATRACHOSAURIA Efremov
- Suborder Anthracosauria Säve-Söderbergh
  - Infraorder Herpetospondyli Panchen
  - Infraorder Embolomeri Cope
  - Infraorder Gephyrostegoidea Carroll
- Suborder Seymouriamorpha Watson

The taxon Herpetospondyli was proposed to include two very early anthracosaurs, *Proterogyrinus* ("*Mauchchunkia*") *scheelei* Romer (Romer, 1970; Hotton, 1970) from the "middle" Mississippian (Viséan) (or Namurian; Holmes, Chapter 14 of this volume) of

West Virginia, U.S.A., and *Eoherpeton watsoni* Panchen (1975) from the probably contemporary Scottish Gilmerton Ironstone of the Edinburgh area. It was also suggested that the isolated femur described by Watson (1914) as *Papposaurus traquairi*, which is probably Namurian, belongs to the same group. *Proterogyrinus* is reconsidered for this volume in Chapter 14, and Mr Holmes tells me that the *Papposaurus* femur is closely similar to that of *Proterogyrinus*. It is clear that *Proterogyrinus* is closely related to the Coal Measure Embolomeri but has vertebrae in which the intercentrum is not fully ossified into the almost complete disc characteristic of the embolomeres (Panchen, 1966).

Unfortunately, however, while the material of *Proterogyrinus* (including "*Mauchchunkia*") consists of good postcranial material but rather poor skulls, the holotype and then only known specimen of *Eoherpeton* consisted of a well-preserved but isolated skull. Discovery of more material of *Eoherpeton* (this time from the Scottish Namurian) makes it clear that it is not closely related to the Herpetospondyli as characterized by *Proterogyrinus* and *Papposaurus*. Fortunately that taxon was proposed on the character of the *Proterogyrinus* vertebrae as described by Hotton (also Holmes and Carroll, 1977) and *Eoherpeton* can be removed painlessly from it. *Eoherpeton* is further considered below.

The Embolomeri of the Coal Measures (Westphalian plus Stephanian) and Lower Permian can be divided into three families (Panchen, 1970, 1975, 1977b). The Anthracosauridae and Archeriidae are both monotypic and characterized by the nominate genera *Anthracosaurus* and *Archeria*, the former known only from the British Middle Coal Measures, the latter, with certainty, only from the American Lower Permian. The third family, the Eogyrinidae, comprises the characteristic embolomeres of the European and North American Coal Measures (Pennsylvanian). All three embolomere families included long-snouted crocodile-like (presumably aquatic) predators. The specimens of *Archeria*, as yet incompletely described, are known to have been long-bodied with some 40 pre-sacral vertebrae (Romer, 1947) and I restored *Eogyrinus attheyi* Watson (Panchen, 1966, 1972a), the largest member of its nominate family, with a similar number (Fig. 3). However, I now regard the evidence on which I did so as inadequate: the closely related *Pholiderpeton scutigerum* Huxley is being redescribed in my laboratory by Miss

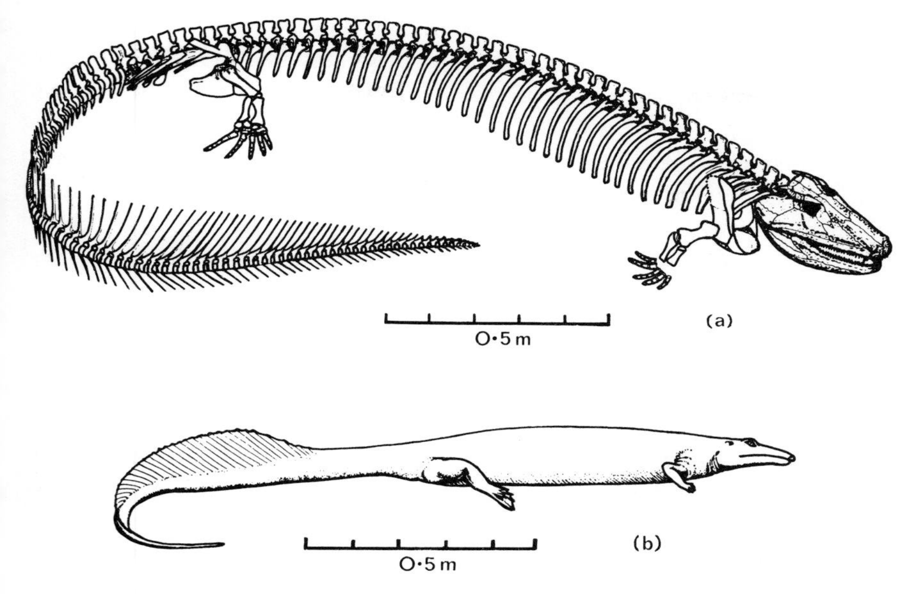

Fig. 3. Whole reconstruction of Embolomeri. (a) *Eogyrinus*; (b) *Archeria* after Romer.

J. A. Agnew and may well settle this problem. Nothing is known about the bodily proportions of *Anthracosaurus russelli* Huxley, the only known anthracosaurid species. This species differs from all other described embolomeres in that the skull-roof is consolidated by junction by firm sutures between the skull-table and the flanking cheek regions. The members of the other families retain a lateral kinetism of the skull-table as a remnant of the kinetically isolated skull-table of their presumed fish ancestors. In *Anthracosaurus* this consolidation is presumably correlated with the massive palatal and marginal dentition (Fig. 6).

Apart from the Herpetospondyli and Embolomeri, a third taxon of undoubted anthracosaurs may be distinguished. The Gephyrostegoidea are a group of relatively small, apparently terrestrial anthracosaurs with a rather reptiliomorph postcranial skeleton but a typical and relatively unspecialized skull and 24 pre-sacral vertebrae, each with an incompletely ossified crescentic intercentrum (Carroll, 1970). Carroll (1969, 1970) regarded the known specimens as a relict of the group ancestral to reptiles but this view has had to be abandoned in the light of more recent work (Panchen, 1972b; see also below). *Gephyrostegus* and possibly related forms all come from the late Upper Carboniferous, but Boy and Bandel (1973) have described an apparently gephyrostegid skeleton from the Namurian of Germany.

It must now be asked what shared derived characters unite all the anthracosaurs so far characterized in contrast to other labyrinthodonts. I think these characters can be seen in terms of three character complexes, the architecture of the back of the skull, the structure of the labyrinthodont teeth and the nature of the vertebrae.

The characteristic tabular–parietal contact in anthracosaurs was certainly produced by two factors. The tabular bone itself was almost certainly a bone of elliptical growth (Parrington, 1956) with its long axis directed antero-mesially and postero-laterally. This produced the characteristic tabular horn, which I regard as the most diagnostic feature of the group, but also produced an extension of the tabular towards the parietals. Backward movement of the parietal–postparietal suture relative to the intertemporal supratemporal suture with which it is in continuity in osteolepiform fish (Romer, 1941; Westoll, 1943) would then establish the parietal–tabular contact.

Two further features seem to be correlated with the presence of tabular horns. Embolomeres lack the posttemporal fossae (fossae Bridgei) present in temnospondyls and their fish ancestors as deep bilateral pockets in the occiput bounded by the tabulars and postparietals dorsally and the paroccipital process (tabular and opisthotic) ventrally (Panchen, 1970). It seems probable that axial muscles from the fossae gained a new insertion on the mesial surface of the horns which also acted as the origin of the depressor mandibulae musculature (Panchen, 1964) (Fig. 5). In *Anthracosaurus* and some eogyrinids the horns are biramous, emphasizing their third function, that of forming the dorsal border of the otic notch and supporting the tympanum.

The teeth and palatal tusks of at least the embolomeres are labyrinthodont in transverse section with the greatest degree of elaboration near the base of the crown. Their structure may be contrasted with that of the contemporary loxommatids (Panchen, 1970) and rhachitomous stereospondyls (Schultze, 1969). In the latter two (and in ichthyostegids) the external primary dentine (Bystrow, 1938 –

Globulärzone of Schultze) which forms the core of each labyrinthine fold has short side-branches at (at least) some of the angles formed by its tortuous course towards the axis of the tooth. In embolomeres these side-branches have disappeared but each fold is more tortuous, a condition paralleled in both respects, certainly convergently, by the advanced capitosaur temnospondyls.

Finally the vertebrae of anthracosaurs are characteristic but not unique in having a cylindrical pleurocentrum which is the principal central element, has a firm junction with the pedicels of the neural arch and, even if not fully ossified, is complete ventrally. The intercentrum in front of it is almost always shorter antero-posteriorly, normally bears the capitular rib-head (as in other labyrinthodonts), and varies in degree of ossification from a small crescentic ventral wedge (in gephyrostegids) to a disc complete but for some dorsal wedging (in typical embolomeres). Both centra have an axial perforation for the notochord, which in large embolomeres may show secondary occlusion by a bony plug, probably as an age character (Panchen, 1977b).

Vertebrae similar to those of anthracosaurs occur in the Microsauria, early reptiles and various "parareptilian" groups discussed in this volume by Heaton (Chapter 18). It is difficult to decide whether the nature of the vertebrae is a character indicating affinity between all these groups and rather similar vertebrae occur in a few aberrant (in this respect) temnospondyls. However, there is some reason for believing that the rhachitomous vertebra typical of most temnospondyls, with dominant intercentrum and small paired dorso-lateral blocks representing the pleurocentrum, is primitive for all tetrapods (Panchen, 1977a).

THE AFFINITIES OF THE ANTHRACOSAURS

The Anthracosauria, as characterized above, certainly form a natural group and it is a reasonable hypothesis that they are monophyletic in the cladist sense, i.e. that they share a unique common ancestry and have in common a series of diagnostic shared derived ("synapomorphic") characters. However their external relationships are much more problematic. Current practice, as noted above, unites them with the Seymouriamorpha, either incorporating the seymouriamorphs within the taxon Anthracosauria (e.g. Romer, 1966) or regarding the two groups as suborders of the order Batrachosauria (Panchen, 1975).

Seymouria (White, 1939) and all other described seymouriamorphs share the diagnostic parietal–tabular contact with anthracosaurs in a skull-roof, which, like that of anthracosaurs, retains a large intertemporal bone and a full complement of the other dermal roofing bones as in "labyrinthodonts". It is worth noting, however, that, while the tabular of anthracosaurs appears to be a bone of elliptical growth, Parrington (1956) was able to reproduce the pattern of the *Seymouria* skull-table with some accuracy on the assumption that the growth of the tabular was even or circular, but that the parietal (also circular) had grown at twice the minimum axis rate of all the surrounding bones. Thus a similar feature in anthracosaurs and seymouriamorphs may have been arrived at by different means.

There is also a well-developed otic notch in seymouriamorphs. However, the otic notch may well have originated independently in at least temnospondyls and anthracosaurs (see below) and is thus not a strong indicator of affinity. Similarly the vertebrae of seymouriamorphs share the character of a dominant pleurocentrum with anthracosaurs, but the vertebral structure is much more like that of the "parareptilian" Diadectidae (Heaton, Chapter 18 of this volume) and some large microsaurs (Carroll and Gaskill, 1978) than like the anthracosaur vertebra. Both *Seymouria* and *Kotlassia* (Bystrow, 1944) have trunk vertebrae in which the pleurocentrum is firmly fused to the neural arch, the neural arch is massively wide and dome-shaped with widely-spaced zygapophyses and the intercentrum is extremely small. In *Discosauriscus* (Špinar, 1952) the vertebrae are poorly ossified even in adults, but have a remarkably long pleurocentrum and a relatively large intercentrum: the neural arches are not conspicuously domed. The only other derived characters of the post-cranial skeleton uniting anthracosaurs and seymouriamorphs are the presence of a parasternal process of the interclavicle extending posteriorly from the body of that bone, and possibly the similar phalangeal formula.

The skull of seymouriamorphs has a number of features which contrast strongly with those in the Anthracosauria. Consolidation of the skull-roof has already been mentioned, but this is paralleled by *Anthracosaurus russelli*. More distinctive is the form of the otic capsule. In both *Seymouria* and *Kotlassia* (the situation is unknown in *Discosauriscus*) the capsule on each side is extended laterally towards the otic notch as a massive periotic tube (lateral otic canal) terminating in the fenestra ovalis. This obviously had the function of reducing the length of the stapes which lay between the fenestra and the tympanum in the notch and is paralleled in *Diadectes* (Olson, 1966; Heaton, Chapter 18 of this volume). In *Seymouria* the tube is floored by a lateral extension of the parasphenoid on each side, but this is not the case in *Kotlassia* (Fig. 4).

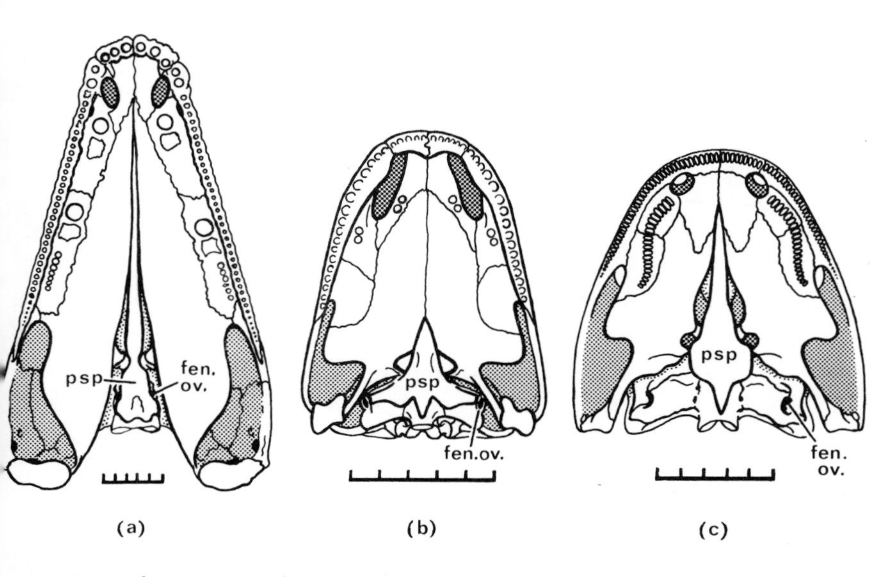

Fig. 4. Anthracosaur palates (scale bars 5 cm). (a) *Eogyrinus*; (b) *Seymouria* (after White); (c) *Kotlassia* (after Bystrow). fen.ov., Fenestra ovalis; psp, parasphenoid.

Heaton also emphasizes the other distinctive features of the seymouriamorph braincase: there is no apparent supraoccipital bone in occipital view, and the synotic tectum, the laterosphenoid region in front of the otic capsule and the neurocranial roof of the sphenethmoid are all said to be unossified. All these features are in striking contrast to the condition known in anthracosaurs.

The anthracosaur braincase has been redescribed in *Palaeoherpeton* ("*Palaeogyrinus*"–Panchen, 1964) and in *Eogyrinus* (Panchen, 1972a). As far as is known, it is essentially similar in both. Firstly it is a unitary structure without the separation into otico-occipital region posteriorly and sphenethmoid region anteriorly produced by loss of ossification of the laterosphenethmoid region. This separation in seymouriamorphs, which is much more radical in *Kotlassia* than in *Seymouria*, is not to be confused with the primary kinetic division of the braincase in crossopterygians (e.g. *Eusthenopteron*—Jarvik, 1954): in the latter the basisphenoid with its paired basipterygoid processes is part of the anterior (ethmosphenoid) region, whereas in the former it is attached to the otico-occipital region. In the two eogyrinid anthracosaurs the sphenethmoid is in continuity with the otic capsule dorsally and the basisphenoid laterally and ventrally. The sphenethmoid has a solid roof perforated only by a fontanelle below the pineal foramen in the skull-roof and was firmly applied to

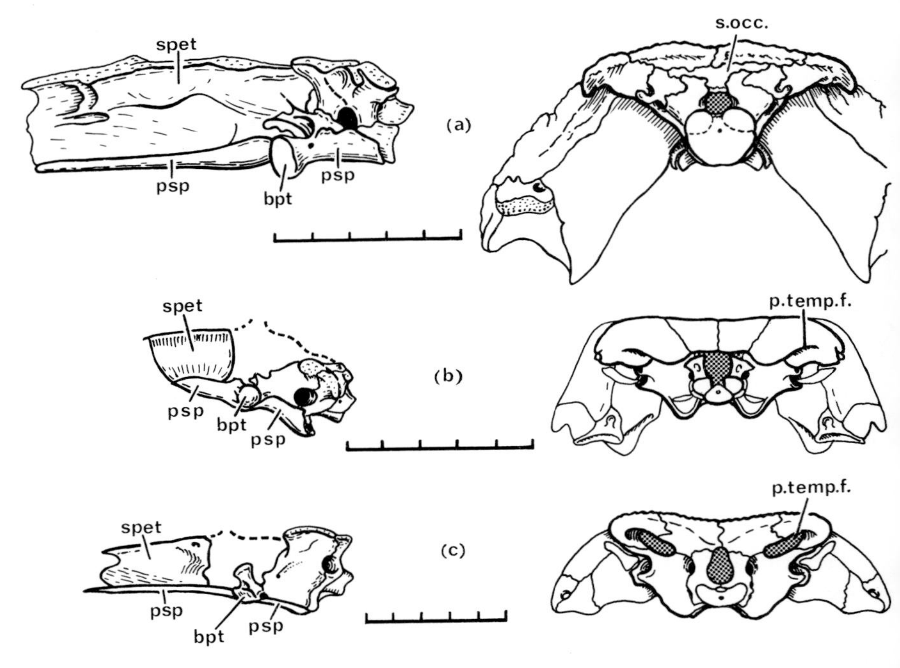

Fig. 5. Braincase in left lateral view (left) and occiput of anthracosaurs (scale bars 5 cm). (a) *Palaeoherpeton*; (b) *Seymouria* (after White); (c) *Kotlassia* (after Bystrow). Broken lines in braincases (b) and (c) show the outline of underside of skull-roof in midline. bpt, Basipterygoid process of basisphenoid; psp, parasphenoid; p.temp.f., post-temporal fossa; s.occ., supraoccipital; spet, sphenethmoid.

the latter. In the eogyrinids the sphenethmoid is a massive block of bone tunnelled by paired canals for the olfactory and possibly vomeronasal nerves. In *Kotlassia* it is also a solid bone but is very short and widely separated from the sphenethmoid. In *Seymouria*, White describes paired sheets of bone ("orbitosphenoids") together forming a V-shaped trough contacting an anterior process ("presphenoid") not apparently separable from the basisphenoid in the ventral midline. There is no dorsal connection between the two regions (Fig. 5).

The region between the paired otic capsules is the synotic tectum. In *Palaeoherpeton* and *Eogyrinus* it is solidly ossified as is the roof of the capsules themselves forming a massive roof to the whole braincase in the otic region which was closely applied to the underside of the skull-roof. In *Palaeoherpeton* an apparent suture was traced on each side of the synotic tectum forming its junction with the otic capsules. In neither anthracosaur was there any visible suture separating the synotic ossification from the sphenethmoid.

It is normally assumed that the synotic tectum is ossified by the supraoccipital bone, if this is present as a separate entity in the occipital plate, but Heaton (Chapter 18 of this volume) makes a distinction between the supraoccipital as an occipital ossification completing the arch above the foramen magnum, and a synotic ossification. If the distinction is valid the two anthracosaurs appear to have had both, but inextricably fused. *Kotlassia* on the other hand had only a small synotic ossification between the pro-otics, with no supraoccipital apparent in occipital view, and *Seymouria* had neither.

Thus the structure of the braincase casts considerable doubt on the relationship of anthracosaurs and seymouriamorphs (as Heaton notes) but also casts doubt on the connection between *Seymouria* and *Kotlassia*. Furthermore, well developed, i.e. primitive, post-temporal fossae are present in *Kotlassia*, whose occiput looks remarkably temnospondyl-like. In *Seymouria* they are described by White as very thin, and are obscured in occipital view by down-growth of the back of the skull-table. In anthracosaurs they are totally absent (Fig. 5). The teeth of *Seymouria* and *Kotlassia* may be similar to those of anthracosaurs (Schultze, 1969) but those of both groups are paralleled by the capitosaur temnospondyls and thus should not be assigned too much taxonomic weight.

Thus in the skull the only derived characters shared by anthracosaurs and seymouriamorphs are the otic notch and the tabular–parietal contact, both quite probably convergent. On the evidence presented so far, therefore, I would agree with Heaton that relationship between the two groups is improbable. However, one line of evidence, not yet considered, may tip the scales in favour of their association.

It was noted above that the unique skull described as *Eoherpeton watsoni* Panchen (Panchen, 1975) did not appear to be closely related to *Proterogyrinus scheelei*, the other described Lower Carboniferous anthracosaur. New material of *Eoherpeton* from the Namurian of Cowdenbeath, Fife, Scotland, collected by Mr Stanley P. Wood (Andrews *et al.*, 1977) is now to hand and is being described by Mr T. R. Smithson (see also Smithson, Chapter 16 of this volume). The material includes at least one more incomplete skull, the articular region of a jaw ramus and, probably, associated post-cranial remains.

In the original description it was noted that *Eoherpeton*, in contrast to other anthracosaurs, shared a number of characters, both primitive and derived, with *Seymouria*. Primitive characters included the form of the lower jaw, the lacrimal bone extending from the orbit to naris and the possible retention of post-temporal fossae. However, the shared derived characters are very striking: the fenestrae ovales are widely separated and floored by lateral extensions of the parasphenoid: the parasphenoid itself is produced into paired basal tuberae for muscle origin as in *Seymouria*, and the configuration of the septomaxillary bone is similar. Furthermore the reptiliomorph shape of the skull is reminiscent of *Seymouria*.

The new skull material emphasizes the resemblance. It is clear that the bony boss identified as supraoccipital in the holotype is extraneous and that a discrete supraoccipital ossification was probably lacking, also that the synotic tectum was probably unossified. Post-temporal fossae were almost certainly present and the seymouriamorph nature of the otic capsule and parasphenoid in front of the basipterygoid articulation are startlingly *Seymouria*-like: the processus cultriformis of the parasphenoid is short and tapers to a point, just as in *Seymouria*, and is overlain by similar basisphenoid ("presphenoid" of White) lying entirely below the site of the interorbital vein. All these features are to be fully described by Mr Smithson.

Despite these resemblances *Eoherpeton* may still be classified as an anthracosaur (s.s.). It shares both primitive and derived characters with *Proterogyrinus* and the Coal Measure forms. Primitive features include the lateral skull-table kinetism, large intertemporal bones and naso-labial groove (Panchen, 1967). Derived characters are the tabular horn and associated otic notch and the (batrachosaur) tabular–parietal contact.

I suggest that *Eoherpeton* is an early gephyrostegid, although corroboration of that hypothesis depends on greater knowledge than we now have of the gephyrostegid skull. The overall shape of the skull and the configuration of the constituent dermal roofing bones is strikingly similar in *Gephyrostegus* (Carroll, 1970) and *Eoherpeton*. Most of the resemblances are probably primitive for all anthracosaurs, but the blunt and rounded form of the tabular horns and configuration of the otic notches are probably useful diagnostic characters.

The otic notch is very similar in the two genera. It is clear that in the original reconstruction the skull-table of *Eoherpeton* should be moved somewhat posteriorly relative to the cheek region (Fig. 6). The back of the grooved squamosal, which forms the anterior border of the notch, then continues a ventral hollowing of the skull-table at the tabular–supratemporal suture to give a smooth profile and very gephyrostegid appearance to the notch. The *Eoherpeton/Gephyrostegus* notch is large, dorso-ventrally long, and shallow, in contrast to the condition in embolomeres and seymouriamorphs and, importantly, to the apparent primitive condition (see below). The large size of the notch in *Gephyrostegus* might be related to the relatively small size of the skull (*c*. 65 mm long) (Parrington, 1959) but the larger skull of *Eoherpeton* (*c*. 150 mm) is of very similar size to that of *Palaeoherpeton*, yet the contrast in shape and size of the otic notch is very striking.

Little information is available on the braincase of *Gephyrostegus* and none on any other gephyrostegid. Carroll (1970) reconstructs a "conventional" otic region in contrast to the extended periotic tube and parasphenoid of seymouriamorphs but the evidence does not seem strong and something like the *Eoherpeton* condition is not ruled out.

Thus, to summarize this section, the hypothesis of close relationship between anthracosaurs (i.e. Herpetospondyli, Embolomeri and

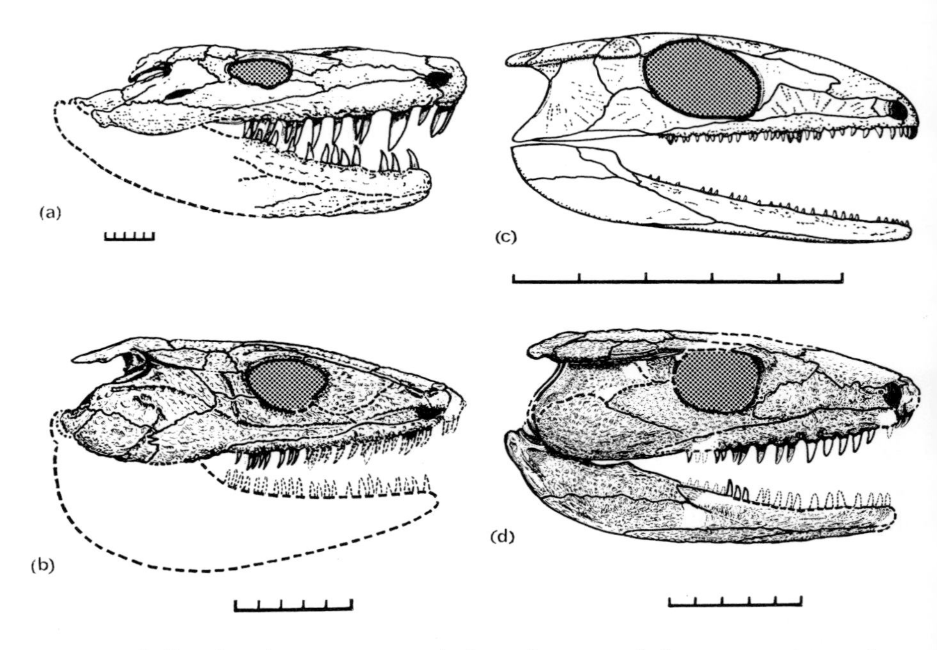

Fig. 6. Skulls of anthracosaurs in right lateral view (scale bars 5 cm). (a) *Anthracosaurus russelli*; (b) *Palaeoherpeton decorum* (lower jaw based on that of *Eogyrinus*); (c) *Gephyrostegus bohemicus* (after Carroll); (d) *Eoherpeton watsoni* (after Panchen (1975) but with position of skull-table modified and new data on jaw after T. Smithson (unpublished)).

Gephyrostegoidea) and seymouriamorphs depends on (1) the acceptance of the significance of the shared derived characters between *Eoherpeton* and *Seymouria*, (2) the acceptance of *Eoherpeton* as a gephyrostegid (rather than a primitive seymouriamorph or abberrant non-gephyrostegid anthracosaur) which in turn depends on (3) the prediction of discovery of significant *Eoherpeton*-like characters in the braincase and elsewhere in *Gephyrostegus*. I feel confident of the first assumption, less so of the second and third.

If the latter two are wrong then I agree with Heaton (Chapter 18 of this volume) that seymouriamorphs, diadectids and other "parareptilian" groups discussed by him, whatever their internal relationships, have nothing to do with the Anthracosauria, and *Eoherpeton* then becomes a primitive seymouriamorph if the first assumption is correct. However, if all three assumptions are correct then a dichotomous arrangement of the Batrachosauria (in my sense) becomes possible, with the primary division between Herpetospondyli and Embolomeri on one hand and Gephyrostegoidea and Seymouriamorpha on the other. This will be discussed further after the next section.

THE ANCESTRY OF ANTHRACOSAURS

When *Crassigyrinus scoticus* Watson, from the Viséan of Scotland was redescribed (Panchen, 1973) the very primitive nature of the side of the skull-roof was apparent. A large preopercular bone was retained; the antero-posterior proportions and configuration of the dermal bones lay between those of a fish such as *Eusthenopteron* and those of *Ichthyostega*; the course of the infraorbital lateral line canal was fish-like and the bones surrounding the naris were possibly the fish anterior tectal and lateral rostral, lost or modified in all other tetrapods (Fig. 7). Nothing, however, was known of the skull-table or any other part of the skull, or of the postcranial skeleton.

The most important find made by Mr Wood at the Namurian site at Cowdenbeath was the almost complete skeleton of a large amphibian illustrated but not identified by Andrews *et al.* (1977). Development of the skull demonstrates that it is *Crassigyrinus scoticus* and also that, contrary to my previous opinion (Panchen, 1973), the Viséan lower jaw ramus "*Macromerium*" *scoticum* Lydekker also pertains to this species. The skull of the Cowdenbeath skeleton is now completely cleaned and a full description is in preparation: but development of the postcranial skeleton will take one or two years.

The skull-table is perfectly preserved and demonstrates the relationship of *Crassigyrinus* to the anthracosaurs. Lateral kinetism is retained as is a large intertemporal bone and in these respects it is similarly primitive. However, large tabular horns, closely similar to those of *Eoherpeton*, are present and, immediately in front of them at the level of the supratemporal–tabular suture, deep rounded emarginations show that large circular otic notches were present. Dermal ornament is strongly developed and of the anthracosaur type (Fig. 9).

The most unexpected discovery is that the configuration of the dermal bones of the table is technically temnospondyl, i.e. primitive. The parietal–postparietal suture is far anterior to that between the supratemporal and tabular, so that postparietal–supratemporal

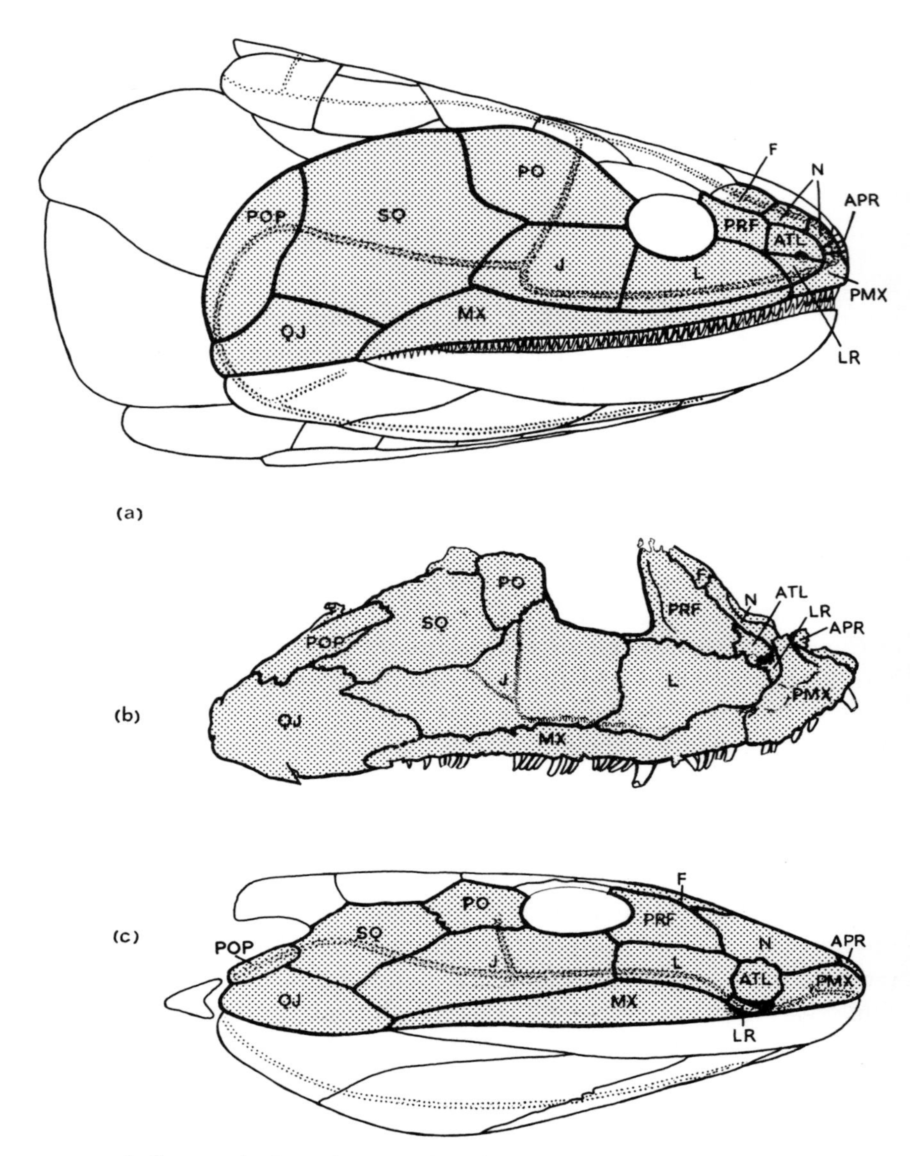

Fig. 7. Skulls in right lateral view reduced to same quadrate length. (a) *Eusthenopteron* (after Jarvik); (b) *Crassigyrinus*; (c) *Ichthyostega* (after Jarvik). (Region preserved in *Crassigyrinus* is stippled.) APR, Anterior postrostral; ATL, anterior tectal; F, frontal; J, jugal; L, lacrimal; LR, lateral rostral; MX, maxillary; N, nasal; PMX, premaxillary; PO, postorbital; POP, preopercular; PRF, prefrontal; QJ, quadratojugal; SQ, squamosal.

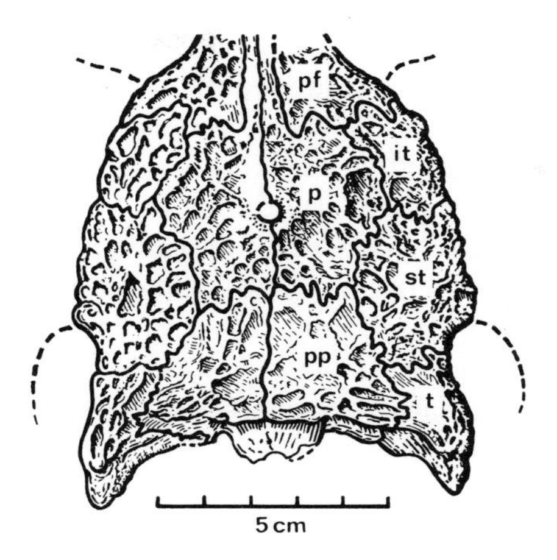

Fig. 8. Skull-table of *Crassigyrinus scoticus* from the Cowdenbeath specimen (BMNH, R10 000). Outline of otic notches and part of posterior border of orbit is shown by the broken line. It, Intertemporal; pf, postfrontal; p, parietal; pp, postparietal; st, supratemporal; t, tabular.

contact is retained. Furthermore the sharply finished posterior edge of each postparietal laterally and tabular mesially make it almost certain that post-temporal fossae were also retained.

Thus in all these features, as well as the primitive cheek and snout already described from the holotype, *Crassigyrinus* demonstrates the probable ancestral condition to all anthracosaurs. However, even as a non-cladist, I have to admit that it cannot be admitted to ancestral status as it comes from the late Viséan and early Namurian. Further, it possesses unique diagnostic features in the large rhomboidal orbits already seen in the holotype and, in consequence, a remarkable constriction of the interorbital region, seen in the Cowdenbeath specimen, which results in the reduction of the parietals anteriorly to a narrow paired medial process, squeezed between wide postfrontals.

Discovery of the skull-table of *Crassigyrinus* illuminates a number of problems in a remarkable way. Firstly it further diminishes the possibility of any close relationship between anthracosaurs and reptiles. The problem of relationship of anthracosaurs and seymouriamorphs to reptiles was reviewed a few years ago and it was concluded that the otic notch and associated features barred any anthracosaur, known or unknown, from reptile ancestry (Panchen, 1972b). Later (Panchen, 1975), at Dr Carroll's suggestion, I considered the possibility of a "notchless anthracosaur" as a reptile ancestor. I concluded that, with independent origin of the otic notch within the "Labyrinthodontia" then accepted, that such organisms may have existed. If they did, I further concluded, there was rather less than a fifty–fifty chance that one of them gave rise to the reptiles. If the otic notch of *Crassigyrinus* represents the primitive condition for all anthracosaurs (and other batrachosaurs), then a primitive notchless anthracosaur did not exist and batrachosaur ancestry of reptiles may be dismissed out of hand.

Secondly, the combination of characteristic tabular horn and otic notch seem almost certain to have been the unique common heritage of all anthracosaurs, and, if the *Eoherpeton*–gephyrostegid–seymouriamorph link is correct, to link the seymouriamorphs and anthracosaurs together and to a "palaeostegalian" ancestor with a *Crassigyrinus*-like skull-table.

Thirdly, there is the potential, in the as yet undeveloped postcranial skeleton of *Crassigyrinus*, to solve the vexed problem of the origin of the characteristic anthracosaur vertebra (Panchen, 1977a). The rhachitomous vertebra of early temnospondyls is probably derived from the type seen in the fish *Osteolepis*, but it is grossly unsatisfying to be told by disciples of Hennigian (cladist) taxonomy that the gastrocentrous vertebra of all but the embolomerous anthracosaurs (plus seymouriamorphs, plus "cotylosaurs", microsaurs and reptiles) is a true "synapomorphy" uniting all these and that one should rejoice in this fact, while leaving all speculation as to its origin alone.

I thus attempted, with some difficulty, to remove two partially exposed apparently adjacent centra from the tangled mass of bone comprising the postcranial remains of the Cowdenbeath specimen. The structure of the two when removed and completely cleaned was a surprise on several counts. Firstly it had been hoped that they would represent intercentrum and pleurocentrum but, while one is slightly longer than the other, they are closely similar in structure (Fig. 9). Each is a half-hoop in end view with almost parallel anterior and posterior edges in lateral view, except for the dorsal third of their height. This tapered region bears (? anteriorly or posteriorly) a facet, presumably for the articulation of the neural arch.

The centra are mere shells and it is evident that either the intact centrum consisted mostly of cartilage or the notochord was virtually unconstricted. My first inclination was to assume that both were intercentra but Mr Smithson pointed out that the number of centra visible in the trunk region of the specimen was approximately double the number of pairs of ribs so the vertebrae may have been diplospondylous and possibly had closely similar intercentra and pleurocentra. There is, furthermore, a slight difference between the two centra in that the one tentatively identified as intercentrum in Fig. 9

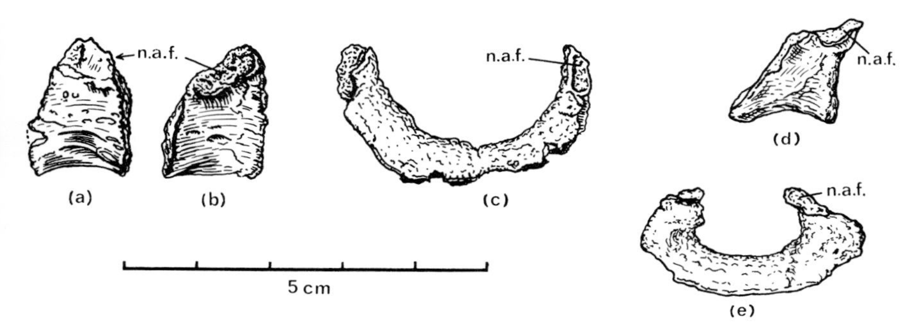

Fig. 9. Vertebral centra as preserved. (a)–(c) *Crassigyrinus scoticus* from the Cowdenbeath specimen (BMNH, R10 000). (a) ?Intercentrum in left lateral view; (b) and (c) ?pleurocentrum in left lateral and anterior view. (d) and (e) One of a group of centra from the Hancock Museum, ?*Megalocephalus pachycephalus*, Low Main Seam, Newsham. (d) (?Left) lateral; (e) ?anterior (not corrected for distortion). n.a.f., Neural arch facet.

has its apex midway between the anterior and posterior border whereas the other (?pleurocentrum) has the apex at one (? the posterior) border. The former also has a neural arch facet directed sagittally and dorsally, whereas in the latter it also has a lateral inclination.

If the vertebrae of *Crassigyrinus* had only one centrum per segment the only remotely similar early tetrapod vertebra is that described by Jarvik (1952) in *Ichthyostega*. If, however, they were diplospondylous with two similar centra, evolution into the embolomerous or gastrocentrous type would be an acceptable hypothesis.

Another intriguing hypothesis is suggested by the *Crassigyrinus* centra. Mr Michael Boyd drew my attention to a group of similar but very poorly ossified centra in the Hancock Museum preserved in coal shale. There are no data, but they are almost certainly from the

Northumberland Low Main Seam (Modiolaris Zone, Westphalian B—Land, 1974). Apart from the virtual lack of periosteal bone they are closely similar to those of *Crassigyrinus* (Fig. 9). They are not attributable to any known fish and may pertain to the common Low Main loxommatid *Megalocephalus pachycephalus* (Barkas). Minute, and dissimilar, intercentra associated with *Megalocephalus* were described by Baird (1957) and drawn by Beaumont (1977) but these, preserved only as natural moulds, could represent the intercentra to the Hancock pleurocentra. This suggests reassessment of the taxonomic position of the Loxommatoidea, which is further considered below.

CONCLUSIONS: PHYLOGENY OF THE BATRACHOSAURIA

All of the hypotheses considered above permit one to construct a classification and a tentative phylogeny of all the amphibians discussed, but before doing so some problems of nomenclature must be resolved.

The relationship of *Crassigyrinus* to the anthracosaurs and seymouriamorphs necessitates the subordination of the taxon Palaeostegalia to the taxon Batrachosauria, which now includes *Crassigyrinus* together with anthracosaurs and seymouriamorphs. A dichotomy of batrachosaurs (cladist or otherwise) is then reasonable with Palaeostegalia as one group and the remaining taxa comprising the other. In proposing a name for the latter it would seem perverse not to revert to Romer's (1947) denotation of the term Anthracosauria, so that this now includes the Seymouriamorpha. However, the connotation of Anthracosauria is still not the same as Romer's in that Batrachosauria, not Anthracosauria, is the collateral group of, and equal in rank to, the taxon Temnospondyli.

Heaton (Chapter 18 of this volume) uses Batrachosauria in another sense to include seymouriamorphs, diadectids and other "parareptilian" groups. However, this conflict of usage will automatically be resolved if one or other of our alternative hypotheses is accepted. The major taxon (subclass or order) Batrachosauria is that which includes the Seymouriamorpha and their relatives.

With recognition that none of the subclasses Labyrinthodontia, Lespospondyli and Lissamphibia can be regarded with any confidence as monophyletic groups, the Class Amphibia can only be divided into ten orders (Ichthyostegalia, Temnospondyli, Batrachosauria, Microsauria, Aïstopoda, Nectridea, Anura, Urodela, Apoda, ?Cotylosauria (*sensu* Heaton)) which suggests an urgent research programme (see Gaffney, 1979a).

The subdivision of the order Batrachosauria is as follows (diagnoses and content of taxa, unless amended here, as in Panchen, 1975, 1977b):

Order BATRACHOSAURIA Efremov

Mainly Palaeozoic Amphibia with skulls retaining a primitive pattern of dermal bones including intertemporals. Interpterygoid vacuities narrow or absent, basal articulation retained, retroarticular process small or absent, a single occipital condyle from basioccipital and exoccipitals, opisthotic forming an exposed para-occipital process between tabular and exoccipital. Vertebrae with compound centra.

Shared derived characters include paired tabular horns extending posteriorly from the corners of the skull-table (in all but the Seymouriamorpha) and a rounded otic notch formed by the emargination of the postero-dorsal margin of the squamosal and the ventro-lateral margin of the skull-table just anterior to the tabular horn. Pleurocentra a complete or incomplete cylinder, always complete ventrally.

Suborder PALAEOSTEGALIA Panchen

Batrachosaurs retaining a large preopercular bone. Postorbital length markedly longer than preorbital, extensive postparietal–supratemporal suture retained, extreme constriction of parietal and ?frontals by flanking postfrontals between orbits. ?Vertebrae equally diplospondylous with intercentra and pleurocentra ventral hemicylinders.

Family Crassigyrinidae Panchen: *Crassigyrinus*

Suborder ANTHRACOSAURIA Säve-Söderbergh

Batrachosaurs lacking a preopercular bone, preorbital length greater than post-orbital, tabular-parietal contact achieved with the elimination of postparietal-supratemporal suture, broad frontal–parietal contact. Vertebrae with pleurocentrum a complete cylinder, usually with ossification of the floor of the neural canal and normally longer than the intercentrum.

Infraorder Herpetospondyli Panchen

About 30 presacral vertebrae, intercentra ossified to about half the height of pleurocentra, the latter highly ossified. (Other characters in Panchen (1975) refer to *Eoherpeton* and are no longer valid: more complete diagnosis awaits redescription by Mr Holmes.)

Family Proterogyrinidae Romer: *Proterogyrinus* ("*Mauchchunkia*"), ?*Papposaurus*

Infraorder Embolomeri Cope
Family Eogyrinidae Watson
Subfamily Eogyrininae Panchen
Subfamily Leptophractinae Panchen
Family Archeriidae Kuhn
Family Anthracosauridae Cope

Infraorder Gephyrostegoidea Carroll
Small to medium terrestrial anthracosaurs. Quadrate condyles at about same level as occipital condyles, otic notch shallow antero-posteriorly but deep dorso-ventrally and well defined, kinetism extending forward to the orbit. No sign of lateral line sulci. Lacrimal extending from near or at orbit to naris. Post-temporal fossae (?always) retained. Incipient basal tubera and periotic tube. Supraoccipital, synotic tectum and laterosphenoid region all unossified. Lower jaw, shallow, primitive. About 25 presacral vertebrae. Pleurocentra incomplete dorsally.
Family Gephyrostegidae Romer
Small lightly built gephyrostegoids. Namurian to end of Carboniferous. Processus cultriformis of parasphenoid long. Ornament pustular on skull-table. Jaw without retroarticular process, lacking coronoid teeth. Intercentra ossified to about one-third height of pleurocentra: *Gephyrostegus, Eusauropleura ?Brukterpeton.*
Family Eoherpetontidae *nov.*
Relatively large, massively built gephyrostegoids. Viséan and Namurian. Processus cultriformis of parasphenoid short. Small shallow pit and ridge ornament on skull-table. Jaw with retroarticular process and coronoid teeth. Intercentra (to be described) similar height to pleurocentra: *Eoherpeton.*

Infraorder Seymouriamorpha Watson
Subordinate taxa as Carroll and Winer (1977) but may also include Diadectidae.

One primitive feature noted in the diagnosis of the Batrachosauria is the absence or slight development of interpterygoid vacuities in front of the basal articulation. In temnospondyls these are palatal fenestrae lying between the sphenethmoid (and the underlying processus cultriformis of the parasphenoid) and the mesial edges of the pterygoids flanking it. Enlargement of the vacuities was one of the most important trends thought by Watson (1919, 1951) to characterize temnospondyl evolution. In 1926 he extrapolated this trend back to the closed palates of Carboniferous anthracosaurs and loxommatids.

The condition of the closed palate is variable in anthracosaurs (s.l.). In eogyrinid and probably archeriid embolomeres the pterygoids are closely applied to the parasphenoid but probably not fused to it or to one another, while in *Anthracosaurus*, no doubt correlated with the general consolidation of the skull, the pterygoids are sutured or fused together anteriorly to give a palatal plate. The same contrast may be made respectively between *Gephyrostegus* and, to judge from the short processus cultriformis, *Eoherpeton*, and again (significantly) between *Kotlassia* and *Seymouria* (Fig. 4).

Returning to the temnospondyls, only one group has members with a closed palate, the Loxommatoidea. With consolidation of the skull-roof but retention of a mobile basal articulation, this probably led to a unique form of skull kinetism (Beaumont, 1977).

It has been noted above that, if correctly characterized in both cases, the vertebrae of loxommatids could be derived from the *Crassigyrinus* condition. I want to suggest, very tentatively, that well-developed interpterygoid vacuities, present even in *Greererpeton* (Carroll, Chapter 12 of this volume), constitute a shared derived ("synapomorphous") character of all members of the Temnospondyli. It is not unique to the temnospondyls, being paralleled by some nectrideans and microsaurs, but I know of no other.

This removes the loxommatids from the temnospondyls. However, there is no feature of either which would debar the origin of loxommatids from a batrachosaur at the *Crassigyrinus* (palaeostegalian) level of organization. Derivation would have involved (1) loss of the skull-table kinetism, (2) loss of the supraoccipital and ossification of the synotic tectum from the opisthotics, (3) reduction of the tabular horns (small occipital "buttons" persist), (4) development of the characteristic preorbital vacuities. The last is a unique feature, all the others are paralleled, at least in part, by other batrachosaurs.

Finally, my phylogenetic hypotheses are presented as a cladogram in which the features of the hypothetical ancestral batrachosaur stock are listed and the reconstructed evolutionary changes in each line are also shown (Fig. 10). The presentation is similar to that favoured by followers of the system of "phylogenetic taxonomy" (cladism) originated by Hennig (1966). I regard the cladogram as an excellent means of exposing a phylogenetic hypothesis. I further consider that Hennig's insistence on the use of shared derived ("synapomorphous") characters as evidence of monophyly as a valuable, though not original, contribution to systematics.

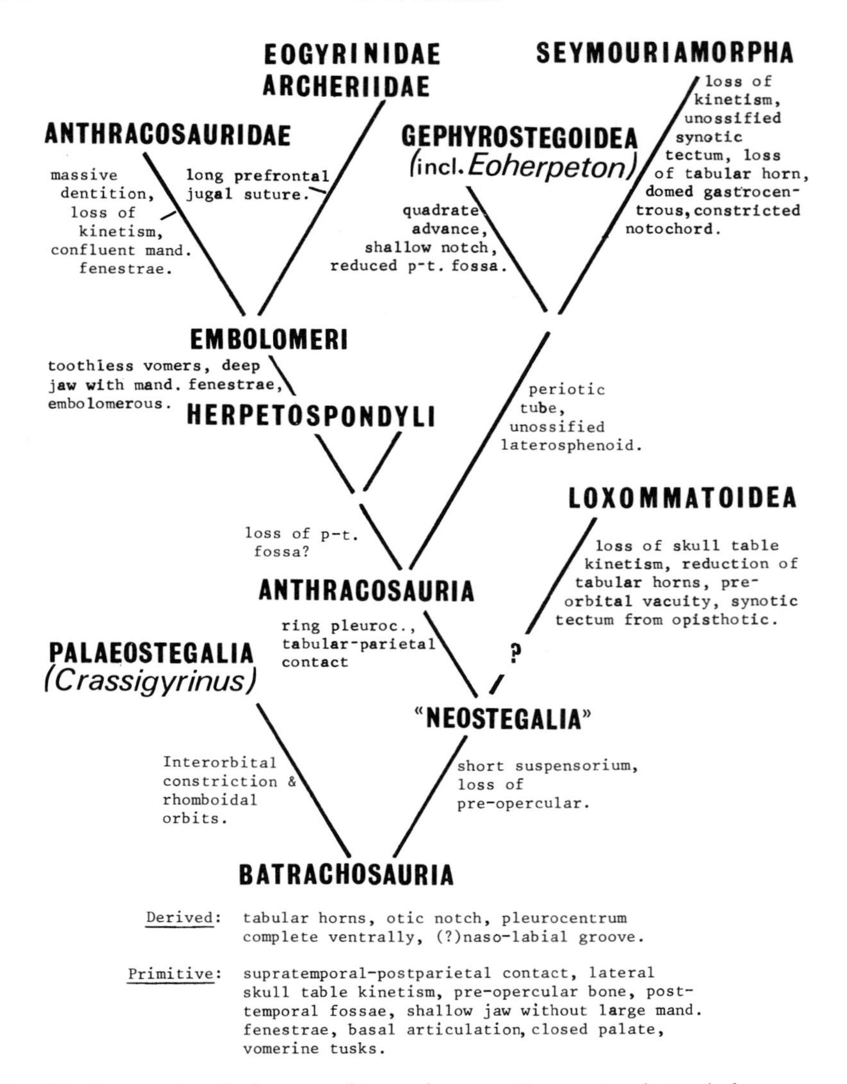

Fig. 10. A tentative phylogeny of batrachosaurs. Successive derived characters of each lineage are indicated. "Neostegalia" is not proposed as a formal taxonomic name and will lapse if the tentative hypothesis of the derivation of loxommatids from batrachosaurs is disproved.

However, contrary to the assertions of many cladists (e.g. Miles, 1975; Gaffney, 1979b) cladistic hypotheses are untestable if the criterion of predictability in the resulting classification is abandoned

(Panchen, 1979; but see Farris, 1977). Also, classifications which are redundant images of Hennigian cladograms are unsatisfactory for a number of reasons (Mayr, 1974; Jardine and Sibson, 1971).

ACKNOWLEDGEMENTS

I am indebted to several participants at the meeting whose ideas influenced the final form of this paper. I would particularly like to acknowledge unpublished information from Dr Carroll, Mr Holmes and Dr Heaton. It is also a pleasure to record the influence of frequent informal discussion on problems of amphibian phylogeny with Mr Timothy Smithson in our joint work on the amphibia of Cowdenbeath. Mr Gordon Howson's photographic work helped in the rapid preparation of the figures and the manuscript was typed by my wife, Rosemary.

REFERENCES

Andrews, S. M., Browne, M. A. E., Panchen, A. L. and Wood, S. P. (1977). Discovery of amphibians in the Namurian (Upper Carboniferous) of Fife. *Nature, Lond.* **265**, 529–532.

Baird, D. (1957). Rhachitomous vertebrae in the loxommid amphibian *Megalocephalus. Bull. geol. Soc. Am.* **68**, 1698.

Beaumont, E. H. (1977). Cranial morphology of the Loxommatidae (Amphibia: Labyrinthodontia). *Phil. Trans. R. Soc.* **B280**, 29–101.

Bolt, J. R. (1969). Lissamphibian origins: possible protolissamphibian from the Lower Permian of Oklahoma. *Science* **166**, 888–891.

Boy, J. A. and Bandel, K. (1973). *Bruktererpeton fiebigi* n. gen. n. sp. (Amphibia; Gephyrostegida) der erste Tetrapode aus dem rheinisch-westfälischen Karbon (Namur B; W-Deutschland). *Palaeontographica* **A145**, 39–77.

Bystrow, A. P. (1938). Zahnstructur der Labyrinthodonten. *Acta zool., Stockh.* **19**, 387–425.

Bystrow, A. P. (1944). *Kotlassia prima* Amalitzky. *Bull. geol. Soc. Am.* **55**, 379–416.

Carroll, R. L. (1969). Problems of the origin of reptiles. *Biol. Rev.* **44**, 393–432.

Carroll, R. L. (1970). The ancestry of reptiles. *Phil. Trans. R. Soc.* **B257**, 267–308.

Carroll, R. L. and Currie, P. J. (1975). Microsaurs as possible apodan ancestors. *Zool. J. Linn. Soc.* **57**, 229–247.

Carroll, R. L. and Gaskill, P. (1978). The Order Microsauria. *Mem. Am. phil. Soc.* **126**, 1–211.

Carroll, R. L. and Winer, L. (1977). Appendix to accompany Chapter 13 (Carroll, R. L., Patterns of amphibian evolution: an extended example of the incompleteness of the fossil record). *In* "Patterns of Evoluton" (A. Hallam, ed.). Elsevier, Amsterdam. (Privately circulated.)

Efremov, J. A. (1946). On the subclass Batrachosauria, a group of forms intermediate between amphibians and reptiles. *Izv. Akad. Nauk SSSR (Biol.)* **6**, 616–638. (In Russian, English summary.)

Farris, J. S. (1977). On the phenetic approach to vertebrate classification. *In* "Major Patterns in Vertebrate Evolution" (M. K. Hecht, P. C. Goody and B. M. Hecht, eds), pp. 823–850. Plenum, New York.

Gaffney, E. S. (1979a). Tetrapod monophyly: a phylogenetic analysis. *Bull. Carnegie Mus. nat. Hist.* No. 13, 92–105.

Gaffney, E. S. (1979b) An introduction to the logic of phylogeny reconstruction *In* "Phylogenetic Analysis and Paleontology" (J. Cracraft and N. Eldredge, eds), pp. 79–111. Columbia, New York.

Hennig, W. (1966). "Phylogenetic Systematics." University of Illinois Press, Urbana.

Holmes, R. and Carroll, R. L. (1976). A temnospondyl amphibian from the Mississippian of Scotland. *Bull. Mus. comp. Zool. Harv.* **147**, 489–511.

Hotton, N. (1970). *Mauchchunkia bassa* gen. et sp. nov. an anthracosaur (Amphibia, Labyrinthodontia) from the Upper Mississippian. *Kirtlandia* No. 12, 1–38.

Jardine, N. and Sibson, R. (1971). "Mathematical Taxonomy." Wiley, London.

Jarvik, E. (1952). On the fish-like tail in the ichthyostegid stegocephalians. *Meddr Grønland* **114** (12), 1–90.

Jarvik, E. (1954). On the visceral skeleton in *Eusthenopteron* with a discussion of the parasphenoid and palatoquadrate in fishes. *K. svenska VetenskAkad. Handl.* (4) **5** No. 1, 1–104.

Kuhn, O. (1965). "Die Amphibien: System und Stammesgeschichte." Oeben, Munich.

Land, D. H. (1974). "Geology of the Tynemouth District." (Explanation of One-Inch Geological Sheet 15, New Series.) *Mem. geol. Surv. U.K.*

Mayr, E. (1974). Cladistic analysis or cladistic classification? *Z. zool. Syst. Evolut.-forsch.* **12**, 94–128.

Miles, R. S. (1975). The relationships of the Dipnoi. *Colloques int. C.n.R.S.* No. 218, 133–148.

Nicholson, H. A. and Lydekker, R. (1889). "A Manual of Palaeontology . . . . ", 3rd Edn, Vol. II. Blackwood, London.

Olson, E. C. (1947). The family Diadectidae and its bearing on the classification of reptiles. *Fieldiana: Geology* **11**, 1-53.

Olson, E. C. (1962). Late Permian terrestrial vertebrates, U.S.A. and U.S.S.R. *Trans. Am. phil. Soc.* (N.S.) **52**, Pt 2, 1-224.

Olson, E. C. (1966). Relationships of *Diadectes*. *Fieldiana: Geology* **14**, 199–227.

Panchen, A. L. (1964). The cranial anatomy of two Coal Measure anthracosaurs. *Phil. Trans. R. Soc.* **B247**, 593–637.

Panchen, A. L. (1966). The axial skeleton of the labyrinthodont *Eogyrinus attheyi*. *J. Zool., Lond.* **150**, 199–222.

Panchen, A. L. (1967). The nostrils of choanate fishes and early tetrapods. *Biol.*

Panchen, A. L. (1970). Teil 5a Anthracosauria. "Handbuch der Paläoherpetologie." Fischer, Stuttgart.

Panchen, A. L. (1972a). The skull and skeleton of *Eogyrinus attheyi* Watson (Amphibia: Labyrinthodontia). *Phil. Trans. R. Soc.* **B263**, 279–326.

Panchen, A. L. (1972b). The interrelationships of the earliest tetrapods. *In* "Studies in Vertebrate Evolution—Essays Presented to Dr F. R. Parrington F. R. S." (K. A. Joysey and T. S. Kemp, eds), pp. 65–87. Oliver and Boyd, Edinburgh.

Panchen, A. L. (1973). On *Crassigyrinus scoticus* Watson, a primitive amphibian from the Lower Carboniferous of Scotland. *Palaeontology* **16**, 179–193.

Panchen, A. L. (1975). A new genus and species of anthracosaur amphibian from the Lower Carboniferous of Scotland and the status of *Pholidogaster pisciformis* Huxley. *Phil. Trans. R. Soc.* **B269**, 581–640.

Panchen, A. L. (1977a). The origin and early evolution of tetrapod vertebrae. *In* "Problems in Vertebrate Evolution" (S. M. Andrews, R. S. Miles and A. D. Walker, eds), pp. 289–318. Academic Press, London and New York.

Panchen, A. L. (1977b) On *Anthracosaurus russelli* Huxley (Amphibia: Labyrinthodontia) and the family Anthracosauridae. *Phil. Trans. R. Soc.* **B279**, 447–512.

Panchen, A. L. (1979). [*In* "The Cladistic Debate Continued"] *Nature, Lond.* **280**, 541.

Parrington, F. R. (1956). The patterns of dermal bones in primitive vertebrates. *Proc. zool. Soc. Lond.* **127**, 389–411.

Parrington, F. R. (1959). A note on the labyrinthodont middle ear. *Ann. Mag. nat. Hist.* (13) **2**, 24–28.

Parrington, F. R. (1967). The identification of the dermal bones of the head. *J. Linn. Soc. (Zool.)* **47**, 231–239.

Parsons, T. S. and Williams, E. E. (1962). The teeth of Amphibia and their relation to amphibian phylogeny. *J. Morph.* **110**, 375–389.

Parsons, T. S. and Williams, E. E. (1963). The relationships of the modern Amphibia: a re-examination. *Q. Rev. Biol.* 38, 26–53.

Romer, A. S. (1941). Notes on the crossopterygian hyomandibular and braincase. *J. Morph.* **69**, 141–160.

Romer, A. S. (1947). Review of the Labyrinthodontia. *Bull. Mus. comp. Zool. Harv.* **99**, 1–368.

Romer, A. S. (1966). "Vertebrate Paleontology", 3rd Edn. University Press, Chicago.

Romer, A. S. (1970). A new anthracosaurian labyrinthodont, *Proterogyrinus scheelei*, from the Lower Carboniferous. *Kirtlandia* No. 10, 1–16.

Säve-Söderbergh, G. (1932). Preliminary note on Devonian stegocephalians from East Greenland. *Meddr Grønland* **94**(7), 1–107.

Säve-Söderbergh, G. (1934). Some points of view concerning the evolution of the vertebrates and the classification of this group. *Ark. Zool.* **26A** (17), 1–20.

Schultze, H. P. (1969). Die Faltenzähne der rhipidistiiden Crossopterygier, der Tetrapoden und der Actinopterygier-Gattung *Lepisosteus. Palaeontogr. ital.* **65** (n.s. 35), 59–137.

Špinar, Z. V. (1952). Revise některých moravských Diskosauriscidů (Labyrinth-

odontia). *Rozpr. ústred. Úst. geol.* **15**, 1–160.
Watson, D. M. S. (1914). On a femur of reptilian type from the Lower Carboniferous of Scotland. *Geol. Mag.* (6) **1**, 347–348.
Watson, D. M. S. (1917). A sketch classification of the pre-Jurassic tetrapod vertebrates. *Proc. zool. Soc. Lond.* 1917, 167–186.
Watson, D. M. S. (1919). The structure, evolution and origin of the Amphibia—The "Orders" Rachitomi and Stereospondyli. *Phil. Trans. R. Soc.* **B209**, 1–73.
Watson, D. M. S. (1926). Croonian Lecture—The evolution and origin of the Amphibia. *Phil. Trans. R. Soc.* **B214**, 189–257.
Watson, D. M. S. (1929). The Carboniferous Amphibia of Scotland. *Palaeont. hung.* **1**, 219–252.
Watson, D. M. S. (1951). "Paleontology and Modern Biology." Yale University Press, New Haven.
Watson, D. M. S. (1954). On *Bolosaurus* and the origin and classification of reptiles. *Bull. Mus. comp. Zool. Harv.* **111**, 297–450.
Westoll, T. S. (1938). Ancestry of the tetrapods. *Nature, Lond.* **141**, 127–128.
Westoll, T. S. (1943). The origin of the tetrapods. *Biol. Rev.* 18, 78–98.
White, T. E. (1939). Osteology of *Seymouria baylorensis* Broili. *Bull. Mus. comp. Zool. Harv.* 85, 325–409.

14 | *Proterogyrinus scheelei* and the Early Evolution of the Labyrinthodont Pectoral Limb

ROBERT HOLMES

Redpath Museum, McGill University, Montreal, Canada

Abstract: The pectoral limb of *Proterogyrinus scheelei* (an Upper Mississippian anthracosaur) is very similar to that of *Archeria* (a Lower Permian anthracosaur). Both *Proterogyrinus* and a contemporary temnospondyl *Greererpeton* possess two foramina piercing the coracoid plate, suggesting that this is a primitive condition for labyrinthodonts. The interclavicle of *Proterogyrinus* is similar to that of *Eogyrinus*. The primitive clavicular stem is more massive than that of later labyrinthodonts. The anterior flange of the humerus, probably homologous with that of the rhipidistian humerus and present in both *Proterogyrinus* and *Greererpeton*, is a primitive feature of labyrinthodonts.

The structure of both the elbow and shoulder joints of *Proterogyrinus* indicates that, compared with later labyrinthodonts and reptiles, the ranges of movement were very restricted. However, the limbs could not have been suitable for aquatic locomotion. Despite the great age of *Proterogyrinus*, its already well developed pectoral limb shows little that is transitional between that of known rhipidistians and later labyrinthodonts, and contributes little to our understanding of the origin and early evolution of the tetrapod pectoral appendage.

INTRODUCTION

The labyrinthodont amphibians were an important component of the terrestrial fauna during the late Palaeozoic. The group is divided into two major lineages (the problematical ichthyostegids and *Crassigyrinus* are placed in isolated orders) based on vertebral structure and several

---

Systematics Association Special Volume No. 15, "The Terrestrial Environment and the Origin of Land Vertebrates", edited by A. L. Panchen, 1980, pp. 351–376, Academic Press London and New York.

important features of the skull (Romer, 1947, 1966). The temnospondyls, although including some terrestrial forms such as the eryopids and dissorophids, consists largely of aquatic and semi-aquatic genera. Although very common in the late Palaeozoic, temnospondyls became extinct in the early Jurassic. The anthracosaurs (*sensu* Romer, 1966) or batrachosaurs (*sensu* Panchen, 1970), although far less common, have received considerable attention because of their supposed position as reptile ancestors (Watson, 1917; Romer, 1966). This relationship has been challenged by Panchen (1972a) and more recently by Heaton (1978), although no other early tetrapod group has been identified as a more probable ancestor of amniotes.

Although the earliest labyrinthodonts, the ichthyostegids (Jarvik, 1952), date from the uppermost Devonian, their remains are otherwise rare below the Pennsylvanian. The few described Mississippian genera are: *Loxomma* (Beaumont, 1977), *Crassigyrinus* (Panchen, 1973), *Eoherpeton* (Panchen, 1975), *Caerorhachis* (Holmes and Carroll, 1977), *Greererpeton* (Romer, 1969) and the closely related *Pholidogaster* (Panchen, 1975), *Proterogyrinus* (Romer, 1970) as well as undescribed limb elements from Horton Bluff, Nova Scotia, (Carroll *et al.*, 1972). Only *Ichthyostega, Caerorhachis, Pholidogaster, Greererpeton* and *Proterogyrinus* show well-articulated postcranial remains. Although a preliminary study of the structure of the vertebral column in *Ichthyostega* has been made (Jarvik, 1952), we are still awaiting a definitive study of the appendicular skeleton of these important tetrapods. The rear limb of *Caerorhachis* has been described, but the anterior limb is virtually unknown. The limbs of the aquatic *Greererpeton* and *Pholidogaster* are weakly developed. *Proterogyrinus* is the only known pre-Pennsylvanian labyrinthodont in which well-developed, articulated limb elements are available for three-dimensional study, rendering it a valuable source of information on the level of development of limbs at this early stage in the evolution of tetrapods.

Proterogyrinus is a medium-sized anthracosaur with a snout-vent length of approximately 60 cm in the largest individual (Fig. 1) and approximately 40 cm in the smallest articulated specimens. The large, well-ossified limbs are very similar to those of *Archeria* (Romer, 1957). There are approximately 31 presacral vertebrae, much lower than the count for *Eogyrinus* (Panchen, 1966) or *Archeria*. Unlike embolomeres, but like the anthracosaur *Gephyrostegus*, the vertebrae

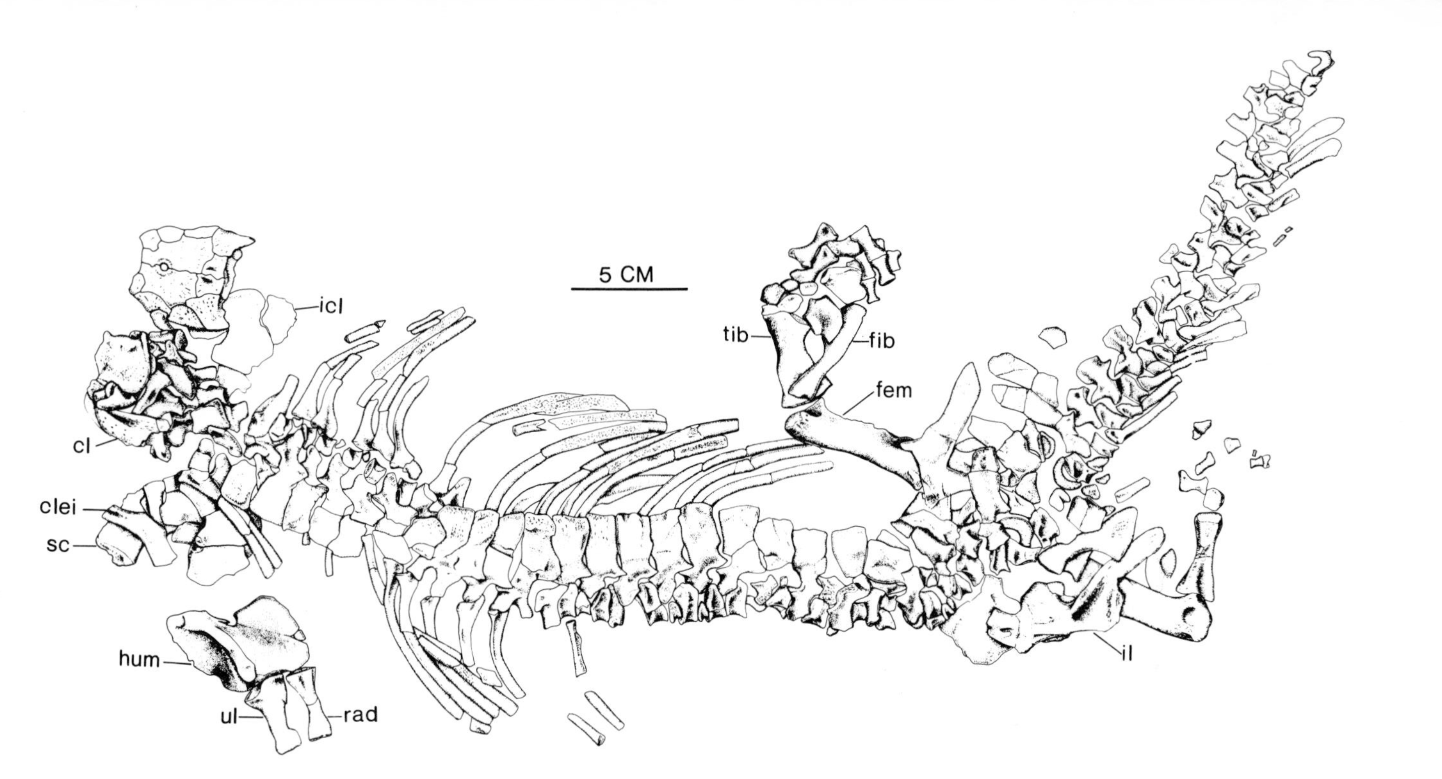

Fig. 1. Articulated specimen of *Proterogyrinus scheelei* (CMNH 11067), dorsal view. cl, clavicle; clei, cleithrum; fem, femur; fib, fibula; hum, humerus; icl, interclavicle; il, ilium; rad, radius; sc, scapula; tib, tibia; ul, ulna.

have a "protoreptilian" structure (Holmes and Carroll, 1977), with a small crescentic intercentrum and a larger, horseshoe-shaped pleurocentrum. The ossification of the central elements is more complete than in *Gephyrostegus* and the pleurocentrum closes over the constricted notochordal canal, although a suture is still visible. The skull is more closely comparable to that of eogyrinids than to the primitive anthracosaur *Eoherpeton* in having a well-developed otic notch and a large number of marginal teeth (approximately 45 on the maxilla), but shares with the latter the presumably primitive kinetic junction that extends the entire distance from the posterodorsal corner of the orbit to the otic notch.

The results given here are a small part of an ongoing study. Space limitations prevent a complete description of the anatomy of *Proterogyrinus*. Rather than presenting a brief survey of the entire skeleton it was felt that it would be more useful to describe one aspect of the anatomy in detail.

SPECIMENS EXAMINED

All specimens are from the collections of the Cleveland Museum of Natural History, the Museum of Comparative Zoology (Harvard), and the U.S. National Museum.

(1) CMNH 10938: Articulated, reasonably complete postcranial skeleton of a small individual.
(2) CMNH 10950: Type specimen. Small individual preserved in three blocks containing a skull and scattered postcranial elements.
(3) CMNH 11067: Magnificent, complete articulated skeleton of a large individual.
(4) MCZ 4537: Reasonably complete, although somewhat disarticulated, large individual.
(5) USNM 26368: Type of "*Mauchchunkia bassa*" (Hotton, 1970). A large individual with a good skull and numerous postcranial elements.

All specimens come from one lenticular ?stream channel deposit from the Upper Mississippian Bickett Shale, Bluefield Formation, Mauch Chunk Group, which is generally considered to be equivalent in age to the Upper Viséan of Europe. However, according to Busanus (1974), although the base of the Mauch Chunk Group of northern

West Virginia is equivalent in age to the Upper Viséan, the upper part of the Group, and hence the Bickett Shale, is contemporaneous with the Lower Namurian A of the European series. The deposit was uncovered at the Greer Quarry, 10·5 km south-east of Morgantown, Monongalia County, West Virginia (for detailed provenance see Hotton, 1970).

In addition to eight specimens of *Proterogyrinus*, the deposit produced a large number of specimens of the temnospondylous amphibian *Greererpeton*, numerous *Gyracanthus* spines, scales of paleoniscoids and crossopterygians, and articulated skeletons of the lungfish *Tranodis* (Busanus, 1974).

DESCRIPTION OF ELEMENTS

1. The Scapulocoracoid (Fig. 2)

The lateral surface of the left scapulocoracoid, lacking the anterior portion of the coracoid plate, is exposed in CMNH 11067. The scapular portion clearly ossified first in ontogenetic development, with only this portion preserved in a smaller specimen (CMNH 10938). As in other labyrinthodonts (and rhipidistians), there is no evidence of more than one centre of ossification. In most features it is quite similar to that of *Archeria* (Romer, 1957). The unfinished dorsal margin of the short scapular blade indicates the presence of a large cartilagenous suprascapula in life. A triangular depression, probably the site of the origin of the subcoracoscapularis muscle, begins immediately dorsal to the glenoid and runs a considerable distance up the thick posterior margin of the blade. The large supraglenoid foramen passes obliquely anteromedially as well as ventrally. Anterior to the greatly thickened posterior border, the blade rapidly tapers to a very thin plate at its anterior margin. Immediately below the supraglenoid foramen, the bone thickens considerably to provide a massive laterally directed process supporting the posterolaterally directed anterior glenoid surface. Little fine detail of the glenoid surface is preserved, and it was probably covered in life by a considerable layer of cartilage. It is clear, however, that there was little twisting of its surface, which turns approximately 15° from a ventrolateral orientation anteriorly to an essentially lateral orientation at its

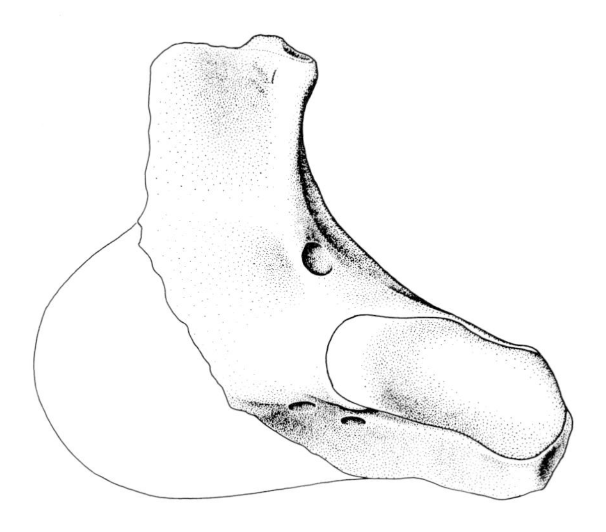

Fig. 2. Left scapulocoracoid of *Proterogyrinus scheelei*, lateral view. Drawin in plane parallel to that of the scapular blade. × 1.

posterior termination. The glenoid is only slightly longer than the corresponding surface on the humerus, suggesting relatively little translation during the locomotor cycle.

Two foramina, both somewhat smaller than the supraglenoid foramen, pierce the coracoid plate. The supracoracoid foramen is located anteoventral to the anterior termination of the glenoid. The glenoid foramen, posterior and slightly ventral to the preceding, is more dorsal in position than its equivalent in *Archeria*. This foramen is present in *Greererpeton* immediately ventral to the middle of the glenoid and, although not characteristic of later temnospondyls, has been reported in *Dissorophus* (Carroll, 1964). Two foramina are present in the coracoid plate of microsaurs (Carroll and Gaskill, 1978) but they are more anterior in position and the homologies are uncertain.

2. *The Dermal Girdle*

The reconstruction of the interclavicle (Fig. 3) is based primarily on CMNH 11067 with the outline for the parasternal process from MCZ 4537. Within the Anthracosauria, its general shape compares most closely with that attributed to *Eogyrinus* (Panchen, 1972b) and

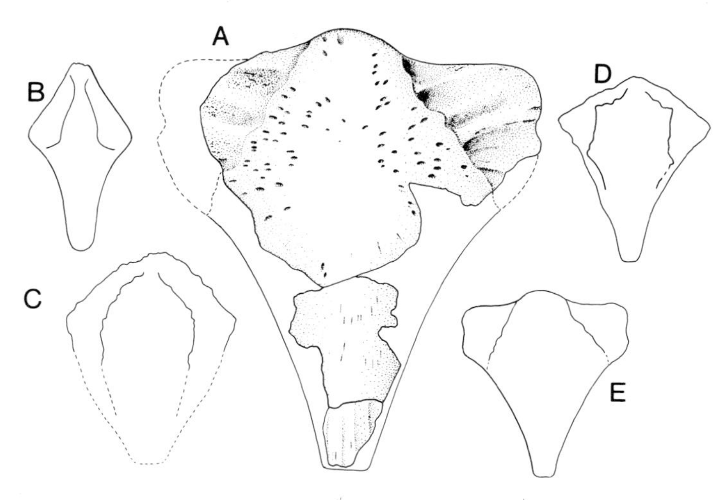

Fig. 3. Anthracosaur interclavicles, ventral views. (A) *Proterogyrinus scheelei*, × 1. (B–E) Outlines of anthracosaur interclavicles drawn to same length: (B) *Archeria* (after Romer); (C) attributed to *Anthracosaurus*; (D) attributed to *Eogyrinus* (C and D after Panchen); (E) *Proterogyrinus*.

differs from that of *Archeria* (Romer, 1957) in being shorter and narrower posteriorly. Viewed ventrally, its outline approaches that of an isosceles triangle with the apex directed posteriorly. The anterior margin bears a broadly rounded convexity. The recessed, overlapping surfaces that received the ventral plates of the clavicles are marked by broad, anterolaterally directed grooves, and form an angle of approximately 25° with the central plate of the interclavicle, giving the dermal girdle a pronounced ventrally convex outline when viewed anteriorly. These surfaces are separated from the raised central area of the interclavicle by an irregular border that makes an angle of approximately 35° with the anteroposterior axis of the bone. Anteriorly, the medial margins of the clavicles were well separated, and the anterior convexity of the interclavicle would be visible ventrally as in *Archeria* (Fig. 4). The diamond-shaped central area of the interclavicle, much wider than in *Archeria*, is covered with widely spaced pits radiating from its centre. The ornamentation is somewhat less well developed than on the larger bone attributed to *Eogyrinus* (Panchen, 1972). As in *Eogyrinus* and *Archeria*, the pits become less distinct posteriorly, and are reduced to shallow striations on the parasternal process. The posterior part of the process is fluted. The anterior portion of the dorsal surface of the interclavicle, exposed in CMNH 11067, is smooth except for a few indistinct, shallow grooves probably representing blood channels.

The interclavicle of *Proterogyrinus* is generally similar to that of *Gephyrostegus* (Carroll, 1970), although the parasternal process of the latter is longer and thinner. It is distinctly different from that of *Seymouria* (White, 1939) and *Tseajaia* (Moss, 1972), in which the process is relatively thin and greatly elongated.

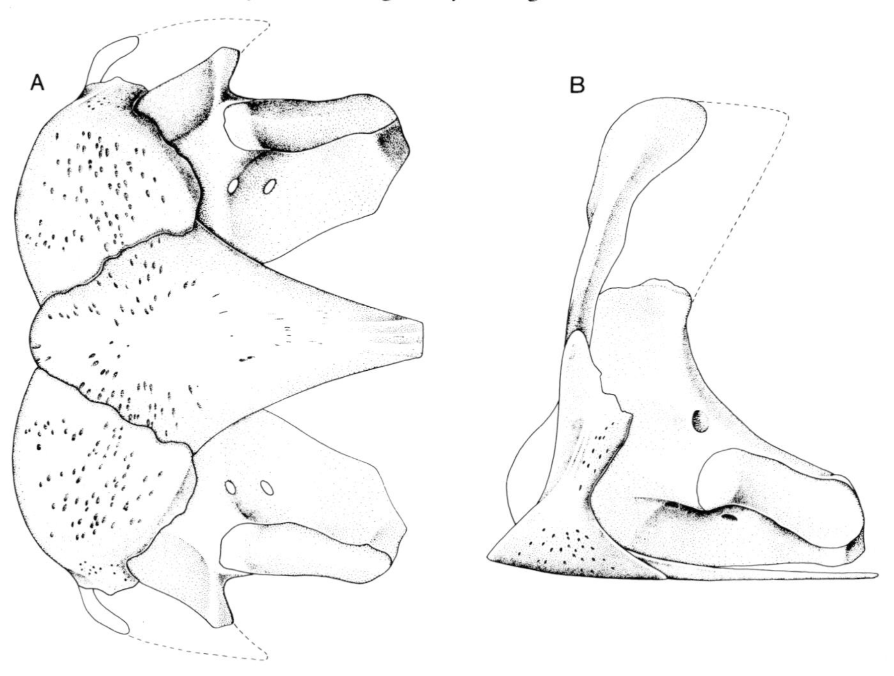

Fig. 4. Reconstruction of the pectoral girdle of *Proterogyrinus scheelei*. (A) Ventral view; (B) lateral view. × $\frac{2}{3}$.

The clavicle (Fig. 5) is preserved in CMNH 11067 and MCZ 4537. The ventral surface of the triangular ventral plate bears the same punctate ornamentation seen on the interclavicle. When the bone is placed in articulation with the interclavicle, it becomes apparent that the lateral angle of the triangle is directed anterolaterally rather than posterolaterally as is the case in *Archeria*. Since the clavicular stem turns dorsally from this point on the ventral plate, the stem is in a much more anterior position relative to the ventral plate (Fig. 4).

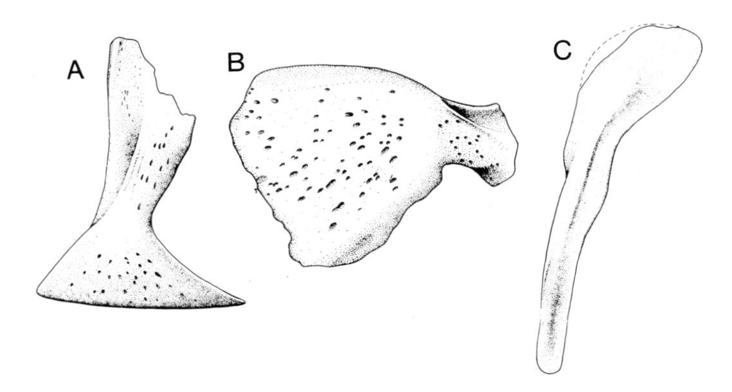

Fig. 5. Dermal elements of the pectoral girdle of *Proterogyrinus scheelei*. (A) Left clavicle, lateral view; (B) left clavicle, drawn in a plane parallel to the ventral plate; (C) left cleithrum, lateral view. All × 1.

The curved posterolateral surface of the stem, exhibiting fine pitting, is essentially a dorsal continuation of the ventral surface. It is separated from the concave anterolateral surface of the stem by a distinct ridge that begins at the anteroventral corner of the base of the stem and traces a broad, posterodorsal curve, losing prominence dorsally. The ridge probably marked the posterior extent of the origin of the ventral throat musculature. The clavicular stem, although similar to that of *Archeria* in its shortness, is much broader anteroposteriorly and is generally more massive. These characteristics are presumably primitive.

The cleithrum (Fig. 5) is preserved in CMNH 11067 and CMNH 10938. It is closely comparable to that of *Archeria*. The thick shaft is triangular in cross-section. As the shaft expands dorsally, it thins to a spoon-shaped plate. The anteromedial surface is grooved, unlike that of *Archeria*. Fluting on the lateral margin of this surface marks the upper part of the origin of the lateral throat musculature. On the lateral surface, a groove begins near the centre of the dorsal expansion, and as it passes ventrally it widens and deepens. The ventral portion of this groove received the dorsal tip of the clavicular stem. The posterior surface is not exposed in any specimen.

The sternum, if present, was unossified.

All articulated specimens as preserved show a distinct neck region between the pectoral girdle and the occiput of the skull. This is in marked contrast to the condition seen in articulated skeletons of

Greererpeton in which the anterior margin of the ventral dermal girdle actually projects anterior to the level of the jaw articulation.

3. The Humerus

The humerus of *Proterogyrinus* (Fig. 6) compares closely with that of *Archeria*, which is of a tetrahedral type common in early tetrapods (Romer, 1957). Both have expanded entepicondyles, an entepicondylar foramen and a prominent lateral (anterior) keel. As in *Archeria*, there is little helical twisting of the proximal articular surface, nor is the "double surface" characteristic of primitive tetrapods (Romer, 1922) apparent. However, ossification, even in large specimens, is incomplete. The anterior-most part of the proximal articular surface faces proximally, forming an angle of 90° with the long axis of the humerus. As it passes posteriorly, the surface turns ventrally, forming an angle of approximately 70° with the long axis of the humerus at its posterior termination. There is no evidence of significant dorsal exposure anteriorly.

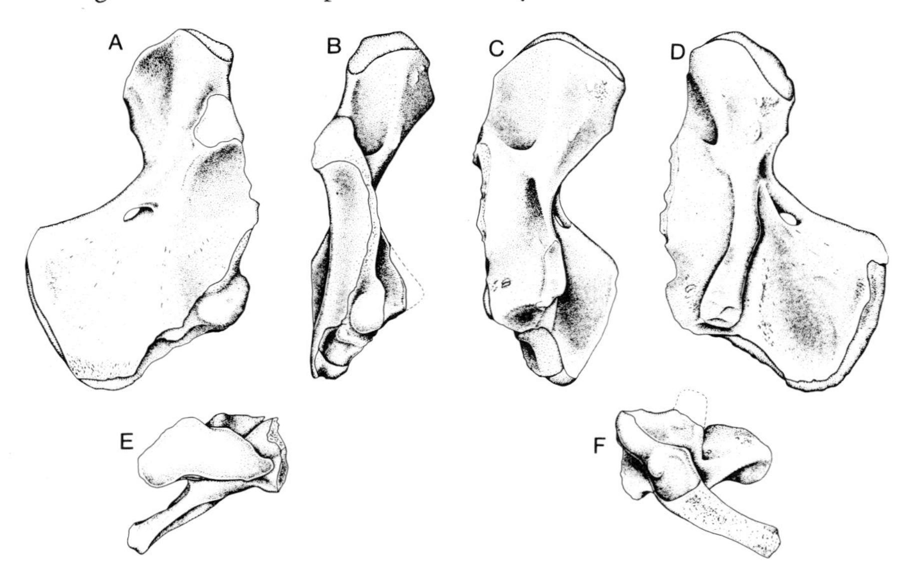

Fig. 6. The left humerus of *Proterogyrinus scheelei.* (A) In the plane of the distal ventral surface; (B) anterior view; (C) in the plane of the proximal dorsal surface; (d) in the plane of the distal dorsal surface; (E) proximal view; (F) distal view.

The arrangement of muscle insertions on the proximal dorsal surface is generally comparable to that described by Romer for *Archeria*. At the posterior corner near the articular surface is an extensive rugose area bearing several small tuberosities that mark the insertion of the subcoracoscapularis muscle. Distal to this is a tuberosity, very prominent in the small type specimen (CMNH 10950), marking the insertion of the latissimus dorsi muscle. A similar process exists in *Eryops* (Miner, 1925). Anterior to these, a concavity marks the insertion of the scapulohumeralis muscle. In one large specimen (CMNH 11067), the concavity deepens distally to form a proximally facing pocket with an overhanging shelf.

The ectepicondylar ridge begins just distal to the insertion of the latissimus dorsi and passes diagonally to the anterodistal corner of the humerus where it forms the body of the ectepicondyle. Distally, the dorsal margin of the ridge did not ossify completely in any of the specimens.

The stout deltopectoral crest is very short. Its flat, unfinished termination faces anteriorly and slightly medially. Continuous with the deltopectoral crest distally is the large dorsally convex anterior flange, probably homologous with the preaxial process of rhipidistians such as *Sterropterygion* (Rackoff, 1976) and *Sauripterus* (Andrews and Westoll, 1970b), that continues distally to the end of the bone. It is best developed proximally where it probably provided much of the insertional area for the deltoideus musculature. This flange, with its inferred primitive muscle arrangement, also appears in the early anthracosaurs *Bruktererpeton* (Boy and Bandel, 1973) and *Gephyrostegus* (Carroll, 1970), as well as in the Upper Mississippian *Greererpeton* (Fig. 7), a temnospondyl only distantly related to *Proterogyrinus* and *Archeria*. It is probable that the flange and associated muscle attachments were primitive features of the labyrinthodont pectoral limb that were retained in anthracosaurs, but lost in most later temnospondyls – Romer (1957) cites *Trimerorhachis* and *Parioxys* as possible exceptions to this – and the line of labyrinthodonts leading to reptiles.

The entepicondylar foramen passes from the dorsal surface of the humerus in an anteroventral direction to the ventral surface. The entepicondyle is similar to that of *Archeria*, being roughly rectangular in outline. Although large, it is quite thin, particularly along the line running from the foramen to its posterodistal corner.

As in *Archeria*, the degree of twisting or "torsion" is rather low as compared with most lower tetrapods. The long axis of the proximal articular surface makes an angle of 37° with the plane of the entepicondyle as compared with 88° for *Eryops*, and 49° with the bicondylar axis as compared with 98° in *Eyrops* – see Andrews and Westoll (1970a) for detailed discussion of humeral axes. Romer stated that the angle between the planes of the entepicondyle and proximal dorsal surface was 20°–25° in *Archeria*, but Figs 5D, E of Romer's paper clearly show that the long axis of the proximal articular surface (essentially parallel with the proximal dorsal surface) makes an angle of approximately 30° with the plane of the entepicondyle and approximately 50° with the bicondylar axis.

Eusthenopteron shows little torsion of the humerus (Andrews and Westoll, 1970a), as compared with that of labyrinthodonts, suggesting that the low degree of torsion seen in *Proterogyrinus* and *Archeria* is transitional between fish and tetrapods. However, the torsion in the humerus of *Sterropterygion*, a rhipidistian with tetrapod-like limbs, is strongly developed, approaching the condition seen in advanced labyrinthodonts such as *Eryops* (Rackoff, 1976, and Chapter 11 of this volume). Whatever the general phylogenetic significance of torsion in tetrapods, it appears that the low degree of torsion in *Proterogyrinus* (and probably in *Archeria*) is related to the orientation of the ulna and radius on the humerus.

The proximal ventral surface of the humerus of *Proterogyrinus* shows two concavities separated by a low, rounded ridge. The anterior concavity begins distally immediately posterior to the deltopectoral crest and passes obliquely anteromedially to merge with the anterior margin of the bone. Roughened bone surface indicates the insertion of the supracoracoideus muscle. Posterior to this, and separated from it by a gentle ridge, is a second, deeper concavity beginning as a deep pocket distally at about the level of the deltopectoral crest and becoming wider and shallower proximally. The roughened, and in some specimens fluted surface of this large concavity, marks the insertion of powerful coracobrachialis brevis musculature. As in *Archeria*, a ridge passes diagonally posterodistally from the deltopectoral crest to the proximal edge of the entepicondyle, where it broadens and becomes confluent with a broad, low thickening that curves anteriorly as it passes distally, terminating just distal to the ulnar articulation.

Although this ridge may be a remnant of the ventral diagonal ridge of the rhipidistian humerus, it bears no rugosities or processes for muscle attachment, and it no doubt only served to strengthen the shaft of the relatively lightly built bone against upward forces produced during the power stroke. Anterior to the ridge is the gently concave ventral surface of the anterior flange, which terminates distally in the radial and ulnar facets. Posterior to the ridge is a low, broad concavity that begins proximally at the entepicondylar foramen and broadens posterodistally. The medial and posterior borders of the entepicondyle are slightly thickened, and the latter is roughened, indicating the origins of the lower limb flexors.

The radial articular surface, although small in *Proterogyrinus*, is more like the capitulum of other primitive tetrapods in its bulbous shape and ventral orientation than is that of *Archeria*. This apparent difference may be due to poor preservation or low degree of ossification in *Archeria*. The ulnar articular surface faces ventrolaterally, as in other primitive tetrapods. A ridge, beginning as a conspicuous swelling on the anteroventral corner of the surface and rapidly losing prominence as it passes posterodorsally, matches well with a depression on the corresponding surface of the ulna. Immediately posterior to this ridge is a deep concavity that accommodates a ridge on the articular surface of the ulna.

Although the radial and ulnar facets are positioned adjacent to one another, they are not as closely associated as they are in most lower tetrapods such as pelycosaurs (Jenkins, 1973) or captorhinids (Holmes, 1977). In these forms, the ulnar articular surface is divided into two parts by a ridge. The posterior surface of the radial condyle (capitulum) serves as the opposing surface for the anterior facet of the ulna. The ridge on the ulna fits into a "trochlear" notch that separates the capitulum from the rest of the distal surface that opposes the posterior facet of the ulna. In *Proterogyrinus*, the anterior ulnar facet does not articulate with the capitulum, but with the swelling on the ulnar articular surface proper. This is presumably a primitive condition associated with the rather restricted range of movement of the lower limb on the humerus. As the capitulum expanded in more advanced forms to allow greater freedom of movement between the radius and humerus, it encroached on the ulnar articular surface and eventually incorporated the swelling. The capitulum, once serving as a radial articulation only, then provided an opposing surface for the anterior facet of the ulna. The large capitulum and deep "trochlear" notch provided much more resistance to torsional forces than the relatively loose joint present in *Proterogyrinus* and *Archeria*, and probably evolved in response to the higher torque imposed on the elbow during the power stroke by the more strongly helical proximal humeral surface of more advanced Palaeozoic tetrapods. The inherent weakness in the joint appears to have been partially offset by a reorientation of the ulna on the humerus. The long axis of the ulna makes an oblique angle with the facet of the humerus, providing a greater area of contact and adding strength to the joint. However, this gives the ulna a distinctly anterior orientation when the long axis of the distal humeral expansion is horizontal with its long axis roughly parallel to the plane of the entepicondyle (Fig. 9). This unusual orientation necessitated low humeral torsion so that the hand could be maintained in a low enough position to make contact with the ground at the beginning of the power stroke.

A curious feature of the distal articular surfaces is the anterior inclination of the ulnar surface, in dorsal view forming an angle of approximately 60° with the long axis of the humerus. In most primitive tetrapods the plane of the ulnar facet is more nearly 90° to the long axis of the humerus.

4. *The Distal Elements*

The ulna and radius (Fig. 7) of *Proterogyrinus* differ from those of *Archeria* in being relatively shorter and stouter. There is considerable variation in proportions among specimens (Table 1).

The proximal articular surface of the ulna faces more nearly proximally than in *Archeria*, forming an angle of 60° with the long axis of the shaft as compared with 32° for *Archeria*. The olecranon, even in large specimens, is not as well ossified. Rugosities and striations, indicating the insertion of the triceps muscle, extend well down onto the shaft. A small swelling on the posteromedial surface just distal to the proximal articular facet marks the insertion of the biceps muscle. This facet is wider than in *Archeria*. It is crossed by a ridge running from the posteroventral to the anterodorsal corner, dividing the surface into posterior and anterior facets. The distal articular surface is indistinctly divided into a large facet for the

Table I. Limb bone measurements in *Proterogyrinus* (mm)[a]

Bone	Length	Proximal width	Shaft width	Distal width
Humerus				
CMNH 10938	40	21	18	29
CMNH 10950	30	15	13	22
CMNH 11067	77	30	20	42
MCZ 4537	62	26	20	38
Ulna				
CMNH 10938	28	13	5	10
CMNH 10950	19	11	4	8
CMNH 11067	30+	18	8	14
MCZ 4537	36	15	7	14
USNM 26368	30	16	6	12
Radius				
CMNH 10938	27	10	5	10
CMNH 10950	19	7	4	9
CMNH 11067	31	11	6	14
MCZ 4537	35	13	7	16
USNM 26368	34	13	6	13

[a]All width measurements are taken perpendicular to the long axes and parallel to the extensor surfaces of the bones. All measurements are maximum except for width of shaft, which is the minimum, and length of ulna, which is measured from the base of the proximal articular surface (olecranon not included).

intermedium and a smaller facet for the ulnare. Viewed from its lateral aspect, the bone has a gentle sigmoidal shape, with its distal end directed ventrally.

The radius is essentially similar to that of *Archeria*. The concave proximal articular surface is subcircular in outline. The slightly convex distal surface is roughly triangular in outline, with the apex directed posteriorly. A ridge probably marking the boundary between the extensor and flexor muscle groups (Romer, 1957), runs diagonally from the middle of the lateral surface of the shaft to the distal dorsolateral corner of the bone. A similar, albeit more prominent, ridge appears in *Archeria*. On the medial surface, a ridge begins immediately proximal to the distal expansion, gains height and thickens as it passes proximally, and terminates in an unfinished prominence that faces medioproximally and slightly towards the

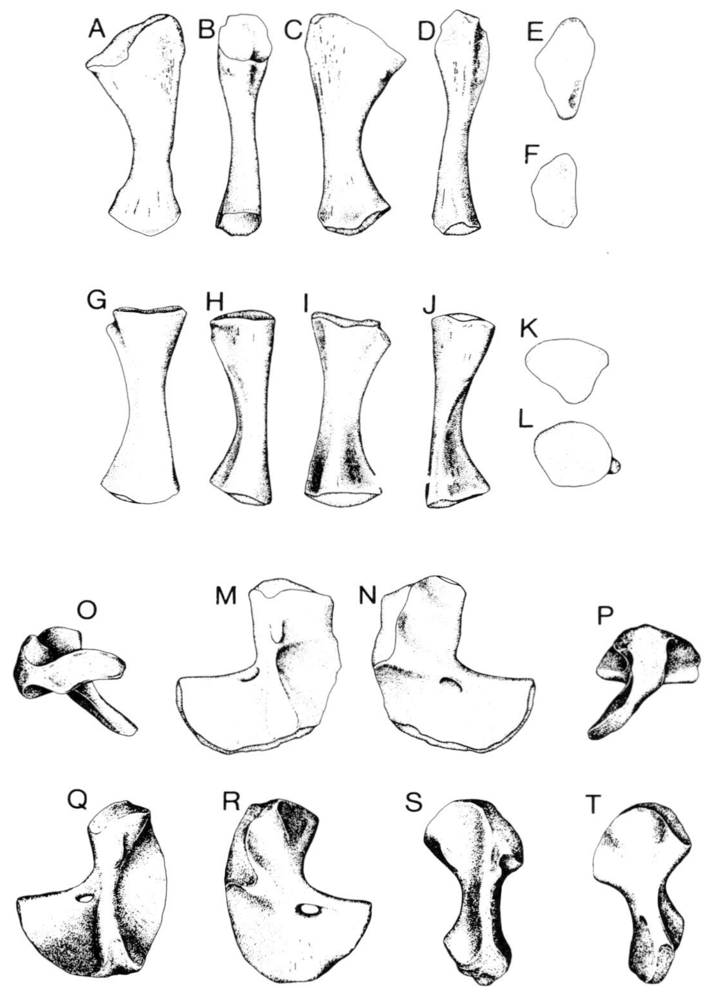

Fig. 7. Limb elements of the Greer labyrinthodonts. (A–F) Left ulna of *Proterogyrinus scheelei*: (A) extensor view; (B) medial view; (C) flexor view; (D) lateral view; (E) proximal view; (F) distal view.

(G–L) left radius of *Proterogyrinus scheelei*: (G) extensor view; (H) medial view; (I) flexor view; (J) lateral view; (K) proximal view; (L) distal view.

(M and N) Right humerus of *Greererpeton burkemorani* (CMNH 11090); (M) in plane of distal dorsal surface; (N) plane of distal ventral surface.

(O–T) Humerus of immature *Proterogyrinus scheelei* (CMNH 10950), type specimen: (O) proximal view; (P) distal view; (Q) in a plane parallel with the distal dorsal surface; (R) in a plane parallel with the distal ventral surface; (S) anterior view; (T) posterior view. All × 1.

flexor surface. The prominence provided an insertion for the tendon of the biceps muscle. The extensor surface is convex in cross-section proximally, but gradually becomes flattened distally. As in *Archeria*, a low ridge begins from the proximal portion of the shaft and as it passes distally curves towards the lateral border, where it subsides. The proximal flexor surface is unremarkable. However, a strong keel develops distally that divides the distal flexor surface into a broad, concave ventromedial surface and a smaller, concave ventrolateral surface producing the same triangular outline of the distal articular surface seen in *Archeria*.

The hand is preserved in MCZ 4537 and CMNH 10938. Since the fore limbs are well articulated in both specimens, it is unlikely that any carpal elements were lost. Nevertheless, except for distal carpals I and II and a bit of bone that has been identified tentatively as a fragment of the intermedium in MCZ 4537, the carpus is unossified (Fig. 8).

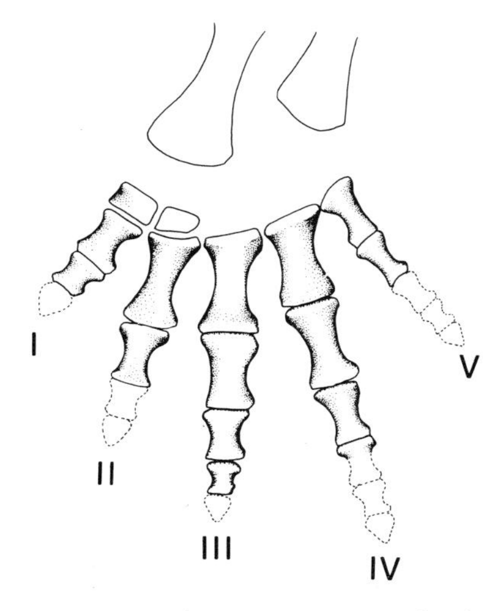

Fig. 8. Restoration of left manus of *Proterogyrinus scheelei*. × 1.

The hands in both specimens, although reasonably well articulated, are incomplete. There are five digits as in *Archeria* (Fig. 8). The phalangeal count is uncertain, for no unguals are preserved. However, judging from the size of the most distal elements preserved, it is not unreasonable to reconstruct the manus with a phalangeal formula of 2, 3, 4, 5, 4 as in *Archeria*. The preserved elements are closely comparable to those of *Archeria*. The metacarpal III is the longest, followed by IV and then II. Metacarpal I, although short is very stout, and metacarpal V, the smallest, is both short and thin. Each metacarpal is only slightly longer than its proximal phalanx. Where preserved, the more distal phalanges are slightly more elongated than the corresponding element in *Archeria*.

THE MECHANICS OF THE FOREARM

1. The Glenoid

Although ossification of the articular surface of the glenoid and proximal humeral head is not complete, some analysis of glenoid functions is possible with the aid of indirect evidence.

Romer (1957) suggested that since the strap-shaped articular surface of the glenoid and proximal humeral head of *Archeria* shows little spiral curvature, the humerus must have moved straight back and forth, presumably with the long axes of the two surfaces aligned producing little rotation of the humerus during the power stroke. Since essentially the same glenoid structure is exhibited in *Proterogyrinus*, a similar function might be inferred. However, with the ulna and radius placed in articulation with the humerus, the above arrangement of glenoid and humeral surfaces places the lower limb in an anterodorsal orientation. If the surfaces are kept in contact as the limb is retracted, the ectepicondyle of the humerus will rotate ventrally, causing the lower limb to approach a horizontal attitude, and allowing the hand to make contact with the substrate only when the humerus if fully retracted. The limb clearly could not have functioned effectively in such a manner. Since the distal humeral articular surfaces are well preserved, the problem must be with the interpretation of the glenoid. The problem can be solved by assuming that the cartilaginous proximal articular surface of the humerus was composed in life of two surfaces, as was the case in most primitive tetrapods (Romer, 1922) and the cartilaginous surface of the glenoid possessed a spiral ridge. As in *Captorhinus* (Holmes, 1977) and *Eryops* (Miner, 1925), the groove would cross the humeral surface from its anterodorsal corner to the midventral point. Because the groove is diagonal, the long axis of the humeral head would form a considerable angle with that of the glenoid surface when fitted into

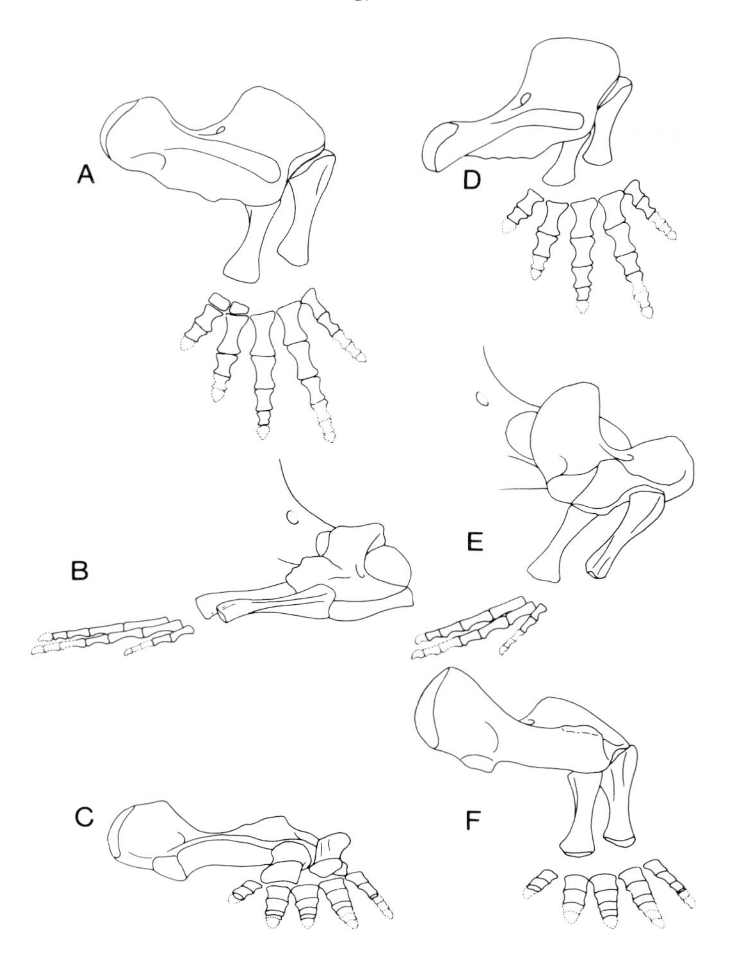

Fig. 9. Restoration of the pectoral limb of *Proterogyrinus scheelei*. (A–C) Beginning of power stroke; (D–F) end of power stroke. (A and D) Dorsal view; (B and E) lateral view; (C and F) anterior view.

the spiral ridge of the glenoid, resulting in an essentially anterior orientation of the lower limb when the humerus is drawn completely forward, and an increasingly ventral orientation as the groove slips over the ridge during the power stroke (Fig. 9).

Although the humeral groove and glenoid ridge almost certainly were present in *Proterogyrinus*, their detailed morphology is unknown, and cannot be used to estimate the amount of rotation that the humerus underwent during the power stroke. However, if it is assumed that maximum rotation at the termination of the power stroke is reached when the deltopectoral crest achieves a directly ventral orientation, an estimate of total rotation can be made (Fig. 9). The preserved articular surfaces can be used to estimate the extent of retraction. In a more advanced tetrapod, the extensive proximal articular surface covers not only the distal part of the humeral head, but also much of the anterodorsal and posterior margins, turning an angle of close to 180° measured anteroposteriorly in the case of *Captorhinus* (Holmes, 1977, Fig. 8). The humeral surface is much larger than the corresponding glenoid surface, allowing considerable translation and resulting in extended retraction and protraction of the humerus. In *Proterogyrinus*, the humeral articular surface is restricted anteriorly and posteriorly, forming a medially facing convexity that turns an angle of about 80°. Despite poor ossification, it is clear that the humeral surface was not much longer than the glenoid. Since the posterior end of the humeral surface cannot be anterior to the posterior end of the glenoid at the beginning of the recovery stroke (Holmes, 1977), the limited development of articular surface onto the posterior margin of the humeral head precludes marked retraction of the humerus. Similarly, the absence of a distinct anterodorsal lappet of the articular surface restricts anterior movement.

Fully protracted, the long axis of the forward-directed humerus was parallel to the ground and formed an angle of approximately 75° with the sagittal plane (Fig. 9). The plane of the entepicondyle was parallel to the ground. As in *Captorhinus* and pelycosaurs (Jenkins, 1971) some depression probably accompanied retraction. At the end of the power stroke, the long axis of the backward-directed humerus formed an angle of approximately 78° with the sagittal plane. The total arc of approximately 27° is considerably less then 60° for *Captorhinus*. The total rotation of the humerus during the power stroke would have been roughly 20° as compared with 45° for *Captorhinus*. Considerable elevation of the glenoid must have accompanied retraction and rotation of the humerus (Fig. 9C,F). At the end of the power stroke, it would have been impossible for the hand of the opposite limb, just beginning the next power stroke, to make contact with the ground unless protraction of that limb were accompanied by considerable twisting of the trunk, dropping the shoulder on that side sufficiently to bring the hand into contact with the substrate.

2. The Elbow

The surface of the ulna fits closely with its counterpart on the humerus, allowing a reasonably accurate reconstruction of the total extent of movement at this joint. Fully extended, the long axis of the ulna lies very close to that of the entepicondyle and makes a 90° angle with the long axis of the humerus (Fig. 9). Since full extension occurs when the humerus is at its anteriormost reach, the long axis of the ulna and radius are close to horizontal. The slight ventral turn of their distal ends would have directed the carpus slightly downward. The concave proximal surface of the radius articulates with the anterior part of the low hemispherical capitulum of the humerus. Because of the spiral nature of the ridge on its proximal articular surface, the dorsal end of the ulna rotates ventrally, its long axis making an increasingly large angle with the plane of the entepicondyle as the elbow is flexed. As in *Captorhinus*, this increases the stride slightly, although the total rotation is less due to the relatively inflexible nature of the joint. Fully extended, the long axis of the ulna is parallel to the plane of the entepicondyle and makes an angle of 90° with the long axis of the humerus. Fully flexed, it makes an angle of 15° with the entepicondyle (a total posterior rotation of 15°), and 84° with the long axis of the humerus, a total flexion of only 6° (Fig. 9). Clearly the posterior rotation is a more important component in elbow movement than simple flexion. This is so not only because of the spiral shape of the opposing articular surfaces, but also because of the oblique orientation of the distal articular surface relative to the long axis of the humerus. Both features cause medially directed forces generated by the biceps to produce a marked posterior rotation of the lower limb.

DISCUSSION

1. Limb Function

Despite the great difference in age between *Proterogyrinus* (Upper Mississippian) and *Archeria* (Lower Permian), the structure of their pectoral limbs is remarkably similar, testifying to the conservative nature of this aspect of the anatomy of embolomeres. However, the interpretations that have been made of limb function in the two animals are fundamentally different. Romer (1957) suggested that

Archeria was poorly adapted to terrestrial locomotion since the animal was obliged to hold the relatively inflexible forearm in a semi-extended position, and the nature of the glenoid did not allow sufficient rotation of the humerus for effective land progression. Rather, the fully extended limb could be turned back along the body, reducing drag and facilitating swimming. However, the articular surfaces are poorly preserved, and Romer was forced to allow a liberal amount of cartilage between bones in his reconstructions. These surfaces are better preserved in *Proterogyrinus*. They show that the lower limb could not have been extended to form an angle of greater that 90° with the long axis of the humerus. The permanently flexed limb must have been a hindrance during swimming in open water, but much more suitable for terrestrial locomotion. Jenkins (1971) pointed out that the screw-shaped glenoid of primitive tetrapods strictly limited the range of movement of the distal end of the laterally directed humerus to a tightly circumscribed ellipse. This precludes any possibility of sufficient humeral retraction in *Proterogyrinus* to allow the limb to be held against the trunk.

Although the pectoral limb of *Proterogyrinus* was not a very versatile terrestrial limb due to the severely restricted range of movement, the laterally directed humerus also must have offered considerable resistance to forward progression during swimming. This would certainly have put the animal at a competitive disadvantage to any fish and even the small-limbed *Greererpeton* while pursuing food in open water. The limb was clearly adapted for use in a terrestrial environment as well as the very shallow vegetation-choked margins of bodies of water where swimming was impractical.

2. *The Origin and Early Evolution of the Tetrapod Pectoral Limb*

The discovery of the Upper Devonian ichthyostegids suggests that the origin of tetrapods could not have occurred much after the late Middle Devonian. Assuming a common ancestry for tetrapods, this allows approximately 35 million years of tetrapod evolution before *Proterogyrinus* appears in the fossil record. The structure of the pectoral limb of *Ichthyostega* (as presently known), and particularly that of *Proterogyrinus*, indicates that well-developed limbs had evolved at this early point in tetrapod history and, assuming that these limbs indicate terrestrial habits, tends to dispel Romer's picture of a persistently aquatic habitat for primitive tetrapods that was not abandoned until long after the amphibian–reptilian transition (Romer, 1974).

Comparison between *Proterogyrinus* and the contemporary temnospondyl *Greererpeton* provides information on the early evolution of the labyrinthodont pectoral limb. The scapulocoracoid, although much smaller in *Greererpeton* appears to be essentially similar. A glenoid foramen is present in both groups, suggesting that primitively the labyrinthodont coracoid plate was pierced by two foramina, and that the presence of only one foramen on the coracoid plate of later temnospondyls is a derived condition.

Of considerable significance is the great similarity between the humerus of *Proterogyrinus* and that of *Greererpeton* (Fig. 7). The humerus of an immature *Proterogyrinus* is closely comparable to a similar-sized element of a mature *Greererpeton*. Both possess an anterior flange. This feature, also present in *Archeria*, was considered by Romer (1957) to be an aquatic specialization, although its function in aquatic locomotion was not discussed. The appearance of the flange on the humerus of two otherwise very differently adapted, primitive amphibians suggests that it was a primitive characteristic of labyrinthodonts that was retained in embolomerous anthracosaurs and lost in other later forms. There is a comparable degree of torsion. Both possess an entepicondylar foramen. This foramen is also present in some microsaurs (Carroll and Gaskill, 1978). As has been suggested by Romer (1947), the presence of an entepicondylar foramen appears to be primitive not only for labyrinthodonts, but for tetrapods in general.

Many characteristic differences in the limbs of the two groups have already been established by the Upper Mississippian. The massive interclavicle of *Greererpeton* is diamond-shaped, as in later temnospondyls, and lacks the parasternal process present in *Proterogyrinus* and other anthracosaurs. The dermal elements of *Greererpeton* are heavily sculptured in a pit and ridge pattern typical of temnospondyls that is distinct from the finer, punctate ornamentation seen in *Proterogyrinus* and other anthracosaurs. The digit count for the two groups (four in temnospondyls and five in anthracosaurs) appears to have already been established.

Unfortunately, little new can be said concerning the mode of evolution of the pectoral limb from that of rhipidistians. General comparisons of the limb of *Eusthenopteron* (Andrews and Westoll, 1970a) and *Sterropterygion* (Rackoff, 1976) with those of tetrapods reveal many similarities, and humeral features have been homologized with considerable success. Andrews and Westoll hypothesized that a tetrapod humerus could be derived from a suitable rhipidistian element by: (1) increased torsion (rotation of the plane of the entepicondyle to form a greater angle with the long axis of the proximal articular surface); (2) migration of the pectoral and deltoid processes to the anterior proximal surface of the humerus to form a deltopectoral crest; (3) development of a distinct shaft; and (4) anteroposterior expansion of the proximal head, creating a strap-shaped humeral surface that was considerably longer than the corresponding glenoid. It was originally hoped that the humerus of *Proterogyrinus* would show features transitional between rhipidistians and more advanced labyrinthodonts, but, other than possessing an anterior flange, it shows no specific similarity to that of rhipidistians and cannot be considered intermediate in structure. Some features of the pectoral limb of *Proterogyrinus*, such as the relatively small size of the scapular blade, wide clavicular stem and low degree of "torsion" of the humerus may be primitive (although the humerus of *Sterropterygion*, a rhipidistian, shows much more torsion (Rackoff, 1976, and Chapter 11 of this volume), but adds little ot our understanding of its transition from fish or its early evolution. Other features, such as the extended entepicondyle and unusual orientation of the lower limb relative to the humerus may be specializations peculiar to *Proterogyrinus* and its allies, and not generally representative of primitive tetrapods. Further finds in the early Mississippian and Devonian, representing an earlier stage in tetrapod evolution, and additional study of the ichthyostegid material, are necessary before the details of tetrapod limb evolution from fish can be worked out.

ACKNOWLEDGEMENTS

I am greatly indebted to Dr R.L. Carroll at whose suggestion and under whose guidance this work was undertaken. I would also like to thank Drs Dunkle and Williams of the Cleveland Museum of Natural History for not only allowing me to borrow specimens of *Proterogyrinus*, but also for the many kindnesses extended to me during my visits to that institution. Thanks are due to fellow student Donald

Brinkman for many enlightening discussions on limb functions. This work was supported in part by a National Science Foundation grant number DEB 72 – 02280 A02.

REFERENCES

Andrews, S. M. and Westoll, T. S. (1970a). The postcranial skeleton of *Eusthenopteron foordi* Whiteaves. *Trans. R. Soc. Edinb.* **68**, 207–327.

Andrews, S.M. and Westoll, T.S. (1970b). The postcranial skeleton of rhipidistian fishes excluding *Eusthenopteron. Trans. R. Soc. Edinb.* **68**, 391–489.

Beaumont, E.H. (1977). Cranial morphology of the Loxommatidae (Amphibia: Labyrinthodontia). *Phil. Trans. R. Soc.* **B280**, 29–101.

Boy, J.A. and Bandel, K. (1973). *Brukterepeton fiebigi* n. gen. n. sp. (Amphibia: Gephyrostegida) der erste Tetrapode aus dem rheinisch-westfälischen Karbon (Namur B; W-Deutschland). *Palaeontographica* **145A**, 39–77.

Busanas, J. W. (1974). "Paleontology and Paleoecology of the Mauch Chunk Group in Northwestern West Virginia." M.Sc. Thesis, Bowling Green State University.

Carroll, R.L. (1964). Early evolution of the dissorophid amphibians. *Bull. Mus. comp. Zool. Harv.* **131**, 163–250.

Carroll, R.L. (1970). The ancestry of reptiles. *Phil. Trans. R. Soc.* **B257**, 267–308.

Carroll, R.L. and Gaskill, P. (1978). The order Microsauria. *Am. phil. Soc. Mem.* **126**, 1–211.

Carroll, R. L., Belt, E. S., Dineley, D. L., Baird, D. and McGregor, D. C. (1972). Vertebrate paleontology of Eastern Canada. Guidebook, field excursion A 59. 24th *int. geol. Congr., Montreal, 1972* 1–113.

Heaton, M.J. (1978). "Cranial Soft Anatomy and Functional Morphology of a Primitive Captorhinid Reptile." PhD. Thesis, McGill University, Montreal.

Holmes, R.B. (1977). The osteology and musculature of the pectoral limb of small captorhinids. *J. Morph.* **152**, 101–140.

Holmes, R.B. and Carroll, R.L. (1977). A temnospondyl amphibian from the Mississippian of Scotland. *Bull. Mus. comp. Zool. Harv.*, **147**, 489–511.

Hotton, N. (1970). *Mauchchunkia bassa*, gen. et. sp. nov., an anthracosaur from the Upper Mississippian. *Kirtlandia* No. 12, 1–38.

Jarvik, E. (1952). On the fish-like tail in the ichthyostegid stegocephalians. *Meddr Grønland* **114**(12), 1–90.

Jenkins, F.A. (1971). The postcranial skeleton of African cynodonts. *Bull. Peabody Mus. nat. Hist.* **36**, 1–216.

Jenkins, F. A. (1973). The functional anatomy and evolution of the mammalian humero – ulnar articulation. *Amer. J. Anat.* **137**, 281–298.

Miner, R.W. (1925). The pectoral limbs of *Eryops* and other primitive tetrapods. *Am. Mus. nat. Hist. Bull.* **51**, 145–312.

Moss, J.L. (1972). The morphology and phylogenetic relationships of the Lower Permian tetrapod *Tseajaia campi* Vaughn (Amphibia: Seymouriamorpha). *Univ. Cal. Publ. geol. Sci.* **98**, 1–72.

Panchen, A.L. (1966). The axial skeleton of the labyrinthodont *Eogyrinus attheyi. J. Zool., Lond.* **150**, 199–222.

Panchen, A.L. (1970). Teil 5a: Anthracosauria. "Handbuch der Palaeoherpetologie" (O. Kuhn, ed.), 1-84. Fischer, Stuttgart.

Panchen, A.L. (1972a). The interrelationships of the earliest tetrapods. *In* "Studies in Vertebrate Evolution – Essays Presented to Dr. F.R. Parrington F.R.S." (K.A. Joysey and T.S. Kemp, eds), pp. 65–87. Oliver and Boyd, Edinburgh.

Panchen, A.L. (1972b). The skull and skeleton of *Eogyrinus attheyi* Watson (Amphibia: Labyrinthodontia) *Phil. Trans. R. Soc.* **B263**, 279–326.

Panchen, A.L. (1973). On *Crassigyrinus scoticus* Watson, a primitive amphibian from the Lower Carboniferous of Scotland. *Palaeontology* **16**, 179–193.

Panchen, A.L. (1975). A new genus and species of anthracosaur amphibian from the Lower Carboniferous of Scotland and the status of *Pholidogaster pisciformis* Huxley. *Phil. Trans. R. Soc.* **B269**, 581–640.

Rackoff, J.S. (1976). "The Osteology of *Sterropterygion* (Crossopterygii: Osteolepidae) and the Origin of Tetrapod Locomotion." Ph.D. Thesis, Yale University.

Romer, A.S. (1922). The locomotor apparatus of certain primitive and mammal-like reptiles. *Am. Mus. nat. Hist. Bull.* **46**, 517–606.

Romer, A.S. (1947). Review of the Labyrinthodontia. *Bull. Mus. comp. Zool. Harv.* **99**, 1–368.

Romer, A.S. (1957). The appendicular skeleton of the Permian embolomerous amphibian *Archeria. Contr. Mus. Pal. Univ. Michigan* **13**(5), 103–159.

Romer, A. S. (1966). "Vertebrate Paleontology", 3rd Edn. University Press, Chicago.

Romer, A. S. (1970). A new anthracosaur labyrinthodont *Proterogyrinus scheelei* Carboniferous. *Kirtlandia* No. 6, 1–20.

Romer, A.S. (1970). A new anthracosaur labyrinthodont *Proterogyrinus scheeli* from the Lower Carboniferous. *Kirtlandia* No. 10, 1–16.

Romer, A.S. (1974). Aquatic adaptation in reptiles – primary or secondary? *Ann. S. Afr. Mus.* **64**, 221–230.

Watson, D. M. S. (1917). A sketch classification of the pre-Jurassic tetrapod vertebrates. *Proc. zool. Soc. Lond.* 1917, 167–186.

White, T.E. (1939). Osteology of *Seymouria baylorensis* Broili. *Bull. Mus. comp. Zool. Harv.* **85**, 325–409.

15 | A Review of the Nectridea (Amphibia)

ANGELA C. MILNER

Department of Palaeontology, British Museum (Natural History), Cromwell Road, London SW7, England

Abstract: The Order Nectridea comprises a group of Carboniferous and Permian amphibians. They share several primitive tetrapod characters and only one autapomorphic character – a holospondylous centrum with fused neural and haemal arches bearing extra articulations above the zygapophyses between adjacent vertebrae in at least part of the vertebral column.

The Order Nectridea has been previously subdivided into two families: the Urocordylidae – characterized by kinetic skulls and retaining a supratemporal in the skull-roof, and the Keraterpetontidae – characterized by the possession of large posteriorly or posterolaterally directed horns which, in Carboniferous forms, are associated with an anteriorly expanded cleithrum. A functional explanation of this feature in relation to the mode of locomotion is proposed and the Carboniferous keraterpetontids are interpreted as actively swimming predatory animals.

Scincosaurus, generally included within the Keraterpetontidae, is confirmed as a hornless form and the family Scincosauridae is reinstated for this genus and (provisionally) for *Sauravus*. *Scincosaurus* is interpreted as a primarily terrestrial form which may have inhabited a "leaf-litter" environment and was a seasonal breeder in shallow swamp-lakes and ponds.

A phylogeny of the Nectridea is presented, based on a revision of the Carboniferous forms, and Beerbower's hypothesis of convergent evolution of the Permian long-horned keraterpetontids is disputed. The Order Nectridea shares some derived characters uniquely with the Aïstopoda. The diversity of the group at its earliest known occurrence (Westphalian A) suggests that it is geologically much older.

Systematics Association Special Volume No. 15, "The Terrestrial Environment and the Origin of Land Vertebrates", edited by A. L. Panchen, 1980, pp. 377–405, Academic Press, London and New York.

INTRODUCTION

The Nectridea comprises an order of superficially newt-like Carboniferous and Permian amphibians. The known animals were small, attaining a maximum length of about 500 cm, with short trunks and very long, usually laterally flattened tails making up to two-thirds of their total length. At their earliest known occurrence, in the Lower Coal Measures (Westphalian A), they are already a diverse and specialized group easily distinguishable from contemporary labyrinthodont Amphibia.

The most characteristic feature of the Nectridea is the vertebral construction. Each vertebra consists of a single holospondylous centrum with a fused neural arch. Holospondylous vertebrae are also found in the Palaeozoic orders Aïstopoda and Microsauria and, on the basis of their vertebral structure, all three orders were grouped together in the Subclass Lepospondyli by Zittel (1888). However, the vertebrae of nectrideans are distinctive in that each neural arch bears an expanded, usually fan-shaped neural spine, grooved and crenellated along its distal border. In the caudal vertebrae these are opposed ventrally by expanded haemal spines of similar shape. The vertebrae bear one or more pairs of accessory articulations above the zygapophyses of adjacent vertebrae in at least part of the vertebral column.

In recent classifications (e.g. Romer, 1966), the Order Nectridea Miall 1875 is divided into three families – Urocordylidae Lydekker 1889, Keraterpetontidae (Jaekel, 1903) and Lepterpetontidae Romer 1945. Two of these families are immediately distinguishable on the basis of skull morphology. Keraterpetontids are characterized primarily by the possession of horn-like extensions of the tabulars projecting back from the skull-table. This feature is most extremely developed in the well-known North American Permian genera *Diplocaulus* and *Diploceraspis*. Urocordylids have a relatively high-sided arrow-shaped kinetic skull. The third family, the Lepterpetontidae, contains the single species, *Lepterpeton dobssii* Wright and Huxley 1866 which is, in fact, a poorly preserved urocordylid. The diversification within the Nectridea at its earliest occurrence, into two distinct skull morphologies, suggests that it is geologically much older than a literal interpretation of the fossil record would suggest.

The Coal Measure Nectridea of both families have been almost completely neglected since their original descriptions in the last century, apart from brief consideration by Steen (1931, 1938) in her revisions of the Coal Measure tetrapod assemblages from Linton, Ohio, and Nýřany, Czechoslovakia, respectively. The later long-horned keraterpetontids have been studied more recently. The systematics and relative growth of *Diplocaulus* were reviewed in a series of papers by Olson (1951, 1952, 1956); the phylogeny of the Keraterpetontidae was discussed by Beerbower (1963) in his thorough study of the Permo-Carboniferous *Diploceraspis*. The only new nectridean taxon to have been described since the widely adopted classification of Romer (1966) is *Peronedon*, first described from the Lower Permian of Oklahoma by Olson (1970) and later revised by Haglund (1977) on the basis of further material from Oklahoma and Texas. The Coal Measure Nectridea have been the subject of two very recent unpublished studies, namely the Urocordylidae by Bossy (1976) and the Keraterpetontidae by Milner (1978). The object of this review is to present a brief summary of some aspects of recent work on Carboniferous nectrideans. Detailed studies of the comparative anatomy, functional morphology and evolutionary relationships of the keraterpetontids are to be published elsewhere by the author.

A complete list of the taxa comprising the Nectridea, together with site records and stratigraphical distribution is given in the Appendix at the end of this chapter.

FUNCTIONAL MORPHOLOGY

1. Skull Structure

The pattern of dermal skull-roofing bones includes a tabular–parietal contact, a condition which also occurs in microsaurs (Carroll and Gaskill, 1978) and in most batrachosaurs (Panchen, Chapter 13 of this volume), and lacks an intertemporal in the skull-table in common with microsaurs. No definite otic notch is present and no stapes has been described in any nectridean. There is, therefore, no basis for determining the position of a tympanum, if indeed one was present.

Most nectrideans, as will be discussed later in this review, are considered to be secondarily aquatic forms. Within other amphibian orders, secondarily aquatic forms such as the trimerorhachids and brachyopids (Temnospondyli) and the pipids (Anura) have lost the

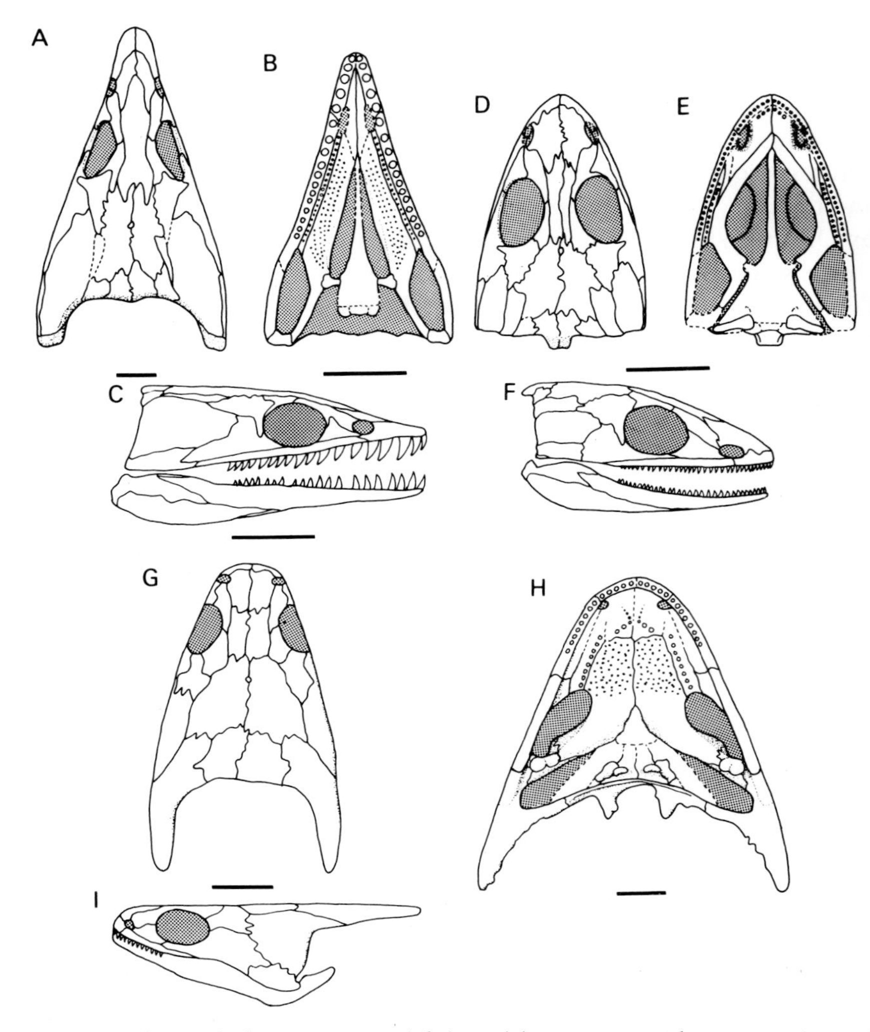

Fig. 1. Skull morphology in urocordylids and keraterpetontids. Restorations of (A) *Sauropleura scalaris* adult, dorsal aspect; (B and C) *Sauropleura* adult, dorsal, palatal and lateral aspects; (G) *Keraterpeton galvani* adult, dorsal aspect; (H) *Batrachiderpeton reticulatum* adult, palate; (I) *Keraterpeton galvani* adult, lateral aspect. (A – F) Modified after Bossy (1976); (G – I) original. Scale bar is 5 mm on all figures.

tympanum and it seems plausible that nectrideans lacked this structure.

The form and function of the keraterpetontid and urocordylid skulls are immediately distinguishable and follow widely differing adaptive trends. The keraterpetontid skull (Fig. 1G) is short-snouted with a broad akinetic skull-table sutured firmly to the cheek region along an undulating line ensuring no plane of weakness. Both supratemporal and intertemporal are absent and a broad parietal flange sutures with the squamosal. The cheek region is primitively quite deep (Fig. 1), but in later Permian forms the skull is flattened. The most characteristic feature of keraterpetontids is the long posteriorly directed horn-like extensions of the tabulars – the so-called tabular horns. Primitively the tabular horns are parallel or slightly divergent. In the later, post-Westphalian "long-horned" genera, they have developed into massive posterolateral extensions of the skull (Fig. 2).

Known urocordylid skulls are, with one exception, elongate and arrow-shaped (Fig. 1A–F). The skull is relatively deep-sided with a narrow kinetic skull-table retaining a supratemporal in all genera. The jaw articulation lies on a level with, or is posterior to, the occiput, in

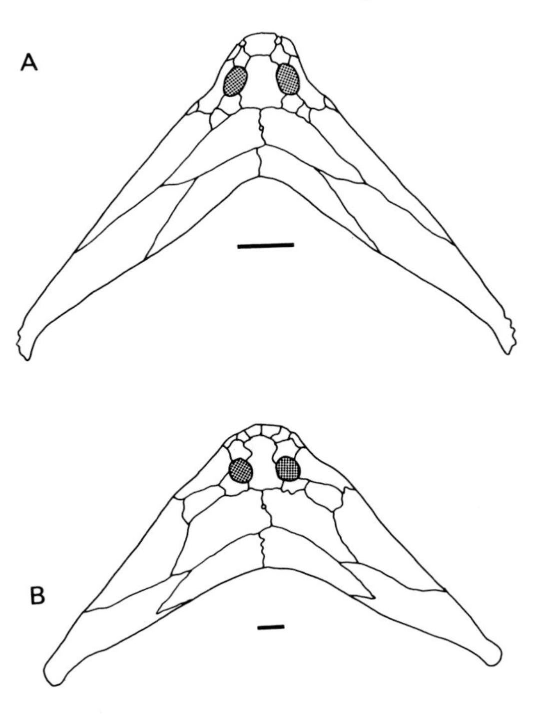

Fig. 2. "Long–horned" nectridean skulls. (A) *Diploceraspis burkei;* (B) *Diplocaulus magnicornis*. (A) After Beerbower (1963); (B) after Olson (1951). Scale bars are 2 cm.

contrast to keraterpetontids in which the jaw articulation varies from being level with the occiput in the primitive genera to far anterior of the occiput in the later genera.

The palate is primitively closed in the early keraterpetontids (Fig. 1H) with large denticle-covered palatine rami of the pterygoids suturing in the mid-line. The quadrate is braced by a broadly rounded internal flange of the squamosal which descends ventrally onto the palate. In later keraterpetontids and in urocordylids, palatal vacuities are present. A well-developed basal articulation is found in all urocordylids, but in *Keraterpeton* and *Batrachiderpeton* mobility of the palate was very restricted and in the late Westphalian *Diceratosaurus* and later genera the palate is firmly sutured to the parasphenoid.

The keraterpetontid mandible has an extremely well-developed retroarticular process and, in *Keraterpeton* and *Batrachiderpeton*, the coronoid is packed with a battery of teeth only slightly smaller than the marginal ones. The marginal teeth in both upper and lower jaws are conical and are angled slightly outwards with incurved tips. Urocordylids have a very slender mandible with a moderately developed retroarticular process. The dentition consists of widely spaced, long, conical teeth in *Sauropleura*, the shorter-skulled *Ptyonius* having a much greater number of small teeth in the marginal rows.

2. *Feeding Adaptations*

(a) Keraterpetontids. The inferred jaw mechanics incorporate a rapid opening mechanism, depression of the mandible being effected by contraction of a powerful depressor mandibulae muscle. The depressor was apparently well developed with an extensive insertion on the ventral surface of the mandible and on the retroarticular process, and a prominent tendon origin preserved as a double pit on the ventral side of the base of the tabular horn. The firmly braced quadrate is adapted to withstand any backwardly directed stress induced by the somewhat oblique placement of the muscle suggested by the angle of the retroarticular process. Any stress on the base of the horn, brought about by rapid depressor contraction, which might tend to tilt the skull would be dampened by the connection of the tabular horn to the cleithrum (discussed under Locomotor

Adaptations). The main adductor musculature was orientated near-vertically in the deep posterior region of the short subtemporal fossa immediately anterior to the jaw hinge, suggesting the possibility of development of occlusal pressure when the jaws were partially closed. The marginal dentition, composed of interlocking teeth angled slightly outward and with incurved tips, is adapted for gripping prey rather than for impaling; the batteries of coronoid teeth oppose part of the pterygoid battery from which it may be inferred that the normal mode of feeding incorporated a crushing action. It is suggested that Carboniferous keraterpetontids fed on invertebrates such as crustaceans and insect nymphs and that the development of occlusal pressure aided retention of prey in the mouth and facilitated the crushing action necessary to break up cuticles and carapaces. The powerful depressor action is suggested as a mechanism necessary to effect rapid and positive opening of the jaws while swimming, to facilitate the capture of active prey. In summary, the keraterpetontid jaw mechanism tends towards the "static pressure" type as characterized by Olson (1961).

(b) Urocordylids. The dentition and jaw mechanics of most urocordylids shows adaptation to snapping at and impaling relatively large active prey, and may be characterized as a specialized "kinetic inertial" system paralleling, in part, the anthracosaur system as described by Panchen (1964, 1970).

The long-snouted *Sauropleura*-type skull was particularly mobile. The palate was hinged anteriorly at a pterygoid-vomer contact corresponding to an external hinge point, namely a lachrymal-prefrontal contact (Bossy, 1976). This anterior hinge allowed much greater mobility of the palate against the braincase and therefore movement of the cheek against the skull-table than occurred in anthracosaurs. On contraction of the depressor mandibulae muscles, movement along the kinetic line between the skull-table and cheek was transmitted anteriorly to the mobile lachrymal which moved vertically downwards on the prefrontal shelf. As a result of this movement, the snout tip anterior to the naris would have been elevated. The frontals have overlapping and sliding junctions with the parietals and nasals which may have served to accommodate changes in the skull-line as the snout was raised. This "dorsal snout shift", as it has been termed by Bossy (1976), allowed reorientation of the tips of the long, recurved premaxillary teeth to a position suitable for impaling prey. On closure of the jaws the tips were reorientated posteriorly as the snout was lowered, preventing the escape of prey. The range of vertical movement of the snout tip is determined by the length of the upper jaw anterior to the lachrymal, and a tendency to elongation of the prelachrymal region of the skull is apparent in the *Sauropleura*-like forms.

The feeding system appears to have been different in *Ptyonius*. In this broad-skulled, short-snouted form, the cranial kinesis was restricted to the supratemporal-squamosal line allowing limited flaring of the cheek (Bossy, 1976). There is no evidence for a dorsal snout shift and the teeth are small and not recurved. The subtemporal fossa is short as in the keraterpetontids and Bossy has suggested that *Ptyonius* tended towards a "static pressure" system of jaw mechanics.

3. Locomotor Adaptations

(a) Swimming. Both the keraterpetontids and the urocordylids were principally if not exclusively aquatic, and Parrington (1967) described the locomotor pattern of the Carboniferous forms as sub-anguilliform, the animals using their long, deep tails for swimming with the short trunk held more or less rigid. They were adapted for anguilliform swimming by polyisomerism within the tail.

In urocordylids and early keraterpetontids, swimming was undoubtedly by sinusoidal flexure of the body and tail using the axial musculature, with the limbs having only a supplementary role. The nectridean vertebral column was uniquely adapted for the lateral bending movements of the trunk and tail associated with anguilliform locomotion. Parrington (1967) demonstrated, using mechanical models, that an arrangement of simple spool centra is the most effective structure for resisting small angles of torsion. This function is enhanced where successive neural arches are articulated by zygapophyses, the facets of which are roughly radial to the axis of torsion. Nectridean trunk vertebrae had almost horizontal zygapophyses which must have minimized dorso-ventral flexure and precluded torsion about the longitudinal axis of the vertebral column. The presence of extra articulations would also have helped to prevent torsion and minimize the possibility of dislocations between adjacent vertebrae. These features, together with the tall neural

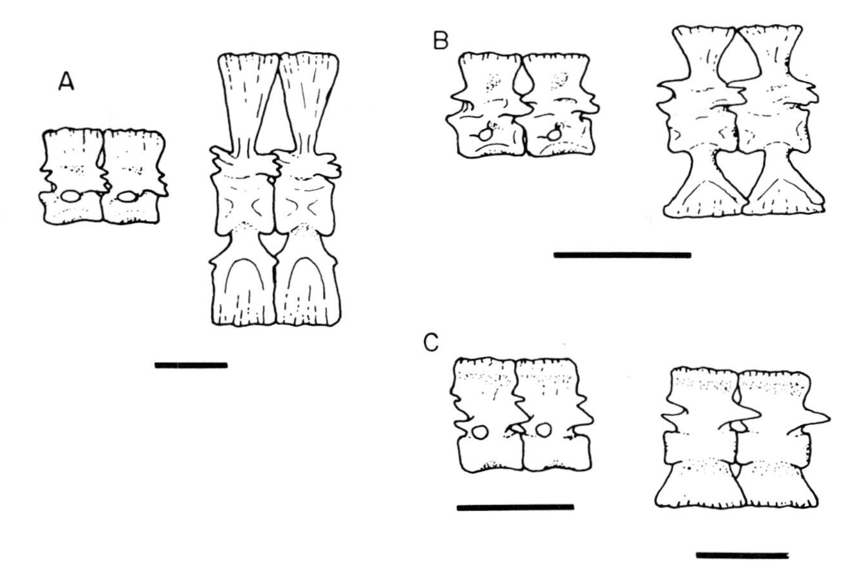

Fig. 3. Vertebral form in urocordylids and keraterpetontids. Pairs of anterior presacrals and caudals of (A) *Urocordylus wandesfordii*; (B) *Sauropleura scalaris*; (C) *Keraterpeton galvani*. (A and B) Modified after Bossy (1976); (C) original. Scale bar is 5 mm on all figures.

spines providing a wide surface area for insertion of the propulsive epaxial trunk musculature, suggest that the trunk, although short, may have been involved in the initiation of the propulsive wave rather than remaining rigid as Parrington suggested. The construction of the caudal vertebrae allowed for progressively greater flexibility in the tail consistent with increasing amplitude and length of the sine wave generated during anguilliform swimming, and the expanded neural and haemal spines confer a deep, vertically-flattened tail profile.

The numbers and morphology of trunk and tail vertebrae vary within the group (Fig. 3); the number of pairs of extra articulations above the zygapophyses in the trunk and tail, the shape and size of the neural and haemal spines, and the presence and position of haemal articulations all suggest slight variation in the locomotor patterns in nectrideans. The *Sauropleura*-type urocordylids seem to have been capable of a considerable degree of lateral flexibility, conferred by a long (for nectrideans) trunk of up to 26 presacral vertebrae, a single pair of extra articulations above the zygapophyses

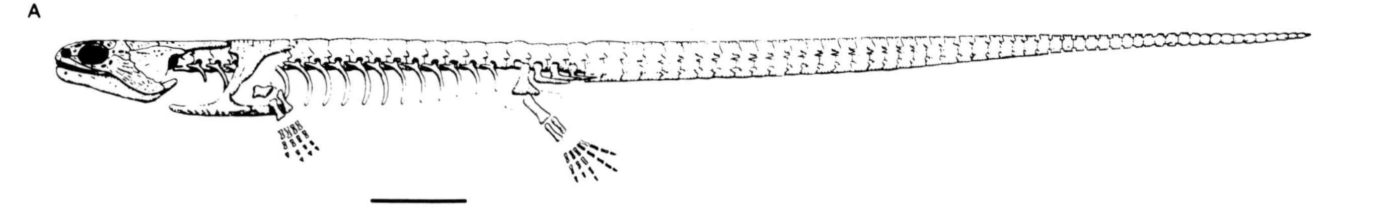
A

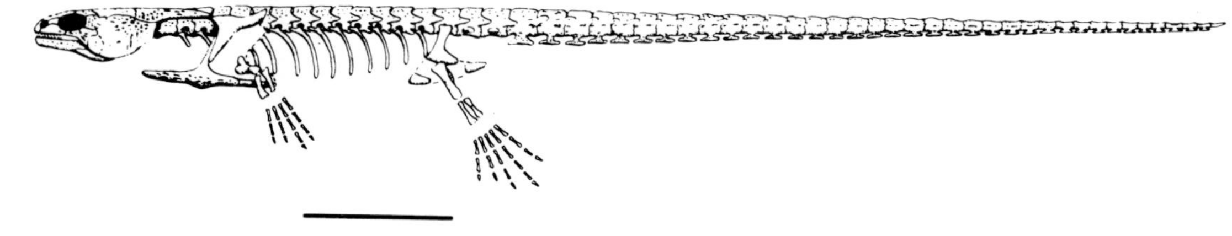
B

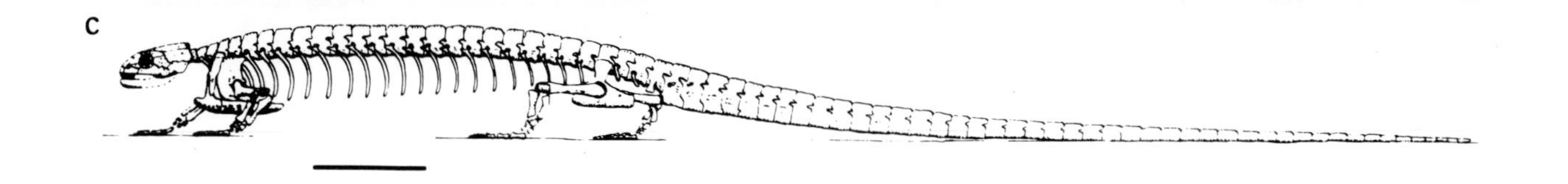

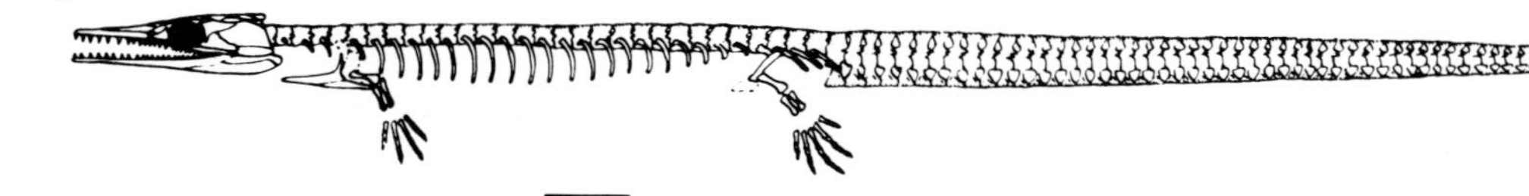

Fig. 4. Composite reconstructions of nectridean skeletons. (A) *Keraterpeton galvani*; (B) *Diceratosaurus brevirostris*; (C) *Scincosaurus crassus*; (D) *Sauropleura pectinata*. All original. Scale bar is 2cm on all figures.

and a tail tapering gradually in height throughout its length with haemal articulations distal to the centrum (Fig. 4D). *Urocordylus* and *Ptyonius*, with shorter trunks (a maximum of 22 presacrals), an increased number of extra neural and haemal articulations, and a very deep tail of constant depth for two-thirds of its length and then diminishing rapidly, were less flexible but efficient swimmers (Bossy, 1976). Bossy suggests that they may have had lower initial acceleration than the *Sauropleura*-type urocordylids but were adapted for sustained powerful swimming. The keraterpetontids were all shorter bodied, with a maximum of 17 presacral vertebrae, and the tail was less deep and relatively shorter than that of urocordylids. No haemal articulations were present but the proximal haemal spines overlapped in a set sequence (Fig. 4A) to facilitate lateral bending. They are envisaged as having been flexible swimmers like *Sauropleura* but, being comparatively less elongate, able to generate fewer undulations in the body at any given moment.

(b) The skull - pectoral girdle connection in keraterpetontids. The general characteristics of anguilliform swimming are such that the head and body oscillate relative to the longitudinal axis of motion, a consequence of the moments induced by changing lateral forces (Gans, 1974). The degree of oscillation, for a given rate of propulsion, is inversely proportional to the number of undulations in the body at any one instant. As it is unlikely that *Keraterpeton* (or *Batrachiderpeton*) was sufficiently elongate to generate more than two or three undulations at any one time, the head may have been subject to marked oscillations, particularly during fast swimming.

Lateral oscillation of the head appears to be sufficiently disadvantageous to efficient lateral-line system function and to the approach to and capture of prey that, in fishes, many compensatory devices have evolved to dampen this oscillation of the head and thus permit a straight-line approach to prey. In osteichthyan fishes, the dermal pectoral girdle contributes to the stiffening of the anterior trunk. Among other functions, it mediates lateral swinging of the head relative to the body through the articulated system of bones linking the back of the skull to the pectoral girdle. In some ostariophysine fishes, the post-temporal and/or the supracleithrum becomes rigidly attached to the skull, precluding movement (Gosline, 1977). In swimming newts the head oscillates slightly, but they do not hunt while actively swimming and so dampening for prey approach is not a necessity.

An analogous adaptation to the skull–pectoral girdle linkage of fishes is apparent in *Keraterpeton*. The posterior end of the tabular horn lies just in front of the anterior arm of the "T"-shaped cleithrum (Fig. 4A). Ridged ornamented surfaces particularly on the anterior arm of the cleithrum suggest a muscular or, more plausibly, a ligamentous connection between it and the tabular horn. Such a system would effectively keep lateral movements of the head relative to the body within controlled limits thus allowing a straight-line approach to prey during active swimming. *Batrachiderpeton* shows evidence of an identical adaptation. The connection of the tabular horn to the dermal pectoral girdle in these animals, as opposed to other contemporary presumed active swimmers (the anthracosaurs and urocordylid nectrideans), is associated with the possession of a broad akinetic skull-roof. Dampening of the oscillation about the skull – atlas fulcrum can be achieved with minimal energy expenditure by maximizing the distance between the fulcrum and the points of leverage of the controlling muscles and ligaments. A broad, flat, rigid skull lends itself to the development of outgrowths serving as levers, whereas the deep, narrow, kinetic skulls of anthracosaurs and urocordylids do not. Such a system inhibits movement of the skull so much that it would be of little value in association with a kinetic system of feeding.

The tabular horn—cleithrum connection is maintained in *Diceratosaurus* which has only 14 presacral vertebrae, the shortest known presacral column of any Palaeozoic amphibian (Fig. 4B). This genus appears to have been less well adapted to anguilliform locomotion; the neural and haemal spines are less expanded and consequently the tail is more rounded in cross-section. The pelvic girdle is comparatively well ossified indicating a greater supportive or propulsive function for the hind limb than in *Keraterpeton*. The femur is distinctly longer than that of *Keraterpeton* and bears a well-developed adductor ridge, the distal limb elements are comparatively short, and the pes is large and lacks an ossified tarsus. The structure of the hind limb suggests that it was employed as the main propulsive organ in swimming. The length of the femur relative to the tibia and fibula is indicative of a powerful rather than a fast power stroke. The non-ossification of the tarsus suggests that it did not have a load-bearing function, again more consistent with swimming. *Diceratosaurus* appears to have swum by paddling with the hind limbs while maintaining the head and trunk in a more or less rigid configuration, an analogous method of locomotion to that of modern frogs which have even shorter bodies. The tabular horn – cleithrum connection is presumed to have functioned in the same manner as inferred for *Keraterpeton*. In *Diceratosaurus*, rigidity of the anterior trunk is ensured by the large, flat, dermal girdle plates (including an extra pair of dermal ossifications ventral to the coracoids), together with the connection of the tabular horn to the cleithrum. Maintenance of a rigid trunk plus head is important in limb-paddling, giving the limbs a rigid structure to push through the water without energy being wasted in lateral undulation.

The connection between the tabular horn and the cleithrum is lost in the later long-horned keraterpetontids which show adaptations implying a completely different locomotor pattern. A. R. I. Cruikshank (personal communication) ascribes a hydrodynamic function to the boomerang-shaped skulls of *Diplocaulus* and *Diploceraspis*.

SCINCOSAURUS – AN APPARENTLY TERRESTRIAL NECTRIDEAN

Scincosaurus has been included in the family Keraterpetontidae by Romer (1945, 1966) and subsequent authors, although Andrews (1895), Woodward (1897) and Jaekel (1903) all observed that *Scincosaurus* was a "hornless' form which was distinct from the genera of horned nectrideans. In a later paper on the classification of tetrapods, Jaekel (1909) proposed the family Scincosauridae, presumably for *Scincosaurus*, although his classification did not list taxa below family level. This family name was not adopted by subsequent authors, but recent work (Milner, 1978) on *Scincosaurus* indicates that it shares few of the derived characters which characterize the Keraterpetontidae and is sufficiently distinct in its structure to merit removal from this family. It is proposed that the family name Scincosauridae be reinstated for *Scincosaurus* and provisionally for the similar *Sauravus*, a genus based on two specimens from the Stephano-Permian of France.

The overall proportions of *Scincosaurus* are different from those of other nectrideans. The skull is very small, between 0·16 and 0·2 of the trunk length compared to a ratio of 0·3 to 0·4 in keraterpetontids

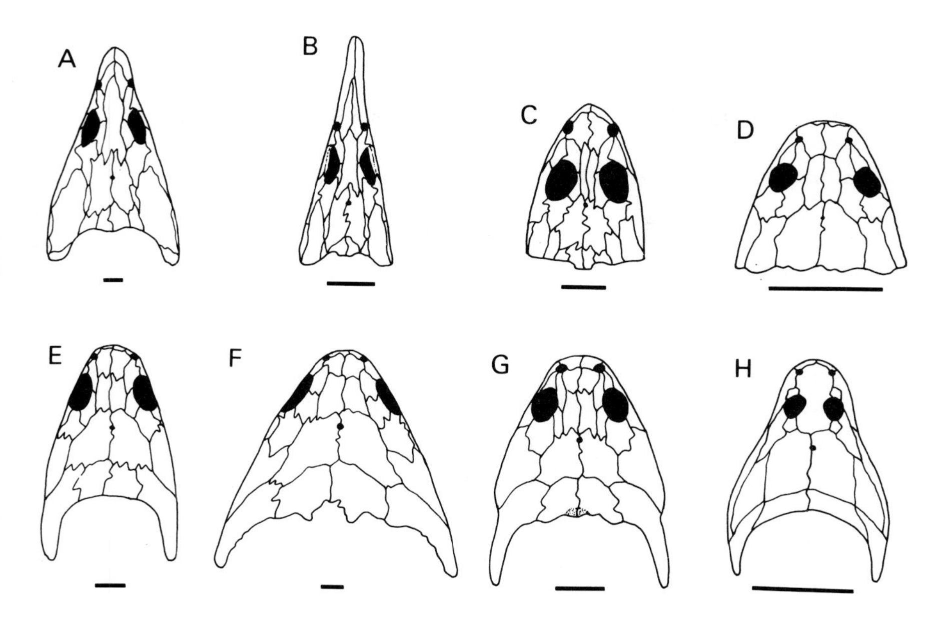

Fig. 5. The diversity of nectridean skulls. (A) *Sauropleura scalaris;* (B) *Sauropleura pectinata*; (C) *Ptyonius marshii*; (D) *Scincosaurus crassus*; (E) *Keraterpeton galvani*; (F) *Batrachiderpeton reticulatum*; (G) *Diceratosaurus brevirostris*; (H) *Peronedon primus*. (A – C) After Bossy (1976); (D – G) original; (H) after Haglund (1977). Scale bar is 5 mm on all figures.

and urocordylids. The tail length is about twice that of the trunk as in most Carboniferous nectrideans.

The ornamentation of the dermal bones of the skull and pectoral girdle is characteristic and consists of densely distributed, deep, circular pitting. The arrangement of the dermal roofing-bones of the skull (Fig. 5) is similar to that in keraterpetontids in that both the supratemporals and intertemporals are absent, but in addition, *Scincosaurus* also lacks postparietals. The posterior margin of the skull-table is bordered by the fluted unornamented edges of the parietals. There is no horn-like process on the tabular although Fritsch (1881, Taf.28) depicted such a structure. The source of this error lies in the preservation of some specimens in a configuration whereby the tabular region of the skull overlays the dorsal end of the detached cleithrum which is unornamented and lacks an expanded head. The shaft of the cleithrum, projecting from the back of the skull, was interpreted by Fritsch as the tabular horn. A sclerotic ring is retained in the orbit as in the urocordylids. No keraterpetontid appears to possess a sclerotic ring and this absence appears to be a derived condition for that family. The palate of *Scincosaurus* shows bracing of the quadrate by an internal squamosal shelf, almost the only derived character shared with the Keraterpetontidae. The marginal tooth row of *Scincosaurus* is very short and the animal apparently had a small gape. The dentition consists of small pedicellate teeth, similar in structure to the monocuspid teeth of some dissorophoid temnospondyls (Bolt, 1977) but without a line of abscission. The small gape and dentition suggest that the animal preyed upon small soft-bodied invertebrates for which a gripping dentition is most suitable for manipulation within the mouth.

The postcranial skeleton is unusually robust and heavily ossified for such a small animal, the largest specimen being 14·5 cm long, over half of which is tail. The appendicular skeleton is typical of terrestrial forms and shows a number of terrestrial adaptations which appear to be primitive tetrapod features relative to the condition in keraterpetontids and some urocordylids. The carpals and tarsals are ossified in *Scincosaurus*, uniquely among the Nectridea, and the pelvis is fully co-ossified. This indicates that these structures are all load-bearing in *Scincosaurus* in contrast to the condition in other nectrideans. The humerus is a robust bone with broad distal and proximal heads set at 90° to one another, and retaining an entepicondylar foramen. The humeri of other nectrideans appear to be simpler in construction. The interclavicle is expanded and heavily ornamented, and may have functioned as a protective shield for the trunk, together with the ventral scales which are arranged in a chevron pattern between the posterior edge of the interclavicle and the pelvic girdle. The trunk vertebrae have large, flat zygapophyses and no extra articulations. The caudal vertebrae bear short, broad, dorso-ventrally flattened ribs giving the tail a relatively rounded cross-section. The vertebrae are locked together by neural and haemal articulations throughout the length of the tail, which must have been relatively inflexible in contrast to the flexible laterally flattened tails of the other nectrideans.

Scincosaurus appears to have been adapted to a terrestrial existence, perhaps living in the equivalent of a "leaf-litter" layer where the heavily ossified pectoral girdle and stiff tail may have functioned in enabling the animal to push or lever its way under litter, stones or logs. Its short limbs and small feet indicate that it was not an active runner and unlikely to have been a surface dweller.

The only evidence for aquatic adaptation in *Scincosaurus* is the presence in some skulls of a lateral-line system represented by a supraorbital sulcus only. This does not conflict with the overall terrestrial adaptation of the skeleton, but indicates that the animal may have been facultatively or seasonally aquatic, as are the primarily terrestrial adults of many living amphibians.

The known specimens of *Scincosaurus*, at least 66 in number, are from the single geographically restricted locality of the Humboldt Mine at Nýřany. They fall within a restricted size range of 12·0 – 14·5 cm total length with the exception of one small specimen 2·5 cm long. This size distribution is unique at Nýřany (A. R. Milner, Chapter 17 of this volume) and represents a single age-class of animals preserved in a single event or more than one similar events. They appear to represent a population(s) of terrestrial or semi-terrestrial adults which had returned to a shallow swamp-pool to breed and had suffered a mass-mortality possibly caused by extreme anoxic conditions developing temporarily in the water body (A. R. Milner, Chapter 17 of this volume).

RELATIONSHIPS

1. Relationships within the Nectridea

Nectrideans are defined principally on vertebral form, and recent work has shown few other characters which define the group. Other shared characters suggest that the primitive nectridean condition, in most respects, reflects what is believed to be the primitive tetrapod condition.

(a) The Order Nectridea. Derived characters defining the Order Nectridea are as follows. They are shown at node 1 on Fig. 6 (this and all other nodes refer to Fig. 6):

(1) One or more extra articulations above the zygapophyses of adjacent vertebrae in at least part of the trunk and tail.

(2) The haemal spine is fused to the centrum in the midline opposite the neural spine.
(3) Neural and haemal spines always expanded in at least part of the vertebral column.
(4) Single coronoid in the mandible where condition is known.

Other derived characters exist which are shared with few non-nectrideans, but these will be discussed in the section on relationships of the group. In order to clarify the nature of the derived features which characterize the families and lower taxa within the Nectridea, a summary is given here of the characters which are believed to represent the plesiomorphic condition for the Nectridea. These are based on the most primitive known characters scattered across the nectridean taxa and many are believed to be primitive tetrapod characters.

(1) Short snout, lachrymal contacts naris and orbit.
(2) Paired frontals.
(3) Intertemporal lacking, postorbital contacts parietal.
(4) Tabular–parietal contact.
(5) Kinetic skull, cheek movable with respect to skull-table.
(6) Large opisthotic forming a paroccipital process to the tabular.
(7) Sclerotic ossifications present.
(8) No otic notch and no stapes known.
(9) No palatal fangs.
(10) Retroarticular process present on mandible.
(11) Double-headed ribs.
(12) Humerus with an entepicondylar foramen.
(13) Wheat or oat-shaped ventral scales arranged in chevrons.

(b) Relationships of nectridean families. As noted in the Introduction and itemized in the Appendix, the Nectridea are here divided into three families, each with a distinctive morphology. Each of these families is believed to be monophyletic and can be defined on unique (apomorphic) characters as follows:

Urocordylidae (node 2)

(1) Elongate neural and haemal spines with crenellated edges.
(2) Long posterodorsal iliac blade.

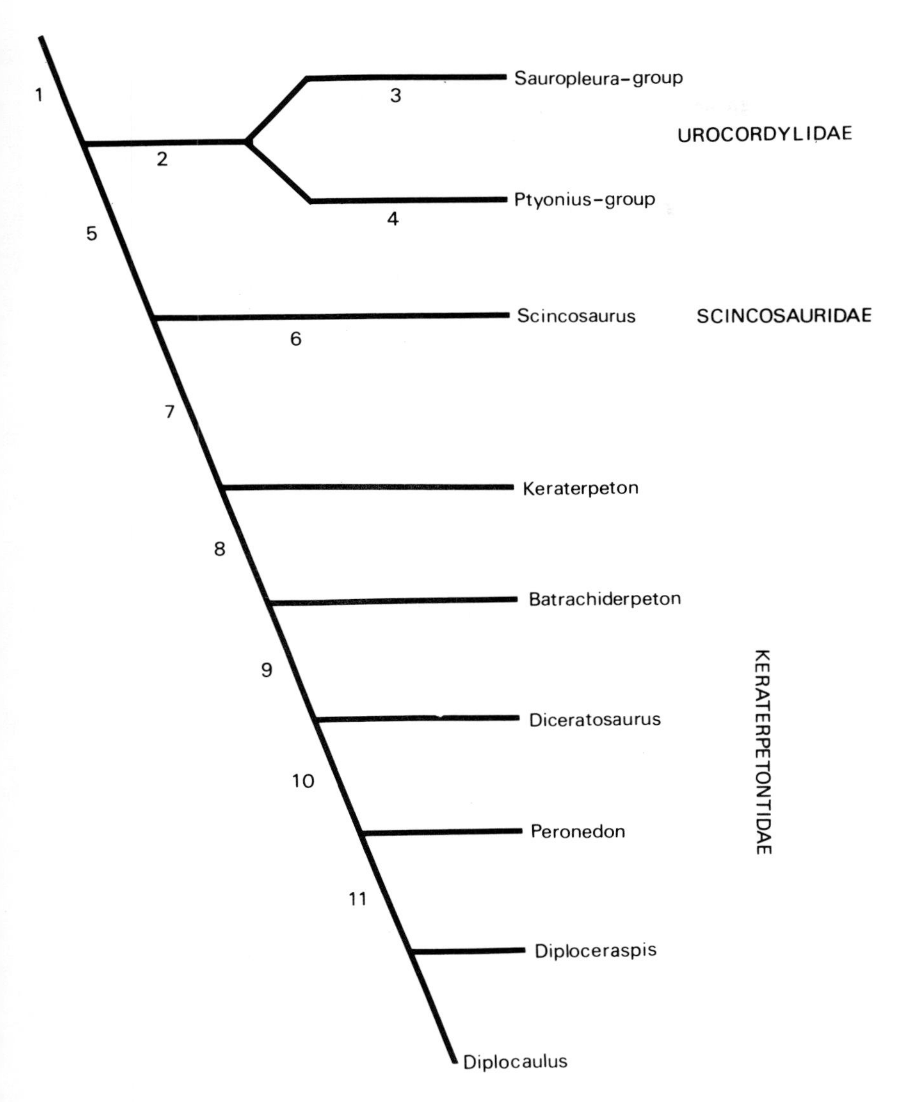

Fig. 6. The relationships of nectrideans depicted in cladogram form.

(3) Characteristic shaped lachrymal with inferred kinetic sliding function (Bossy, 1976).
(4) Kinetic joint between postorbital and jugal where known.

Scincosauridae (node 6)

(1) Loss of postparietals, skull-table bordered by fluted parietals.
(2) Enlarged tabular.
(3) Spatulate teeth.
(4) Neural and haemal spines abut throughout their depth along the whole length of the vertebral column.

Keraterpetontidae (node 7)

(1) Enlarged tabular horns.
(2) Loss of sclerotic ossifications.
(3) Double occipital condyles.
(4) Short presacral column, 17 vertebrae or less.
(5) Cleithrum with "T"-shaped expansion of the head where known.
(6) No extra articulations between adjacent haemal spines.

The following derived characters are shared by two out of the three families, and hence are those which can be used to form a basis for considering relationships between them. The Keraterpetontidae and Scincosauridae share the following:

(1) Akinetic skull.
(2) Loss of supratemporal.
(3) Internal shelf of squamosal braces quadrate.
(4) Short mandible and tooth row.

Keraterpetontidae and Urocordylidae share the following:

(1) Loss of entepicondylar foramen.
(2) Non-ossification of carpals, tarsals and incomplete ossification of pelvis.

The Urocordylidae and Scincosauridae share no derived characters which the keraterpetontids lack. This distribution of shared characters suggests that the Scincosauridae is more closely related to the Keraterpetontidae than either family is to the Urocordylidae (node 5). Characters shared between the Keraterpetontidae and the Urocordylidae are fewer and can be interpreted as simple reductions relating to an aquatic mode of life, trends which also occur in other groups.

(c) Relationships within the Urocordylidae. Within the Urocordylidae there appear to be two distinct morphological types, which have been studied by Bossy (1976). The two groupings are apparently

definable on derived characters; they are not named here but are described by reference to the best preserved genus of each type. The *Sauropleura*-group is taken to include, in addition to *Sauropleura*, *Lepterpeton* and *Crossotelos*; and the *Ptyonius*-group is taken to include, in addition to *Ptyonius*, *Urocordylus* and *Ctenerpeton*.

The *Sauropleura*-group can be defined by the following characters (node 3):

(1) Single frontal bone where known.
(2) Tendency to snout elongation, principally of the premaxillae, the nasals – extending between the premaxillae well anterior to the nares and the vomers.
(3) Tendency to increasingly posterior position of the jaw articulation relative to the occiput.
(4) Enlarged premaxillary teeth.
(5) Haemal spines articulate at their distal tips.

The *Ptyonius*-group can be defined by the following characters (node 4); however *Ctenerpeton* is known only from the posterior trunk and tail and *Urocordylus* is known only from very poorly preserved material:

(1) Increase in the number of extra articulations between neural arches.
(2) Haemal spine articulation(s) close to centrum.
(3) Increase in length of neural and haemal spines.
(4) Tail height constant for two-thirds of the length of the tail and then diminishes rapidly.

(d) Relationships within the Scincosauridae. There are only two genera attributable to this family, *Scincosaurus* and *Sauravus*. The latter is represented by only two specimens which require detailed study before discussion of the relationships within the family can be attempted.

(e) Relationships within the Keraterpetontidae. A phylogeny of the Keraterpetontidae was presented by Beerbower (1963), based on first-hand work on *Diploceraspis*, and he concluded that the two "long-horned" genera had evolved by convergence. Beerbower's conclusions are discussed after the phylogeny of the Keraterpetontidae which is proposed here.

Keraterpeton is the earliest known and most primitive member of the family. Characters which define this genus are those which define the primitive family condition (node 7). It has no specialized characters which debar it from ancestry of the other genera within the family.

Batrachiderpeton and *Diceratosaurus* share the following characters which neither share with *Keraterpeton* (node 8):

(1) Maxilla borders the orbit margin separating lachrymal and jugal.

(2) Emargination of the medial region of the postparietals.

This latter character is very pronounced in both genera, although the shape of the posterior margin of the skull-table in *Batrachiderpeton* is masked by the presence of postparietal horns which is an apomorphic character for the genus.

Diceratosaurus shares a complex of characters with the post-Westphalian forms that it does not share with *Batrachiderpeton* (node 9). Some of these characters are:

(1) Interpterygoid vacuities.
(2) Parasphenoid with a cultriform process.
(3) Squamosal enlarged.
(4) Loss of palatal denticles and teeth on posterior region of vomer and reduction or absence of coronoid teeth.
(5) Vertebral centra bear striate ornament.
(6) Neural arches of caudal vertebrae never waisted.
(7) All haemal arches very waisted.

Thus *Diceratosaurus* appears to be more closely related to the post-Westphalian keraterpetontids than to either *Keraterpeton* or *Batrachiderpeton* on the basis of the above derived characters.

Both *Diplocaulus* and *Diploceraspis* share a number of derived characters with *Peronedon* (Olson, 1970), a Lower Permian short-horned form (node 10):

(1) Very large interpterygoid vacuities.
(2) Fused frontals.
(3) Quadratojugal rolled ventrally to form a lateral palatal exposure.
(4) Orbits dorsally placed with loss of prefrontal–postfrontal contact.
(5) Comparatively enlarged parietals.
(6) Nasals absent.

Diplocaulus and *Diploceraspis* share the following derived characters with respect of *Peronedon* (node 11):

(1) Development of long horns by posterolateral expansion of the tabulars, squamosals, parietals and postparietals.
(2) Long double transverse processes on the centre, hence vertical rib heads.

Beerbower's (1963) phylogeny was based on the then current state of knowledge of the keraterpetontids. Watson (1913) had interpreted the condition of the skull-roof in *Batrachiderpeton* as intermediate between *Keraterpeton* and *Diplocaulus*; his key characters were the enlargement of the squamosal and extreme reduction of the nasals. However, detailed study of *Batrachiderpeton* (Milner, 1978) has shown that the pattern of dermal roofing-bones is typical of primitive keraterpetontids, thus the characters cited by Watson and subsequent works as evidence of close relationship between *Batrachiderpeton* and *Diplocaulus* are not valid.

Keraterpeton and *Diceratosaurus* were regarded as closely related contemporaneous genera. However, work by Eagar (1961, 1964) has demonstrated that *Keraterpeton* is considerably earlier than *Diceratosaurus* – Westphalian A as opposed to Westphalian D. Beerbower assumed, because *Diceratosaurus* and *Keraterpeton* were then regarded as neocontemporary, that the palatal configuration of both must have been similar. On the basis of further study of both genera (Milner, 1978), Beerbower's assumptions are shown to be incorrect. He suggests that *Diploceraspis* evolved from a lineage ("lineage A", p. 102) including *Keraterpeton* and *Diceratosaurus*, and that *Diplocaulus*, the other "long-horned" genus evolved from *Batrachiderpeton* ("lineage B", p. 102). Compared to the relationships presented here, Beerbower's phylogeny is, with respect to the "long-horned" forms, unparsimonious in relation to the distribution of derived characters in these genera.

2. Relationship of the Nectridea to Other Early Tetrapods

Nectrideans have been classified together with the microsaurs and aïstopods in the Subclass Lepospondyli by previous workers. The characters cited to unite the group are the possession of holospondylous vertebrae with spool-shaped centra, the absence of palatal fangs, non-labyrinthine teeth and small size of the animals.

The phyletic unity of the Lepospondyli has been questioned recently by Thomson and Bossy (1970) and Bossy (1976) who have suggested that the holospondylous lepospondyl vertebrae are, in part, the product of functional convergence. Nectrideans and aïstopods share several derived vertebral characters: elongate centra; tall neural spines (with a crenellated border in some) fused to the neural arch; intervertebral spinal nerve foramina (Bossy, 1976) and extra articulations above the zygapophyses (Bossy, 1976) that are not found in microsaurs. Microsaur vertebrae are quite different to those of nectrideans; the neural arch may be fused or loosely sutured but the haemal arches are never fused as they are in nectrideans. Not all microsaurs are holospondylous, for trunk intercentra have been reported in several genera (Carroll and Gaskill, 1978).

The absence of palatal fangs is a shared derived character occurring also in early captorhinomorph reptiles and, therefore, has no phylogenetic significance as a lepospondyl character. Non-labyrinthine teeth are size-related – small temnospondyls have non-labyrinthine teeth and those of large microsaurs show labyrinthine infolding (Carroll and Gaskill, 1978). This feature, together with the size of the animals, appears to have little phylogenetic significance with respect of lepospondyls.

On the basis of shared derived characters as currently understood it is, therefore, demonstrable that the Microsauria cannot be inferred as being immediately related to the Nectridea and that the Lepospondyli is an unnatural polyphyletic group.

Bossy (1976) proposes a new superorder, "Holospondyli", for the Nectridea and the Aïstopoda on the basis of shared vertebral characters and also derived skull characters. In both groups the opisthotic forms a very large paroccipital process to the tabular and the post-temporal fossae are small. A similar opisthotic configuration and small post-temporal fossae also occur in the Batrachosauria (*sensu* Panchen). Bossy suggests this may be a shared derived condition indicating possible relationship of the Holospondyli to the Batrachosauria.

The earliest known probable aïstopod is from the Viséan Lower Oil Shale Group at Wardie, near Edinburgh, Scotland (Baird, 1964). If the aïstopods are the sister-group of the nectrideans, then the dichotomy between them must have occurred at the basal Carboniferous at the latest or, more plausibly, in the late Devonian.

ACKNOWLEDGEMENTS

I should like to thank all those colleagues in various institutions who made specimens available to me during the course of my studies for my thesis, information and reconstructions from which are incorporated in this review. In particular, I should like to thank Dr A. L. Panchen for advice during the course of the work and Dr K. V. H. Bossy for permission to quote from her unpublished thesis.

The help of the British Museum (Natural History) Photographic Unit in the preparation of the text-figures is gratefully acknowledged.

Finally, I should like to express my sincere gratitude to my husband, Andrew, for invaluable discussions, help and criticism in the preparation of this manuscript, and for preparing the reconstructions shown in Fig. 4A–C.

REFERENCES

Andrews, C. W. (1895). Notes on a specimen of *Keraterpeton galvani* from Staffs. *Geol. Mag.* (4) **2**, 81–84.

Baird, D. (1964). The aïstopod amphibians surveyed. *Breviora* No. 206, 1–17.

Beerbower, J. R. (1963). Morphology, palaeoecology and phylogeny of the Permo-Pennsylvanian amphibian *Diploceraspis*. *Bull. Mus. comp. Zool. Harv.* **130**, 31–108.

Bolt, J. R. (1977). Dissorophoid relationships and ontogeny, and the origin of the Lissamphibia. *J. Paleont.* **51**, 235–249.

Bossy, K. V. H. (1976). Morophology, paleoecology and evolutionary relationships of the Pennsylvanian urocordylid nectrideans (Subclass Lepospondyli, Class Amphibia). *Diss. Abstr.* **B37**, 2731. Ph.D. Thesis, Yale University.

Carroll, R. L. and Gaskill, P. (1978). The Order Microsauria. *Mem. Am. phil. Soc.* No. 126, 1–211.

Case, E. C. (1902). On some vertebrate fossils from the Permian beds of Oklahoma. *In* "Second Biennial Report" (A. H. Van Fleet, ed), pp. 62–68. Department of Geology and Natural History, Territory of Oklahoma.

Cope, E. D. (1868). Synopsis of the extinct Batrachia of North America. *Proc. Acad. nat. Sci. Philad.* 1868, 208–221.

Cope, E. D. (1875). Supplement to the extinct Batrachia, Reptilia and Aves of North America. 1. Catalogue of the air-breathing Vertebrata from the Coal Measures of Linton, Ohio. *Trans. Am. phil. Soc.* **15**, 261–278.

Cope, E. D. (1877). Description of extinct vertebrates from the Permian and Triassic Formations of the United States. *Proc. Am. phil. Soc.* **17**, 182–193.

Cope, E. D. (1882). Third contribution to the history of the Vertebrata of the Permian Formation of Texas. *Proc. Am. phil. Soc.* **20**, 447–461.

Eagar, R. M. C. (1961). A note on the non-marine lamellibranchs of Leinster, Slieveardagh and Kanturk coalfields. *In* "The Westphalian of Ireland" (W. E. Nevill, ed.) *C.r. 4 Congr. Avanc. Etud. Stratigr. carb., (Heerlen)* **2**, 453–460.
Eagar, R. M. C. (1964). The succession and correlation of the Coal Measures of South Eastern Ireland. *Cr. 5 Congr. Avanc. Etud. Strarigr. carb., (Paris)* **2**, 359–374.
Fritsch, A. (1876). Über die fauna der Gaskohle des Pilsner und Rakonitzer beckens. *Sber. K. böhm. Ges. Wiss. Math. -nat. Kl.* 1875, 70–78.
Fritsch, A. (1883). "Fauna der Gaskohle und der Kalksteine der Permformation Böhmens", Band 1 (1879–1883). Prague.
Gans, C. (1974). "Biomechanics: An Approach to Vertebrate Evolution." Lippincott, Philadelphia.
Gosline, W. A. (1977). The structure and function of the dermal pectoral girdle in bony fishes with particular reference to ostariophysines. *J. Zool. Lond.* **183**, 329–338.
Haglund, T. R. (1977). New occurrences and paleoecology of *Peronedon primus* Olson (Nectridea). *J. Paleont.* **51**, 982–985.
Hancock, A. and Atthey, T. (1869). On a new labyrinthodont amphibian from the Northumberland coalfield, and on the occurrence in the same locality of *Anthracosaurus russelli. Ann. Mag. nat. Hist.* (4)4, 182–189.
Jaekel. O. (1903). Ueber *Ceraterpeton, Diceratosaurus* u. *Diplocaulus. Neues Jb. Miner. Geol. Paläont.* **1**, 109–134.
Jaekel, O. (1909). Ueber die Klassen der tetrapoden. *Zool. Anz.* **34**, 193–212.
Lydekker, R. (1889). *In* "A Manual of Palaeontology" (H. A. Nicholson and R. Lydekker, eds), 3rd Edn. Blackwood, Edinburgh.
Miall, L. C. (1875). Report of the committee . . . on the structure and classification of the labyrinthodonts. *Rep. Br. Ass. Advmt Sci.* 1874, 149–192.
Milner, A. C. (1978). "Carboniferous Keraterpetontidae and Scincosauridae (Nectridea: Amphibia) – a review." Ph.D. Thesis, University of Newcastle upon Tyne.
Olson, E. C. (1951). *Diplocaulus*. A study in growth and variation. *Fieldiana: Geology*. **11**, 55–154.
Olson, E. C. (1952). Fauna of the Upper Vale and Choza: 6. *Diplocaulus. Fieldiana: Geology*. **10**, 147–166.
Olson, E. C. (1956). Fauna of the Vale and Choza: 11. *Lysorophus*: Vale and Choza; *Diplocaulus, Cacops* and Eryopidae: Choza. *Fieldiana: Geology*. **10**, 313–322.
Olson, E. C. (1961). Jaw mechanisms: rhipidistians, amphibians, reptiles. *Am. Zool.* **1**, 205–215.
Olson, E. C. (1970). New and little known genera and species of vertebrates from the Lower Permian of Oklahoma. *Fieldiana: Geology* **18**, 359–426.
Olson, E. C. (1972). *Diplocaulus parvus* n.sp. (Amphibia: Nectridea) from the Chickasha Formation (Permian: Guadaloupian) of Oklahoma. *J. Paleont.* **46**, 656–659.
Panchen, A. L. (1964). Cranial anatomy of two Coal Measure anthracosaurs. *Phil. Trans. R. Soc. Lond.* **B247**, 593–637.

Panchen, A. L. (1970). Teil 5a Anthracosauria. "Handbuch der Paläoherpetologie." Fischer, Stuttgart.
Parrington, F. R. (1967). The vertebrae of early tetrapods. *Colloques int. C.n.R.S.* No. 163, 269–279.
Romer, A. S. (1930). Pennsylvanian tetrapods of Linton, Ohio. *Bull. Am. Mus. nat. Hist.* 59, 77–147.
Romer, A. S. (1945). "Vertebrate Paleontology", 2nd Edn. University Press, Chicago.
Romer, A. S. (1952). Late Pennsylvanian and early Permian vertebrates of the Pittsburgh–West Virginia region. *Ann. Carnegie Mus.* 33, 47–112.
Romer, A. S. (1966). "Vertebrate Paleontology", 3rd Edn. University Press, Chicago.
Steen, M. C. (1931). The British Museum collection of Amphibia from the Middle Coal Measures of Linton, Ohio. *Proc. zool. Soc. Lond.* 1930, 849–891.
Steen, M. C. (1938). On the fossil Amphibia from the Gas Coal of Nýřany and other deposits in Czechoslovakia. *Proc. zool. Soc. Lond.* **B108**, 205–283.
Thevenin, A. (1906). Amphibiens et reptiles de terrian houiller de France. *Annls Paléont.* **1**, 145–163.
Thevenin, A. (1910). Les plus anciens quadrupèdes de France. *Annls Paléont.* 5, 1–63.
Thomson, K. S. and Bossy, K. H. (1970). Adaptive trends and relationships in early Amphibia. *Forma Functio* 3, 7–31.
Watson, D. M. S. (1913). *Batrachiderpeton lineatum* Hancock and Atthey, a Coal Measure stegocephalian. *Proc. zool. Soc. Lond.* 1913, 949–962.
Woodward, A. S. (1897). On a new specimen of the stegocephalan [*sic*] *Ceraterpeton galvani* Huxley from the Coal Measures of Castlecomer, Kilkenny, Ireland. *Geol. Mag.* (4)4, 293–298.
Wright, E. P. and Huxley, T. H. (1866). On a collection of fossils from the Jarrow Colliery, Kilkenny. *Geol. Mag.* 3, 165–171.
Zittel, K.von (1888). "Handbuch der Palaeontologie", 1 Abt., "Palaeozoologie", III Band, "Amphibia". Oldenbourg, Munich and Leipzig.

APPENDIX

The following is a checklist of the families, genera and valid species within the Order Nectridea as used in this paper. It is based on the work of Milner (1978), Bossy (1976) and on published sources.

Order NECTRIDEA Miall 1875

Family Urocordylidae Lydekker 1889

Urocordylus wandesfordii Wright and Huxley 1866
Four specimens from Jarrow Colliery, Castlecomer, Co. Kilkenny, Eire. From the Jarrow Coal, Westphalian A.

Lepterpeton dobbsii Wright and Huxley 1866
Two specimens, Jarrow Colliery, Co. Kilkenny, Eire. Jarrow Coal, Westphalian A.

Sauropleura pectinata Cope 1868
Many specimens from "Diamond" Mine, Linton, Ohio, U.S.A. From the "Number 6" coal, Allegheny Series, Westphalian D.

S. scalaris (Fritsch) Baird 1964
At least 32 specimens from the Humboldt Mine, Nýřany, Czechoslovakia. From the Gaskohle of the Plzeň Basin, Westphalian D.

S. unnamed species.
Specimen from the Archer City Bone Bed, Archer Co., Texas, U.S.A. Putnam Formn, Wichita Gp, Lower Permian (Bossy, 1976).

Ctenerpeton remex (Cope) Romer 1930
At least 8 specimens, "Diamond" Mine, Linton, Ohio, U.S.A. Westphalian D.

Ptyonius marshii (Cope 1875)
At least 15 specimens, "Diamond" Mine, Linton, Ohio, U.S.A. Westphalian D. Also a specimen from Mazon Creek, Illinois, U.S.A. Francis Creek Shales, Carbondale Formn, Westphalian D.

Crossotelos annulatus Case 1902
Material from Orlando, Noble Co., Oklahoma. From the Wellington Formn, Lower Permian.

Family Keraterpetontidae (Jaekel 1903 as Ceraterpetontidae, emended by Romer 1945).

Keraterpeton galvani Wright and Huxley 1866
At least 24 specimens from Jarrow, Co. Kilkenny, Eire. Westphalian A.

K. unnamed species
A single specimen from Longton Hall Colliery, Longton, Staffordshire, England. Ash Coal Shale, *similis-pulchra* zone, Westphalian C (A. C. Milner, 1978).

Batrachiderpeton reticulatum (Hancock and Atthey 1869) comb. nov.
12 specimens from Newsham Colliery near Blyth, Northumberland, England. Low Main Seam, *modiolaris* zone, Westphalian B.

Diceratosaurus brevirostris (Cope) Romer 1930
At least 8 specimens from Linton, Ohio, U.S.A. Allegheny Series, Westphalian D.

Diplocaulus salamandrioides Cope 1877
Material from Danville, Illinois, U.S.A. Upper Pennsylvanian.

D. magnicornis Cope 1882
Much material from Texas and Oklahoma. Arroyo Formn and earlier, Lower Permian.

D. brevirostris Olson 1951
Material from Texas, Arroyo Formn, Clear Fork Group, Lower Permian.

D. recurvatus Olson 1952
Material from Texas, Vale and Choza Formn, Upper Clear Fork, Lower Permian.

D. parvus Olson 1972

One specimen from Blaine Co., Oklahoma. Chickasha Formn, Guadaloupian, Upper Permian.

Diploceraspis conemaughensis Romer 1952

From Soho St Quarry, Pittsburgh, Penn., U.S.A. Conemaugh Gp, Upper Pennsylvanian.

D. burkei Romer 1952

From many localities in Ohio, West Virginia and Pennsylvania, U.S.A. Washington and Greene Formn, Dunkard Gp, Lower Permian.

Peronedon primus Olson 1970

At least ten specimens from Norman and Grandfield, Oklahoma and from Texas. Arroyo to Choza Formn equivalents in Oklahoma, Clear Fork Gp, Lower Permian.

Family Scincosauridae Jaekel 1909

Scincosaurus crassus Fritsch 1876

Represented by at least 66 specimens from the Humboldt Mine, Nýřany, Czechoslovakia. From the Gaskohle of the Plzeň basin, Westphalian D.

Sauravus costei Thevenin 1906

Represented by a single specimen from Blanzy, Saône et Loire, France. Upper Stephanian, Upper Carboniferous.

S. cambrayi Thevenin 1910

Represented by a single specimen from Les Télots, near Autun, Saône et Loire, France. Autunian, Lower Permian.

Scincosauridae *incertae sedis*

An undescribed specimen (Museum of Comparative Zoology, Harvard, MCZ 2164) from the "Diamond" Mine at Linton, Ohio, U.S.A. From the "Number 6" coal, Allegheny Series, Westphalian D.

16 | An Early Tetrapod Fauna from the Namurian of Scotland

T. R. SMITHSON

Department of Zoology, The University, Newcastle upon Tyne, England. (Present Address: Redpath Museum, McGill University Montreal, Canada)

Abstract: Amphibia from the Namurian (basal Upper Carboniferous) are only known from the Midland Valley of Scotland, Nova Scotia, West Virginia and the Ruhr of Germany. The amphibian fauna from the Namurian of Scotland is reviewed here. The fauna includes a new (?)trimerorhachid (to be described elsewhere) together with adelogyrinid microsaurs, loxommatids, anthracosaurs (*sensu* Panchen), *Crassigyrinus scoticus* Watson, the edopoid *Caerorhachis bairdi* Holmes and Carroll, and possibly *Acherontiscus calaedoniae* Carroll.

A number of palaeoecological aspects of the fauna are discussed. It is concluded that aquatic and semi-terrestrial forms were present together with well-developed terrestrial members. Much of the fauna fed primarily on fish, but the adelogyrinids were possibly aquatic herbivores and the terrestrial anthracosaur *Eoherpeton* was probably a predator of the terrestrial invertebrate fauna.

A comparison of the Namurian and Westphalian tetrapod assemblages has shown them to be almost totally independent of each other. Only three families of pre-Coal Measure Amphibia are certainly found in the Westphalian. It is concluded that until new localities of pre-Coal Measure Amphibia are found outside the areas from which they are already known the possibility of finding the ancestors or close relatives of the majority of Westphalian Amphibia is remote.

INTRODUCTION

The majority of Carboniferous amphibia from Europe and North America have been discovered in beds equivalent to the British Coal

---

Systematics Association Special Volume No. 15, "The Terrestrial Environment and the Origin of Land Vertebrates", edited by A. L. Panchen, 1980, pp. 407–438, Academic Press, London and New York.

Measures (Westphalian and Stephanian). Far fewer have been found from strata below the Coal Measures (Tournaisian, Viséan and Namurian). Until recently, only two areas were known from which pre-Coal Measure forms had been found: the Scottish Midland Valley and Hinton, West Virginia (Watson, 1929; Romer, 1941). However, in the last two decades the number of pre-Coal Measure amphibian localities has almost doubled. New discoveries have been made in Nova Scotia, West Virginia, the Scottish Midland Valley and the Ruhr of Germany (Andrews *et al.*, 1977 and references). Also an undisputed pre-Carboniferous amphibian has been reported from the Devonian of Australia (Campbell and Bell, 1977) to add to the classic ichthyostegid finds from the Devonian of East Greenland.

These new discoveries have been important in two respects. Firstly they have provided the opportunity to find primitive representatives of some of the families of amphibians which up until now have appeared so suddenly in the Coal Measures, and secondly they have taken an important step nearer to the time of the fish-amphibian transition.

The first of the discoveries was reported by Romer (1958a). He described the cast of an anthracosaur lower jaw which had been made from a natural mould found in Namurian strata at Point Edward, Nova Scotia. This was the first of a number of reports on the amphibian fauna from Point Edward which includes the aberrant loxommatid *Spathicephalus* and the earliest known anthracosaur with embolomerous vertebrae (Baird, 1962; Romer, 1963; Carroll *et al.*, 1972). A second pre-Coal Measure locality was discovered in Nova Scotia by Dr Donald Baird in 1966 (Carroll *et al.*, 1972). Amphibian remains, including femora thought to be from a form well adapted for terrestrial locomotion, were found in Tournaisian strata at Horton Bluff.

In 1969 Romer reported the discovery of an important new amphibian locality at Greer in the Viséan of West Virginia. In the first of three papers on the Greer amphibian fauna, Romer (1969) described a new species of rhachitome, *Greererpeton burkemorani*, which established the presence of a Mississippian representative of the family Colosteidae, a group which had previously only been found in the Pennsylvanian. In the following year Romer (1970) and Hotton (1970) described new species of anthracosaur from Greer.

The discovery of the first new European pre-Coal Measure locality in the Namurian of the Ruhr, Germany, was reported by Boy and Bandel (1973). From it they described a new gephyrostegid anthracosaur, *Brukterepeton fiebigi*, and in doing so extended the range of the gephyrostegids down from the upper Coal Measures into the Namurian.

These new discoveries have in all cases extended the range of a number of amphibian families down into the pre-Coal Measures. However, the small faunal diversity of each locality had meant that the majority of Coal Measure amphibian families were still unrepresented in earlier sediments.

In 1974 amphibian remains were found by Mr Stanley Wood in a richly fossiliferous bone-bed at the Dora Opencast Site near Cowdenbeath, in the Namurian of Fife, Scotland (Andrews *et al.*, 1977). Much of the bone-bed was collected before open-cast operations removed the surrounding and underlying strata. Most of this has now been prepared and a diverse and extensive amphibian fauna revealed.

In this paper the Amphibia from Cowdenbeath, together with those from the other two Scottish Midland Valley Namurian localities, Loanhead and Niddrie, are briefly reviewed. A number of palaeoecological aspects of the fauna are discussed, and the extent and diversity of the pre-Coal Measure tetrapod assemblage is considered.

THE LOCALITIES

1. Loanhead

Four vertebrate-bearing horizons have been recognized in Namurian strata at Loanhead (Henrichsen, 1970). Of these, two have yielded amphibian remains: the Burghlee Ironstone and the No. 2 Ironstone (Paton, 1975). The fossiliferous horizons were exposed during underground coal working in the last century and all the tetrapod remains were collected during the 1880s.

Much of the material from Loanhead is well preserved. A number of specimens are partially or completely articulated, notably the smaller amphibians *Adelospondylus* (Carroll, 1967), *Caerorhachis* (Holmes and Carroll, 1977) and possibly *Acherontiscus* (Carroll, 1969).

2. *Niddrie*

Three vertebrate-bearing horizons have been recorded in the Namurian at Niddrie (Henrichsen, 1970). Of these, two have yielded amphibian remains: the shale overlying the Blue Coal Seam and the shale overlying the South Parrott Coal Seam (Paton, 1975). An isolated amphibian scute is the only recognizable amphibian specimen from the shale overlying the Blue Coal Seam. As at Loanhead, the fossiliferous horizons at Niddrie were exposed during underground coal working in the last century and all the tetrapod remains are thought to have been collected towards the end of the 1890s.

The fossiliferous matrix which overlies the South Parrot Coal Seam is a black shale which closely resembles that overlying the Low Main Seam at Newsham, Northumberland.

3. *Cowdenbeath*

Vertebrate-bearing sediments were originally discovered at the Dora Opencast Site, Cowdenbeath, in 1974 (Andrews *et al.*, 1977). The most productive horizon lay beneath a coal seam below the Lochgelly Blackband Ironstone and is referred to throughout this paper as the Dora Bone Bed. Several other vertebrate-bearing beds have been recognized but only the Lochgelly Blackband Ironstone has yielded amphibian remains. The Dora Bone Bed had been disrupted by two faults. After the overburden had been removed, three bone-bed platforms were exposed: the Top, Middle and Bottom platforms. These were gridded into 0·33m squares to preserve relationships before the bone-bed was removed.

The Dora Bone Bed comprised a localized patch of seat rock (Andrews *et al.*, 1977). In this respect it is unusual. Most fossil vertebrates discovered in the Carboniferous of Scotland have been found either in shales overlying oil shales or coal seams, or in ironstones associated with oil shales or coal seams.

THE FAUNA: "LEPOSPONDYLI"

1. *Order Microsauria?*

(a) Family Adelogyrinidae. The adelogyrinids are currently being

prepared and described by Dr S. M. Andrews at the Royal Scottish Museum.

Of the few groups of Amphibia found in the pre-Coal Measures of Scotland, the adelogyrinids are one of the most common. They have been found at five localities in the Viséan and all three localities in the Namurian of the Midland Valley. Four genera have been described: *Dolichopareias* and *Adelogyrinus* (Watson, 1929), *Palaeomolgophis* (Brough and Brough, 1967) and *Adelospondylus* (Carroll, 1967). Since all the material has been found within a very limited area and from a comparatively short geological time span, there is now some question as to whether all four genera are really

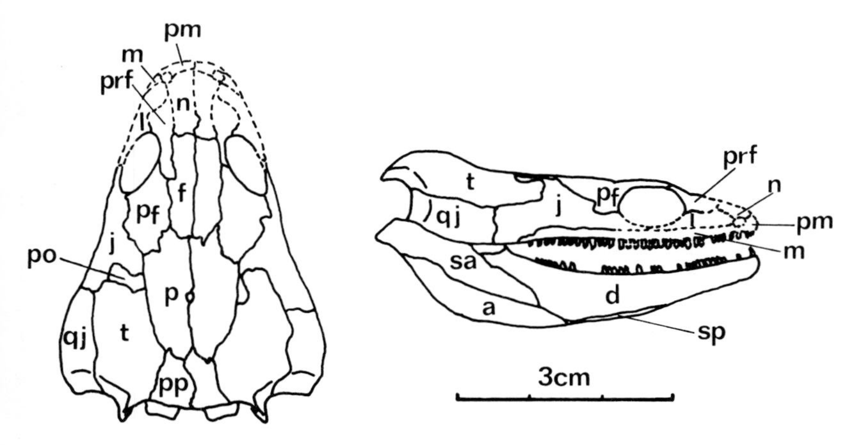

Fig. 1. *Adelospondylus watsoni* Carroll, skull in dorsal and lateral view, after Carroll.

Key to abbreviations: a, angular; art, articular; co, coronoid; d, dentary; f, frontal; it, intertemporal; j, jugal; l, lacrimal; m, maxillary; n, nasal; p, parietal; pf, postfrontal; pm, premaxillary; po, postorbital; posl, postsplenial; pra, prearticular; prf, prefrontal; qj, quadratojugal; sa, surangular; sp, splenial; st, supratemporal; t, tabular.

distinct (Carroll and Gaskill, 1978; S. M. Andrews, personal communication).

In general morphology all are very similar. They have a reduced complement of cranial bones with only the postparietals, quadratojugals and a third pair which might be termed either squamosal or tabular bordering the back of the skull (Fig. 1). The orbits are far forward with the bone normally described as the postorbital excluded from the orbital margin.

The pattern of bones in the lower jaw follows that characteristic of microsaurs but the adductor fossa is unusually short and there is a large retroarticular process. The coronoids are devoid of teeth or denticles. The structure of the palate is unknown.

Associated with the skull of *Palaeomolgophis* are a number of dissociated bones which Brough and Brough (1967) identify as elements of a well-developed branchial arch skeleton.

The teeth of the premaxilla, maxilla and dentary are very distinctive. In lateral view they are chisel-shaped with slightly recurved tips closely resembling those of the Permian embolomere *Archeria* and the aberrant loxommatid *Spathicephalus* (Panchen, 1970; Carroll and Gaskill, 1978). However, the teeth of *Archeria* and *Spathicephalus* are bullet-shaped in posterior view. This is quite unlike the condition found in the adelogyrinids in which the teeth are gently incurved and present a profile similar to that seen in the incisor teeth of mammals or the marginal dentition of iguanid lizards (Fig. 2). The apical edge of each tooth does not lie parallel to the jaw margin but is set at an oblique angle which runs from the lingual to the buccal surfaces of the jaw.

Postcranial remains have been described in *Palaeomolgophis* and *Adelogyrinus* (Watson, 1929; Brough and Brough, 1967) and *Adelospondylus* (Carroll, 1967). The vertebrae consist of holospondylus centra with suturally attached neural arches. No intercentra are associated with the dorsal vertebrae but they have been reported in the tail of *Palaeomolgophis*. The pre-sacral vertebral column is long: 37 presacral vertebrae were included in the restoration of *Palaeomolgophis* by Brough and Brough. The ribs are

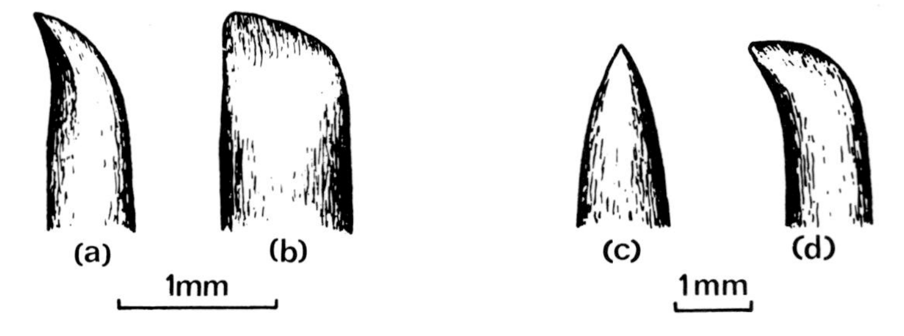

Fig. 2. Comparison of the teeth of the adelogyrinids with those of *Spathicephalus*. (a and b) Adelogyrinid tooth in posterior and lateral view; (c and d) *Spathicephalus* tooth in posterior and lateral view.

double-headed and a flange of bone extends from the tuberculum along the rib shaft. The lateral extent of the flange along the rib varies with respect to rib position.

The appendicular skeleton is inadequately known. The pectoral girdle is represented on all specimens by elements of the dermal girdle. Other appendicular bones are found only on the type of *Palaeomolgophis*. A well-ossified scapulo-coracoid, together with the distal end of a humerus, and epipodial bones and distal elements of a small fore limb are preserved. Impressions of a femur, part of an ischium and some phalanges have also been identified (Brough and Brough, 1967). An element lying behind the skull of the type of *Adelogyrinus* identified by Brough and Brough as a humerus may possibly be an anterior axial skeletal element similar to the "neck rib" in lungfish (S. M. Andrews, personal communication).

The bodily form of the adelogyrinids is very similar to that of the microbrachid microsaurs. Both have a long presacral column and small limbs. The two forms may have retained functional gills and it is probable that the adelogyrinids like the microbrachids were strictly aquatic Amphibia.

However, amongst the adelogyrinids only specimens attributed to the genus *Dolichopareias* have any trace of a lateral-line canal system. In almost all other obligate aquatic Amphibia an extensive lateral-line canal system is a prominent feature. Its absence in the adelogyrinids may be the result of modifications associated with their feeding habits.

The unusual dentition of the adelogyrinids is one of their most characteristic features. Their gently incurved chisel-shaped teeth with obliquely arranged apical edges formed a long, partially serrated, cutting surface. During jaw closure the incurved apical edges of the upper teeth passed over the buccal surfaces of the lower teeth in a manner similar to that seen in certain mammals or in the marginal dentition of iguanid lizards. This unusual arrangement has not been described before in Palaeozoic Amphibia and is of fundamental importance in any consideration of the feeding habits of the adelogyrinids.

Chisel-shaped teeth have only been found in two other Palaeozoic amphibian genera, *Spathicephalus* and *Archeria*. A. R. Milner (1978) has suggested that the modified dentition in *Archeria* enabled it to provide a continuous surface for gripping or nipping soft-bodied animals, whilst Tilley (in Panchen, 1970) has tentatively proposed that the chisel-shaped teeth in *Spathicephalus* were associated with filter-feeding. However, in neither form are the apical edges of the teeth incurved and it is unlikely that the shearing action of the upper teeth over the lower teeth, seen in the adelogyrinids, took place to such a degree.

The small adductor fossa of the lower jaw of the adelogyrinids suggests that their jaw-closing muscles were poorly developed (Carroll, 1967). This contrasts sharply with the condition seen in *Archeria* where a deep adductor fossa together with a high surangular crest (Case, 1915; Stovall, 1948; Panchen, 1970) would have provided an extensive area for the attachment of large adductor muscles. The kinetic inertial jaw-closing mechanism usually found in primitive Amphibia (Olson, 1961) would have allowed *Archeria*, with its powerful adductor muscles, to close its jaws with considerable force. The small adductor muscles of the adelogyrinids would have been incapable of such action and almost certainly prevented them from exploiting a similar food resource to that proposed for *Archeria*.

The filter-feeding mechanism tentatively proposed for *Spathicephalus* is associated with certain aberrant features of the skull, including its spade-like shape, the presence of double rows of teeth on the lower jaws and skull, and the restricted gape of the mouth (Tilley, 1971). The rather elongate shape of the adelogyrinid skull and the absence of a row of teeth on the coronoids, together with jaws which open normally, probably precluded them from adopting a filter-feeding habit.

The presence of a large retroarticular process on the lower jaw of the adelogyrinids led Carroll (1967) to suggest that they may have adopted a bottom-living habit with the lower jaws resting on the sediment in a manner similar to that proposed by Watson (1919, 1951) for forms with very flat skulls, for example the capitosaurs. However, the validity of Watson's hypothesis has recently been challenged by Chernin and Cruickshank (1978). In a re-analysis of the feeding habits of the large Triassic stereospondyls, they rejected the idea that the capitosaurs were primarily bottom-dwellers and suggested that they were mid-water feeders.

It therefore appears that the adelogyrinids represent a group of Amphibia utilizing a food resource not normally exploited by aquatic tetrapods. They show none of the specializations which might be associated with filter-feeding or those associated with a durophagous habit. Their weak adductor muscles probably prevented them feeding on soft-bodied animals in the manner proposed for *Archeria*, and their modified teeth were unsuitable for the piscivorous feeding habits normally found in aquatic Amphibia. They do, however, show features which suggest that they may represent the earliest recorded herbivorous Amphibia. Their shearing dentition was sufficiently well developed to cut through the stems and leaves of aquatic algae. The non-fibrous nature of the plant tissues would have provided little resistance to the partially serrated dentition and a strong jaw-closing musculature was therefore unnecessary. This non-fibrous tissue would probably have required little mastication before being ingested and the elaborate grinding teeth associated with many herbivorous forms were also unnecessary.

The forward position of the eyes probably indicates the importance of vision in food detection. The reduced lateral-line canal system may also be the result of the herbivorous nature of their diet. Extensive lateral-line canal systems are normally found in aquatic predators and it is probable that one of their principal functions is prey detection. If this is the case, the absence of a lateral-line canal system in the adelogyrinids can be taken as further evidence of their non-predatory habits.

The success of the group may also, in part, be due to their herbivorous diet. They have been found at all but three of the pre-Coal Measure localities in the Scottish Midland Valley, often from localities at which other Amphibia are unknown.

(b) Acherontiscus caledoniae *Carroll. Acherontiscus* was originally described by Carroll (1969). He based his description on an isolated specimen from the collections of the Royal Scottish Museum (RSM GY 1967.13.1). The provenance of the specimen was unknown and attempts to determine its geological age by analysis of ostracods and plant spores in the matrix led to two different conclusions. On the basis of the ostracod fauna the specimen was most likely to be of Westphalian A or B age, but analysis of the spore assemblage placed it from strata within the range Upper Viséan to Middle Namurian. Since the specimen is possibly from the Scottish Namurian it is briefly reviewed here.

The specimen comprises a skull and associated postcranial skeleton. It is poorly preserved and details of its postcranial anatomy were obtained by etching away the bone with acid and taking a silicone rubber cast from the resulting mould. Only the right side of the skull is preserved and what is visible resembles that of tuditanomorph microsaurs (Carroll and Gaskill, 1978), although it has a complex of distinct lateral-line canals which are not normally found in microsaurs. Two distinct types of teeth are found in the jaw. The anterior teeth are slim, sharply pointed cones, the teeth in the posterior portion of the jaws are bulbous with shearing crowns (Carroll, 1969; Carroll and Gaskill, 1978). Immediately behind the skull are a number of bones which are probably remains of the visceral arch apparatus.

The postcranial skeleton is unlike that normally found in microsaurs. The presacral vertebral column is moderately long and comprises between 26 and 31 vertebrae. The vertebrae are diplospondylous and closely resemble those of the embolomerous anthracosaurs. The appendicular skeleton is poorly preserved, but the limbs appear to have been rather small.

THE FAUNA: LABYRINTHODONTIA

1. Order Temnospondyli

(a) Caerorhachis bairdi *Holmes and Carroll.* The remains of a small temnospondyl amphibian *Caerorhachis bairdi* (Museum of Comparative Zoology No. 2271), almost certainly from Loanhead, were recently prepared and described by Holmes and Carroll (1977).

The skull resembles that of primitive temnospondyls. The bones of the skull-table are arranged in typical temnospondyl fashion with the supratemporal bones contacting the postparietals. No lateral-line canal grooves are present and there is no otic notch. The palate is perforated by small interpterygoid vacuities and, like the mesial surface of the lower jaw, extensively covered with denticles. Holmes and Carroll suggested that in its cranial anatomy *Caerorhachis* was similar to edopoid temnospondyls and in particular *Dendrerpeton*.

The postcranial anatomy of *Caerorhachis* differs from that normally expected in primitive temnospondyls. Unlike most early temnospondyls, which have rhachitomous vertebrae, the vertebrae of

Caerorhachis are gastrocentrous and very similar to those of the anthracosaur *Gephyrostegus*. The posterior elements of the appendicular skeleton are relatively well preserved but little is known of the fore limb and pectoral girdle. The pelvic girdle and hind limb are relatively large and in overall structure are similar to those of early temnospondyls. The epipodial bones of the hind limb are relatively short and measure less than half the length of the femur.

The habits of *Caerorhachis* were discussed by Holmes and Carroll. On the basis of the absence of lateral-line canals and relatively large limbs, they concluded that it was a semi-terrestrial form living much of its life around the margins of ponds and streams, feeding principally on stranded fish and only occasionally venturing into the water.

(b) New trimerorhachid? The most common amphibian at Cowdenbeath is an undescribed possibly trimerorhachid temnospondyl. Large numbers of fragmentary specimens have been discovered but no complete skulls are known and no postcranial remains can be associated with certainty. Specimens have also been found at Niddrie and Pitcorthie (Mid-Viséan, Fife). A lower jaw from Niddrie (RSM GY 1898.107.51) is the most complete specimen known.

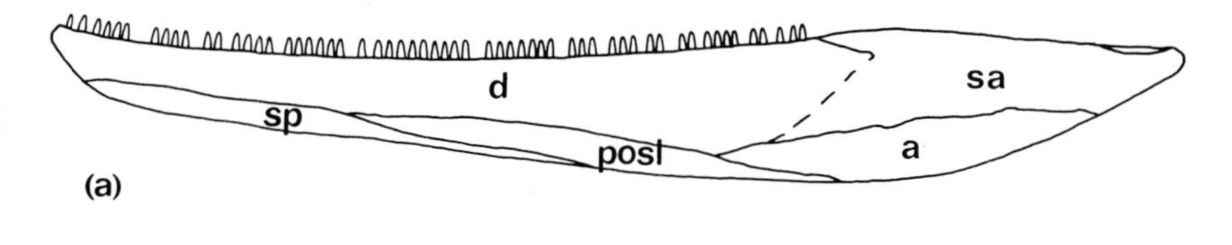

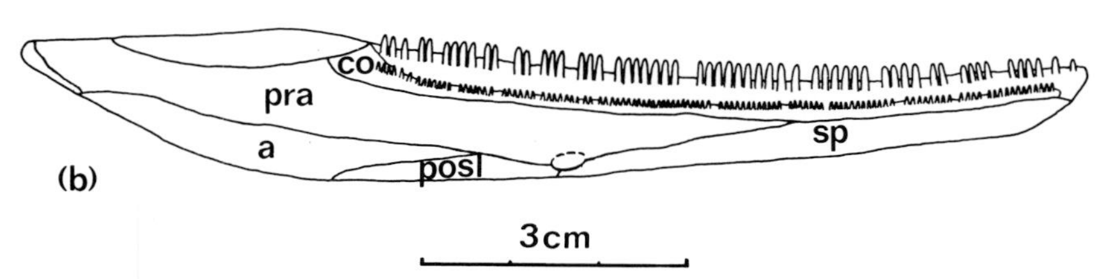

Fig. 3. Preliminary reconstruction of the lower jaw of a new ?trimerorhachid (see text). (a) Lateral view; (b) mesial view. See key to abbreviations in legend to Fig. 1.

The jaw is long and shallow, and similar in its overall construction to the lower jaws of trimerorhachids. (Fig. 3). The majority of the jaw specimens are from animals with lower jaws approximately 12cm long, but a number of specimens have been found at Cowdenbeath which are considerably larger. Each lower jaw terminates posteriorly with a distinct retroarticular process. A deep lateral-line groove, occasionally bridged with bone in a manner similar to that found in *Pholidogaster* (Panchen, 1975), runs along the ventral margin of the jaw. The dentition is similar in a number of respects to that of the Australian Devonian amphibian *Metaxygnathus denticulus* Campbell and Bell (1977). There are a large number of closely spaced, strongly incurved dentary teeth. A row of uniform, slim, conical teeth occurs along the coronoid series. There is no parasymphisial tusk.

A full description of the new material is in preparation.

(c) Loxomma *sp.* A lower jaw from the Burghlee Ironstone, Loanhead (RSM GY 1967.201.277), was tentatively referred to the genus *Loxomma* by Beaumont (1977, p.45). The specimen was briefly described but not illustrated. It consists of a left lower jaw which is clearly that of a labyrinthodont amphibian and has a suture pattern which is similar to that of known loxommatids. The jaw cannot be attributed to any other known labyrinthodont from the Namurian of Scotland.

(d) Spathicephalus mirus *Watson*. Watson (1929) based the original description of *Spathicephalus* on a number of partially complete skulls from Loanhead. This material, together with new fragmentary cranial remains from Cowdenbeath, are currently being redescribed by Dr E. H. Beaumont (*née* Tilley).

Spathicephalus is a moderately large, aberrant loxommatid. It has retained the "keyhole" orbits characteristic of the group but expanded its snout laterally to produce a peculiar spade-shaped skull. The orbits lie very close to the mid-line of the skull and are separated by the frontals and prefrontals only. They are posteriorly positioned on the skull-roof and considerably reduce the forward extent of the skull-table. The intertemporal bones are absent.

The marginal dentition of *Spathicephalus* is very distinctive (Fig. 2). In lateral view the teeth are chisel-shaped with slightly recurved tips and closely resemble those of *Archeria* and the adelogyrinids. In posterior view they are bullet-shaped like those of *Archeria*, but differ from those of adelogyrinids which have an incurved dentition (see above).

Apart from an interclavicle which has been attributed to *Spathicephalus* (Tilley, 1971) the postcranial skeleton is unknown.

The shape of the teeth, their large number (about 80 in each maxilla – Watson, 1929) and the arrangement of the jaw musculature led Tilley (in Panchen, 1970) to propose tentatively that *Spathicephalus* was a filter-feeding form. The arrangement and shape of the teeth would have precluded *Spathicephalus* from adopting the piscivorous habits of normal loxommatids (Beaumont, 1977) or the herbivorous diet of the adelogyrinids.

*2. Order Batrachosauria (*sensu *Panchen, Chapter 13 of this volume)*

(a) Crassigyrinus scoticus *Watson. Crassigyrinus* was originally described by Watson in his review of Scottish Carboniferous Amphibia (Watson, 1929). The holotype, thought to be from the Gilmerton Ironstone, was recently redescribed by Panchen (1973). New material from Cowdenbeath, in particular the almost complete skeleton BM(NH)R 10000 illustrated by Andrews *et al.* (1977), is currently being prepared and described by Dr Panchen.

Crassigyrinus exhibits a number of features which Panchen (Chapter 13 of this volume) suggests point to it being a relict member of the stock from which the anthracosaurs (*sensu* Panchen, Chapter 13 of this volume) and possibly the loxommatids evolved. It has retained the preopercular bones, the lateral kinetism between the skull-table and cheek region, and post-temporal fossae. The bones of the skull-table are arranged in the temnospondyl pattern with the supratemporals contacting the postparietals. Well-developed otic notches are present together with tabular horns. The lower jaws are similar in many features to those of the loxommatid *Megalocephalus*.

The cheek of *Crassigyrinus* is very deep and the orbits are situated high on the skull-roof: behind the orbits is a long suspensorium. The lateral-line canal system is moderately well developed. The tooth row in each jaw is long and includes a large number of simple conical teeth. On the lower jaw large teeth are present at intervals and probably correspond with the large palatal fangs in a manner similar to that seen in *Megalocephalus* (Beaumont, 1977).

Most of the postcranial skeleton is currently awaiting preparation.

*Suborder Anthracosauria (*sensu *Panchen, Chapter 13 of this volume).* At the two principal Namurian amphibian localities in Scotland, Loanhead and Cowdenbeath, anthracosaur remains have been discovered. Cowdenbeath has by far the largest number and four discrete assemblages have been recognized:

(1) The cranial remains of *Eoherpeton watsoni* Panchen, noted by Andrews *et al.* (1977).
(2) An incomplete postcranial skeleton.
(3) The skull-table (× 12) illustrated by Andrews *et al.* (1977) and a number of associated postcranial remains.
(4) A lower jaw "Jaw 90" and a number of associated postcranial elements.

At Loanhead three discrete assemblages of anthracosaur remains have been recognized:

(1) The isolated femur of *Papposaurus traquairi* Watson.
(2) The anterior portion of a headless trunk mentioned by Panchen (1970, 1975) but not described. This material is almost certainly to be attributed to the amphibian species represented by assemblage 2 from Cowdenbeath.
(3) Two jaw specimens with teeth very similar to those found in "Jaw 90" (Cowdenbeath assemblage 4).

(b) Eoherpeton watsoni *Panchen: Assemblage 1.* The recent preparation and redescription of the skull of *Pholidogaster pisciformis* revealed that the animal is not a primitive anthracosaur as had previously been thought (e.g. Watson, 1929; Romer, 1964) but is a colosteid temnospondyl closely related to *Greererpeton* (Panchen, 1975). The isolated anthracosaur skull from the Gilmerton Ironstone which had erroneously been attributed to *Pholidogaster* thus became the type specimen of a new genus and species, *Eoherpeton watsoni* Panchen.

In its general morphology *Eoherpeton* is unmistakably, a primitive anthracosaur (*sensu* Panchen, Chapter 13 of this volume). The large tabular bones contact the parietals antero-medially and they are extended posteriorly to produce characteristic horns. The intertemporal is a persistent bone in the skull-roof and the lateral kinetism between the skull-table and the squamosal bones of the cheek region is retained.

However, Panchen (1975) also noted that a number of cranial characters closely resembled those found in the Seymouriamorpha. These similarities are particularly noticeable in the braincase. The fenestrae ovales are widely spaced and floored by lateral processes of the parasphenoid. The posterior border of the parasphenoid develops into two deep concavities, one on either side of the mid-line which are probably homologous with the basal tubera in *Seymouria*. Post-temporal fossae are retained.

The marginal teeth of *Eoherpeton* are stout, slightly recurved, bluntly pointed cones. They are well spaced in the jaws and relatively few in number: Panchen (1975) restores 18 marginal teeth in the upper jaw and 20 in the lower jaw. In typical embolomeres there is usually a much larger number of marginal teeth. Only *Anthracosaurus russelli* is unusual in this respect, having fewer marginal teeth than *Eoherpeton* (Panchen, 1977).

The length of the dentary tooth row in *Eoherpeton* is slightly more than half the total length of the lower jaw. In typical embolomeres (e.g. *Eogyrinus* – Panchen, 1972) the tooth row is relatively much longer, occupying about two-thirds of the total length of the lower jaw. The adductor fossa in *Eoherpeton* is therefore relatively longer than that found in the embolomeres. A specimen of the lower jaw of *Eoherpeton* from Cowdenbeath has provided additional information on the bones surrounding the adductor fossa (Fig. 4). The surangular crest is not developed to such a degree as typical embolomeres but closely resembles that found in *Seymouria* (White, 1939). The dorso-lateral surface of the surangular crest is highly rugose, probably for the insertion of a muscle mass of the posterior adductor muscle. The mesial-ventral border of the adductor fossa is developed into a relatively long overturned shelf which probably served for the insertion of middle and posterior adductor muscle masses.

The arrangement of adductor muscles around the adductor fossa of *Eoherpeton* is similar to that found in *Seymouria* and suggests that *Eoherpeton*, like *Seymouria*, probably adopted the static pressure system of jaw closure. In this type of jaw-closing mechanism, pressure is developed when the jaws are not in motion, usually at or near the occlusal postion (Olson, 1961). This contrasts sharply with the kinetic inertial system normally found in anthracosaurs (Panchen, 1970). Here, the jaws are accelerated from an open position and the kinetic energy acquired is sufficient to ensure complete closure. Forms with the kinetic inertial system have little ability to develop the occlusal pressure required for chewing.

Eoherpeton watsoni was clearly a terrestrial anthracosaur. The absence of lateral-line sulci and the presence of a well-developed auditory apparatus are normally associated with animals which spent most of their lives on land. The presence of basal tubera suggests that the dorsal axial musculature was well developed to hold up the head, a feature which would be unexpected in an aquatic form where buoyancy effects would reduce the load on the neck muscles.

The dentition and jaw-closing mechanism of *Eoherpeton* differ significantly from those of embolomeres. The majority were aquatic piscivorous carnivores which used an extreme case of the kinetic inertial system when catching prey (Panchen, 1970). *Eoherpeton* probably used a poorly developed static pressure system which Olson (1961) suggested was "first developed among tetrapods as land habituation was being exploited and was probably an adaptation to feeding upon invertebrates, insects, molluscs and similar forms".

(c) Assemblage 2. The incomplete remains of an anthracosaur skeleton were found in the Dora Bone Bed at Cowdenbeath, scattered along the lower platform within an area 6 m long and 1 m wide (Fig. 5). Very few remains which can be associated with certainty were found outside this area. The partial skeleton includes a right humerus, right pelvic girdle, left femur and tibia, right and left fibulae, and a number of vertebral elements and ribs. This material is currently being prepared and described by the author. The anterior portion of a headless trunk from the Burghlee Ironstone, Loanhead, is attributed to the Cowdenbeath form principally on the basis of similarities in vertebral structure (see below). This specimen comprises a left scapulo-coracoid and clavicle, and a number of partially articulated vertebrae and ribs. The specimen was originally prepared by Dr Eileen Beaumont and was to be described jointly by her and Dr A. L. Panchen (note Panchen, 1975, p. 606).

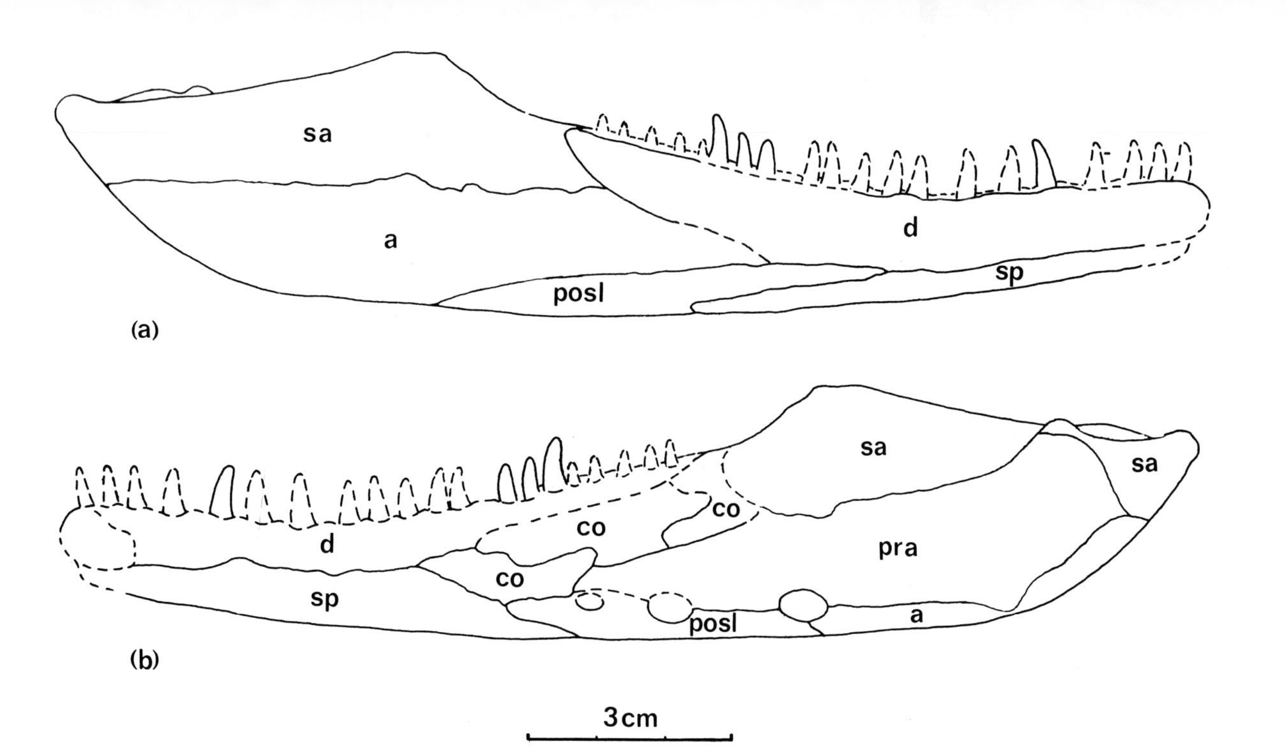

Fig. 4. New reconstruction of the lower jaw of *Eoherpeton watsoni* Panchen, partly after Panchen. (a) Lateral view; (b) mesial view. See key to abbreviations in legend to Fig. 1.

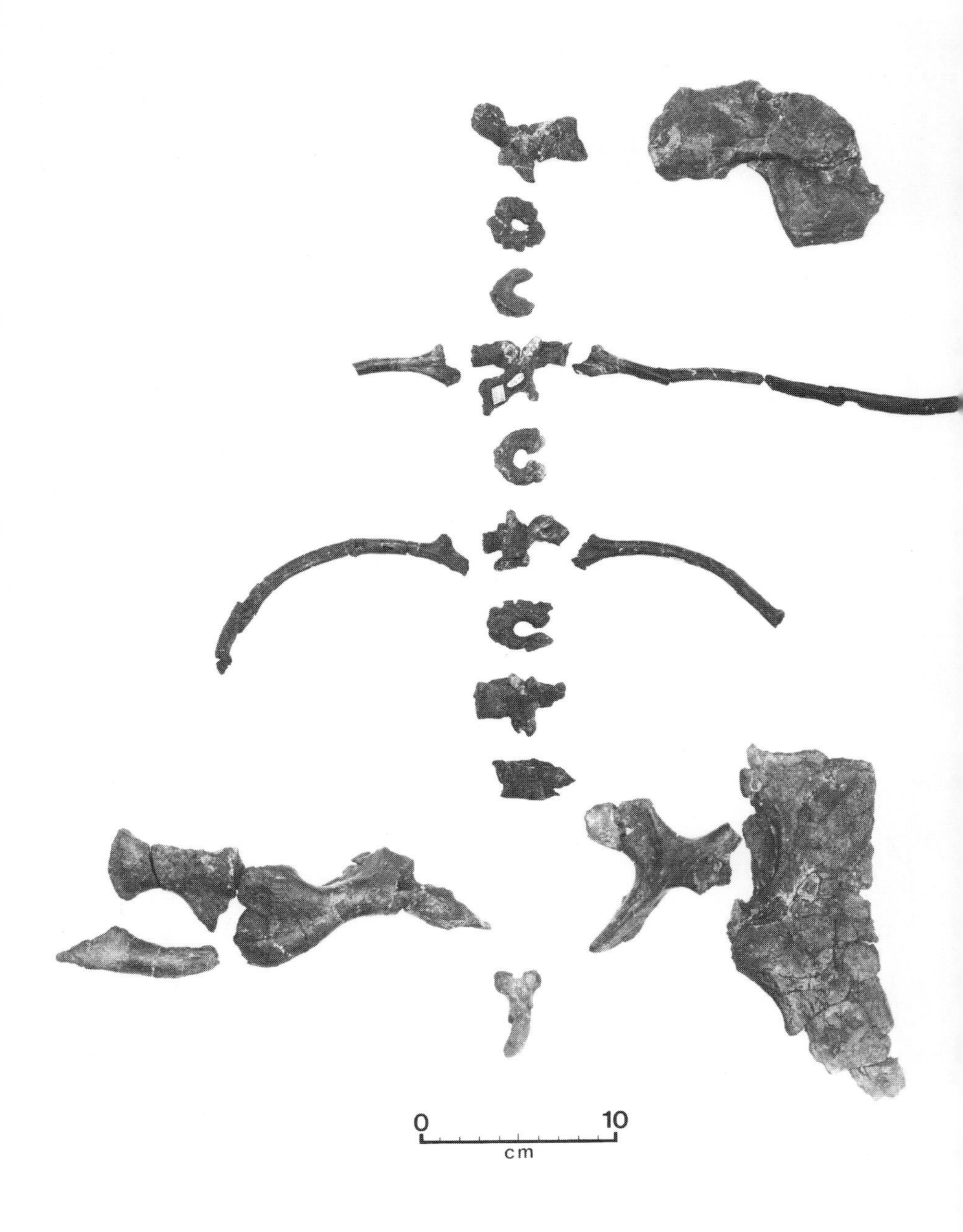

Fig. 5. Incomplete anthracosaur postcranial skeleton from Cowdenbeath ("assemblage 2").

Axial skeleton. The vertebrae are very similar to those described from *Gephyrostegus* (Carroll, 1970). The pleurocentra are incomplete dorsally and the notochordal perforation is very large. The intercentra are high horseshoe-shaped elements. The pleurocentrum and intercentrum belonging to the same vertebra have the same antero-posterior thickness. The neural arches are similar to those found in the embolomeres and gephyrostegids. The zygopophyses are well spaced and stoutly buttressed. The inward inclination of the pre-zygopophyses is approximately 40°, an arrangement which would have successfully restricted torsion. The neural spines are roughly rectangular in lateral view and possess prominent rugosities for the attachment of the dorsal axial musculature. Each spine is pierced from its anterior to its posterior edge by a very high supraneural canal. The canal is floored ventrally by the roof of the neural canal and extends dorsally to a position more than half-way up the spine. The sides of the canal taper dorsally and are lined with smooth periosteal bone.

The trunk ribs associated with assemblage 2 are double-headed in typical early tetrapod fashion. They are similar to those of *Eogyrinus* and show no lateral expansion of the rib shaft characteristic of *Ichthyostega* (Jarvik, 1955).

Appendicular skeleton. A large left scapulo-coracoid is preserved on the specimen from the Burghlee Ironstone. It closely resembles the scapulo-coracoid of *Seymouria* but the scapula and coracoid elements are co-ossified. The glenoid fossa is elongated antero-posteriorly with a strap-shaped surface which is partially twisted to form an articular surface often described as screw-shaped. It is orientated along the antero-posterior line of the scapulo-coracoid in a manner similar to that seen in *Seymouria*. It differs in this respect from *Archeria* which has a glenoid orientated obliquely downward.

A large left clavicle is also preserved with the scapulo-coracoid. The clavicular plate is roughly triangular and has a relatively long antero-posterior axis. The clavicular shaft rises almost perpendicularly from the postero-lateral corner of the clavicular plate. The area of the clavicular plate is relatively much larger than that found in either *Seymouria* or *Archeria*.

The fore limb is represented by a complete right humerus from Cowdenbeath. In its overall construction it is very similar to the humerus of *Seymouria*. It shows the same degree of "torsion" (whichever method of measurement is applied: c.f. Andrews and Westoll, 1970, p. 248), which is considerably greater than that measured in *Archeria* (Romer, 1957) or *Proterogyrinus* (Holmes, Chapter 14 of this volume). The origins and insertions of the principal locomotory muscles are well developed and clearly visible. A small supinator process is present. The capitellum is primitive in occupying a terminal-ventral position. It does not wholly lie on the ventral surface of the humerus and is not the rounded boss-shaped capitellum found in many primitive tetrapods (e.g. *Eryops* – Miner, 1925). The entepicondyle is relatively larger than that found in *Seymouria* and, like all batrachosaur humeri, is pierced by the entepicondylar foramen.

A large right pelvic girdle has been isolated from the Dora Bone Bed. In its overall construction it resembles that of *Archeria*. However, the ilium is a much stouter element. It is not so waisted above the acetabulum, the iliac blade is more strongly developed and the postiliac process is relatively short. The transverse line separating the axial musculature from the limb musculature is primitively low.

The pubo-ischiadic plate is essentially like that of *Archeria*. Beneath the acetabulum the pubis is perforated by the obturator foramen. Posterior to the obturator foramen on the mesial surface the pubis is thickened and forms a prominent ridge running from the ilium to the pubic symphysis. The symphysial surface appears to be restricted to the base of the pubic ridge, contrasting sharply with the condition in *Seymouria* where the whole length of the ventral edges of the pubo-ischiadic plates are thickened and ruguse and closely interlock.

The sutures delimiting the three bones of the pelvic girdle cannot be determined.

An incomplete left femur, a left tibia and fibula, and an incomplete right fibula constitute the hind limb elements prepared from Cowdenbeath. The femur is similar in its general construction to those described in *Archeria* and *Papposaurus*. However, the arrangement of the distal condyles differs significantly from the situation normally found in batrachosaurs. The tibial condyle is ventrally situated in a manner similar to that found in the pelycosaurs, for example *Ophiacodon*. This contrasts with the condition normally found in the batrachosaurs where the tibial condyle is terminal.

The tibia and fibula are much stouter elements than those found in primitive batrachosaurs, but in their general morphology they resemble those of *Archeria*. The tibia is primitively short and is approximately half the length of the femur.

Relationships of assemblage 2. The postcranial remains of assemblage 2 are clearly those of a moderately terrestrial anthracosaur. The vertebrae are constructed to withstand torsion, the limb girdles are large and the limbs, in particular the propodials, are relatively long.

The occurrence at Cowdenbeath of the postcranial remains of a terrestrial anthracosaur, and the cranial remains of *Eoherpeton*, also a terrestrial anthracosaur, raises the question: should the two assemblages be united? Currently there is insubstantial evidence to support this, although it is hoped that a specimen from Cowdenbeath, deposited in the Royal Scottish Museum and awaiting preparation, will resolve these difficulties. The specimen comprises a neural arch of the type described from assemblage 2 together with part of a braincase which may belong to *Eoherpeton*.

(d) Assemblage 3. The skull-table "X12" (Newcastle University Department of Zoology 75.11.2) illustrated by Andrews *et al.* (1977) is clearly that of an anthracosaur. The horned tabular bones contact the postparietals, the intertemporal bones are retained and the lateral kinetism between the skull-table and cheek is persistent (Fig. 6). In most respects "X12" is very similar to the skull-table of the British eogyrinids *Eogyrinus* and *Palaeoherpeton* and that of the American form *Proterogyrinus*. However, lateral-line canal sulci are absent from the frontal and postfrontal bones and the kinetic margin of the skull-table with the cheek extends anteriorly almost to the orbital margin in a manner similar to that described for *Eoherpeton* (Panchen, 1975). The sutures between a number of bones at the back of the table are raised on the dorsal surface into sutural ridges. This condition is also found in *Proterogyrinus* (R. Holmes, personal communication) and in an undescribed skull-table from the Namurian of Point Edward, Nova Scotia (A. L. Panchen, personal communication).

A number of postcranial bones have been found in the area of bone-bed surrounding the skull-table and they are tentatively associated with it. They comprise a neural arch, an incomplete left clavicle and an incomplete element which may be a humerus.

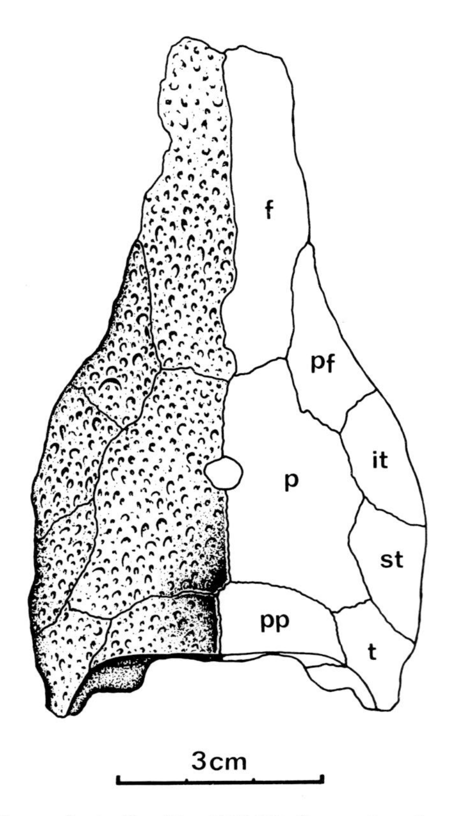

Fig. 6. Reconstruction of skull-table "X12" from Cowdenbeath. See key to abbreviations in legend to Fig. 1.

Relationships of assemblage 3. Skull-table "X12" differs markedly from that of *Eoherpeton*. The raised sutures and structure of the tabular horn are notable differences. The few postcranial remains available for comparison also show differences to those from assemblage 2. It is therefore probable that assemblage 3 represents the remains of an anthracosaur quite distinct from those of assemblages 1 and 2.

(e) Assemblage 4. A poorly preserved anthracosaur lower jaw "Jaw 90" (RSM GY 1977.46.23) was found in bone-bed from the top platform at Cowdenbeath. The jaw was originally exposed in lateral view but careful preparation has revealed much of the mesial surface. Found alongside "Jaw 90" were an incomplete vertebra and rib, and

also associated with it are a number of isolated vertebrae and an incomplete interclavicle. Two jaw specimens from Loanhead are cautiously included with assemblage 4.

In profile "Jaw 90" is very similar to the lower jaw from Point Edward, Nova Scotia, described by Romer (1963) (Fig. 7). The structure of its mesial surface is very similar to that described for *Eogyrinus* (Panchen, 1972). There are two large meckelian fenestrae characteristic of the embolomere lower jaw. The surangular crest is strongly developed but overall the jaw is shallower than those of typical embolomeres, for example *Eogyrinus* (Panchen, 1972) and *Eobaphetes* (Panchen, 1977). The teeth in "Jaw 90" are moderately tall and slender and taper very slightly over their exposed length. The tip of the crown tapers to a sharp point which is both hooked back and incurved.

The vertebrae associated with "Jaw 90" are very similar to those of *Proterogyrinus* drawn by Hotton (1970) (as *Mauchchunkia*) and Holmes and Carroll (1977). However, the pleurocentra are fused dorsally and a persistent supraneural canal pierces the neural spine. The overall construction of the vertebrae is of a type normally associated with a partially terrestrial form. The neural spines are high and rugose and suggest an extensive dorsal axial musculature. The inward inclination of the pre-zygopophyses is approximately 30°, an arrangement which would have successfully resisted torsion. The pleurocentra are well ossified with a small notochordal perforation and prominent ventral keel. The intercentra are low horseshoe-shaped elements.

The incomplete interclavicle closely resembles that attributed by Panchen (1972) to *Eogyrinus*.

Relationships of assemblage 4. In its overall construction "Jaw 90" closely resembles the lower jaw of *Eogyrinus*. It differs significantly from that of *Eoherpeton* particularly in the character of its dentition, the degree of development of the surangular crest and the size of the meckelian fenestrae (c.f. Fig. 4 and Fig. 7). The vertebrae of assemblage 4 are also markedly different from those of assemblage 2. The pleurocentra are more robust elements, fused dorsally and with a small notochordal perforation. The intercentra are low and the supraneural canal is not so extensive.

At this stage it is impossible to say for certain whether assemblage 4 should be associated with assemblage 3. I have included the two

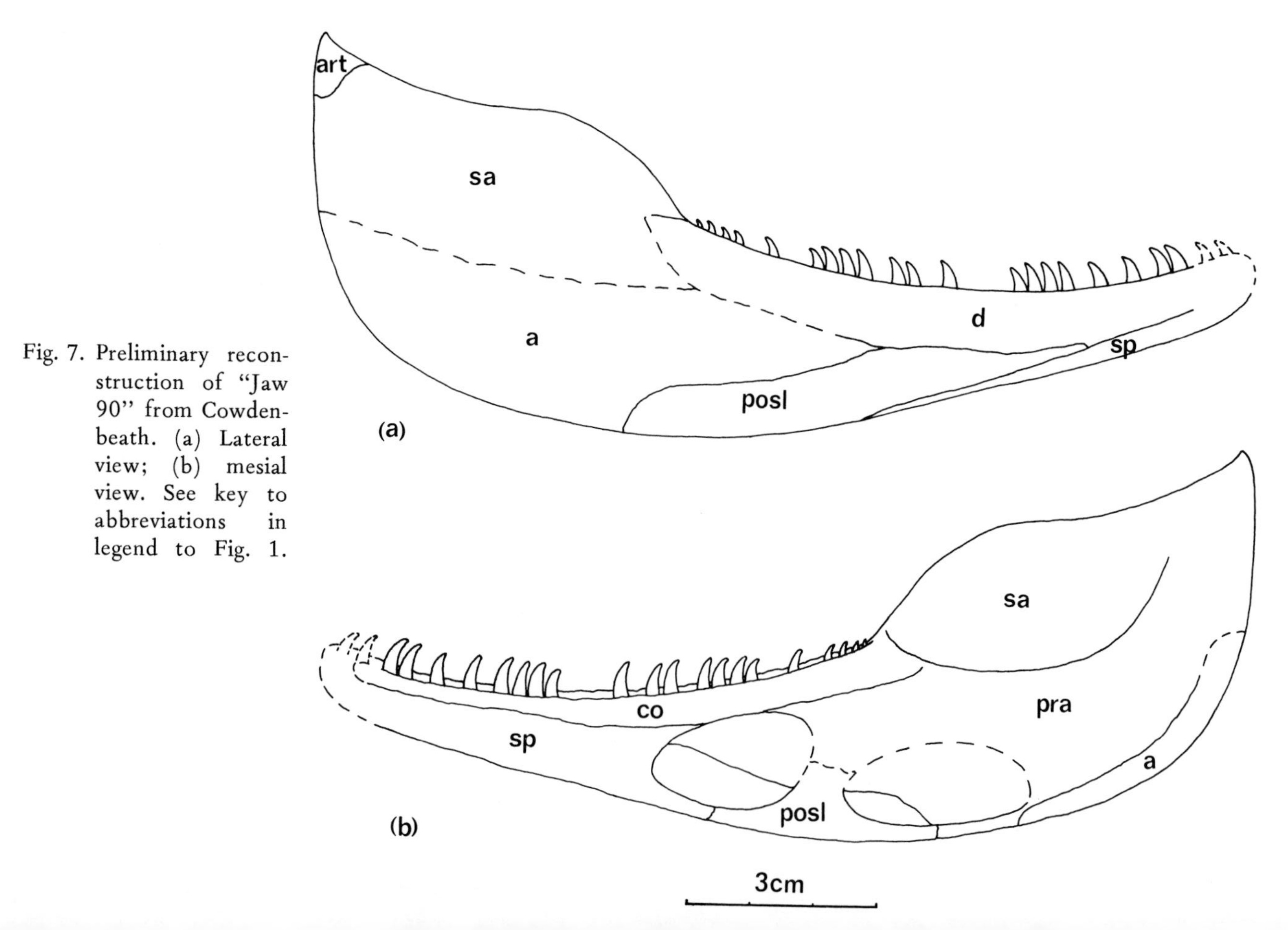

Fig. 7. Preliminary reconstruction of "Jaw 90" from Cowdenbeath. (a) Lateral view; (b) mesial view. See key to abbreviations in legend to Fig. 1.

assemblages in the infraorder Herpetospondyli (see Panchen, Chapter 13 of this volume) and tentatively placed them in the family Proterogyrinidae.

(f) Papposaurus tranquairi *Watson*: *Assemblage 5*. An isolated femur thought to be from Loanhead was described and illustrated by Watson (1914). It closely resembles the femur of *Proterogyrinus* (R. Holmes, personal communication). It is less than half the length of the femur from assemblage 2, the distal condyles are orientated differently (see above) and the adductor crest is not so strongly developed.

The disparity in size and the different orientation of the distal condyles in the femur from assemblage 2 and that of *Papposaurus* probably prevents the association of the two assemblages. The femur is also probably too small to be associated with *Eoherpeton*. It is possible, however, that it should be associated with either assemblage 3 or 4 (or both) and I have therefore also included *Papposaurus* in the family Proterogyrinidae.

ECOLOGY AND LOCOMOTION

The anthracosaurs from the pre-Coal Measures of Scotland are particularly important in one respect: they provide evidence to show that the successful exploitation of the terrestrial environment by tetrapods occurred much earlier than is currently believed. Until now, Romer's (1958b) suggestion, that fully terrestrial tetrapods did not develop until the late Carboniferous, has remained unchallenged. He proposed that amphibians exploited the terrestrial environment by two distinct and separate steps: firstly by the development of limbs in the Devonian, giving the potential for terrestrial existence, and secondly by the utilization of these limbs for life on land during the late Carboniferous. To support his hypothesis Romer cited a number of examples from the fossil record. He suggested that the Devonian ichthyostegids with their small limbs and fish-like tail were primarily water-dwelling piscivores. He also suggested that the climatic conditions in the Devonian were unsuitable for the development of terrestrial tetrapods. He considered most early Carboniferous amphibians to be aquatic forms and emphasized that until the development of such forms as *Eryops* and *Cacops* towards the end of the Carboniferous and early Permian no amphibian could be "seriously considered as living to any extent on land".

The cranial anatomy of *Eoherpeton* exhibits a number of features normally associated with terrestrial amphibians: the absence of lateral-line canal sulci, a well-developed auditory apparatus, the static pressure jaw-closing mechanism and a strong axial musculature for holding up the head. The postcranial remains of assemblage 2 and those associated with "Jaw 90" also exhibit features normally found in terrestrial forms: vertebrae well developed to resist torsion, strong limb girdles and relatively long limbs.

The terrestrial features of a number of pre-Coal Measure Amphibia fit into a trend of developing terrestriality. The ichthyostegids currently exemplify the base of this development. Their strong, stumpy limbs and modified axial skeleton (Panchen, 1967) are suggestive of animals capable of prolonged progression on land. Limb bones from the Tournaisian (basal Lower Carboniferous) of Horton Bluff, Nova Scotia, thought to be those of a well-developed terrestrial amphibian (Carroll *et al.*, 1972), and the anthracosaur *Proterogyrinus* from the Viséan of West Virginia (Holmes, Chapter 14 of this volume), are evidence of forms capable of land progression in the pre-Coal Measures of North America. *Eoherpeton*, the Namurian herpetospondyls and possibly *Caerorhachis* are European examples of pre-Coal Measure terrestrial forms.

The early appearance of an otic notch and auditory apparatus in the Batrachosaurs (*sensu* Panchen, Chapter 13 of this volume), for example *Crassigyrinus*, is an indication of the development of aerial hearing, a mechanism of little or no use to animals which spend all their time in water.

The arid climatic conditions of the Devonian were, Romer thought, unsuitable for prolonged excursions on land by early amphibians. He argued that the damp humid environment of the Coal Measure swamps were the conditions during which tetrapods first began to exploit the terrestrial environment. However, damp conditions were sufficiently common in the Devonian to allow land floras to flourish (Edwards, Chapter 4 of this volume) and associated with this terrestrial vegetation was a diverse community of land invertebrates (Rolfe, Chapter 6 of this volume).

By the beginning of the Carboniferous the "arid" conditions of the Devonian had ceased. In all areas where pre-Coal Measure tetrapods have been discovered, lacustrine deposits and in some cases coal swamp deposits – for example in the Horton Group, Horton Bluff (Carroll *et al.*, 1972), and the Limestone and Limestone Coal Groups of the Scottish Midland Valley – are the dominating facies. Thus at certain times in the pre-Coal Measures and in certain areas conditions similar to those commonly found in the Coal Measures were also in operation. The conditions which Romer probably rightly proposed were necessary for the successful exploitation of land by amphibians were therefore not restricted to the Coal Measures but are found intermittently throughout the Carboniferous.

Terrestrial locomotion prior to the Coal Measures, however, was almost certainly inefficient. In all known forms the epipodial limb bones were relatively small and rarely exceeded half the length of the propodials. During locomotion the torso would probably have remained close to the substrate surface and in certain forms continue to contact the ground: for example the tetrapod responsible for trackway II from the Devonian of Australia (Warren and Wakefield, 1972). The well-developed scutes and gastralia of early Amphibia may have been necessary to protect the ventral body surface from harmful abrasion during land-walking. Efficient locomotion was only possible with the development of a strong vertebral column and the reduced size of the head (Olson, 1976). Following this the body was "jacked up" off the ground by the elongation of the epipodial limb bones.

DISCUSSION

In the Lower Carboniferous of Scotland eight amphibian localities are known. Of these only three (Burdiehouse, Gilmerton and Pitcorthie) have yielded more than a single specimen, and only at Gilmerton are more than two species represented. By the beginning of the Upper Carboniferous, however, a diverse amphibian fauna had become established. At the two principal Namurian localities, Loanhead and Cowdenbeath, a combined total of at least eight species is recognized. Most Namurian forms have been recorded in the Lower Carboniferous but a number, notably *Caerorhachis*, *Spathicephalus* and the herpetospondyls, have not been recorded from earlier horizons.

The pre-Coal Measure amphibian fauna from North America is very similar to that found in Scotland. The colosteid *Greererpeton*, a form closely related to *Pholidogaster* (Panchen, 1975), and the herpetospondyl *Proterogyrinus* have been described from the Upper Viséan of West Virginia (Romer, 1969, 1970). *Spathicephalus* has been found in the Namurian of Nova Scotia (Baird, 1962) together with the earliest embolomeres and an undescribed trimerorhachoid (Romer, 1958, 1963; Carroll *et al.*, 1972).

By the Westphalian the pre-Coal Measure amphibian fauna of North America and Europe has largely been replaced. Of the 12 amphibian families currently known in the Viséan and Namurian, only three are certainly present in the Westphalian: the Colosteidae, Loxommatidae and Gephyrostegidae. Two Namurian forms, *Caerorhachis* and "*Pholiderpeton*" *bretonense*, are less certainly included in families found in the Westphalian (Panchen, 1977; A. R. Milner, 1980).

In Britain the structure of the pre-Coal Measure amphibian assemblage differs in a number of respects from that found in the Westphalian. The few data available from the pre-Coal Measures, and in particular the Namurian, suggest that the principal components of the fauna were the adelogyrinds, the undescribed trimerorhachid and the herpetospondyls, plus *Crassigyrinus* and *Spathicephalus*. The characteristic fauna associated with the lacustrine deposits of the British Westphalian Coal Measures includes the eogyrinids, loxommatids, keratopetontid nectrideans and the aïstopod *Ophiderpeton* (A. R. Milner, 1978). The pre-Coal Measure piscivores *Crassigyrinus*, the herpetospondyls and the undescribed trimerorhachid were replaced in the Westphalian by the eogyrinids and loxommatids. There are, however, no Westphalian counterparts for the adelogyrinids and *Spathicephalus*. This is particularly surprising in view of the similar environmental conditions operating in the Limestone Coal Group of the Namurian (in which the fossiliferous deposits at Loanhead and Cowdenbeath were discovered) and the Coal Measures (Goodlet, 1959). It is possible that the niches vacated by the adelogyrinids and *Spathicephalus* were filled by members of the diverse Westphalian fish assemblage, but the absence of detailed accounts of many of the Coal Measure fishes prevents a satisfactory comparison.

The discovery of nectrideans and aïstopods at the earliest Westphalian locality, Jarrow, Eire (Huxley and Wright, 1867; Eager, 1964; A. C. Milner, Chapter 15 of this volume) highlights one of the major unexplained problems of the faunal transition between the Namurian and Westphalian. Of the 30 or so tetrapod families currently known in the Westphalian, only three are certainly found in the pre-Coal Measures. A number of higher labyrinthodont taxa have been found in the pre-Coal Measures, however, notably the Embolomeri and possibly the Edopoidae (Holmes and Carroll, 1977) and the Trimerorhachoidea, but the majority have no known antecedents. The recent discovery at Cowdenbeath of a Namurian amphibian fauna has in no way helped to resolve this problem. Indeed the discovery has confirmed the integrity of the Scottish pre-Coal Measure tetrapod assemblage. Until new faunas are found outside the areas from which pre-Coal Measure Amphibia are already known, the possibility of finding the ancestors or close relatives of the majority of the Westphalian Amphibia is remote.

ACKNOWLEDGEMENTS

I am grateful to the staffs of the Royal Scottish Museum, Edinburgh, and the British Museum (Natural History), London, for permission to borrow and study specimens in their care. I have benefited from discussion with colleagues, particularly Dr S. M. Andrews and Dr A. L. Panchen both of whom kindly read and commented on the manuscript. I gratefully acknowledge the important work of Mr S. P. Wood who discovered fossil Amphibia at the Dora Opencast Site, Cowdenbeath, and was responsible for their collection. Mr G. Howson advised me on the photography.

My thanks are also due to Dr Andrews for permitting me to include and comment on some of her unpublished data on the Adelogyrinidae, and to Dr Eileen Beaumont and Dr Panchen for passing on the anthracosaur skeleton from the Burghlee Ironstone to enable me to describe it alongside the other material from Cowdenbeath.

This work was completed while I was a Junior Research Associate financed by a Natural Environmental Research Council grant (No. GR3/2983) awarded to Dr A. L. Panchen.

REFERENCES

Andrews, S. M. and Westoll, T. S. (1970). The postcranial skeleton of *Eusthenopteron foordi* Whiteaves. *Trans. R. Soc. Edinb.* 68, 207–329.

Andrews, S. M., Browne, M. A. E., Panchen, A. L. and Wood, S. P. (1977). Discovery of amphibians in the Namurian (Upper Carboniferous) of Fife. *Nature, Lond.* **265**, 529–532.

Baird, D. (1962). A rhachitomous amphibian, *Spathicephalus*, from the Mississippian of Nova Scotia. *Breviora* No. 157, 1–10.

Beaumont, E. H. (1977). Cranial morphology of the Loxommatidae (Amphibia: Labyrinthodontia). *Phil. Trans. R. Soc.* **B280**, 29–101.

Boy, J. A. and Bandel, K. (1973). *Bruktererpeton fiebigi* n. gen. n. sp. (Amphibia: Gephyrostegida) der erste Tetrapode aus dem Rheinisch – Westfälischen Karbon (Namur B; W.-Deutschland) *Palaeontographica* **145A**, 39–77.

Brough, M. C. and Brough, J. (1967). Studies on early tetrapods. I. The Lower Carboniferous microsaurs. II *Microbrachis* the type microsaur. III. The genus *Gephyrostegus. Phil. Trans. R. Soc.* **B252**, 107–165.

Campbell, K. S. W. and Bell, M. W. (1977). A primitive amphibian from the late Devonian of New South Wales. *Alcheringa* **1**, 369–381.

Carroll, R. L. (1967). An adelogyrinid lepospondyl amphibian from the Upper Carboniferous. *Can. J. Zool.* **45**, 1–16.

Carroll, R. L. (1969). A new family of Carboniferous amphibians. *Palaeontology* **12**, 537–598.

Carroll, R. L. (1970). The ancestry of reptiles. *Phil. Trans. R. Soc.* **B257**, 267–308.

Carroll, R. L. and Gaskill, P. (1978). The Order Microsauria. *Mem. Am. phil. Soc.* **126**, 1–211.

Carroll, R. L., Belt, E. S., Dineley, D. L., Baird, D. and McGregor, D. C. (1972). "Excursion A59 Vertebrate Palaeontology of eastern Canada Guidebook", *24th int. Geol. Congr., Montreal, 1972.*

Case, E. C. (1915). The Permo-Carboniferous red beds of North America and their vertebrate fauna. *Publs Carnegie Instn* No. 207, 1–176.

Chernin, S and Cruickshank, A. R. I. (1978). The myth of the bottom-dwelling capitosaur amphibians. *S. afr. J. Sci.* **74**, 111–112.

Eager, R. M. C. (1964). The succession and correlation of the Coal Measures of South Eastern Ireland, *C.r. 5 Congr. Avanc. Étud. Stratigr. carb. Paris* **2**, 359–374.

Edwards, D. (1980). Early land floras. *In* "The Terrestrial Environment and the Origin of Land Vertebrates" (A. L. Panchen, ed.), pp. 55–85. Academic Press, London and New York.

Goodlet, G. A. (1959). Mid-Carboniferous sedimentation in the Midland Valley of Scotland. *Trans. Edinb. geol. Soc.* **17**, 217–240.

Henrichsen, I. G. C. (1970). "A Catologue of Fossil Vertebrates in the Royal Scottish Museum, Edinburgh. Part one/Actinopterygii". Information Series, Royal Scottish Museum.

Holmes, R. (1980) *Proterogyrinus scheelei* and the early evolution of the labyrinthodont pectoral limb. *In* "The Terrestrial Environment and the Origin of Land Vertebrates" (A. L. Panchen, ed.), pp. 352–376. Academic Press, London and New York.

Holmes, R. and Carroll, R. (1977). A temnospondyl amphibian from the Mississippian of Scotland. *Bull. Mus. comp. Zool. Harv.* **147**, 489–511.

Hotton, N. (1970). *Mauchchunkia bassa* gen. et. sp. nov. an anthracosaur (Amphibia: Labyrinthodontia) from the Upper Mississippian. *Kirtlandia* No. 12, 1–38.

Huxley, T. H. and Wright E. P. (1867). On a collection of fossil Vertebrata, from the Jarrow Colliery, County of Kilkenny, Ireland. *Trans. R. Irish Acad.* **24**, 351–369.

Jarvik, E. (1955). The oldest tetrapods and their forerunners. *Scient. Mon. N.Y.* **80**, 141–154.

Milner, A. C. (1980). A review of the Nectridea. *In* "The Terrestrial Environment and the Origin of Land Vertebrates" (A. L. Panchen, ed.), pp. 379–405. Academic Press, London and New York.

Milner, A. R. (1978). A reappraisal of the early Permian amphibians *Memonomenos dyscriton* and *Cricotillus brachydens*. *Palaeontology* **21**, 667–686.

Milner, A. R. (1980). The temnospondyl amphibian *Dendrerepeton* from the Upper Carboniferous of Ireland. *Palaeontology* **23**, 125–141.

Miner, R. W. (1925). The pectoral limb of *Eyrops* and other primitive tetrapods. *Bull. Am. Mus. nat. Hist.* **51**, 145–312.

Olson, E. C. (1961). Jaw mechanisms: rhipidistians, amphibians, reptiles. *Am. Zool.* **1**, 205–215.

Olson, E. C. (1976). The exploitation of land by early tetrapods. *In* "Morphology and Biology of Reptiles" (A. d'A. Bellairs and C. B. Cox, eds), pp. 1–30. Academic Press, London and New York.

Panchen, A. L. (1967). The homologies of the labyrinthodont centrum. *Evolution* **21**, 24–33.

Panchen, A. L. (1970). Teil 5a Anthracosaura. "Handbuch der Paläoherpetologie". Fischer, Stuttgart.

Panchen, A. L. (1972). The skull and skeleton of *Eogyrinus attheyi* Watson (Amphibia: Labyrinthodontia). *Phil. Trans. R. Soc.* **B263**, 279–326.

Panchen, A. L. (1973). On *Crassigyrinus scoticus* Watson, a primitive amphibian from the Lower Carboniferous of Scotland. *Palaeontology* **16**, 179–193.

Panchen, A. L. (1975). A new genus and species of anthracosaur amphibian from the Lower Carboniferous of Scotland and the status of *Pholidogaster pisciformis* Huxley. *Phil. Trans. R. Soc.* **B269**, 581–640.

Panchen, A. L. (1977). On *Anthracosaurus russelli* Huxley (Amphibia: Labyrinthodontia) and the family Anthracosauridae. *Phil. Trans. R. Soc.* **B279**, 447–512.

Panchen, A. L. (1980). The origin and relationships of the anthracosaur amphibia from the late palaeozoic. *In* "The Terrestrial Environment and the Origin of Land Vertebrates (A. L. Panchen, ed.), pp. 319–350. Academic Press, London and New York.

Paton, R. L. (1975). "A Catalogue of Fossil Vertebrates in the Royal Scottish Museum, Edinburgh. Part Four/Amphibia and Reptilia". Information series, Royal Scottish Museum.

Rolfe, W. D. I. (1980). Early invertebrate terrestrial faunas. *In* "The Terrestrial Environment and the Origin of Land Vertebrates" (A. L. Panchen, ed.), pp. 117–157. Academic Press, London and New York.

Romer, A. S. (1941). Earliest land vertebrates of this continent. *Science N.Y.* **94**, 279.
Romer, A. S. (1957). The appendicular skeleton of the Permian embolomerous amphibian *Archeria. Contr. Mus. Geol. Univ. Mich.* **13**, 103–159.
Romer, A. S. (1958a). An embolomere jaw from the Mid-Carboniferous of Nova Scotia. *Breviora* No. 87, 1–8.
Romer, A. S. (1958b). Tetrapod limbs and early tetrapod life. *Evolution* **12**, 365–369.
Romer, A. S. (1963). The larger embolomerous amphibians of the American Carboniferous. *Bull. Mus. comp. Zool. Harv.* **128**, 415–454.
Romer, A. S. (1964). The skeleton of the Lower Carboniferous labyrinthodont *Pholidogaster pisciformis. Bull. Mus. comp. Zool. Harv.* **131**, 129–159.
Romer, A. S. (1966). "Vertebrate Paleontology", 3rd Edn. University Press, Chicago.
Romer, A. S. (1969). A temnospondylous labyrinthodont from the Lower Carboniferous. *Kirtlandia* No. 6, 1–20.
Romer, A. S. (1970). A new anthracosaurian labyrinthodont. *Proterogyrinus scheelei*, from the Lower Carboniferous. *Kirtlandia* No. 10, 1–16.
Stovall, J. W. (1948). A new species of embolomerous amphibian from the Permian of Oklahoma. *J. Geol.* **56**, 75–79.
Tilley, E. H. (1971). "Morphology and Taxonomy of the Loxommatoidea (Amphibia)". Ph.D. Thesis, University of Newcastle upon Tyne.
Warren, J. W. and Wakefield, N. A. (1972). Trackways of tetrapod vertebrates from the Upper Devonian of Victoria, Australia. *Nature, Lond.* **238**, 469–470.
Watson, D. M. S. (1914). On a femur of reptilian type from the Lower Carboniferous of Scotland. *Geol. Mag.* (6) **1**, 347–348.
Watson, D. M. S. (1919). The structure, evolution and origin of the Amphibia – The "orders" Rhachitomi and Stereospondyli. *Phil. Trans. R. Soc.* **B209**, 1–73.
Watson, D. M. S. (1929). The Carboniferous Amphibia of Scotland *Palaeont. hung.* **1**, 219–252.
Watson, D. M. S. (1951). "Paleontology and Modern Biology". Yale University Press, New Haven.
White, T. E. (1939). Osteology of *Seymouria baylorensis* Broili. *Bull. Mus. comp. Zool. Harv.* **85**, 325–409.

17 The Tetrapod Assemblage from Nýřany, Czechoslovakia

ANDREW R. MILNER

Department of Zoology, Birkbeck College (University of London), Malet Street, London WC1, England

Abstract: The tetrapod assemblage from the late Westphalian (Upper Carboniferous) Plattelkohle from the Humboldt mine at Nýřany, Czechoslovakia, is ‹ of the largest and most diverse accumulations of Carboniferous amphibians and reptiles. The material accumulated, possibly within a century, in a small shallow, poorly aerated swamp-lake in a vegetation-rich intermontane basin with a high inflow of organic and inorganic sediment. A census of over 450 tetrapod specimens permits the abundant endemic forms to be distinguished from the rare erratic elements from neighbouring environments. By comparison with other Westphalian assemblages, associations of open-water, shallow swamp-lake and terrestrial/marginal tetrapods can be identified.

The open-water amphibians are a small number of specialized lineages which are associated with fish-dominated assemblages and which were either large piscivores or were modified for anguilliform swimming. The shallow-water/swamp-lake assemblage appears to include both permanently aquatic tetrapods and seasonally aquatic forms, the latter including larvae or juveniles of amphibious or terrestrial adults and also a breeding association of terrestrial adults (*Scincosaurus*). The content of swamp-lake assemblages may be influenced by factors such as season of deposition/preservation and degree of accessibility to fishes. The terrestrial/marginal tetrapods all appear to have been insectivores or carnivores preying on other vertebrates, and the terrestrial tetrapod community must have been dependent on leaf-litter and forest-floor invertebrates as primary consumers. The absence of terrestrial herbivorous vertebrates in Westphalian lowland assemblages is attributed to the coal-forest being made up of arborescent lycopods or, in dryer conditions, gymnosperm trees, giving a concentration of primary production and edible plant material in a canopy layer. Such

Systematics Association Special Volume No. 15, "The Terrestrial Environment and the Origin of Land Vertebrates", edited by A. L. Panchen, 1980, pp. 439–496, Academic Press, London and New York.

a forest, like modern canopy-forest, could not have supported a high density of ground-dwelling herbivores, while the large (2–4 m) Stephano-Permian herbivores (the Diadectidae and Edaphosauridae) with primitive tetrapod gait were unable to adopt an arboreal existence.

INTRODUCTION

The small mining town of Nýřany (previously) Nürschan) is situated about 13 km south-west of Plzeň in western Bohemia, now in north-west Czechoslovakia. Carboniferous vertebrates from the Humboldt mine at Nýřany were first reported by Antonín Frič (Anton Fritsch) in 1870 and some material was named and given a preliminary description a few years later (Frič, 1876). Between 1879 and 1901, Frič published his monumental 15-part study "Fauna der Gaskohle und der kalksteine der Permformation Bohmens" in which he described 41 species of amphibian from Nýřany, mostly from collections in Bohemia. Further tetrapod material was dispersed in collections over Europe and was subsequently described or referred to by Jaekel (1902, 1909, 1911), Broili (1905, 1908, 1924), Schwarz (1908), Hummel (1913), Watson (1913, 1926), Pearson (1924), Stehlik (1924), Bulman and Whittard (1926) and Augusta (1940). The assemblage was reviewed and further material described by Steen (1938) and the systematics of the labyrinthodonts was rationalized by reduction in synonymy of several nominal taxa by Romer (1947). Subsequent work has been orientated towards more detailed studies of individual taxa, in particular, the loxommatid *Baphetes* (Beaumont, 1977), the batrachosaurs (Carroll, 1969, 1970a, 1972; Panchen, 1970), the microsaurs (Carroll, 1966; Carroll and Gaskill, 1978), the captorhinomorphs (Carroll and Baird, 1972) and the pelycosaur *Archeothyris* (Reisz, 1975). Briefer reviews or significant mention of Nýřany tetrapods are made by Carroll (1964) and Boy (1972) on temnospondyls, Brough and Brough (1967) and Carroll and Baird (1968) on microsaurs, and Baird (1964), McGinnis (1967) and Lund (1978) on aïstopods.

Currently unpublished work incorporating study of Nýřany taxa includes theses by A. R. Milner (1974) on "branchiosaurs" including *Branchiosaurus* and *"Limnerpeton"*, Bossy (1976) on urocordylid nectrideans including *Sauropleura*, and A. C. Milner (1978, also Chapter 15 of this volume) on non-urocordylid nectrideans including *Scincosaurus*. Thus most of the Nýřany tetrapods have been revised in the last decade or are being revised, and it is now possible to present a faunal list of 22 species (Table I) which should remain relatively stable. With this fundamental prerequisite achieved, an assessment of the palaeoecology of the Nýřany tetrapods becomes practicable for the first time.

The significance of the Nýřany assemblage is that it is one of only two diverse assemblages of well-preserved Carboniferous tetrapods represented by large numbers of individual specimens (at least 500). Only the contemporary, late Westphalian swamp-lake assemblage from Linton, Ohio, is represented by as much material, comparably preserved. Other assemblages are either poorly preserved (Jarrow, Eire), consist of disarticulated fragments (Joggins, Nova Scotia) or are restricted to a few (2–5) taxa (Greer, West Virginia, and Newsham, Northumberland). It may further be noted that the difference between the comparatively good representation of tetrapods in Westphalian deposits (*c*. 30 families) and the poorer representation in earlier Carboniferous deposits (*c*. 10 families) is largely attributable to the contents of just three ponds or small lakes (Jarrow, Linton and Nýřany) and several hollow lycopod stumps in two localities (Joggins and Florence). Hence, despite being of relatively late Carboniferous age, the well-preserved assemblages from Nýřany and Linton remain significant sources of information about many aspects of Carboniferous tetrapod biology.

The Nýřany assemblage marks the latest appearance of many amphibian taxa in Europe (Loxommatidae, Cochleosauridae, Edopidae, Gephyrostegidae, Phlegethontiidae, Urocordylidae and Tuditanidae), most Stephanian and Permian assemblages being characterized by a restricted range of families (Actinodontidae, Micromelerpetontidae, Branchiosauridae and Discosauriscidae). The Nýřany assemblage would therefore appear to be a late example of its kind but is probably representative for Westphalian faunas in so far as many of the component families have been described from early Westphalian localities such as Jarrow (Westphalian A) and Joggins and Newsham (Westphalian B). This implies that similar faunas were present throughout the Westphalian in Euramerica, but not necessarily earlier. The Nýřany assemblage is clearly too late to be of direct relevance to the origin and early radiation of tetrapods, but late Westphalian assemblages such as that from Nýřany are the earliest which can be used to make inferences about the ecological relationships of Palaeozoic tetrapods. This exercise is an attempt to identify some associations within a major assemblage of Carboniferous tetrapods, to link these associations to different environments and to reconstruct the roles which those tetrapods played.

STRATIGRAPHY AND PALAEOENVIRONMENT

Almost all the tetrapods from Nýřany were collected from a 30 cm thick sequence of laminated canneloid shales and mudstones, the Plattelkohle, near the base of the Nýřany Gaskohle series, principally from the Humboldt Mine at Nýřany. The Nýřany Gaskohle series is part of the Nýřany Member of the Lower Grey Beds of the Plzeň limnic basin and was dated as being of uppermost Westphalian D age within the Upper Carboniferous by Nemejč (1952). This stratigraphical position has been followed by subsequent authors (e.g. Pešek, 1968; Holub and Tásler, 1978).

The late Palaeozoic intermontane sedimentary basins of the Bohemian Massif region of central Europe developed, as a result of the Variscan Orogeny, as depressions on a folded Proterozoic and early Palaeozoic basement (Holub and Tásler, 1978). The Plzeň basin first came into existence as a depositional area in the Westphalian B when the Radnice Member was laid down (Pešek, 1968, Abb. 10 transposed from Abb. 7). After a discontinuity, the Nýřany Member was laid down over a much greater area during the Westphalian D (Pešek, 1974, Fig. 4). Most of the Nýřany Member consists of flood-plain deposits and alluvial fans from rivers flowing off the massif and crossing the basin. Only a restricted area to the north of Nýřany bears lacustrine and swamp deposits including the Gaskohle series. The lacustrine and swamp facies occupy an area about 8 km east to west and 2 km north to south, with river-bed facies extending further north-east from this area (Pešek, 1974, Fig. 4) indicating that a river flowed through the swamp. This was one of the very few perennial lake sequences in the Bohemian Massif area at this time (Holub *et al.*, 1975) and represents an intermittently overgrown lake (Pešek, 1974).

The sequence of facies within the lacustrine Gaskohle series was described by Frič (1879) for the Humboldt mine locality. They represent a phase of local transition from a swamp to a large stratified lake. The lowest horizon in Frič's section is a green shale bearing *Calamites* with occasional thin layers of coal. *Calamites* was a characteristic swamp inhabitant (Scott, Chapter 5 of this volume) and the deposit was laid down under reducing conditions, both of which suggest a swamp environment. Above this is the 30 cm band of Plattelkohle, the finely laminated coals, shales, mudstones and ironstones which contain virtually all the Nýřany tetrapods (Frič, 1879). From its fine lamination and undisturbed bedding, the Plattelkohle was evidently laid down under current-free conditions and in the absence of a benthic fauna. As described in more detail below, much of the included skeletal material is preserved in articulation and was evidently undecomposed or little decomposed, implying minimal transport prior to burial and an anaerobic, disturbance-free substrate. Even some of the terrestrial erratics are so preserved indicating close proximity to dry land. The water body at this time was probably a small, shallow, frequently stagnant pond or lake situated in an open part of a swamp-forest (Westoll, 1944; Boy, 1977). The swamp-forest was probably composed of *Calamites* in up to 1 m of standing water. The mudstone and shale layers in the Plattelkohle indicate that an occasional high inflow of inorganic silt occurred, as well as a sustained deposition of carbon from allochthonous sources (algae, spores and plant material from the vegetation-rich surroundings) (Boy, 1977). The interpretation of the Plattelkohle swamp-lake as a shallow, poorly aerated body, less than 1 km^2 in area and cut off from other water bodies except by percolation through the *Calamites* swamp, is also supported by the very presence of diverse small aquatic tetrapods combined with a low diversity of fish and the absence of both large fish and large aquatic tetrapods (i.e. more than 50 cm long). In contrast, at the Newsham locality, many of the remains are of vertebrates from 3 m to 5 m long.

Skoček (1968), working on contemporary, seasonally laminated perennial lake deposits from the Bohemian Massif area, concluded that 100 years of deposition was represented by from 4·4 cm to 10·8 cm of consolidated mudstones. If this is applied to the Plattelkohle, this suggests a deposition time of 300 – 700 years for the entire 30 cm sequence. The tetrapods are found in at least three distinct horizons within the Plattelkohle (Frič, 1879). One labyrinthodont specimen was collected from the base of the Plattelkohle but most of the several hundred small tetrapods and some of the fish come from within 1 cm of a white or pale grey band of mudstone, 2 – 5 mm thick, in the lower Plattelkohle (Frič, 1879). This band could be seen in the sides of the slabs of many of the specimens examined in the course of this study. The general proximity of most of the specimens to this band indicates, from Skoček's above-mentioned estimates, that the main assemblage was initially preserved in a period of not more than a hundred years and probably substantially less. In specimens where the white band is a uniform 3 – 4 mm deep, the distance from specimen bedding plane to band varies from 3 mm to 10 mm indicating that specimens were not the product of a single catastrophic event and were not restricted to a single bedding plane. The specimens are always within a coal layer and preserved as pale brown bone. Associated with the small tetrapods are small xenacanths (*Orthacanthus*) up to about 30 cm long, and small acanthodians (*Traquairichthys*) and palaeoniscoids (principally *Pyritocephalus*) both less than 10 cm long. Also present are abundant crustaceans (*Palaeocaris*) sometimes packed in single bedding planes, and frequent terrestrial arthropods, particularly a wide range of myriapods from 1 cm to 30 cm long. A third productive layer within the Plattelkohle at a higher horizon is a shaly coal containing some of the large *Cochleosaurus* specimens (Frič, 1879). These can be distinguished from *Cochleosaurus* from the main assemblage by the darker grey-brown colour of the preserved bone and the lighter grey matrix.

Above the Plattelkohle are 30 cm of papershales followed by 30 cm of canneloid Breitelkohle. These layers contain much fish material and few tetrapod bones. The papershales are rich in small palaeoniscoids and acanthodians while the Breitelkohle contains isolated teeth up to 2 cm long of xenacanth sharks. These deposits were evidently laid down in a large permanent, stratified lake, predominantly inhabited by fish, some large, and finally becoming deep enough to permit formation of cannel coal from the sapropel on the lake bed under the hypolimnion (Moore, 1968).

It appears therefore that the main Nýřany assemblage originates from a band in the Plattelkohle laid down over a relatively short period of time, not more than a hundred years and possibly substantially less. It represents a phase in the evolution from a swamp to a perennial stratified lake, when the water body was a small, shallow lake in a swamp-forest transected by a river-system in an intermontane basin. Although the tetrapods were not in a single bedding plane, the assemblage accumulated in a single environment in a comparatively short period of time and can be considered to be derived from a set of contemporary communities occurring within one geographical area, namely the Plzeň basin.

TAPHONOMY

The fine preservation of the main assemblage in the vicinity of the white band is the product of an unusual combination of factors and provides information about the circumstances in which the animals died and were preserved. The several fortuitous factors which have combined to permit the preservation of this material are as follows.

(i) The presence of unoxidized carbon (coal), of iron fixed as siderite and of intact animal remains demonstrates that the substrate was a reducing one and must have been completely anaerobic to prevent decomposition or any form of oxidation. The presence of finely laminated interbedded coals and shales shows that the lake bed was laid down in current-free conditions and was biologically inert, i.e. not supporting a benthic or burrowing fauna which would have disturbed the sediment. If the lake bed was current-free and anaerobic, this indicates the presence of a stagnant, poorly aerated hypolimnion within the lake, or even a lake which was entirely stagnant and anoxic at times. Streams bringing in silt and dissolved salts must have flowed into the upper stratum of the lake on occasion, but without generating a complete circulation. Vertebrates dying in the lake and sinking to the bottom would have been unscavenged, undisturbed by currents and would have decomposed slowly by autolysis or not at all.

(ii) Westphalian lake-bed facies rarely contain bone, being derived from sapropels produced in anaerobic conditions. In such conditions, non-decomposition of the plant acids in the buried plant debris or derivatives produces an acid substrate in which bone dissolves completely or occasionally incompletely as at Jarrow, Eire (Rayner, 1971, p.444). At Nýřany the presence in the Plattelkohle of bands of siderite and a greyish mudstone testifies to the existence of a base-rich inflow of silt or dissolved salts from the surrounding massif. This has neutralized the acids in the substrate and permitted bone to survive. The proximity of much of the skeletal material to the white mudstone band is probably not a coincidence.

(iii) The articulated condition of much of the tetrapod material not only demonstrates that post-burial decomposition did not occur, but also that the animal cadavers must have sunk immediately after death and settled on or in the substrate without transport having occurred. Some terrestrial erratics (*Archeothyris*, *Sparodus*) are only represented by isolated bones and probably were transported, but these are exceptional. Most of the assemblage is autochthonous, composed of individuals which were swimming in the lake prior to death. The intact forms which appear to be marginal or terrestrial animals are rare erratics and presumably individuals which either fell in and drowned or were facultative swimmers which occasionally crossed the lake.

(iv) Many of the abundant taxa are preserved as individuals over a wide size-range, most of which are presumably juveniles or larvae. These juveniles were evidently not dying of old age nor were they being predated. Either they were killed by minor local catastrophes such as the lake periodically becoming anoxic up to the surface in dry, wind-free conditions when circulation was minimal, or they may represent an accumulation of individuals swimming into the lower anaerobic region of the water body and dying there.

(v) As noted above, many of the smaller tetrapods are completely in articulation with structures such as sclerotic plates, branchial ossicles and dermal scales in place and showing no evidence whatsoever of decomposition. However, some of the small specimens and most of those more than 30 cm long show evidence of varying degrees of decomposition and disintegration manifested by progressive disarticulation of the skeleton, usually *in situ*. The pattern of this disintegration suggests that the cause is autolysis of the hind-gut contents occurring before, during or after burial and generating gas bubbles inside the corpse with various possible effects. The work of Schäfer (1972, pp.53–61) on the patterns and rates of decomposition of small marine fish at 15–18°C gives some indication of the nature of this process as it occurred in the Nýřany tetrapods. Schäfer found that, in some types of fish, the gas generated was sufficient to refloat the body which then disintegrated while floating in less anoxic conditions or, having ruptured and lost the decomposition gas bubbles, re-sank in a more poorly articulated condition. Other fish, having initially sunk, were not refloated by gas which was either inadequate to float a dense, bony corpse (e.g. *Callionymus*, *Trigla*) or which diffused or ruptured its way out of the abdominal wall without disturbing the rest of the corpse.

Those small tetrapods from Nýřany which are intact have evidently not been subject to significant gas build-up or flotation, but there are also isolated fragments of small animals suggesting that some did refloat and disintegrate over a wider area. A first stage in this process is shown by the *Microbrachis* specimen shown in Fig. 1. Here the anterior region is intact (the anterior skull is off the slab) but autolysis of the gut contents has generated gas which has ruptured the body wall at the bases of the right posterior ribs which, together with the attached skin and dermal scales, have been displaced to one side. Further disintegration of the posterior region has occurred and it is possible that this end of the animal was partly floated by the gases of decomposition and was buried more slowly and disintegrated further as a consequence. In other *Microbrachis*

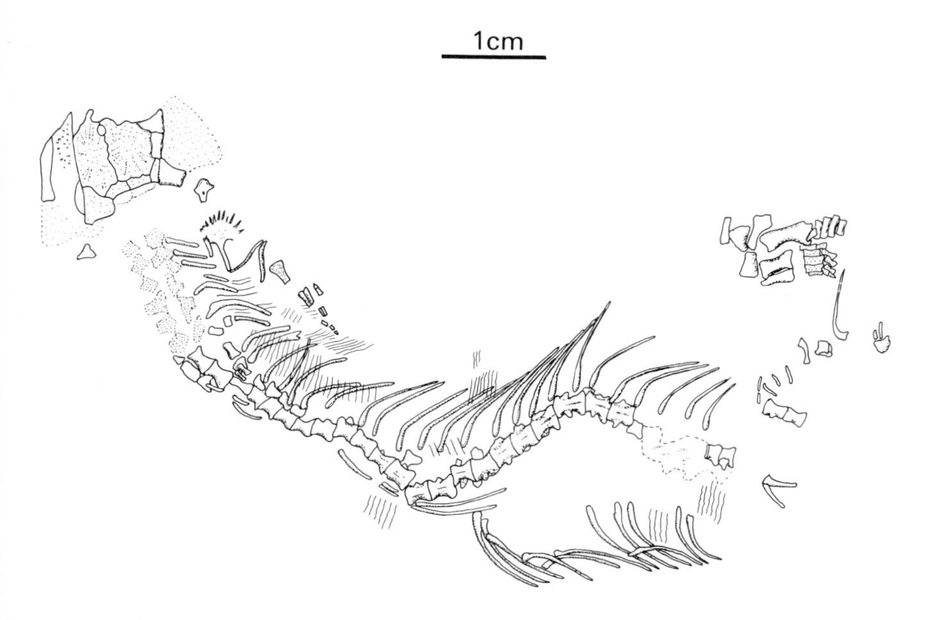

Fig.1 *Microbrachis pelikani* Fr. (uncatalogued specimen, Národní Muzeum, Praha) showing effects of partial decomposition in the abdominal region. A build-up of gas in the hind-gut has ruptured the partly decomposed body wall dorsally and spread the nearly detached skin and ribs out to one side.

specimens the posterior abdomen is completely missing although the anterior half of the skeleton is still intact; presumably the former has floated elsewhere. This variation in preservation of small tetrapods is comparable to the individual variation that Schäfer found in some fish and may depend on no more than the nature and quantity of the intestine contents at the time of death. A further category of disintegration found in many medium to large specimens (e.g. most *Cochleosaurus, Amphibamus*, the *Baphetes* skull and some large microsaurs), is characterized by the intact skull being present in isolation or associated only with disarticulated pectoral elements. The combination in these forms, particularly the temnospondyls, of large, dense skull and pectoral girdle anteriorly and a potentially autolysing hind-gut seems to have regularly resulted in the bodies and tails of the corpses floating away or disintegrating while the denser anterior ends remained where they settled and were buried. There is a general correlation between size and degree of disintegration, few large tetrapods being represented by intact skeletons. This might be due to the larger corpses taking longer to be buried by incoming silt or it may be a consequence of larger animals having a larger potentially decomposable volume of hind-gut, capable of generating more gas before decomposition was inhibited by the surroundings. Schäfer's results suggest that undisintegrated specimens must have experienced less than two days' decomposition unless the lake-bed was particularly cool. It is unlikely that sufficient sediment to completely bury corpses could be laid down in two to three days without the transporting currents stirring the lake detectably. This suggests that the lake had an anaerobic hypolimnion and that the size-linked decomposition generally related to the degree of autolysis and the volume of generated gas developed prior to the inhibition of decay by the anaerobic surroundings. Occasionally, a specimen may completely disintegrate *in situ* without any elements being floated off, some *Gephyrostegus* and *Cochleosaurus* occurring in this condition.

The Nýřany assemblage exists because of the unusual permutation of basic and anaerobic conditions on the bed of a Westphalian swamp-lake. These conditions foreshadow those of the Stephanian and Autunian lake-beds of Europe in which there was much less carbon but in which preservational conditions were similar, as exemplified by the many "branchiosaur" assemblages found in them.

These Autunian assemblages are species-depauperate, usually made up of only 2–3 types of tetrapod, and occur in oil-shales, mudstones and freshwater limestones. The uniqueness of the Nýřany locality lies in the fact that it is one of the earliest Variscan intermontane limnic basins inhabited by one of the latest Carboniferous coal-swamp faunas.

METHOD OF STUDY – A CENSUS

In order to identify associations with the Nýřany assemblage, I have carried out a quantitative study on as much of the Nýřany material as I have been able to examine. About 450 determinate or semi-determinate tetrapod specimens from the Plattelkohle were recorded on card with as much of the following information as possible for each specimen: (i) identity; (ii) size usually skull length or trunk length but sometimes humerus length was used, and for aïstopods, individual vertebral length was used; (iii) preservation, whether articulated, partly articulated, disarticulated or fragmentary. The sample is composed mainly of material in the collections at the Národní Muzeum, Praha, and the Západočeské Muzeum, Plzeň. It also includes material from the British Museum (Natural History); Cambridge University Museum of Zoology; the Royal Scottish Museum; the Hunterian Museum, Glasgow; the Geologisches Museum, Marburg; the Paläontologisches Institut, Johannes Gutenberg-Universität, Mainz; the Senckenberg Museum, Frankfurt; and the Bayerische Staatssammlung, München. It includes described specimens only, from the Humboldt Museum, Berlin, but not the main collection there. This might have the effect of rendering some of the rare erratics appear to be slightly more frequent than they should be, but not significantly so.

Why attempt a palaeoecological quantitative study on the basis of collections made a century ago? The unfortunate situation is that all the major compact assemblages incorporating Westphalian tetrapods, namely those from Jarrow, Newsham, Joggins, Linton and Nýřany, were discovered and collected between 1850 and 1900. Pending discovery of a new Westphalian tetrapod locality which could be subject to first-hand palaeoecological study, one is reduced to attempting interpretation of these century-old collections of which Nýřany assemblage is one of the largest and most diverse. The original systematic descriptions of Nýřany vertebrates concentrated on recording from one to several of the best preserved specimens of each taxon and until now no attempt has been made to assess the relative frequency of all of the tetrapod taxa at Nýřany. Only Frič (1879–1901) made any assessment of the numbers of specimens of some of the taxa which he described, but he was so profligate with his taxa that his figures cannot be translated into currently used taxa.

The principal potential losses of information in a century-old collection are in the precise geography and horizon of collection and in the possible sampling bias of selection of specimens during collection. For Frič's material, the bedding data for specimens are not precise, and although it is certain that the Plattelkohle material surveyed here came from a narrow band, it cannot now be established whether the assemblage was the product of a few minor catastrophes or up to a century's worth of continuous accumulation of occasional specimens as described under Taphonomy (iv). Likewise, the different taxa might be uniformly distributed or occur locally in different bedding planes. These possibilities do not significantly affect the interpretation of the rare erratic elements in the assemblage, but there is a loss of information concerning the coexistence or non-coexistence of the various abundant forms which is discussed later when the shallow-water association is described. It is considered that the geographically, chronologically and lithologically restricted nature of the Plattelkohle is such that the enclosed fossil assemblage can be considered as a single entity for the purposes of establishing general associations. More seriously problematical is the possibility that the material passed through the filter of morphologically orientated collectors looking for well-preserved individual specimens rather than unbiased samples of material. Examination of the catalogued and uncatalogued material at Prague, which forms the bulk of the census material, convinces me that Frič and his contemporaries were conscientious in their collection of much unpromising and scrappy material. The main limitation to sample size has been my inability to determine many of the scraps and fragments rather than any apparent selection on Frič's part, and any more poorly preserved material which Frič may have failed to collect would have been largely indeterminate. The other, smaller collections incorporated here all consist of selected well-preserved specimens and show considerable variation in the representation of taxa. Only by combining all the smaller collections with each other and with the larger collection at the Národní Muzeum, could I procure a sufficiently large sample to minimize the possibility of bias through inadequate sample size. The possibility remains that the abundant small aquatic forms may be under-represented if a proportion of the original cadavers floated and disintegrated into isolated bones which are either indeterminate or were uncollected.

TAXONOMY OF NÝŘANY TETRAPODS

A census of an assemblage of organisms is only as valid as the taxa to which the material is referred. The taxa used in this census mostly reflect the published situation, except for a few synonymies and attributions based on my unpublished work on the temnospondyls. The following notes serve to clarify some of the nomenclature and attributions of material. All genera from Nýřany are, in the most recent literature, monospecific and therefore my attribution of material to taxa has effectively been to genera. No attempt was made to identify multiple species within a genus. Thus I have followed Carroll and Gaskill (1978) in synonymizing all *Microbrachis* species and have not attempted to confirm or refute this synonymy. The currently valid nomenclature with authorship is incorporated in Table 1 which gives the quantitative results of the census. The sources are as follows:

Temnospondyli. The nomenclature of the loxommatid follows Beaumont (1977). The only diagnostic skull is not the original type and I include here with doubt the type mandibular ramus and another similar jaw ramus. The edopoid systematics is based mainly on Romer (1947) but also includes some unpublished attributions. The *Edops*-like *Gaudrya* includes the "*Capetus*" specimen of Steen (1938) and the "*Sclerocephalus*" skull of Jaekel (1908), but not any of the "*Nyrania*" material. *Cochleosaurus* incorporates the various specimens in Steen's growth series and many more, plus the "*Nyrania*" type skull which possesses the base of a postparietal lappet, one of the diagnostic features of *Cochleosaurus*. There are three dissorophoids present. The micromelerpetontid has been confused with *Branchiosaurus* in the past but is a longer-bodied, shorter-limbed form like *Micromelerpeton* from the Autunian of the Rheinpfalz (Boy, 1972). The binomen *Limnerpeton laticeps* has priority but as

Steen (1938, p.263) made the indeterminate *L.modestum* into the lectogenotype of *Limnerpeton*, a new generic name will be required. *Branchiosaurus salamandroides* is similar to the Permian *B. petrolei* (for description see Boy, 1972) but more primitive in several characters. The dissorophid has been most recently described as *Amphibamus calliprepes* by Carroll (1964) but the unlocated specimen of *Potomochoston limnaios* (Steen, 1938) may be the same species. Several undescribed specimens exist.

Batrachosauria (sensu *Panchen, 1970*). The systematics follows Carroll (1970a, 1972) and Panchen (1970). Following Panchen (1972b), *Solenodonsaurus* is referred to the Batrachosauria and no specimens other than those described are referred to this genus. *Gephyrostegus*, however, is represented by several specimens other than those described by Carroll (1970a). The scattered material of "*Diplovertebron punctatum*" which Carroll removed from type status is a *Gephyrostegus* as is some of the *Sparagmites lacertinus* material. Several undescribed specimens belong here including one (N.M.P. M398) which appears to be a *Gephyrostegus* with embolomerous caudal vertebrae. The lectotype slab of *Diplovertebron* (see Carroll, 1970a; Panchen, 1970) and one other specimen are the only possible eogyrinid specimens and attribution to that family is doubtful. The lectotype could equally be a poor association of gephyrostegid fragments.

Aïstopoda. I have followed McGinnis (1967) in synonymizing the Nýřany *Phlegethontia* with that from Linton, Ohio. As noted by Lund (1978), the specific name of the Linton animal has several years' priority. The nomenclature of *Ophiderpeton* follows Frič (1880) and Baird (1964). No comparative study of *Ophiderpeton* species has been undertaken and it is not certain that all the Nýřany material of *Ophiderpeton* represents one species. There is a considerable size-range in the preserved material.

Nectridea. The nomenclature of the urocordylid follows Baird (1964, p.14) and Bossy (1976). Also following Bossy (1976), only a single species of *Sauropleura* is recognized in the Nýřany fauna. The re-establishment of the family Scincosauridae follows A.C. Milner (Chapter 15 of this volume). *Scincosaurus crassus* is sufficiently distinctive that its taxonomy has remained straightforward.

Microsauria. The systematics and taxonomy follows Carroll and Gaskill (1978) in all respects with the addition of the little-known

Microbrachis fritschi Augusta (1940) to the synonymy of *M. pelikani*.

Reptilia. The taxonomy of the Nýřany reptiles follows the published work of Carroll and Baird (1972) and Reisz (1975).

CENSUS RESULTS

The results of the census are shown in Table I and depicted in histogram form in Fig. 2. A total of 457 specimens was incorporated in the census, but it may be noted that several European collections were not examined and that the total number of specimens collected at Nýřany is greater. Of the 457 specimens, 401 are believed to be determinate, 9 doubtfully so, and 47 belong to various categories of

Table I. Results of census of Nýřany tetrapods

	AMPHIBIA		
Temnospondyli			
Loxommatidae	*Baphetes bohemicus* (Frič) Beaumont 1977	1 (+2?)	
Edopidae	*Gaudrya latistoma* Frič 1885	4 (+2?)	+15[a]
Cochleosauridae	*Cochleosaurus bohemicus* (Frič) 1885	47[e]	
Micromelerpetontidae	"*Limnerpeton*" *laticeps* Frič 1881	48	
Branchiosauridae	*Branchiosaurus salamandroides* Frič 1876	39	+24[b]
Dissorophidae	*Amphibamus calliprepes* (Steen) Carroll 1964	9	
Batrachosauria (*sensu* Panchen)			
Gephyrostegidae	*Gephyrostegus bohemicus* Jaekel 1902	11 (+2?)	
Eogyrinidae?	*Diplovertebron punctatum* Frič 1885	2	
Solenodonsauridae	*Solenodonsaurus janenschi* Broili 1924	3	
Aiistopoda			
Phlegethontiidae	*Phlegethontia* cf. *linearis* Cope 1871	8	
Ophiderpetontidae	*Ophiderpeton granulosum* Frič 1880	23 (+2?)	
Nectridea			
Urocordylidae	*Sauropleura scalaris* (Frič) Baird 1964	32	
Scincosauridae	*Scincosaurus crassus* Frič 1876	66	
Microsauria			
Microbrachidae	*Microbrachis pelikani* Frič 1876	82	
Hyloplesiontidae	*Hyloplesion longicostatum* (Frič) 1883	12	
Gymnarthridae	*Sparodus validus* Frič 1876	8	
Tuditanidae	*Crinodon limnophyes* (Steen) Carroll and Gaskill 1978		+ 7[c]
Hapsidopareiontidae	*Ricnodon copei* Frič 1883	11 (+1?)	
	REPTILIA		
Captorhinomorpha			
Romeriidae	*Brouffia orientalis* Carroll and Baird 1972	1	+ 1[d]
	Coelostegus prothales Carroll and Baird 1972	1	
Pelycosauria			
Ophiacodontidae	*Archeothyris* sp. Reisz 1975	1	
Reptilia *incertae sedis*	Unnamed reptile Carroll and Baird 1972	1	
	Complete total =	457	

[a]Fragments of indeterminate large temnospondyls. [b]Indeterminate small dissorophoid temnospondyls. [c]Indeterminate non-microbrachid microsaurs. [d]small reptile jaw. [e]Excludes *Cochleosaurus* from "upper band".

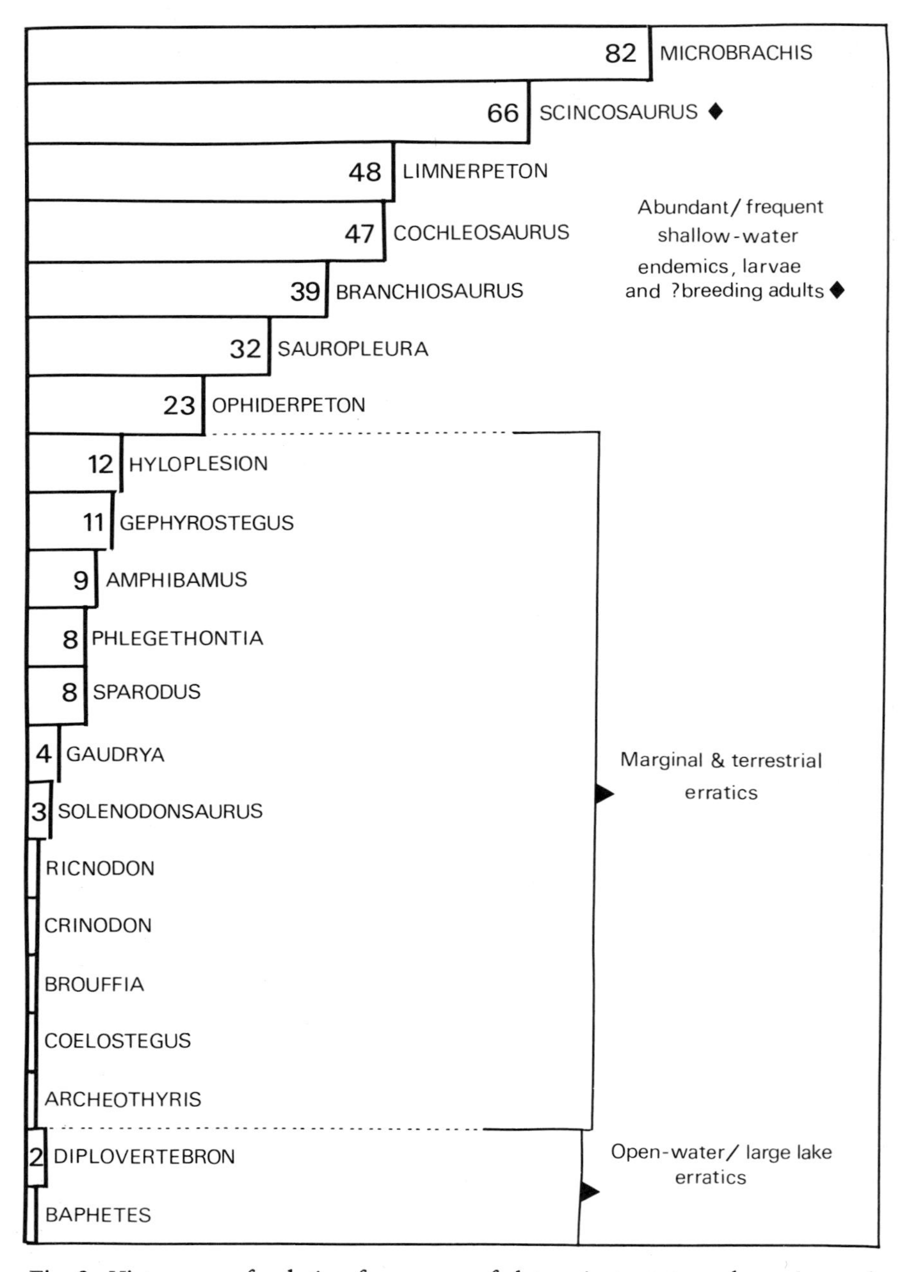

Fig. 2. Histogram of relative frequency of determinate tetrapod specimens in the studied sample of the Nýřany assemblage, based on the information in Table I.

indeterminacy. The many indeterminate temnospondyl specimens result in the representation of determinate temnospondyls being lower than its potential compared to, for example, the nectrideans *Sauropleura* and *Scincosaurus* which are determinable from tiny fragments. The indeterminate specimens do not, however, affect the general relationship of abundant to rare taxa in the assemblage, and it is this broad distinction between abundant endemics and rare erratics which is being sought in this study.

Table II. Maximum sizes of Nýřany tetrapods. (Dimensions are all snout-tip to pelvis (s–p) unless otherwise stated.)

	OPEN-WATER/LACUSTRINE	
Baphetes	Only specimen – 18cm skull (28cm)[a]	
Diplovertebron	Estimated 30cm s–p,	against *Eogyrinus*[g] (2m)[a]
	TERRESTRIAL/MARGINAL	
Gaudrya	Estimated 2m s–p,	against *Eryops*[g]
Solenodonsaurus	Estimated 60cm s–p[f]	
Archeothyris	Estimated 48cm s–p,	against Florence specimen[g]
Indeterminate reptile	Estimated 28cm s–p[b]	
Gephyrostegus	21cm s–p[f]	
Phlegethontia	Estimated 1m total length[e]	
Amphibamus	17cm s–p	
Sparodus	Estimated 16cm s–p,	against *Cardiocephalus*[g,c]
Crinodon	Estimated 16cm s–p,	against *Tuditanus*[g,c]
Coelostegus	16cm s–p,	juvenile?[b]
Brouffia	13cm s–p,	juvenile?[b]
Hyloplesion	8cm s–p[c]	
Ricnodon	Estimated 7cm s–p,	against *Saxonerpeton*[g,c]
Scincosaurus	5·5 cm s–p	
	SHALLOW-WATER/SWAMP-LAKE	
Ophiderpeton	Estimated 1·5m total length (based on "*O.forte*")	
Cochleosaurus	Estimated 80cm s–p,	against juvenile[g]
Microbrachis	17cm s–p	
Sauropleura	15cm s–p[d]	
"*Limnerpeton*"	11cm s–p,	against *Micromelerpeton*[g]
Branchiosaurus	5cm s–p	

[a]Comparable dimension in largest specimen from British coal measures. [b]Carroll and Baird, 1972. [c]Carroll and Gaskill, 1978. [d]Bossy, 1976. [e]McGinnis, 1967. [f]Carroll, 1972. [g]Largest skull or fragments scaled against intact small specimens or intact specimens of related forms from other localities as indicated.

Some taxa proved to be more abundant than a study of the literature suggested, in particular *Scincosaurus* (66 specimens), "*Limnerpeton*" (48 specimens) and *Gephyrostegus* (11 specimens). Others such as *Phlegethontia* (8) and *Hyloplesion* (12) proved to be represented by little more than the described material.

Where possible, specimens were measured and although much of the resulting information is not incorporated here, I have included in Table II a series of measurements or estimates of the maximum size which each taxon attains at Nýřany. Many of these are, of necessity, estimates based on the largest known skulls or fragments which have been scaled against intact specimens, either smaller conspecifics from Nýřany or confamilial or similar forms from elsewhere. In discussion of the roles played by the various taxa in their respective communities the size attained by each taxon is an important factor. Those species represented by only one or two specimens can only be represented by the dimensions of those specimens and many of the rare elements could be represented only by juvenile individuals. Carroll and Baird (1972) suggest this possibility for the Nýřany romeriids and it is a probable explanation of the small size of the known loxommatid and presumed eogyrinid material.

IDENTIFICATION OF ASSOCIATIONS AND COMMUNITIES

The tetrapod material from Nýřany and other Westphalian localities contains several types of information which can be combined to permit identification of associations and the interpretation of them as communities.

(i) Relative frequency. As can be seen from Table I and Fig. 2, there is a spectrum of abundance within the Nýřany assemblage. There are abundant to frequent forms represented by 20 – 80 individuals (5 – 20% of the sample), some occasional forms represented by 5 – 20 individuals (roughly 1 – 5% of the sample) and some rare erratics (1 – 4 individuals, fewer than 1%). On this alone, the abundant forms are more likely to represent major components of the endemic aquatic community, the occasional forms could be rare aquatics or abundant lake-margin forms appearing as erratics, the rare elements are likely to be accidental specimens brought in (or straying in) from environments other than the Nýřany water body or immediate margin.

(ii) Associations elsewhere. Comparison with other, more restricted, Westphalian assemblages, and with some early Permian assemblages, indicates association of certain tetrapod types with particular environments and with each other. The lycopod-stump localities of Joggins and Florence, Nova Scotia, are generally agreed to contain assemblages consisting predominantly of terrestrial tetrapods and arthropods (Carroll *et al.*, 1972), while the assemblage from Newsham is believed to be derived from a large body of open water (Panchen, 1970). The Lower Permian fissure fauna from Fort Sill, Oklahoma, is generally agreed to be made up of upland terrestrial tetrapods and, although the material is believed to be several million years younger than the Nýřany fossils, many of the tetrapods are closely related to Nýřany taxa. The associations from all these localities show some correspondence to the occasional and rare elements in the Nýřany assemblage.

(iii) Size distribution and degree of articulation. Some of the abundant forms are represented by a wide size-range of individuals whilst one, *Scincosaurus*, is represented by specimens which are almost uniform in size and represent a single age-class. Those which show a wide size-range may or may not show some type of modality in the preserved size-distribution. The maximum preserved size for a given genus at Nýřany (or, for a few genera, elsewhere) gives an indication of the possible adult size and, in several abundant forms, the largest individuals are rare, implying that most preserved individuals were juveniles (e.g. *Cochleosaurus*, "*Limnerpeton*"). Such juveniles are sufficiently smaller than the largest known specimens that they need not be assumed to have occupied the same niche, nor were they necessarily part of the same association. Some of the rare elements in the assemblage are only known from disarticulated fragments indicating the possibility of transport from another environment involving a longer pre-burial phase when decomposition could have occurred (e.g. *Archeothyris*, *Sparodus*).

(iv) Functional morphology. Many inferences about mode of life and life-habitat can be derived from the preserved skeletal components of the feeding, locomotor, respiratory and receptor systems. In this review there is insufficient space to discuss the morphology of each form, but reference is made to recent published work and some morphological features of interest are noted.

Of the above types of information (i), (ii) and sometimes (iii) serve to demarcate the major associations, while (ii), (iii) and (iv) can all be used in the interpretation of associations as living communities. The associations, as identified and itemized here, are testable in that they can be refuted or supported by the structures of further assemblages as and when they are discovered; the community structures suggested here are speculative in so far as they are not directly testable and are based on plausible interpretation of the associations. In the following sections, those associations which are rare components of the Nýřany assemblage are dealt with first, the endemic association and *Scincosaurus* being discussed subsequently.

1. Open-water/Lacustrine Association

The two tetrapods in the Nýřany assemblage identified as belonging to this association are the loxommatid *Baphetes* and the presumed eogyrinid *Diplovertebron*, both of which are extremely rare at Nýřany (Fig. 2).

In the earlier Westphalian of the British Coal Measures, the Loxommatidae and Eogyrinidae (restorations in Fig. 3) are the two most frequent tetrapod families. At Newsham, Fenton, Swanwick and Pirnie, members of the two families occur as the most frequent or only amphibians (Panchen and Walker, 1961). The Newsham assemblage is one of the largest of this type, and in it three amphibian families are represented by several tens of specimens, the Loxommatidae (*Megalocephalus*), the Eogyrinidae (*Eogyrinus* and *Pteroplax*) and the Keraterpetontidae (*Batrachiderpeton*). *Ophiderpeton* occurs as a single erratic. They are associated with a rich and diverse fish fauna, over 20 genera being recorded by Andrews (in Land, 1974) and two haplolepid genera are also present (Westoll, 1944). Other fish in the assemblage include osteolepidids, dipnoans, coelacanths, hybodonts, xenacanths, acanthodians and several palaeoniscoids. At Pirnie, *Baphetes* occurs with eogyrinid material; at Fenton, *Megalocephalus* and eogyrinids occur (Panchen and Walker, 1961); and a fish-rich assemblage from Longton Hall, Staffs., includes a keraterpetontid specimen. These associations appear to have been the inhabitants of large fish-dominated areas of open water, either large, permanent lakes as suggested by Panchen (1970), or large river channels in a deltaic system. The fossils from the Low Main Seam at Newsham are found in association with a "swelly" or local thickening

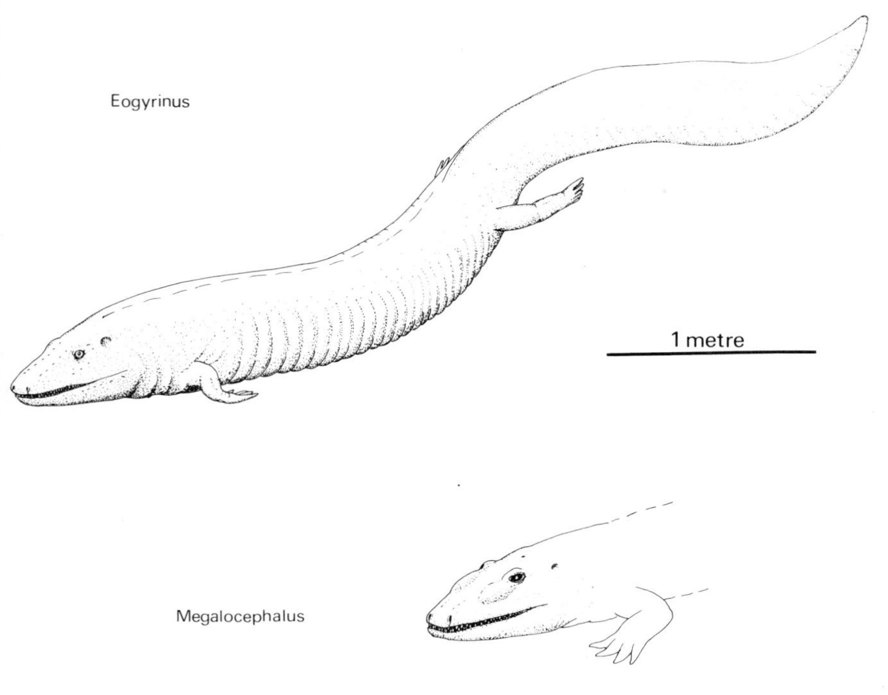

Fig. 3. Restorations of amphibians representing the families Eogyrinidae and Loxommatidae, the two principal members of the "open-water" association. The restorations are based on large individuals of the better-known Newsham genera *Eoryrinus* (modified after Panchen, 1972a) and *Megalocephalus*. The loxommatid postcranial skeleton is almost completely unknown.

of the seam which is several miles long (Land, 1974) and may have been a silted-up deltaic channel. As *Eogyrinus* attained a length of 4m (Panchen, 1972a) and some of the rhipidistian fragments belonged to fish up to 6m long, the original water body must have been open and large enough to support a fauna with such large predators as the ultimate consumers. From their morphology, both eogyrinids and keraterpetontids were specialized swimmers (Panchen, 1970; A. C. Milner, Chapter 15 of this volume). The loxommatid postcranium is unknown but their skulls are either crocodile-like (*Megalocephalus*) or broader and alligator-like (*Baphetes*) in general shape with long, slender teeth suggestive of piscivory, and they were presumably aquatic or amphibious piscivores. The morphology of the

Newsham tetrapods is consistent with the interpretation that they are part of an open-water association.

The above tetrapod families and many of the associated fish are rare in, or absent from, the Nýřany assemblage. As recorded in this census, *Baphetes* is represented by one certain and two doubtful specimens, while eogyrinids are doubtfully represented by the lectotype of *Diplovertebron* and one other specimen. The only dipnoan specimens are three scales. Unrecorded from the Plattelkohle are *Megalocephalus*, keraterpetontids, osteolepidids, coelacanths, hybodonts or large xenacanths. The *Baphetes* specimen is an isolated cranium which may or may not have been transported, most large Nýřany tetrapods being partly broken up, as discussed under Taphonomy. The presumed eogyrinid material is made up of isolated or disarticulated associated bones. The specimens of both families are small examples of their kind (see estimated sizes in Table II) and could have been either juveniles or "dwarf" adults of smaller intermontane basin species. The absence at Nýřany of forms such as *Megalocephalus* or *Keraterpeton* is most likely to be ecological and cannot be taken to have chronological or geographical significance.

The contemporary swamp-like assemblage from Linton, Ohio, is larger and contains a greater diversity of fish than Nýřany, but the comparable open-water elements are equally rare if allowance is made for the larger number of specimens collected from Linton. The osteolepidid *Megalichthys* is absent, *Baphetes* is represented by one specimen, *Megalocephalus* by four specimens (Beaumont, 1977), the eogyrinid *Leptophractus* by six specimens (Panchen, 1977) and the keraterpetontid *Diceratosaurus* by eight specimens recorded in the literature (A. C. Milner, personal communication). The Linton fauna is thus comparable to the Nýřany fauna in the rarity of this association within the larger assemblage.

In conclusion, the loxommatids and eogyrinids from Nýřany are not typical members of the dominant community, but are erratics from an association from a neighbouring environment, either from a deeper, more open region of the swamp-lake or, perhaps more likely, from the river-system crossing the Plzeň basin.

2. Terrestrial/Marginal Association

The 13 tetrapods in the Nýřany assemblage which most obviously belong to this association are the temnospondyls *Gaudrya* (Edopidae)

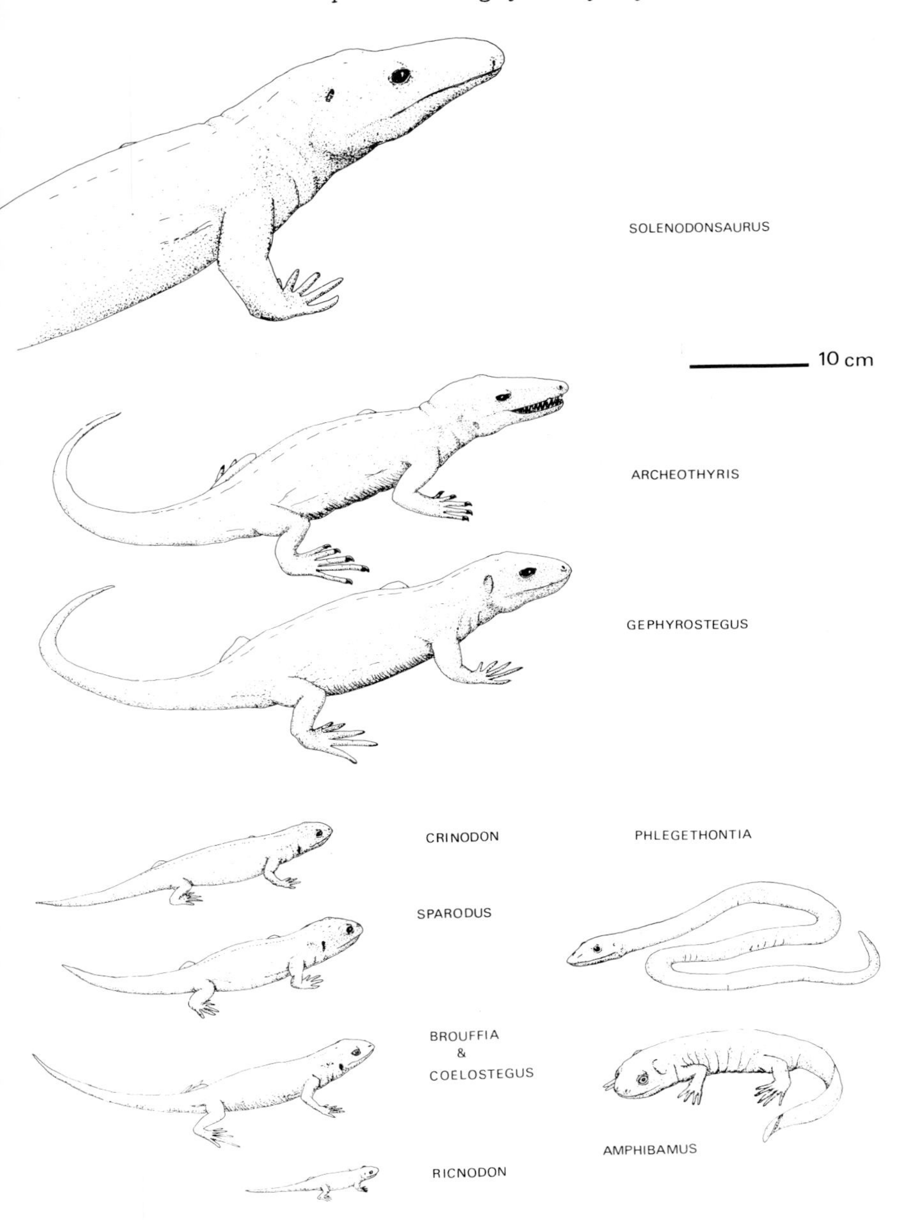

Fig. 4. Restorations of some of the tetrapods comprising the terrestrial/marginal association at Nýřany. Omitted forms include the small amphibians *Hyloplesion* and *Scincosaurus* (shown in Fig. 7), and the large temnospondyl *Gaudrya* which is known only from skull fragments.

and *Amphibamus* (Dissorophidae), the batrachosaurs *Gephyrostegus* and *Solenodonsaurus*, the aïstopod *Phlegethontia*, the microsaurs *Hyloplesion* (Hyloplesiontidae), *Sparodus* (Gymnarthridae), *Ricnodon* (Hapsidopareiontidae) and *Crinodon* (Tuditanidae), and the reptiles *Brouffia* and *Coelostegus* (Romeriidae), *Archeothyris* (Ophiacodontidae) and one undetermined but distinct reptile. Most of these genera are illustrated in Fig. 4. All are represented by 12 specimens or less and most by one to two specimens, and, as can be seen in Fig. 2, the association comprises all such rare elements other than those already characterized in the previous section. Four abundant Nýřany tetrapods may also form part of this association for part of their life-history (*Limnerpeton*, *Cochleosaurus*, *Ophiderpeton*, *Scincosaurus*) but these are discussed separately in following sections.

This association is considered to comprise those tetrapods which spend most of their adult life (and possibly juvenile life) above the air/water interface and which are primarily adapted to feeding and locomotion on land. They can be characterized as such by their rarity in swamp-lake assemblages such as Nýřany, Linton and Jarrow; by the predominance of confamilial forms in terrestrially derived assemblages such as Joggins, Florence and Fort Sill; and by the mode of life inferred from their anatomy.

The association, as represented by the Joggins and Florence lycopod-stump assemblages, is made up of the temnospondyl *Dendrerpeton*, gymnarthrid, tuditanid, hapsidopareiontid and pantylid microsaurs, a limnoscelidid, romeriid reptiles, and the ophiacodont *Archeothyris*, all of which are generally agreed to have been terrestrial (Carroll *et al.*, 1972). Two possibly anomalous forms also present are the edopoid *Cochleosaurus* at Florence represented by small but adult specimens which may have been terrestrial (see later discussion) and some small individuals of an eogyrinid *Calligenethlon* from Joggins. The lycopod-stump tetrapods are associated, not with fish (except for a few scales) but with myriapods, eurypterids and snails. The origin of these assemblages and their essentially terrestrial nature have been discussed by Carroll (1970b) and Carroll *et al.* (1972). The other comparable assemblage is that from the limestone fissure fillings of Fort Sill, Oklahoma, which are of Lower Permian age and derived from an upland environment (Olson, 1967; Carroll and Gaskill, 1978). Despite its later age, this assemblage is dominated by a similar association of dissorophids, a phlegethontiid, gymnarthrid and hapsidopareiontid microsaurs, captorhinomorphs including undescribed romeriids, and small pelycosaurs. A single xenacanth tooth is the only fish fossil recorded.

The assemblages outlined above share an association (allowing for some temporal variation) made up of the Dissorophidae/Dendrerpetontidae, Phlegethontiidae, Gymnarthridae, Tuditanidae, Hapsidopareiontidae, Romeriidae and Ophiacodontidae. This association and the other Nýřany forms of comparable rarity can now be discussed on the basis of the Nýřany representatives. Where they are known, confamilials from Linton and Jarrow are also noted, together with their status at those localities.

The edopid *Gaudrya* is the form least certainly placed in this association as it appears to have had a crocodile-like skull like *Edops*. However the edopid skull is massively built, heavily ossified with large, broad teeth, bears no detectable lateral-line system and is not obviously built for piscivory. Edopids occur as rare erratics at Linton (three to four specimens) and Jarrow (one specimen currently being described by the author). I interpret *Gaudrya* as a heavily built swamp or lake-margin carnivore but not a habitually aquatic form. The dissorophid *Amphibamus* is identified by Carroll (1964) as one of the more terrestrial coal-swamp inhabitants, and later dissorophids are agreed to have been among the most terrestrial of temnospondyls. *Amphibamus*' short trunk, long limbs and toes, ossified carpals and tarsals, large tympanic notch and absence of lateral-line system are all consistent with feeding and locomotion on land. The Linton species *A. lyelli* is also a rare erratic known from about seven specimens. In the early Westphalian, a more primitive relative of *Amphibamus* filling the same niche is *Dendrerpeton*. This is the most abundant form at Joggins (Carroll *et al.*, 1972) while at Jarrow it is known as an erratic from three specimens (Milner, 1980).

The batrachosaurs *Gephyrostegus* and *Solenodonsaurus* are not known from terrestrially derived assemblages but Carroll's work (1969, 1970a, 1972) indicates that they have a fundamentally terrestrial morphology. The Gephyrostegidae are represented at Linton by *Eusauropleura* (two specimens).

The mode of life of *Phlegethontia* has most recently been discussed by Lund (1978) who interprets the phlegethontiids as being primarily terrestrially adapted snake-like crawlers although they may have been aquatic feeders. Although conspicuous and easily recognized members of coal-swamp faunas, they are relatively rare at Linton (14 specimens – McGinnis, 1967) and Jarrow (two specimens).

The microsaur *Hyloplesion* (Fig. 7) has been interpreted as a potentially aquatic or terrestrial microsaur by Carroll and Gaskill (1978, p.134) but later (p.200) they identify it as being typically aquatic. I here interpret it as being terrestrial when adult. Of the 12 specimens recorded, most have skulls which are 7 mm long or less and could be aquatic juveniles. Only three have 8–12 mm skulls and these have no lateral-line sulci, no branchial ossicles but do possess ossified carpals and tarsals. Carroll and Gaskill (1978) note the relatively large number of *Hyloplesion* specimens (nine in their study) as evidence of its typically aquatic nature but in relation to overall numbers at Nýřany this is not an abundant form. The family is almost unknown outside the Plzeň basin and there is no corroborative information from other assemblages. The remaining microsaurs of the families Tuditanidae, Gymnarthridae and Hapsidopareiontidae are all uncontroversially considered to be terrestrial forms (Carroll and Gaskill, 1978, p.201). Only the Tuditanidae occur as erratics at Linton (two specimens of *Tuditanus*), and no terrestrial microsaurs have been found in the Jarrow assemblage.

The romeriid captorhinomorphs and the ophiacodont *Archeothyris* are also agreed to be terrestrial (Carroll and Baird, 1972, pp. 354–355; Reisz, 1975, p.526) and both families also occur as rare erratics at Linton.

Thus this association is represented by erratic elements, not only in the Nýřany fauna but also at Linton and, from a smaller sample, at Jarrow. At Nýřany the scatter of representation of these forms is from 12 specimens to one which is a factor of over ten and this may represent a significant difference in representation between lake-margin species and those from non-marginal terrestrial environments. *Amphibamus* and *Gephyrostegus* both fall into the former category and may have lived in lake-margin vegetation, not competing directly with more terrestrial microsaurs and reptiles living further from the water.

In conclusion, the association comprising the 11 families of Nýřany tetrapods discussed above represents the terrestrial and marginal tetrapods living in the vicinity of the Nýřany swamp-lake and being preserved as erratic elements in an assemblage dominated by the endemic lake-dwellers.

3. Shallow-water / Swamp-lake Association

This association is made up of six of the seven remaining taxa which are all represented by 20 or more specimens. These are the cochleosaurid *Cochleosaurus*, the micromelerpetontid "*Limnerpeton*", the branchiosaurid *Branchiosaurus*, the aïstopod *Ophiderpeton*, the urocordylid *Sauropleura* and the microbrachid microsaur *Microbrachis*. These forms are illustrated in Figs 5 and 6. The only

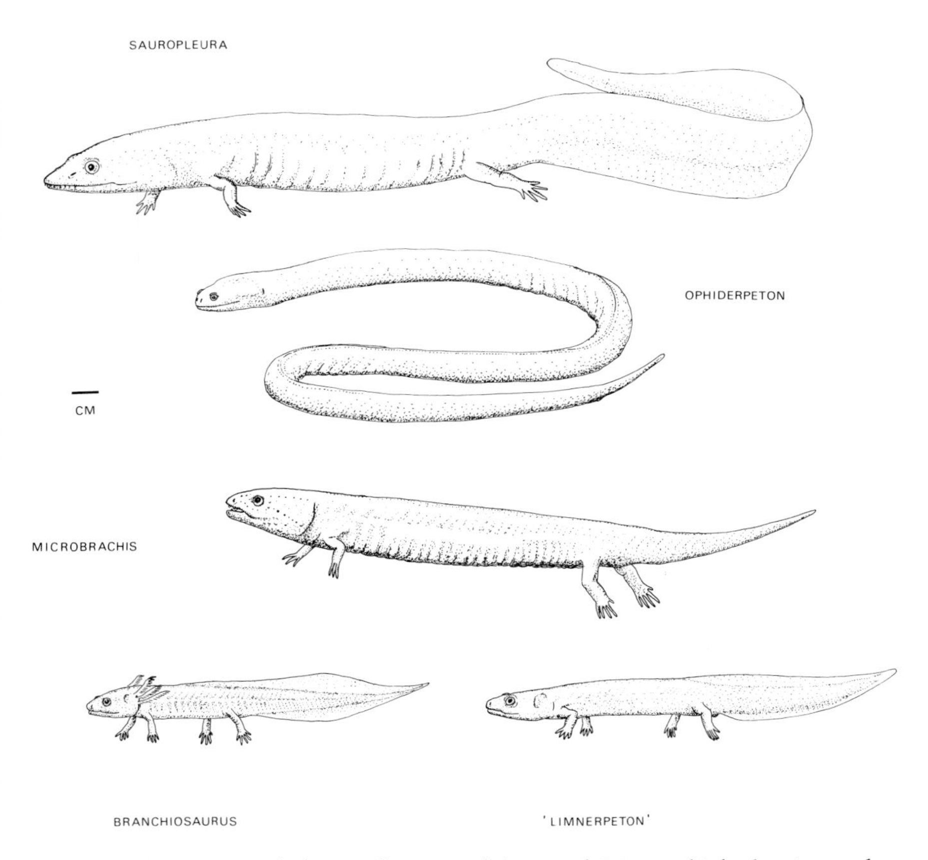

Fig. 5. Restorations of the small swamp-lake amphibians which dominate the Nýřany assemblage. The *Branchiosaurus*, *Microbrachis* and *Sauropleura* restorations correspond in size to the largest known specimens from Nýřany. The "*Limnerpeton*" and *Ophiderpeton* restorations correspond to the largest size which is frequently represented at Nýřany, both being known from a few larger individuals, possibly terrestrial adults.

abundantly represented tetrapod not included here is *Scincosaurus* which is discussed separately in a following section.

Of the six families comprising this association, only the Ophiderpetontidae and Urocordylidae are similarly abundant at Linton and Jarrow. The other four families are represented by single specimens – a possible microbrachid at Jarrow (Carroll and Gaskill, 1978) – or are unrecorded from other swamp-lake assemblages. The members of this association are therefore characterized by their abundance in the Nýřany assemblage and their aquatic adaptations rather than by reference to other assemblages.

The most abundant tetrapod preserved at Nýřany is *Microbrachis* which is an anatomically well known but atypically aquatic microsaur. It possesses lateral-line sulci, gill ossicles, unossified carpals and tarsals, small limbs and an elongate trunk and could have evolved by paedomorphosis (Carroll and Gaskill, 1978, p.200). A few large specimens would have been 300 mm in length but most were half this size or smaller, and probably juvenile. The long trunk, short tail and reduced limbs are indicative of anguilliform swimming. *Microbrachis* had well-developed sight (sclerotic ring present) and this, combined with the lateral-line sulci on the skull suggest an animal feeding on small active invertebrates, either ostracods in the plankton or small crustaceans. The gill ossicles on the internal gill arches could have served as gill-rakers such as are found in many plankton-feeding fish where they serve to prevent small food particles being lost through the gill-slits.

Branchiosaurus and "*Limnerpeton*" are the other abundant small aquatic amphibians. Both are relatively more abundant at Nýřany than the total of determinate specimens in the census suggests, as about one in five of the small dissorophoid specimens were not determinable to family level. They are similar forms and both are characterized by poor ossification including unossified centra, carpals, tarsals, pubes, coracoids and braincase. Both possess ossified gill-rakers, large orbits with a sclerotic ring and a long flattened tail. In *Branchiosaurus* there are no visible lateral-line sulci although the system may have been superficial on the poorly ossified dermal bones; the mouth has a wide shallow gape, simple peg-like marginal teeth and no palatal denticles; there are feathery external gills visible in Permian forms and the trunk is short (20 presacrals). Boy (1972, 1978) has argued that *Branchiosaurus* is a secondarily aquatic, neotenous derivative of the dissorophoid group and that it was a poor swimmer, feeding on small invertebrates. At Nýřany, *Branchiosaurus* does not exceed 12 mm skull length but it may have grown larger. In most Permian "branchiosaur" assemblages, *Branchiosaurus* does not exceed 15 mm skull length but a few specimens with skulls 30 mm long occur (Boy, 1978; A. R. Milner, unpublished). These show no metamorphic trends and appear to be large aquatic adults. The Nýřany *Branchiosaurus* might not have grown to this size but the sample is small enough not to include infrequent large adults. "*Limnerpeton*" has a deeper gape than *Branchiosaurus*, the tympanic notch is larger and deeper, the trunk is longer (26 presacrals) and the limbs and dermal pectoral girdle are more heavily built. These features suggest a compromise to terrestrial feeding and locomotion in the adult. Most of the specimens of "*Limnerpeton*" are less than 22 mm skull length but one, the type, has a 29 mm skull and is a partly disarticulated specimen. This could be a form which is represented by large numbers of aquatic larvae and the occasional terrestrial adult. Like *Branchiosaurus*, the small "*Limnerpeton*" must have eaten small invertebrates.

The other abundant temnospondyl is *Cochleosaurus* (Fig. 6) which in the main Plattelkohle assemblage is represented mostly by small specimens from 25 mm to 120 mm skull length, several but not all of the large specimens coming from a different Plattelkohle horizon and not included in the census. *Cochleosaurus* was a superficially crocodile-like form with slightly recurved, pointed teeth, a sclerotic ring accommodation system, no detectable lateral-line sulci (in the Nýřany material), a large rhomboidal interclavicle (a common feature of aquatic temnospondyls) and short limbs. It was a predator on other smaller vertebrates to judge from the size and dentition, and was clearly abundant in the Nýřany swamp-lake as a small aquatic animal and less frequent as a larger one. This could be attributable to aquatic adults being present in smaller numbers or to the adults being at least semi-terrestrial and using the lake as a breeding area. The presence of a different species of *Cochleosaurus* in the lycopod-stump assemblage at Florence supports the latter hypothesis.

Two frequent forms, both at Nýřany and Linton, are the nectridean *Sauropleura* and the aïstopod *Ophiderpeton*. The former is clearly a specialized swimmer with a massive tail with deep vertebrae modified for movement in a horizontal plane only. It

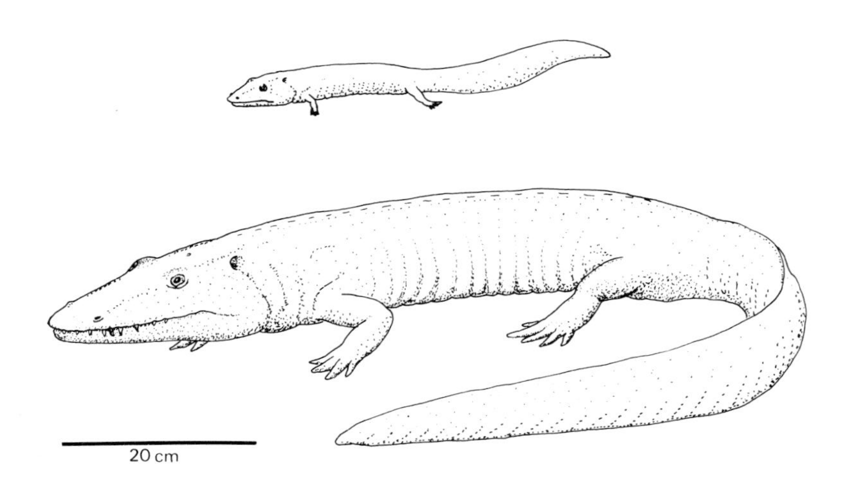

Fig. 6. Restoration of juvenile and presumed adult of *Cochleosaurus*. The larger individuals corresponds to the largest known size of this genus at Nýřany.

occurs over a wide size-range and appears to be a permanently aquatic form. It possesses pointed, recurved teeth and a narrow snout, and like juvenile *Cochleosaurus* appears to have been a predator on small vertebrates. *Ophiderpeton* had a skull of snake-like construction (McGinnis, 1967; Thomson and Bossy, 1970) and appears to have been a "rib-walker" with a short tail (Lund, 1978; Zidek and Baird, 1978) although, like snakes, capable of swimming. The genus has not been reported from any primarily terrestrial assemblage, and not only is it frequent at Nýřany, Jarrow and Linton but it also occurs as an erratic in the fish-dominated assemblages from Cannelton (Baird, 1978), Newsham and Kounová (Baird, 1964). This suggests a habitual swimmer, perhaps a "water-snake"-like form. Most specimens from Nýřany are small, less than 10 mm body width, but a few are large – 20–30 mm body width. Perhaps, as in *Cochleosaurus* and "*Limnerpeton*", this may represent frequent aquatic juveniles and occasional terrestrial-marginal adults. The dentition of *Ophiderpeton* is of small, blunt, peg-like teeth (Steen, 1931) more useful for gripping than puncturing, and suggesting predation on much smaller animals which did not require to be impaled in order to be held.

In conclusion, the association comprised of the six families discussed above is that of the predominant swamp-lake inhabitants at Nýřany. Some members of this association are found in comparable swamp-lake assemblages elsewhere (Urocordylidae and Ophiderpetontidae) but the remainder of the association has not been found in any other Carboniferous assemblage.

4. Scincosaurus, *an Anomaly*

In the attempt to characterize the most abundantly represented Nýřany tetrapods as the aquatic endemic inhabitants of the swamp-lake, it became apparent that *Scincosaurus crassus* was an anomalous

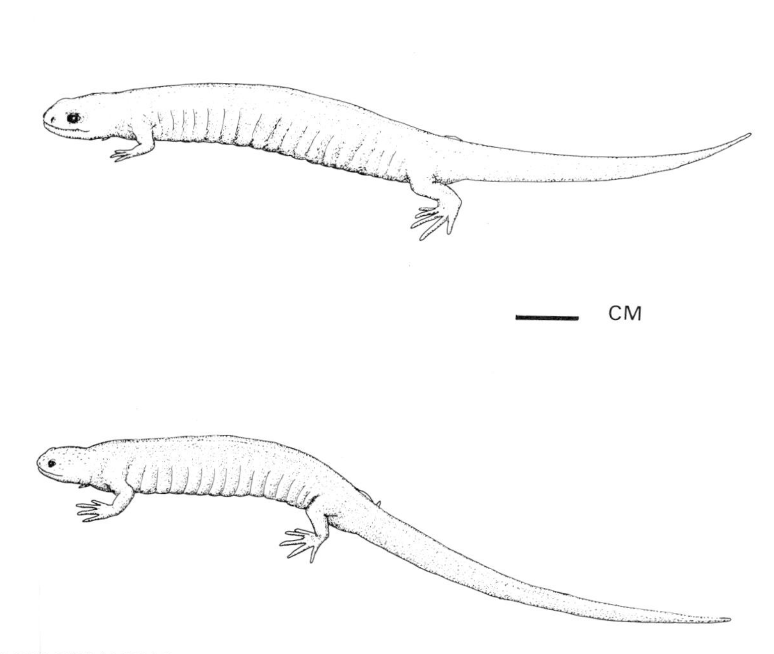

Fig. 7. Restorations of the two small terrestrial/marginal amphibians from Nýřany, *Hyloplesion* and *Scincosaurus*.

form. It is the second most abundant genus present at Nýřany after *Microbrachis*, but elsewhere the Scincosauridae are known only as single erratic specimens from Linton (A. C. Milner, 1978) and, as *Sauravus*, from two Stephano-Permian localities in France (Piveteau, 1925). More problematical is that the morphology of *Scincosaurus* points to it being an entirely terrestrial tetrapod (A. C. Milner, Chapter 15 of this volume). The tiny head, the relatively massive and heavily ossified appendicular skeleton, the reduction of the characteristic nectridean neural arches in the caudal vertebrae and the thick armour of ventral dermal scutes, all indicate a terrestrial form similar in build and size to the living salamander *Ambystoma texanum*, the small-headed salamander, which lives in damp, terrestrial environments and only returns to the water to breed.

The abundance of this apparently terrestrial form at Nýřany may be interpreted by reference to the size-distribution within the census material. The measurable *Scincosaurus* specimens, with one exception, fall within a very narrow size-range. The trunk length varies between 35 mm and 45 mm in intact specimens and most of the fragmentary specimens are of a size consistent with this. A single tiny specimen with a 17 mm trunk was the only measurable specimen outside this range. The preserved material represents a single age-cohort of animals either killed by a single local catastrophe or an accumulation of animals which have died in a regular seasonal pattern. They represent either a generation of terrestrial adults which have returned to the water to breed and died in quantity during this aquatic phase, perhaps killed by temporary anoxia in the lake; or a generation of juveniles killed immediately prior to their emergence onto land. None show any retained aquatic characteristics such as might be expected in at least some larvae, and the latter explanation seems less plausible than the former. Likewise, the five-fold difference between the abundance of *Scincosaurus* and the next most abundant inferred terrestrial/marginal forms renders it extremely unlikely that *Scincosaurus* was simply a lake-margin inhabitant being accidently washed in. *Scincosaurus* seems, then, not to have been a permanent member of the endemic swamp-lake association at Nýřany but a seasonally present form, entering the swamp-lake to breed and probably being present in the form of small aquatic larvae for part of the year as well. *Cochleosaurus*, "*Limnerpeton*" and *Ophiderpeton* are all represented by numerous small individuals and few large ones and may represent similar life-cycles preserved at a different phase, with larval generations well represented and adults present only as terrestrial or marginal erratics. This means that the distinction between swamp-lake and terrestrial/marginal communities cannot be made by mutually exclusive associations of genera. Some genera appear to have inhabited two different environments and played different roles in two communities during their life-cycle. The communities discussed in the next section are not, therefore, direct derivatives of the three major associations as defined in this section, but some attempt is made to take account of taxa which alternate between two communities.

5. Summary of Associations

Regardless of how or whether one interprets them as communities, the Nýrany tetrapods appear to segregate into three major associations from three different environments and it is to this that the diversity of tetrapods at Nýřany is attributable. The associations and their relationship to the Nýřany swamp-lake and also to the other major types of preserving environment, are shown in Fig. 8, a diagrammatic transect of Westphalian continental environments which have produced these associations. The families depicted are strictly those described from Nýřany.

The three principal tetrapod associations can be expressed in family terms as follows.

Loxommatid–eogyrinid association. Characteristically associated with large assemblages of fish belonging to diverse taxa, particularly osteolepidids, dipnoans, petalodonts and a wide range of palaeoniscoids, notably the platysomoids. The loxommatids and eogyrinids are the two most characteristic tetrapod families. Keraterpetontids also form part of this association but occur less uniformly and are not known from Nýřany nor from some of the British Coal Measure localities. I briefly mentioned this association in an earlier paper (A. R. Milner, 1978) in which I included *Ophiderpeton* with the above forms but it appears to be only an erratic when it occurs in this association as at Newsham and Kounová.

Dissorophid/dendrerpetontid–gymnarthrid–hapsidopareiontid–tuditanid–romeriid–ophiacodont association. These families are the most

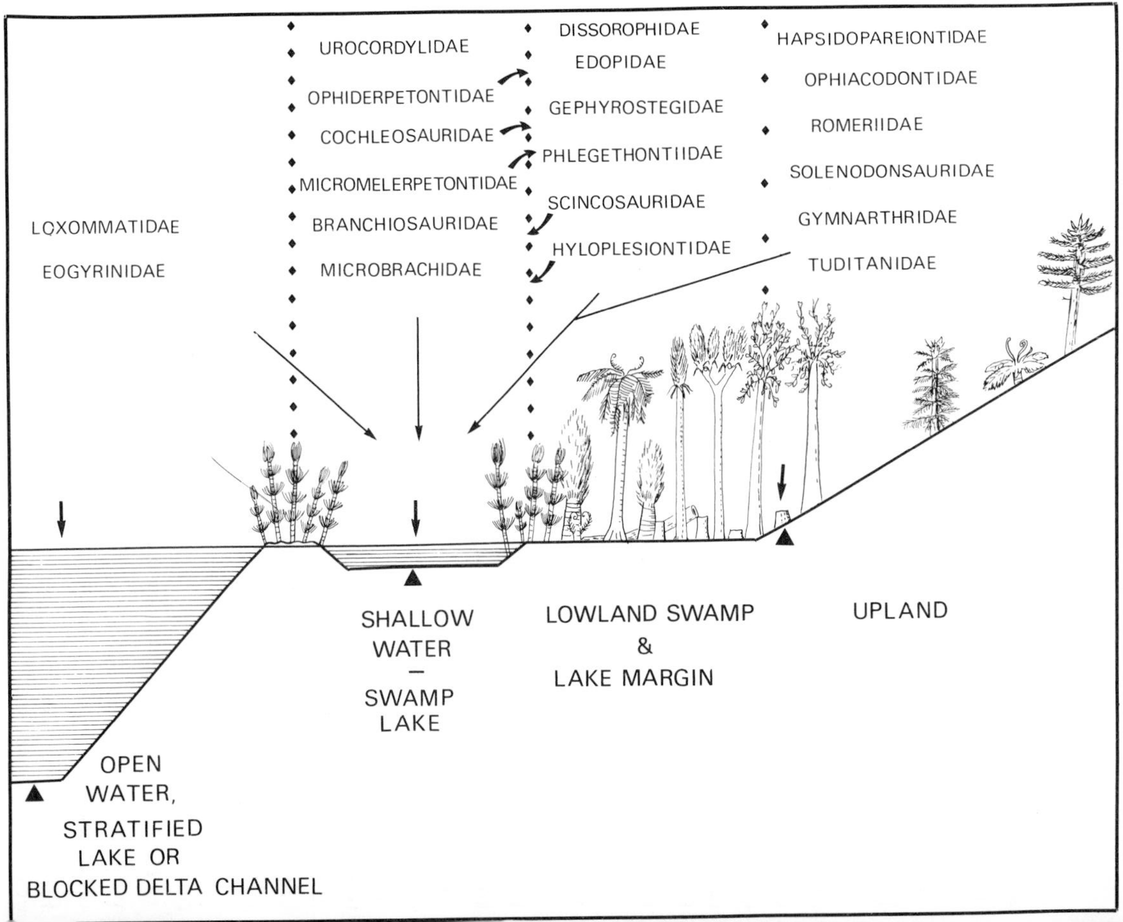

Fig. 8. Generalized transverse section of the major environments contributing elements to Westphalian tetrapod assemblages. The tetrapod families, as described from Nýřany, are depicted in the environment from which they are believed to have been derived. The left triangle symbol indicates the Newsham type of depositional environment, the central triangle indicates the Nýřany/Linton type and the right triangle indicates the Joggins/Florence type. The arrows between family associations indicate forms which may alternate between two environments in their life-cycles, as discussed in the text.

consistently occurring members of this association. The dendrerpetontids and dissorophids seem to be successive temnospondyl families (possibly one lineage) filling the same niche, the former in Westphalian A–B localities, the latter from the Westphalian D onwards. Other families which belong to this association but which occur with less regularity for geographical, chronological, or indeterminate reasons or random circumstances are the edopids, *Stegops*, the gephyrostegids, *Solenodonsaurus*, the phlegethontiids, *Hyloplesion*, the pantylids, the limnoscelidids and probably the earliest diapsids. The adults of *Cochleosaurus* and *Scincosaurus* probably belong here and, less certainly, the adults of "*Limnerpeton*" and *Ophiderpeton*. Parts of this association have been identified by Carroll and other authors in several papers (e.g. Carroll *et al.*, 1972; Carroll and Baird, 1972; Reisz, 1975). Despite the assertions of Romer (1974) that these forms all lived near water and were therefore semi-aquatic, the evidence of morphology, association and circumstances of preservation support the conclusion that these forms lived above the air-water interface at least during their adult existence.

Ophiderpetontid–urocordylid association. The third association can only be diagnosed from two tetrapod families as an ophiderpetontid–urocordylid association. As described by Westoll (1944) the haplolepid palaeoniscoids form part of this association as do small acanthodid acanthodians and some small xenacanth sharks. Other tetrapods occur as part of this association, often in abundance, but not sufficiently uniformly to be diagnostic. These include microbrachids, branchiosaurids and micromelerpetontids (Nýřany), colosteids and lysorophids (Linton and Jarrow), and saurerpetontids (Linton). Also represented by juvenile swarms are *Cochleosaurus* (Nýřany) and *Keraterpeton* (Jarrow).

Previous characterizations of Westphalian associations have been restricted by the reliability or otherwise of the state of the systematics of the various groups and the extent of comparative studies of taxa from different localities. Westoll (1944, pp. 105–108) was aware of this problem, and in characterizing a "facies fauna" incorporating the Haplolepididae he was able to associate amphibians with the haplolepids only at ordinal level, there being no studies of amphibian groups at that time comparable in detail to his haplolepid study. Westoll's haplolepid–aïstopod–nectridean assemblage corresponds to the swamp-lake association diagnosed here for Nýřany and he observes that the included forms occur only as occasional individuals at Newsham and Longton, here identified as open-water associations. The associations presented in this study are dealt with at a family level and the aïstopod–nectridean component of Westoll's "facies fauna" is here restricted to the families Ophiderpetontidae and Urocordylidae. The remaining aïstopods and nectrideans are associated primarily with other tetrapods and other environments.

COMMUNITIES AND THEIR STRUCTURE

The problem of identifying the roles of 22 tetrapods in one association (and by inference, community) at Nýřany has now been reduced to a similar exercise spread over three associations interpretable as communities living in distinct environments. The identification of associations gives insight into the structure of Westphalian animal communities and also into the significance of some of the patterns of distribution of Carboniferous tetrapods in space and time. These aspects are elaborated in the following discussions.

1. General Palaeogeography

All Carboniferous tetrapod fossils discovered to date derive from the late Palaeozoic continent usually referred to as Euramerica which was composed of what now comprises North America and Europe west of the Urals (Smith *et al.*, 1973). Within Euramerica, all tetrapod or footprint assemblages derive from within a few degrees latitude of the palaeomagnetic and palaeoclimatic equator (Panchen, 1973). The regional differentiation of lowland vegetation into several geographical provinces in the Westphalian supports the contention that equatorial Euramerica was a distinct floral province at that time (Chaloner and Lacey, 1973) and therefore probably a distinct faunal province as well. Regardless of whether the absence of tetrapod fossils outside equatorial Euramerica is a genuine representation of the distribution at the time (Panchen, 1973; Cox, 1975) or an artifact of the fossil record (Carroll and Gaskill, 1978), the conclusions concerning communities put forward here can only be taken to apply to equatorial Euramerica. No assumptions can be made concerning the nature of tetrapod communities on other continents or in other climates.

2. Open-water Community

(a) Structure. The assemblage at Newsham is characterized by over 20 types of fish (Andrews, in Land, 1974, p.61). Some are erratics from the swamp-lake association (the haplolepids) but most are frequent and this appears to have been the open-water community characteristic of large, permanent, well-aerated water bodies, either large lakes with circulation at least in the epilimnion, or the lower reaches of large river-systems or deltas. Some fish grew to 5–6 m in length indicating large volumes of water containing the considerable biomass needed to support such top consumers. In such environments, fish would be expected to predominate.

The range of feeding types among the fish can only be partly determined because, although there are some obvious specialists, many fish show no pronounced modifications of the teeth or jaws. Furthermore, many living fish feed in more than one mode on more than one type of food (Olson, 1971, p.476). At Newsham there are clearly medium to large carnivores feeding on other aquatic vertebrates – the rhipidistians, xenacanth sharks and possibly large palaeoniscids; there are durophagous feeders – the dipnoans and petalodonts – feeding on bivalve and gastropod molluscs, arthropods and possibly other vertebrates; there are nektonic plankton feeders – the small acanthodians and probably the juveniles of many palaeoniscids – feeding on ostracods and other small crustaceans; and there are small deep-gaped palaeoniscoids such as *Rhadinichthys* and *Elonichthys* which presumably preyed on other smaller vertebrates and arthropods. Of the range of fish at Newsham, the only forms which show adaptations interpretable as being for herbivorous browsing, either on algae or vascular plants, are some of the platysomoid palaeoniscoids and in particular *Chirodus*. This genus has pterygoid and coronoid tooth-plates, no marginal dentition and a beak-like premaxillary (Moy-Thomas and Miles, 1971). It is not demonstrably herbivorous and could have been an arthropod-feeder, but it is the only form which could have possessed a keratinous rasping beak combined with a crushing dentition such as characterize many living "nibbling" forms (Olson, 1971, p.485). Olson also observes that this type of dentition is commonly associated with deep bodies and high mobility, and the platysomoids are uniquely deep-bodied among Carboniferous fish.

Apart from *Chirodus*, the fish from Newsham and other similar assemblages all appear to have fed on other animals. Most of the energy of fish would have derived from ostracods (detritus and phytoplankton feeders), palaeocarid crustaceans (scavengers and detritus feeders), small chelicerates (carnivores and scavengers), bivalves (detritus/filter feeders), gastropods (algal browsers), and various soft-bodied invertebrates of which we have no record. There is no good evidence for aquatic insect nymphs or adults in the Carboniferous (Wootton, 1972a,b) although carnivorous odonatan nymphs may have been present. The two ultimate sources of energy in Carboniferous open-water bodies seem to have been detritus and algae, both planktonic and encrusting.

The three amphibians found in association with this community were all structurally very specialized forms and some of thes specializations must have enabled them to coexist with fish. The loxommatids are only known from skulls and a few postcranial scraps, and so their postcranial adaptations are unknown. They had crocodile-shaped heads up to 30–40 cm long and their dentition indicates that they were piscivores. They could have been river-margin or lake-margin dwellers like crocodiles, moving into water to feed but otherwise resting at the margin of the water.

The eogyrinids grew to 4 m in length and were large anguilliform piscivores (Panchen, 1970, 1972a). Retaining limbs, they could probably have moved over land, particularly when small, some small individuals (*Calligenethlon*) occurring in the Joggins lycopod-stumps (Carroll, 1967). Dr Ian Rolfe (personal communication) has pointed out that there was a potential niche for a large, long-bodied, lycopod-stump/trunk-inhabiting predator on the giant myriapod *Arthropleura* which appears to have fed on lycopod stems (Rolfe, Chapter 6 of this volume). Eogyrinids might have preyed on *Arthropleura* in marginal vegetation but the large individuals are unlikely to have moved far from water. On the other hand, the presence of 1–5 m rhipidistians and other predatory fish in the British Coal Measures suggests that eogyrinids must have been more than just another group of large aquatic carnivores. As anguilliform amphibians with limbs, they would have been well suited to swimming/crawling among dense *Calamites* stands on the lake margin or in and out of hollow submerged lycopod trunks. There are no really anguilliform fish known from the Carboniferous apart from the tiny *Tarrassius*, and the eogyrinids might have been at a substantial advantage living in dense marginal vegetation above or below the water-line and catching fish on the periphery.

The keraterpetontids were short-bodied, long-tailed aquatic specialists up to 40 cm long. They had akinetic skulls and dentitions of short, blunt, interlocking marginal and palatine teeth and opposing batteries of pterygoid and coronoid denticles, and were apparently adapted to gripping and crushing small invertebrates with exoskeletons or shells (A. C. Milner, 1978). Small crustaceans, gastropods and bivalves are plausible aquatic food animals as are drowned terrestrial arthropods, there being no evidence for aquatic insects as noted above. Like eogyrinids, keraterpetontids were elongate tetrapods (if the tail is included) and retained limbs. They too may have lived in stands of marginal vegetation and fed on invertebrates in the water on the periphery of the vegetation. They were probably capable of swimming without yawing of the head during undulation (A. C. Milner, 1978) which would have enabled them to seize small swimming or floating invertebrates whilst swimming.

In conclusion, it is suggested that the three amphibian families in this open-water community were not occupying feeding niches that fish could not fill, nor can their air-breathing abilities be assumed to be a particularly significant characteristic as the dipnoans and rhipidistians are generally accepted as having been air-breathers. The generally eel-like body-form with retained limbs of the eogyrinids and keraterpetontids suggest a life in dense stands of vegetation where conventionally shaped fish would not have been at any advantage. More speculatively, the loxommatids may have lived a crocodile-like existence in more open marginal areas. Although principally associated with large open-water fish communities, these three amphibian types may have primarily lived on the margin of this community although utilizing it for food. The open-water vertebrate community as a whole was made up of secondary and tertiary consumers, the only plausible primary consumers being the chirodontid palaeoniscoids. Many of the primary consumers appear to have been detritus feeders relying on decomposers. Living plants in the form of algae were consumed and incorporated into the food-web mainly by ostracods and gastropods which were undoubtedly major food-sources for the vertebrates. There is no evidence either for small aquatic vascular

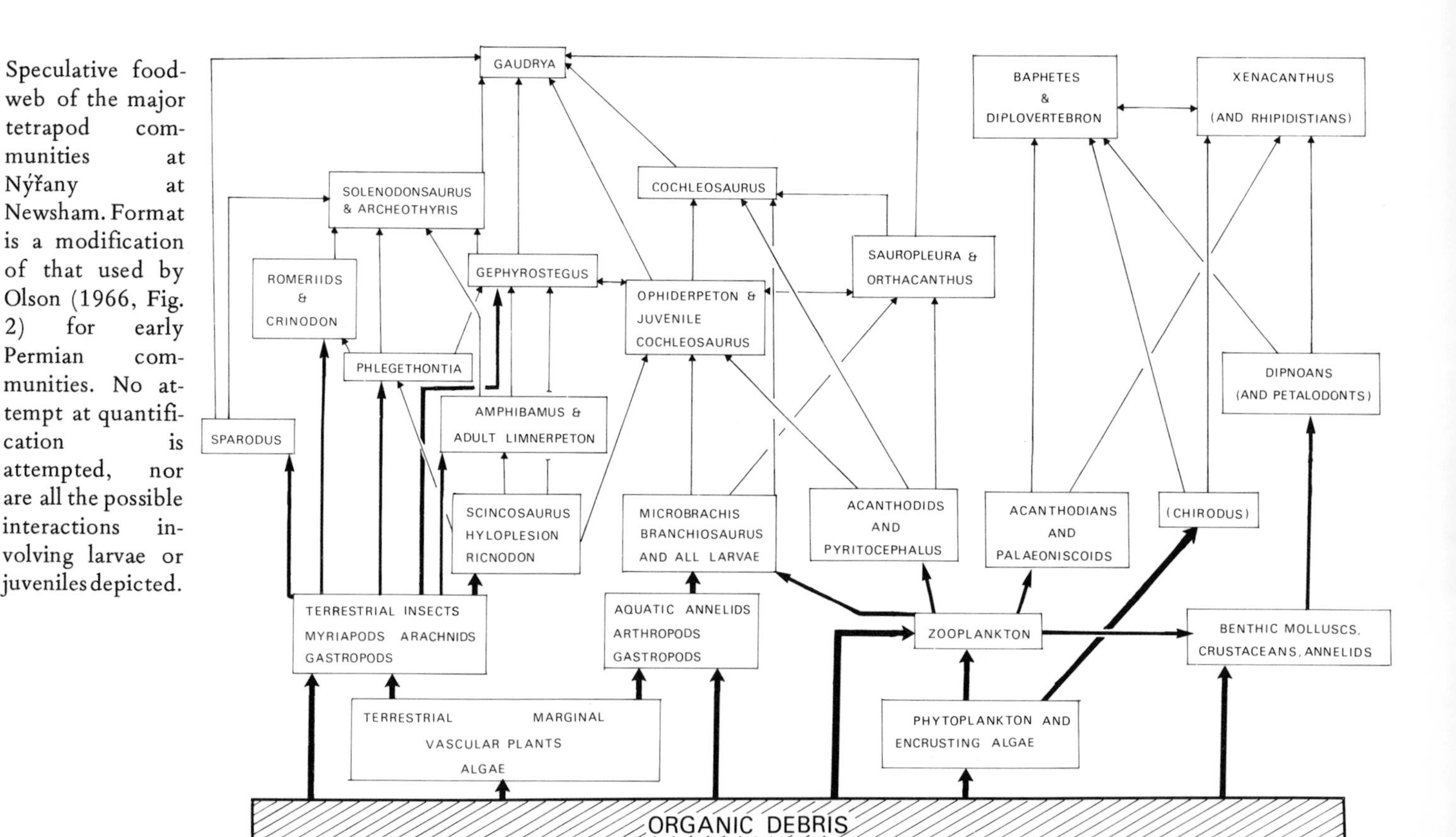

Fig. 9. Speculative food-web of the major tetrapod communities at Nýřany at Newsham. Format is a modification of that used by Olson (1966, Fig. 2) for early Permian communities. No attempt at quantification is attempted, nor are all the possible interactions involving larvae or juveniles depicted.

plants or for invertebrates which might have consumed them. These conclusions are implicit in the right-hand part of the food-web shown in Fig. 9, which is a simplified interpretation of this community.

(b) Palaeogeography. Such elements as are known from Nýřany represent the eastern end of a widespread and conservative community. *Baphetes* has been described from Czechoslovakia, the British Isles, Nova Scotia and Ohio (Beaumont, 1977). If *Diplovertebron* is an eogyrinid, it belongs to a family with a similar distribution. *Megalocephalus* and the Keraterpetontidae are known from the British Isles and Ohio and, as noted previously, their absence from Nýřany is as likely to be ecological as geographical. Panchen (1977) has noted some regional difference between the eogyrinids of the north-east U.S.A. and those of the British Coal Measures, placing these regional groups in the subfamilies Leptophractinae and Eogyrininae respectively. This may be evidence for limited regional endemism but is based on cranial characters and cannot be applied to *Diplovertebron*.

(c) Response to Permian changes. The latest known occurrences of tetrapods referable to this community are Westphalian D for primitive keraterpetontids (*Diceratosaurus* at Linton) and loxommatids (*Megalocephalus* at Linton and *Baphetes* at Linton and Nýřany), and Stephanian for certain or probable eogyrinids (*Neopteroplax* from the Tristate area, U.S.A., and material from Kounová). There is no positive evidence to suggest that these forms persisted into the Permian. In the Permian of North America only, are found diplocauline nectrideans and archeriid anthracosaurs which are highly derived relatives of the Carboniferous keraterpetontids and eogyrinids respectively. They are believed to have been pond- and stream-dwellers and differ considerably from the Carboniferous forms. No comparable related forms are known from the Permian of Europe. It appears that the development of fold mountain systems in central Euramerica in the late Carboniferous and early Permian, combined with the alteration in climate to a highly seasonal wet/dry regime, resulted in the large perennial water bodies and associated swamps becoming too isolated in space and time for the Westphalian open-water lake fauna to maintain itself (A. R. Milner, 1978). The highly specialized tetrapods in this community could not apparently

survive in ephemeral lake systems or in river systems, both characterized by other tetrapods in the Stephanian and Permian.

3. Swamp-lake Community

(a) Structure. The nature of the Westphalian swamp-lake environment as represented at Nýřany and Linton was discussed by Westoll (1944) with particular reference to food relations. He suggested a close analogy to swamps in the Paraguayan Chaco as described by Carter and Beadle (1930) where the water is about 1 m deep, stagnant and current-free because of the abundant standing vegetation and very poorly oxygenated except for a few centimetres near the surface. However, in open pools and lakelets in the swamps, the increased light permits a greater abundance of phytoplankton than in the rest of the swamp. The Chaco pools not only have a more accessible primary food-source in the algae but are also better oxygenated as a result of its presence and are consequently more productive and support a richer fauna. This appears to be a closely comparable environment to the Nýřany swamp-lake (Westoll, 1944) which was shallow, stagnant and closely associated with *Calamites*-dominated swamp.

If phytoplankton was present in the Nýřany swamp-lake, there may have been seasonal blooms and cycles of high productivity. There is no evidence for extreme climatic seasonality in the Westphalian, but local seasonal nutrient cycles in lakes, perhaps triggered by influxes of dissolved salts, may have resulted in cyclic plankton blooms which would have influenced the entire dependent food-web by permitting the existence of ephemeral niches associated with seasons of high productivity. The abundant presence in the Nýřany assemblages of both permanently aquatic tetrapods and larvae of amphibious or terrestrial adults suggests that niches for small, seasonally present, plankton feeders existed. The primary sources of food in the Nýřany swamp-lake, namely algae and detritus, must have initially been consumed by planktonic crustaceans living near the surface such as ostracods and palaeocarids, and possibly by small larval vertebrates. There is no evidence for bivalves in this assemblage, the substrate having been uninhabitable, and, as noted previously, there is no evidence for aquatic insect nymphs in Westphalian coal-swamp communities, only adults and nymphs as terrestrial erratics.

The dominant vertebrate association at Nýřany is consistent with derivation from a community inhabiting an environment as described above, being dominated by small forms which were either air-breathers or which had accessory gills. The presence of terrestrial erratics showing little evidence of transport indicates that the swamp-lake must have been in close proximity to dry land and hence a suitable breeding ground for terrestrial amphibians with aquatic larvae.

The swamp-lake community was made up primarily of a restricted diversity of small aquatic tetrapods together with an even more limited range of fish. The fish present are acanthodids and haplolepids up to 10 cm long and xenacanths up to 30 cm long. The former two families are believed to have fed on planktonic invertebrates in the oxygenated upper layers of the lake (Moy-Thomas and Miles, 1971; Westoll, 1944). The xenacanths were presumably carnivores feeding on other fish and also on small aquatic tetrapods. This limited range of fish may represent the forms most tolerant of the poorly aerated swamp-lake environments or, alternatively, it may represent the degree of colonization of the lake through the filter of the swamp. The possibility that these swamp-lakes may have been imperfectly accessible to potential inhabitants among the fish, is raised by the assemblage from Třemošná which is contemporary to that from Nýřany and occurs a few kilometres away in the Plzeň basin. This assemblage, preserved in very similar conditions, is dominated by a wider range of fish, and most of the dozen or so tetrapod specimens collected here are terrestrial erratics. This suggests that a swamp-lake which is either tolerable by, or accessible to most fish could not be utilized by small or larval aquatic tetrapods. Only xenacanths, acanthodids and certain palaeoniscoids coexist with small aquatic tetrapods in the late Palaeozoic of central Europe, not only at Nýřany but also in many Autunian "branchiosaur" assemblages.

Of the apparently permanent aquatic tetrapods in this community, *Microbrachis* and *Branchiosaurus* were of a size and morphology to be consumers of ostracods, other crustaceans and soft-bodied invertebrates. Both had broad gapes, small teeth and ossified gill-rakers. Both are represented by large numbers of individuals which are smaller than the largest, presumed adults, and both were probably abundant as tiny larvae which are rarely detectable in the Plattel-kohle but which would have been a major food-source for larger vertebrates. There is evidence from Permian *Branchiosaurus* assemblages that the life-cycle of this genus included 2 cm long larvae (Boy, 1974) which are preserved as carbonized films, visible on the Autunian mudstones in which they are preserved but potentially invisible on the black Plattelkohle. Tiny *Microbrachis* down to 22 mm overall length are known from Nýřany. The third permanently aquatic tetrapod at Nýřany was *Sauropleura* which was an active swimmer with a proportionately large skull, slender snout and long recurved teeth, evidently a carnivore feeding on other vertebrates, even when small. The largest individual from Nýřany has a skull 43 mm long, large enough to catch and feed on all known sizes of *Branchiosaurus, Microbrachis* and *Pyritocephalus*.

The remaining tetrapods which are attributed to this community all have adults which are less certainly aquatic and hence less certainly members of the aquatic ecosystem. Although occurring in the assemblage in different overall quantities, "*Limnerpeton*", *Ophiderpeton* and *Hyloplesion* are all represented mostly by small individuals and few possibly terrestrial adults. *Scincosaurus* is represented mostly by terrestrial adults and only a few larvae. *Cochleosaurus* is represented by individuals over a wide size-range with no obvious modality of size-distribution. The larvae of "*Limnerpeton*", *Scincosaurus*, *Ophiderpeton* and *Hyloplesion* could only have fed on tiny invertebrates in feeding niches which may have overlapped those of *Branchiosaurus* and *Microbrachis*. The juvenile *Cochleosaurus*, like *Sauropleura*, were undoubtedly predators on other vertebrates. As discussed earlier, the adult of *Ophiderpeton* could have been terrestrial or a "water-snake"-like form, and may have been a predator on vertebrates in either the swamp-lake system or the terrestrial system or both. The adult of *Cochleosaurus* may likewise have been either terrestrial or aquatic, but its crocodile-like skull suggests that it may have primarily preyed on smaller vertebrates in the lake.

A relatively simple hierarchy of consumers is as much as can be suggested for this community (see Fig. 9, central portion). The algae and detritus were consumed by planktonic invertebrates which, together with larval vertebrates, formed the diet of the acanthodids, haplolepids, *Branchiosaurus* and *Microbrachis* throughout the year, and larval "*Limnerpeton*", *Scincosaurus*, *Hyloplesion* and *Ophiderpeton* seasonally. These were consumed by xenacanths, *Sauropleura*,

Ophiderpeton and middle-sized *Cochleosaurus*. The largest *Cochleosaurus* seem to have been the terminal consumers. The two most imponderable factors are the role of juvenile or larval stages in the system and the extent of seasonal variation. Larvae of all taxa may have been preyed on by larger individuals of all taxa. The seasonal distribution of larvae of different species could have been more or less synchronous exploiting a period of high productivity, or alternatively the relevant species of amphibian could have bred seasonally at 1–2 week intervals so that the larvae would have overlapped minimally as they successively filled the same niche through a longer period of time. The representation of this community in a given assemblage could thus depend on the season of preservation. The presence of *Scincosaurus* in abundance at Nýřany, but as a single specimen at Linton, could be no more than a seasonal difference.

(b) Palaeogeography. The Nýřany swamp-lake association appears to be made up of some widespread genera and two possibly regionally endemic forms which may have arisen at the eastern end of the Euramerican coal-swamp range. Of the widespread genera, *Sauropleura* occurs as a distinct species at Linton (Steen, 1931; Bossy, 1976), *Ophiderpeton* occurs at several Westphalian localities in Europe and North America (Baird, 1964), *Cochleosaurus* occurs as a distinct species at Florence, Nova Scotia, recently described by Rieppel (1980). *Branchiosaurus* is predominantly known from European material and is absent from the Linton assemblage but is known from a single Stephanian specimen from the Tristate area (Romer, 1939). The two possible regionally endemic forms are the microbrachids and the micromelerpetontids ("*Limnerpeton*") which are only known from the Permocarboniferous of Europe.

It is within this association that potentially geographical differences between the Linton and Nýřany faunas may exist (Milner and Panchen, 1973). In the aforementioned study, it was suggested that the developing Appalachian/Caledonian mountain range acted as a filter to the interchange of small aquatic or amphibious tetrapods resulting in regional differentiation in the late Carboniferous and early Permian. However some subsequent discoveries have reduced the number of taxa which this particular filter might have restricted. The colosteids and lysorophids, present at Linton but absent from

Nýřany, have both been identified in the Westphalian A assemblage from Jarrow, Eire (colosteid – A. R. Milner, unpublished; lysorophid – Carroll and Gaskill, 1978). As noted previously, *Scincosaurus* occurs as an erratic at Linton and was hence widespread. Only the Micromelerpetontidae and the Microbrachidae in Europe and the Saurerpetontidae in North America remain plausible examples of geographically localized swamp-lake tetrapods in the Westphalian.

(c) Response to Permian changes. Some elements of the European swamp-lake community not only survived into the Autunian (basal Permian) but also diversified, while others remain unrecorded from European post-Carboniferous deposits. The Branchiosauridae and Micromelerpetontidae are abundant in the European Autunian and represented by diverse species (Boy, 1972, 1978). There are two records of possible Autunian microbrachids (Schroeder, 1939; Kuhn, 1959) although neither has been re-studied recently. *Cochleosaurus* is replaced ecologically by the Actinodontidae, also abundant in the European Autunian, which are not descendants of *Cochleosaurus* but may be closely related to that genus. Neither urocordylids nor ophiderpetontids are known from European Permian deposits. Thus, as noted by Westoll (1944), the widespread Carboniferous swamp-lake tetrapods do not appear to have survived into the Permian, at least in European Euramerica. However, some of the forms uniquely abundant at Nýřany also characterize the following Autunian lacustrine deposits, particularly the small dissorophoids. The European Autunian lake assemblages are characterized by a low diversity of vertebrates, usually all present in abundance. Most contain one to three types of fish and one to three types of tetrapod, and the remains of terrestrial or marginal forms are extremely rare. These lakes fluctuated in level and may represent a less stable and more rigorous environment than the Westphalian swamp-lakes resulting in the survival of a few lineages capable of withstanding a fluctuating environment.

4. Terrestrial / Marginal Community

(a) Structure. The marginal and terrestrial environment in the lowland coal forests is believed to have been warm, humid and stable, rich in vegetation and in invertebrates. The arborescent lycopods, *Cordaites*

and conifers would, like modern forest trees, have generated shaded vegetation-free forest floors and local clearings rich in ground vegetation. In marginal areas, the hollow trunks of fallen lycopods, compressed to varying extents, could have built up to give a complex humid substrate/habitat rich in decomposers, invertebrates and amphibians. The terrestrial invertebrate community is discussed extensively by Rolfe (Chapter 6 of this volume) but it may be noted that, in the Plzeň basin both at Nýřany and other contemporary localities, a wide range of terrestrial arachnids, myriapods and insects have been described, myriapods being notably abundant. Gastropod snails are known from the Joggins tree-stump assemblage and terrestrial annelids can be assumed to have been present. These invertebrates form a spectrum of herbivores, detritus feeders and carnivores, mostly a few centimetres long although some myriapods and chelicerates were considerably larger. Considered as tetrapod food, the coal-forest invertebrates represent a wide range of feeding niches, varying in such factors as agility/mobility, presence and extent of protective exoskeleton, life-habitat in, on, or above forest-floor and in wet or dry areas.

The major category of marginal and terrestrial tetrapods are the small insectivores (taken to include all invertebrate eaters), categorizable as tetrapods up to 30 cm long (or 1 m for the snake-like *Phlegethontia*) with either small, gripping, marginal teeth and palatal denticles, or large, rounded, crushing teeth. The range of dentition types in terrestrial Westphalian tetrapods was noted by Carroll *et al.* (1972, pp. 73–74) in the Joggins microsaurs and reptiles. They noted that the tuditanids, hapsidopareiontids and romeriids were probably feeding on insects and small slow-moving invertebrates, whilst the durophagous gymnarthrids and pantylids may have consumed molluscs and myriapods. The Nýřany terrestrial community is similar to that from Joggins but with a few more types of non-microsaur amphibian, perhaps indicating an association from wetter conditions. *Ricnodon*, *Hyloplesion* and *Scincosaurus* are all small (12 cm or less – see Table II) and small-headed forms with tiny teeth, and must have fed on very small arthropods and soft-bodied invertebrates. In the 10–30 cm category are the romeriids and *Crinodon* (probably fully terrestrial insect-eaters with the romeriids being more agile predators), *Sparodus* (a durophagous mollusc/myriapod eater), and *Amphibamus* and *Gephyrostegus* (possibly water-margin forms as noted earlier, both insectivores/small carnivores with *Gephyrostegus* as a more agile form). *Phlegethontia*, though up to 1m in length was small-headed and probably a forest-floor/leaf-litter insectivore. The next major category of consumers are the 30cm – 1·5m carnivores with large conical teeth and pseudocanines in some instances. *Archeothyris*, *Solenodonsaurus* and the adult *Cochleosaurus* fall into this category, the latter genus being of very crocodile-like configuration while the former two appear to be more fully adapted to terrestrial locomotion. Finally, *Gaudrya* may have grown to over 2m in length and is the largest preserved tetrapod at Nýřany.

If the preserved material is fully representative, this is the earliest example of the type III community of Olson (1966). The major energy flow was from plants and detritus to invertebrates to ten genera of invertebrate-eating tetrapods to four genera of larger predatory tetrapods. As with the other communities, the roles of juvenile tetrapods as predators and prey is a major complexity which cannot be assessed. The general structure of the food-web is depicted in Fig. 9 (left side).

The key feature of the type III community, as defined by Olson, is the complete absence of herbivorous tetrapods. Relatively few post-Carboniferous communities have this structure, herbivorous vertebrates being almost ubiquitous in space and time from the Permian onwards. Olson (1966) discusses the possible exceptions. However, prior to the Stephano-Permian, there are no known tetrapods showing dental adaptation for herbivory and the possibility exists that in the coal-forest terrestrial community we have a relic of the earliest type III community prior to the appearance of herbivorous tetrapods. The earliest terrestrially feeding tetrapods are deduced to have been carnivores (in the broad sense), all known adaptations to herbivory being derived conditions within tetrapod groups. The earliest terrestrial communities incorporating tetrapods must have been type III communities as the early tetrapods would have been incapable of feeding on detritus or plants. As the tetrapods diversified, hierarchies of carnivores would have developed and large vertebrates capable of feeding on smaller vertebrates, as opposed to arthropods, would have evolved. This appears to be the structure of the Nýřany terrestrial community.

When, in the Stephanian of Euramerica, herbivores do appear, namely the diadectids and edaphosaurs, they are clearly lineages without antecedents or close relatives in the known Westphalian faunas. This could be size sampling as both diadectids and edaphosaurs grew to 2–4 m adult size, but no possible juveniles or other small herbivores are known from the Westphalian. The herbivores appear suddenly in the fossil record at the same time as the coal-forest floras are replaced by "upland" floras (Scott, Chapter 5 of this volume) and Olson (1976) has reasonably proposed that these herbivorous lineages evolved in the uplands during the Westphalian or earlier and "followed" this vegetation into the lowlands in the Stephanian. Two interesting problems arise. Why were the earliest herbivores restricted to being such massive forms and what precluded the early evolution of small herbivorous tetrapods? Secondly, why were herbivorous tetrapods either absent from the Westphalian coal-forests or at least sufficiently rare not to be preserved even as erratics? Diadectids and edaphosaurs were evidently abundant in the Stephano-Permian lowlands so what factors prevented them from being similarly abundant in the Westphalian coal-forest?

The consistently large size of the earliest known herbivores may be interpreted as a "short-cut" solution to the problem of utilizing food with a low energy content. Without complex morphological and biochemical adaptations (rumens, coprophagy etc.) a tetrapod can only become a herbivore if it can minimize its energy requirement and maximize its utilization of energy in plant material. In this respect, large size has three advantages: (i) the absolute size increase and elongation of the hind-gut provides a longer period for cellulose breakdown and nutrient absorption; (ii) even with an unspecialized dentition, a large head and jaws provides a more efficient means of crushing plant material in order to break down fibres and cellulose cell-walls; (iii) maintenance energy consumption per unit weight decreases with mass increase (Kleiber, 1932) and has a simple exponential relationship to it within the poikilotherms (Hemmingson, 1960). As a result, absolute body size is an important factor in determining the minimum energy content of food which can be tolerated; and in lineages of primitive tetrapods shifting from a carnivorous to a herbivorous diet, increase in size could be expected as one of the simplest adaptations to a diet with a lower protein and energy content. Not until the appearance of small dicynodonts in the Middle Permian, do we have evidence of sufficient adaptation to a herbivorous diet that large size ceased to be obligatory. Even now, it appears that true herbivory has rarely arisen among small poikilotherms, the most notable exceptions being the tadpoles of several families of Anura. Among the reptiles, the Testudinidae and Iguanidae include many herbivorous forms, but not only have most proved to be omni-herbivores on closer study but the hatchlings/juveniles are insectivores or small carnivores, shifting to herbivory as they grow. Juvenile diadectids and edaphosaurs need not be assumed to have been herbivores or have shared the same dentition-type as the adults.

The large size and primitive tetrapod gait of the earliest herbivorous tetrapods may have been the factors which precluded their abundance in the Westphalian lowland forests. In modern tropical rain-forest and many other categories of forest, shading by the trees results in a vegetation-poor ground layer and the majority of the primary production taking place in the canopy. In modern rain-forests, large ground-dwelling herbivores are rare and thinly distributed, the majority of primary consumers living in the canopy as arboreal forms. The structure of the arborescent plants in the Westphalian lowlands (e.g. *Calamites*, lycopods, *Cordaites*) indicates that they mostly had concentrations of leaves or comparable structures in their upper regions and that, in dense stands, there would have been shading of the ground layer by a canopy layer. The edible photosynthetic and reproductive structures in such a canopy layer would not have been accessible to ground-dwelling primitive tetrapods. The massive but short-limbed diadectids and edaphosaurs could not have adopted an arboreal existence, small agile herbivores had not evolved and only in clearings in the coal-forest would there have been sufficient ground vegetation to support ground-dwelling herbivores. The numbers and concentrations of large herbivorous tetrapods supported by Coal Measure forest would have been low and dependent on the existence of marginal areas such as clearings and river banks where ground layers of vegetation could have survived. It is this circumstance which appears to be responsible for the absence of herbivorous tetrapods in Westphalian assemblages, and which has permitted the primitive hierarchy of insectivorous/carnivorous tetrapods to persist to the end of the Westphalian.

(b) Palaeogeography. The terrestrial and marginal components of the Nýřany, Linton and Mazon Creek faunas show surprisingly good

correspondence to each other considering that they are represented by erratics in all three assemblages. Of the Nýřany reptile families, the romeriids and ophiacodonts occur at Florence and Linton and a romeriid occurs at Mazon Creek. Among the amphibians in this community, several apparent congeners are shared with Linton (*Amphibamus*, *Gaudrya*, *Phlegethontia* and *Scincosaurus*) or Joggins (*Ricnodon*). Other forms are shared between Nýřany and Linton at a family level, the Gephyrostegidae (*Gephyrostegus* and *Eusauropleura*) and the Tuditanidae (*Crinodon* and *Tuditanus*). *Hyloplesion* has an unnamed probable relative from Mazon Creek and the gymnarthrid *Sparodus* has a probable relative represented by a fragment from Florence (Carroll and Gaskill, 1978). Of the Nýřany forms, only *Solenodonsaurus* has no apparent immediate relatives in the Carboniferous of North America. Thus the terrestrial marginal tetrapod associations show remarkable uniformity across the Euramerican coal-swamp localities. Although they could be coincidentally similar random samples from a larger, more diverse, terrestrial fauna, the obvious interpretation is that they represent the actual diversity of tetrapods within the terrestrial coal-swamp community and that this community was uniform over Euramerica as observed by Milner and Panchen (1973).

(c) Response to Permian changes. The Westphalian terrestrial community not only persisted but diversified in the Lower Permian of Euramerica (Olson, 1976). The major diversification was by amniote lineages (Olson, 1976) such as the captorhinomorphs (Clark and Carroll, 1973) and the pelycosaurs (Romer and Price, 1940), although the latter must have been diversifying considerably during, or even prior to, the Westphalian (Reisz, 1972) presumably in more upland areas. One non-amniote group, the Dissorophidae, diversified conspicuously in the Permian of North America (DeMar, 1968; Olson, 1976). Other lineages persisted into the Permian without further apparent diversification, for example Edopidae, Phlegethontiidae and Gymnarthridae in North America, Scincosauridae in Europe, Hapsidopareiontidae in both. No Permian gephyrostegids, solenodonsaurs, ophiderpetontids or tuditanids have been identified. Much of this turnover is attributable to the replacement of the more terrestrial amphibian families by amniote counterparts, as the

constantly humid climate within the lowland coal-forests was succeeded by the highly seasonal wet/dry climate of the Permian lowlands.

GENERAL CONCLUSIONS

(i) By the Westphalian, at least three and almost certainly four communities incorporating tetrapods existed in Euramerica. The fish-dominated open-water community, the swamp-lake community and the lowland terrestrial community are all represented by associations of tetrapod fossils, while the existence of an upland terrestrial community can be inferred from the sudden appearance in the Stephano-Permian of many terrestrial tetrapod groups without apparent antecedents in the Westphalian. The known pre-Westphalian Carboniferous tetrapods are nearly all members of open-water/lacustrine communities, the other environments being unrepresented. Thus, although the known pre-Westphalian Carboniferous tetrapods have the potential to provide us with information about the evolution of lake communities, the antecedents of the Westphalian swamp-lake and terrestrial communities are unknown and the apparent increase in diversity of tetrapods in the Westphalian is an artifact. As a result of this bias in the pre-Westphalian record, the lacustrine tetrapods have frequently had ascribed to them a greater structural, functional and evolutionary significance than they necessarily merit.

(ii) The lowland terrestrial tetrapod community is a late relict of the earliest type of complex terrestrial tetrapod community (Olson's type III) in which a hierarchy of carnivorous vertebrates is entirely dependent on invertebrate groups as consumers of plants and detritus prior to the evolution of herbivory in tetrapods. Antecedents of the Westphalian terrestrial community must have existed since the initial diversification of terrestrial tetrapods in the Devonian.

(iii) Many of the tetrapods in the swamp-lake community at Nýřany were either larvae of terrestrial or amphibious adults, or small aquatic forms which are believed to have been paedomorphic relatives of contemporary groups with terrestrial/amphibious adults (*Microbrachis* and *Branchiosaurus*). Only the urocordylids appear to have been an unambiguously aquatic family which did not have immediate relatives which were non-paedomorphic. Even so, the skeletal morphology indicates descent from a terrestrial tetrapod, albeit more distantly.

(iv) The open-water tetrapods are not obviously paedomorphic but are all highly specialized aquatic forms, some showing structural convergence with fish. The eogyrinids and keraterpetontids have immediate terrestrial relatives in the gephyrostegids and the scincosaurs respectively, and are structurally derivable from terrestrial forms, their aquatic adaptations being most readily interpreted as secondary conditions. The relationships of the loxommatids to other tetrapods is less certain and their postcranial anatomy is almost unknown.

(v) It is inaccurate to characterize any major order, group or lineage of early tetrapods as being simple "aquatic" or "terrestrial" apart from the presumed amniote groups. The temnospondyls, batrachosaurs, aïstopods, nectrideans and microsaurs all include both families with terrestrial adults and families with amphibious or aquatic adults. All orders have at least some representatives with aquatic larvae or small apparently aquatic juveniles. Thus all orders seem to have been able to radiate into both aquatic and terrestrial niches. However, most of the aquatic adults are either paedomorphic, highly specialized or, in the case of the ophiderpetontids, ambiguous in their aquatic adaptation, and it is suggested that all the aquatic tetrapods as represented at Nýřany derive from antecedents with terrestrial adults. It is suggested, albeit on the basis of late Carboniferous communities, that the water–land transition was just that and that the major amphibian orders subsequently diversified on land as amphibians with terrestrial adults, and that, subsequent to this diversification, lineages within each order "returned to the water" by extending their larval phase or by otherwise modifying terrestrial feeding and locomotor systems to produce a series of uniquely specialized aquatic lineages, all evolutionary "cul-de-sacs". Hence none of these forms have any relictual status in relation to the water–land transition but are derived from post-transition forms with terrestrial or amphibious adults.

ACKNOWLEDGEMENTS

For permission to examine material at their institutions, I am indebted to colleagues at the Národní Muzeum, Prague; the Západočeské

Muzeum, Plzeň; the British Museum (Natural History); the Royal Scottish Museum; the Hunterian Museum, Glasgow; Cambridge University Museum of Zoology; the Senckenberg Museum, Frankfurt, and the Paläontologisches Institut, Johannes Gutenberg-Universität, Mainz. Dr P. Wellnhofer kindly provided information about the collections at the Bayerische Staatsammlung, München. A research visit to Germany in 1977 was funded by the British Council and another to Czechoslovakia in 1978 was funded by the University of London Central Research Fund. Many colleagues have influenced this work during discussions and in correspondence, and I would particularly like to acknowledge Dr Don Baird, Dr Jürgen Boy, Dr Robert Carroll, Dr Alec Panchen, Dr Ian Rolfe, Dr Andrew Scott and, most important of all, my wife Angela.

REFERENCES

Augusta, J. (1940). (A new stegocephalian, *Microbrachis friči.* n.sp., from the Upper Carboniferous of Nýřany). *Véda přir.* **20**, 59–61.

Baird, D. (1964). The aïstopod amphibians surveyed. *Breviora* No. 206, 1–17.

Baird, D. (1978). Studies on Carboniferous freshwater fishes. *Am. Mus. Novit.* No. 2641, 1–22.

Beaumont, E. H. (1977). Cranial morphology of the Loxommatidae (Amphibia: Labyrinthodontia). *Phil. Trans. R. Soc.* **B280**, 29–101.

Bossy, K. V. (1976). "Morphology, Paleoecology and Evolutionary Relationships of the Pennysylvanian Urocordylid Nectrideans (Subclass Lepospondyli; Class Amphibia)." Ph.D. Thesis, Yale University.

Boy, J. A. (1972). Die Branchiosaurier (Amphibia) des saarpfalzischen Rotliegenden (Perm, SW-Deutschland). *Abh. hess. L-Amt Bodenforsch.* **65**, 1–137.

Boy, J. A. (1974). Die larven der rhachitomen Amphibien (Amphibia: Temnospondyli; Karbon-Trias). *Paläont. Z.* **48**, 236–268.

Boy, J. A. (1977). Typen und genese jungpaläozoischer tetrapoden-lagerstätten. *Palaeontographica* **156A**, 111–167.

Boy, J. A. (1978). Die tetrapodenfauna (Amphibia, Reptilia) des saarpfalzischen Rotliegenden (Unter-Perm; SW-Deutschland). 1. *Branchiosaurus. Mainzer geowiss. Mitt.* **7**, 27–76.

Boy, J. A. and Bandel, K. (1973). *Bruktererpeton fiebigi* n.gen. n.sp. (Amphibia: Gephyrostegida) der erste tetrapode aus dem Rheinisch-Westfälischen Karbon (Namur B, W-Deutschland). *Palaeontographica* **145A**, 39–77.

Beoili, F. (1905). Beobachtungen an *Cochleosaurus bohemicus* Fritsch. *Palaeontographica* **52**, 1–16.

Broili, F. (1908). Über *Sclerocephalus* aus der Gaskohle von Nürschan und das alter dieser ablagerungen. *Jb. geol. Reichsanst. Wien* **58**, 49–69.

Broili, F. (1924). Ein cotylosaurier aus dem oberkarbonischen Gaskohle von Nürschan in Böhmen. *Sber. bayer. Akad. Wiss.* 1924, 3–11.

Brough, M. C. and Brough, J. (1967). Studies on early tetrapods. I The Lower Carboniferous microsaurs. II *Microbrachis*, the type microsaur. III The genus *Gephyrostegus*. *Phil. Trans. R. Soc.* **B252**, 107-165.

Bulman, O. M. B. and Whittard, W. F. (1926). On *Branchiosaurus* and allied genera. *Proc. zool. Soc. Lond.* 1926, 533-580.

Carroll, R. L. (1964). Early evolution of the dissorophid amphibians. *Bull. Mus. comp. Zool. Harv.* **131**, 161-250.

Carroll, R. L. (1966). Microsaurs from the Westphalian B of Joggins, Nova Scotia. *Proc. Linn. Soc. Lond.* **177**, 63-97.

Carroll, R. L. (1967). Labyrinthodonts from the Joggins Formation. *J. Paleont.* **41**, 111-142.

Carroll, R. L. (1969). Problems of the origin of reptiles. *Biol. Revs.* **44**, 393-432.

Carroll, R. L. (1970a). The ancestry of reptiles. *Phil. Trans. R. Soc.* **B256**, 267-308.

Carroll, R. L. (1970b). The earliest known reptiles. *Yale Sci. Mag.* October, 16-23.

Carroll, R. L. (1972). Gephyrostegida, Solenodonsauridae. *In* Teil 5B, Batrachosauria (Anthracosauria), Gephyrostegida-Chroniosuchida, pp. 1-19. "Handbuch der Paläoherpetologie." Fischer, Stuttgart.

Carroll, R. L. and Baird, D. (1968). The Carboniferous amphibian *Tuditanus (Eosauravus)* and the distinction between microsaurs and reptiles. *Am. Mus. Novit.* No. 2337, 1-50.

Carroll, R. L. and Baird, D. (1972). Carboniferous stem-reptiles of the family Romeriidae. *Bull. Mus. comp. Zool. Harv.* **143**, 321-364.

Carroll, R. L. and Gaskill, P. (1978). The Order Microsauria. *Mem. Am. phil. Soc.* No. 126, 1-211.

Carroll, R. L., Belt, E. S., Dineley, D. L., Baird, D. and McGregor, D. C. (1972). Vertebrate paleontology of Eastern Canada. Guidebook field excursion A59. *24th int. geol. Congr., Montreal, 1972* 1-113.

Carter, G. S. and Beadle, L. C. (1930). The fauna of the swamps of the Paraguayan Chaco in relation to its environment. I. Physico-chemical nature of the environment. *J. Linn. Soc. (Zool).* **37**, 205-258.

Chaloner, W. G. and Lacey, W. S. (1973). The distribution of Late Palaeozoic floras. *In* "Organisms and Continents Through Time" (N. F. Hughes, ed.), *Spec. Pap. Palaeont.* No. 12, 271-290.

Clark, J. and Carroll, R. L. (1973). Romeriid reptiles from the Lower Permian. *Bull. Mus. comp. Zool. Harv.* **144**, 353-407.

Cope, E. D. (1871). Observations on the extinct batrachian fauna of the Carboniferous of Linton, Ohio. *Proc. Am. phil. Soc.* **12**, 177.

Cox, C. B. (1975). Vertebrate palaeodistributional patterns and continental drift. *J. Biogeog.* **1**, 75-94.

DeMar, R. E. (1968). The Permian labyrinthodont amphibian *Dissorophus multicinctus* and adaptations and phylogeny of the family Dissorophidae. *J. Paleont.* **42**, 1210-1242.

Frič (Fritsch), A. (1876). Über die fauna der Gaskohle des Pilsner und Rakonitzer beckens. *Sber. Bohm. Ges. Wiss.* 1875, 70-79.

Frič (Fritsch), A. (1879–1901). "Fauna der Gaskohle und der Kalksteine der Permformation Böhmens", Band I–IV. Prague. (Band 1, Heft 1 (1879); Heft 2 (1880); Heft 3 (1881); Heft 4 (1883). Band 2, Heft 1 (1885); Heft 2 (1885); Heft 3 (1888); Heft 4 (1889). Band 3, Heft 1 (1890); Heft 2 (1893); Heft 3 (1894); Heft 4 (1895). Band 4, Heft 1 (1899); Heft 2 (1899); Heft 3 (1901).

Haubold, H. (1971). Teil 18. Ichnia Amphibiorum et Reptiliorum Fossilium. "Handbuch der Paläoherpetologie." Fischer, Stuttgart.

Hemmingson, A. M. (1960). Energy metabolism as related to body size and respiratory surfaces and its evolution. *Rep. Steno meml Hosp.* **9**, 1–110.

Holub, V. and Tásler, R. (1978). Filling of the late Palaeozoic continental basins in the Bohemian Massif as a record of their palaeogeographical development. *Geol. Rundschau* **67**, 91–109.

Holub, V., Skoček, V. and Tásler, R. (1975). Palaeogeography of the late Palaeozoic in the Bohemian Massif. *Palaeogeogr., Palaeoclimatol., Palaeoecol.* **18**, 313–332.

Hummel, K. (1913). Über *Ricnodon* cf. *dispersus* Fritsch aus dem Böhmischen Oberkarbon. *Z. dt. geol. Ges.* **65**, 591–595.

Jaekel, O. (1902). Ueber *Gephyrostegus bohemicus* n.g.n.sp. *Z. dt. geol. Ges.* **54**, 127–132.

Jaekel, O. (1909). Über die klassen der tetrapoden. *Zool. Anz.* **34**, 193–212.

Jaekel, O. (1911). "Die Wirbeltiere. Eine übersicht uber die fossilen und lebenden formen." Berlin.

Kleiber, M. (1932). Body size and metabolism. *Hilgardia* **6**, 315–353.

Kuhn, O. (1959). Ein neuer microsaurier aus dem deutschen Rotliegenden. *Neues Jb. Geol. Paläont. Mh.* 1959, 424–426.

Land, D. H. (1974). Geology of the Tynemouth district. *Mem. geol. Surv. (N.S.)* No. 15.

Lund, R. (1978). Anatomy and relationships of the family Phlegethontiidae (Amphibia, Aïstopoda). *Ann. Carnegie Mus.* **47**, 53–79.

McGinnis, H. J. (1967). The osteology of *Phlegethontia*, a Carboniferous and Permian amphibian. *Univ. Calif. Publs geol. Sci.* **71**, 1–49.

Milner, A. C. (1978). "Carboniferous Keraterpetontidae and Scincosauridae (Nectridea: Amphibia) – a review." Ph.D. Thesis, University of Newcastle upon Tyne.

Milner, A. R. (1974). "A Revision of the "Branchiosaurs" and some Associated Palaeozoic Temnospondyl Amphibia." Ph.D. Thesis, University of Newcastle upon Tyne.

Milner, A. R. (1978). A reappraisal of the early Permian amphibians *Memonomenos dyscriton* and *Cricotillus brachydens*. *Palaeontology* **21**, 667–686.

Milner, A. R. (1980). The temnospondyl amphibian *Dendrerpeton* from the Upper Carboniferous of Ireland. *Palaeontology* **23**, 125–141.

Milner, A. R. and Panchen, A. L. (1973). Geographical variation in the tetrapod faunas of the Upper Carboniferous and Lower Permian. *In* "Implications of Continental Drift to the Earth Sciences" (D. H. Tarling and S. K. Runcorn, eds), Vol. 1, pp. 353–368. Academic Press, London and New York.

Moore, L. R. (1968). Cannel coals, bogheads and oil shales. *In* "Coal and Coal-bearing Strata" (D. G. Murchison and T. S. Westoll, eds), pp. 19–29. Oliver and Boyd, Edinburgh.

Moy-Thomas, J. A. and Miles, R. S. (1971). "Palaeozoic Fishes." Chapman and Hall, London.

Nemejč, F. (1952). On some more detailed problems in the stratigraphy of limnic Permocarboniferous basins of Bohemia and Moravia. *C.r.3 Congr. Avanc. Étud. Stratigr. carb.* **2**, 475–481.

Olson, E. C. (1966). Community evolution and the origin of mammals. *Ecology* **47**, 291–302.

Olson, E. C. (1967). Early Permian vertebrates of Oklahoma. *Circ. Okla. geol. Surv.* **74**, 1–111.

Olson, E. C. (1971). "Vertebrate Paleozoology." Wiley, New York.

Olson, E. C. (1976). The exploitation of land by early tetrapods. *In* "Morphology and Biology of Reptiles" (A. d'A. Bellairs and C. B. Cox, eds), pp. 1–30. Academic Press, London and New York.

Panchen, A. L. (1970). Teil 5A Anthracosauria "Handbuch der Paläoherpetologie." Fischer, Stuttgart.

Panchen, A. L. (1972a). The skull and skeleton of *Eogyrinus attheyi* Watson (Amphibia: Labyrinthodontia). *Phil. Trans. R. Soc.* **B263**, 279–326.

Panchen, A. L. (1972b). The interrelationships of the earliest tetrapods. *In* "Studies in Vertebrate Evolution" (K. A. Joysey and T. S. Kemp, eds), pp. 65–87. Oliver and Boyd, Edinburgh.

Panchen, A. L. (1973). Carboniferous tetrapods. *In* "Atlas of Palaeobiogeography" (A. Hallam, ed.), pp. 117–125. Elsevier, Amsterdam.

Panchen, A. L. (1977). On *Anthracosaurus russelli* Huxley (Amphibia: Labyrinthodontia) and the family Anthracosauridae. *Phil. Trans. R. Soc.* **B279**, 447–512.

Panchen, A. L. and Walker, A. D. (1961). British coal measure labyrinthodont localities. *Ann. Mag. nat. Hist.* (13)**3**, 321–332.

Pearson, H. S. (1924). *Solenodonsaurus broili*, a seymouriamorph reptile. *Ann. Mag. nat. Hist.* (9)**14**, 338–343.

Pešek, J. (1968). Geologischer bau und entwicklung der Karbon-ablagerungen des Plzeň-steinkohlengebirges. *Acta Univ. Carolinae, Geol.* 1968, 329–357.

Pešek, J. (1974). Lateral passages between variegated and grey Carboniferous sediments. *C.r. 7 Congr. Avanc. Étud. Stratigr. carb.* **4**, 75–83.

Piveteau, J. (1925). Sur la morphologie et la position systematique du genre *Sauravus* Thevenin. *Bull. Soc. geol. France* (4)**25**, 89–96.

Rayner, D. H. (1971). Data on the environment and preservation of late Palaeozoic tetrapods. *Proc. Yorks. geol. Soc.* **38**, 437–495.

Reisz, R. (1972). Pelycosaurian reptiles from the Middle Pennsylvanian of North America. *Bull. Mus. comp. Zool. Harv.* **144**, 27–62.

Reisz, R. (1975). Pennsylvanian pelycosaurs from Linton, Ohio and Nýřany, Czechoslovakia. *J. Paleont.* **49**, 522–527.

Rieppel, O. (1980). The edopoid amphibian *Cochleosaurus* from the Middle Pennsylvanian of Nova Scotia. *Palaeontology* **23**, 143–149.

Romer, A. S. (1939). Notes on branchiosaurs. *Am. J. Sci.* **237**, 748–761.
Romer, A. S. (1947). Review of the Labyrinthodontia. *Bull. Mus. comp. Zool. Harv.* **99**, 1–368.
Romer, A. S. (1974). Aquatic adaptation in reptiles, primary or secondary? *Ann. S. Afr. Mus.* **64**, 221–230.
Romer, A. S. and Price, L. I. (1940). Review of the Pelycosauria. *Spec. Pap. geol. Soc. Am.* No. 28, 1–538.
Schäfer, W. (1972). "Ecology and Palaeoecology of Marine Environments." Oliver and Boyd, Edinburgh.
Schroeder, E. (1939). Ein neuartiger amphibienrest (?*Microbrachis*) aus dem saarländischen Rotliegenden. *Z.dt. geol. Ges.* **91**, 812–815.
Schwarz, H. (1908). Über die wirbelsäule und die rippen holospondyler stegocephalen. *Beitr. Paläont. Geol. Öst.Ung.* **21**, 63–105.
Skoček, V. (1968). (Upper Carboniferous varvites in coal basins of central Bohemia). *Věstn. ústřed. Úst. geol.* **43**, 113–121.
Smith, A. G., Briden, J. C. and Drewry, G. E. (1973). Phanerozoic world maps. *In* "Organisms and Continents through Time" (N. F. Hughes, ed.), *Spec. Pap. Palaeont.* No. 12, 1–42.
Steen, M. C. (1931). The British Museum collection of Amphibia from the Middle Coal Measures of Linton, Ohio. *Proc. zool. Soc. Lond.* 1930, 849–891.
Steen, M. C. (1938). On the fossil Amphibia from the Gas Coal of Nýřany and other deposits in Czechoslovakia. *Proc. zool. Soc. Lond.* **B108**, 205–283.
Stehlík, A. (1924). (New stegocephalians from Moravian Permian formations and additions to the knowledge of stegocephalians from Nýřany). *Acta Soc. Sci. nat. Morav.* **1**, 199–283.
Thomson, K. S. and Bossy, K. V. H. (1970). Adaptive trends and relationships in early Amphibia. *Forma Functio* **3**, 7–31.
Watson, D. M. S. (1913). *Micropholis stowi* Huxley, a temnospondylous amphibian from South Africa. *Geol. Mag.* **10**, 340–346.
Watson, D. M. S. (1926). On the evolution and origin of the Amphibia. *Phil. Trans. R. Soc.* **B214**, 189–257.
Westoll, T. S. (1944). The Haplolepidae, a new family of late Carboniferous bony fishes. *Bull. Am. Mus. nat. Hist.* **83**, 1–121.
Wootton, R. J. (1972a). The evolution of insects in freshwater ecosystems. *In* "Essays in Hydrobiology" (R. B. Clark and R. J. Wootton eds), pp. 69–82. University of Exeter Press.
Wootton, R. J. (1972b). Nymphs of Palaeodictyoptera (Insecta) from the Westphalian of England. *Palaeontology* **15**, 662–675.
Zidek, J. and Baird, D. (1978). *Cercariomorphus* Cope 1885, identified as the aïstopod amphibian *Ophiderpeton*. *J. Paleont.* **52**, 561–564.

18 | The Cotylosauria: A Reconsideration of a Group of Archaic Tetrapods

M. J. HEATON

Department of Earth and Planetary Sciences, Erindale College, University of Toronto, Mississauga, Ontario, Canada

Abstract: The Order Cotylosauria was once thought to consist of three suborders, Seymouriamorpha, Diadectomorpha and Captorhinomorpha, and was regarded as the most primitive group of reptiles. More recently the first suborder and part of the second (including *Diadectes* itself) were removed from the Cotylosauria and placed with the anthracosaurian amphibians, despite the fact that *Diadectes* is the "type" cotylosaur. The remaining "diadectomorphs" (Procolophonia and Pareiasauria) were and are regarded as primitive true reptiles.

It is concluded here that the Cotylosauria (now used in the sense of seymouriamorphs, diadectids and related forms) were not "stem reptiles" but rather were a group of advanced, non-anthracosaurian amphibians closely related to the Reptilia. The Seymouriamorpha (Seymouriidae, Kotlassiidae, and *Nycteroleteridae* and the non-reptilian Diadectopmorpha (Limnoscelidae, Tseajaiidae, and Diadectidae) formed a natural group, the Batrachosauria, with the most primitive Reptilia and their immediate ancestors, that possessed such shared derived characters as: a dorsally open braincase without a tectum synoticum ossification (labyrinthodont braincases were roofed by a synotic ossification); "Y"-shaped sphenethmoid; separate coracoid ossifications; pleurocentrum-dominated vertebrae with laterally swollen neural arches; humeri with both large supinator processes and entepicondylar foramina (labyrintho-

Systematics Association Special Volume No. 15, "The Terrestrial Environment and the Origin of Land Vertebrates", edited by A. L. Panchen, 1980, pp. 497–551, Academic Press, London and New York.

donts and "lepospondyls" never had both present together); and a transverse flange on the pterygoid. These characters were primitive for both reptiles and cotylosaurs but may have been modified in more derived members of either group. Cotylosaurs primitively possessed such shared derived characters as: loss of primitive cranial neurokinesis; development of a single coracoid ossification; and reduced or closed post-temporal fenestrae. These characters were primitive for cotylosaurs and may have been modified in more derived cotylosaurs. Reptiles (including the Captorhinomorpha) primitively possessed such shared derived characters as: an amniotic egg; advanced cranial metakinesis (non-batrachosaurian amphibians were not metakinetic); large post-temporal fenestrae; and two coracoid ossifications. These characters were primitive for reptiles and may have been modified in more derived reptiles.

INTRODUCTION

The Order Cotylosauria was first erected by Cope (1880) based on the peculiar molariform cheek teeth of *Diadectes* and a misinterpretation of the occipital condyle in the genus "*Empedocles*" (*Diadectes*). The name of the order was derived from the appearance of a specimen which was damaged in such a manner that the basioccipital had been lost from the braincase. It thus appeared to him that the basal portion of the braincase (basisphenoid–parasphenoid) extended posteriorly and branched to form two separate, cup-shaped condyles or cotyli (exoccipitals) lateral to the foramen magnum. Cope (1883, 1889) later realized his mistake and eventually corrected it. Nonetheless, the term has remained in the literature. A misunderstanding has arisen over the translation of the Greek root *kotyle* as a plant or stem rather than as a cup-shaped hollow (Jaeger, 1955) as Cope had originally intended. Thus, the term Cotylosauria has been interpreted literally into English as "stem reptiles" or into German as "Stämreptilien". This would not be so unfortunate were it not for the fact that this translation has been reinterpreted to indicate that the Cotylosauria were the "stem reptiles" from which all later reptiles arose (Broili, 1904; Romer, 1956, 1966, 1968). Once this interpretation has been accepted, almost any primitive reptiliomorph tetrapod could be included within the Cotylosauria for lack of a more appropriate assignment. By 1911 when Case's important monograph on the Cotylosauria appeared, the Cotylosauria was composed of such diverse families as: the Diadectidae; Bolosauridae; "Pariotichidae", including *Pariotichus*, a microsaurian amphibian

(Gregory *et al.*, 1956; Carroll and Gaskill, 1978) and *Isodectes*, a trimerorhachoid temnospondyl amphibian (Romer, 1966); Captorhinidae; Seymouriidae; Pareiasauridae; Procolophonidae; and Pantylidae. The Cotylosauria had become a palaeontological waste-basket and a rather large one at that.

Although Case (1911) had succeeded in resolving, at the family level, many of the systematic problems associated with these animals, he was unable to draw any major conclusions at the suprafamilial level. It was left to Watson (1917) to accept this challenge. He divided the order Cotylosauria into three suborders, the Diadectomorpha, the Seymouriamorpha and the Captorhinomorpha. He also recognized the close relationships of the Captorhinomorpha to the synapsid or mammal-like reptiles. Unfortunately Watson interpreted the Diadectomorpha as the ancestors of all "sauropsid" (including all modern) reptiles and included within it the suborders Procolophonia and Pareiasauria.

It is now generally recognized that *Diadectes* is not a true reptile. Romer (1964, 1966, 1968) believed *Diadectes* to be an advanced seymouriamorph, this latter group having been transferred from the Cotylosauria to the Amphibia and specifically to the Anthracosauria. This caused problems since it removed *Diadectes*, the type genus of the Order Cotylosauria, from that order. Olson (1947, 1965, 1966b) likewise believed that *Diadectes* was not a true reptile and attempted to solve the problem of fitting *Diadectes* into its proper systematic position by erecting the reptilian subclass Parareptilia to include the Diadectomorpha (including the Seymouriamorpha), the Procolophonia, the Pareiasauria, and initially (1947) the Chelonia. As evidence has continued to accumulate, it appears more probable that the Procolophonia and the Pareiasauria, like the Chelonia (Olson, 1965), were true reptiles that laid amniotic eggs and that the Diadectomorpha (including the Seymouriamorpha in Olson's sense) were amphibians. As Olson (1971) has indicated, the term Cotylosauria has, within the last 20 years, come to include only the Captorhinomorpha, together with the Procolophonoidea and Pareiasauroidea (Romer's Procolophonia, 1966, 1968), all of which are true reptiles.

It appears necessary as both Olson (1971, 1977) and Jollie (1978) have indicated, to re-define what is meant by the term Cotylosauria. Romer's note (1968) that "for two-thirds of a century the

Cotylosauria has been synonymous with 'stem reptile'", is misleading since it is only within the last decade or two that the Diadectomorpha and Seymouriamorpha have not been considered to be cotylosaurs. Thus, if historical precedent is to be maintained, the use of the term Cotylosauria, as a suborder encompassing Diadectomorpha and Seymouriamorpha, should be continued as it was used rigorously from 1917 (Watson, 1917) to 1961 and somewhat more loosely from 1881 to 1964. Even Romer (1956) included the Seymouriamorpha and the Diadectomorpha within the order Cotylosauria.

Thus the amphibian order Cotylosauria contains two known suborders, the Diadectomorpha and the Seymouriamorpha. The type genus of the order is *Diadectes* ("*Empedocles*") Cope 1878.

THE COMPOSITION OF THE COTYLOSAURIA

In order to follow the discussion of the anatomy and systematics of the Cotylosauria, it is necessary to present a brief outline of the nomenclature used. Although this paper deals primarily with the systematic position of the Cotylosauria, it must be indicated that this order is a member of the amphibian Subclass Batrachosauria as here amended to exclude the Anthracosauria. The batrachosaurs are thought to have given rise to two separate lineages, one the Cotylosauria, the other the Reptilia. No non-cotylosaurian batrachosaurs that might be closely associated with the origin of the Reptilia have yet been identified. The Order Cotylosauria is composed of two groups of archaic tetrapods that are believed to have laid a primitive anamniotic egg and hence may be referred to, loosely, as "amphibians". Each of these groups, the Seymouriamorpha and the Diadectomorpha, is recognized as a suborder. The Suborder Seymouriamorpha is poorly known but is provisionally thought to be composed of the families Seymouriida, Kotlassiidae and possibly the Nycteroleteridae. The Discosauriscidae are here considered to be anthracosaur Amphibia (see below). The Suborder Diadectomorpha is here taken to be composed of three families, the Limnoscelidae, Tseajaiidae and Diadectidae.

For the purposes of the discussion below, the following is the classification and terminology to be used:

Class AMPHIBIA
Subclass Batrachosauria Efremov 1946
Order Cotylosauria Cope 1880

Suborder Seymouriamorpha Watson 1917
Family Nycteroleteridae Efremov 1940
Family Seymouriidae Williston 1911
Family Kotlassiidae Amalitzky 1921
Suborder Diadectomorpha Watson 1917
Family Limnoscelidae Williston 1911
Family Tseajaiidae Vaughn 1964
Family Diadectidae Cope 1880

The most recent characterization of the Order Cotylosauria is that provided by Romer (1956). It is too long to be regarded as a diagnosis but it does indicate the extent to which primitive characters were put in identifying this order. Such characters as sculpture of the skull-roof elements, absence of temporal fenestration, presence of a large pineal foramen, and presence of a tabular, supratemporal and postparietal, to mention a few, are not valid diagnostic characters since they are primitive characters possessed by all manner of primitive tetrapods including temnospondyl and anthracosaurian labyrinthodont amphibians, microsaurian and nectridean lepospondyl amphibians, and captorhinomorph, procolophonian and pareiasaurian reptiles. For a valid diagnosis of the Order Cotylosauria to be produced, specific homologous shared derived characters must be recognized that unite all members of this order and its subdivisions.

Previous discussions of the relationships of the Cotylosauria have typically neglected to explain the nature and significance of the osteological characters used to establish the proposed phylogenies and classifications. For this reason, an extended, but still too brief, discussion of important aspects of cotylosaurian osteology is presented before the discussion of cotylosaurian phylogeny and classification.

COTYLOSAURIAN SKULL

The primitive amphibian posterior skull-roof was composed of a series of paired median elements, frontals, parietals and postparietals. In the postorbital region, the median elements were flanked by the postfrontals, intertemporals, supratemporals and tabulars (Fig. 1). Primitively, the large dorsal postparietal contacted the supratemporal and prevented contact from being made between the

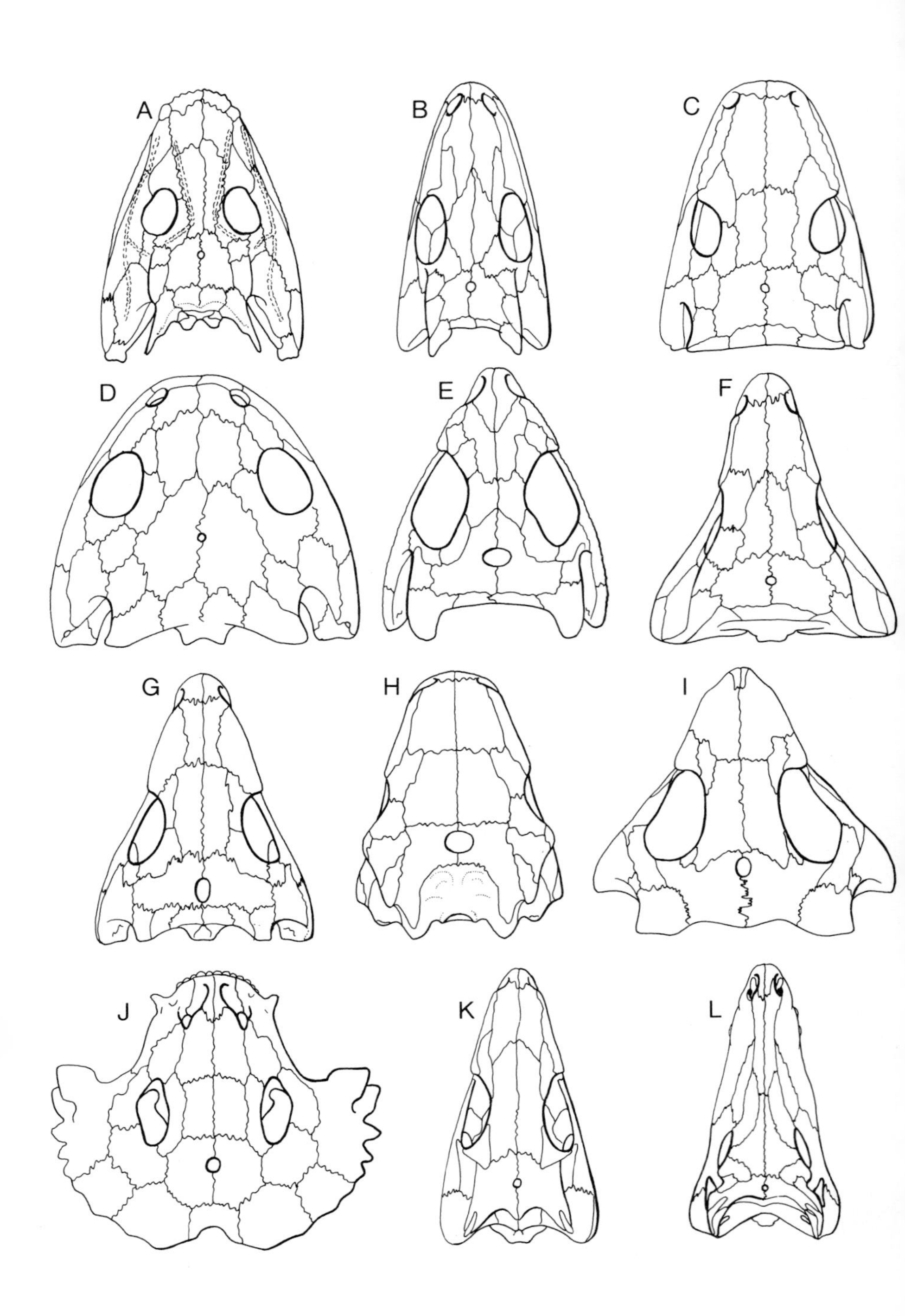
A
B
C
D
E
F
G
H
I
J
K
L

tabular and parietal just as in rhipidistian crossopterygian fish such as *Eusthenopteron* (Jarvik, 1944), edopoid temnospondyl amphibians such as *Caerorhachis* (Holmes and Carroll, 1977), *Dendrerpeton* (Carroll, 1966) and *Edops* (Romer and Witter, 1942).

Progressive shortening of the postorbital region of the skull and particularly the post-pineal region characterized the fish-tetrapod transition (Panchen, 1975). Reduction in size of the postparietal eventually led to the development of the tabular–parietal contact present in anthracosaurs, cotylosaurs and reptiles. This has been regarded as an homologous shared derived character uniting these three major groups, although "lepospondyls" have the same arrangement, and is one of the main pillars of the theory of an anthracosaur–seymouriamorph origin of reptiles. As Panchen (1972, 1975, 1977a) has indicated, most of the characters thought to support this theory, and in particular the Romerian theory for the evolution of the reptilian vertebra (Romer, 1947, 1956, 1966), have been shown to be invalid. Panchen (1975) has noted that it is quite likely that the tabular–parietal contact of reptiles developed convergently and did not arise from the condition present in anthracosaurs. The development of a tabular–parietal contact was the natural consequence of a continued relative reduction of the postorbital region and was not restricted to these groups as it appeared in the clearly temnospondyl intasuchid amphibians (Konjukova, 1953). This obvious convergence – no other derived characters are shared in common with this group – should inspire caution when interpreting the similarity of skull-roof patterns of anthracosaurs and cotylosaurs. If, as with the intasuchids, they did not share other derived characters – and as far as can be determined

---

Fig. 1. Skulls in dorsal aspect. (A) The anthracosaur *Palaeoherpeton*; (B) the anthracosaur *Gephyrostegus*; (C) the seymouriamorph cotylosaur *Seymouria*; (D) the seymouriamorph *Kotlassia*; (E) the seymouriamorph *Nycteroleter*; (F) the diadectomorph cotylosaur *Limnoscelis*; (G) the diadectomorph *Tseajaia*; (H) the diadectomorph *Diadectes*; (I) the procolophonid reptile *Procolophon*; (J) the pareiasaur *Scutosaurus*; (K) the captorhinomorph *Protorothyris*; (L) the pelycosaur *Dimetrodon*. (A) After Panchen, (B) after Carroll, (C) modified after White, (D) after Bystrow, (E) after Efremov, (F) after Huene and Romer, (H) after Olson, (I) modified after Broili and Schroeder, (J) after Vjushkov, (K) after Carroll, (L) after Romer and Price. Not to scale.

they did not – then likewise, the similarity was probably a result of convergence. The seymouriamorph cotylosaurs *Nycteroleter*, *Seymouria* and *Kotlassia* retained the primitive, large, dorsal exposure of the tabulars and postparietals on the dorsal surface of the skull-roof but also extended prominent flanges ventrally onto the occipital surface (Fig. 2). In diadectomorphs, the dorsal exposure of the tabulars and postparietals was greatly reduced, with these elements being restricted primarily to the occipital surface. In most diadectomorphs the occipital surface was oriented vertically while, in the Diadectidae, emargination of the posterior skull-roof to accommodate enlarged cervical musculature has resulted in some anterodorsal sloping of the occiput thus exposing the postparietals and tabulars dorsally.

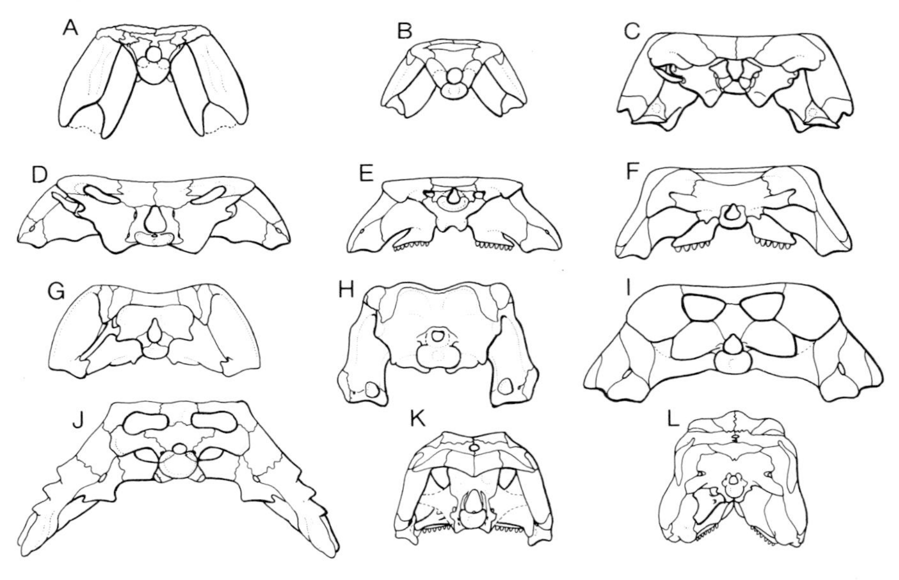

Fig. 2. Skulls in occipital aspect. (A) The anthracosaur *Palaeoherpeton*; (B) the anthracosaur *Gephyrostegus*; (C) the seymouriamorph cotylosaur *Seymouria*; (D) the seymouriamorph *Kotlassia*; (E) the seymouriamorph *Nycteroleter*; (F) the diadectomorph cotylosaur *Limnoscelis*; (G) the diadectomorph *Tseajaia*; (H) the diadectomorph *Diadectes*; (I) the procolophonid reptile *Procolophon*; (J) the pareiasaur *Scutosaurus*; (K) the captorhinomorph *Protorothyris*; (L) the pelycosaur *Dimetrodon*. Not to scale.

Panchen (1975), in proposing the convergent development of the reptilian tabular–parietal contact, assumed, in accord with the theories championed by Carroll (1964, 1969a,b,c, 1970), that captorhinomorphs were the most primitive group of reptiles. However, there is evidence that captorhinomorphs were in some respects quite advanced reptiles, although generalized in structure, and that such groups as millerosaurs, mesosaurs, procolophonids and pareiasaurs, although exhibiting numerous obvious specializations, were generally more primitive. The primitive reptilian skull-roof appears to have included a substantial postorbital–supratemporal contact of the type present in pelycosaurs, millerosaurs, procolophonids, pareiasaurs, bolosaurs and mesosaurs. While it is possible that a convergent development of the postorbital–supratemporal contact occurred once or twice, it seems highly unlikely that it could have occurred as many as five (known) times. The primitive reptilian condition is foreshadowed in at least one cotylosaurian family, the Limnoscelidae. In this group the intertemporal was lost and its position occupied by an extension of the supratemporal. This contact resembles the condition seen in eothyridid pelycosaurs and pareiasaurs but is believed to have been developed convergently. The presence of an intertemporal is believed to be a primitive character among batrachosaurs since it is present in seymouriamorph cotylosaurs and possibly also in the diadectomorph cotylosaur *Diadectes* (Olson, 1947, 1950). If diadectids possess an intertemporal it is more or less co-ossified with the parietal and, to a lesser extent, with the supratemporal, with a loss, particularly on the internal surfaces of the bone, of any identifiable sutures. This gives the impression that a lateral lappet of the parietal extends between the postorbital and the supratemporal to meet the squamosal (Watson, 1954). In the tseajaiid diadectomorph *Tseajaia* there is no identifiable intertemporal (Moss, 1972) with the result that there is a parietal lappet that extends laterally to contact the squamosal just as Watson (1954) described in *Diadectes*. It is not known whether the parietal lappet in tseajaiids developed as a coelescence of the parietal and intertemporal as in diadectids or as a lateral replacement of the intertemporal by the parietal.

Some dermal skull elements in seymouriamorph cotylosaurs, most notably the maxilla, jugal and squamosal, possess the remnants of a lateral-line system (White, 1939). This suggests that at least some of the life-cycle was spent as an aquatic form presumably with external gills. This tadpole or branchiosaur stage, as in most cases it was (a neotenic "axolotl" form was a possible alternative), is indicative of development from a primitive anamniotic egg. The absence of lateral-line canals or pits in diadectomorphs – *Tseajaia* is a possible exception – was not indicative of development from an amniotic egg but rather is only an indicator of advanced terrestriality at least in the known adult forms. The presence of a lateral-line

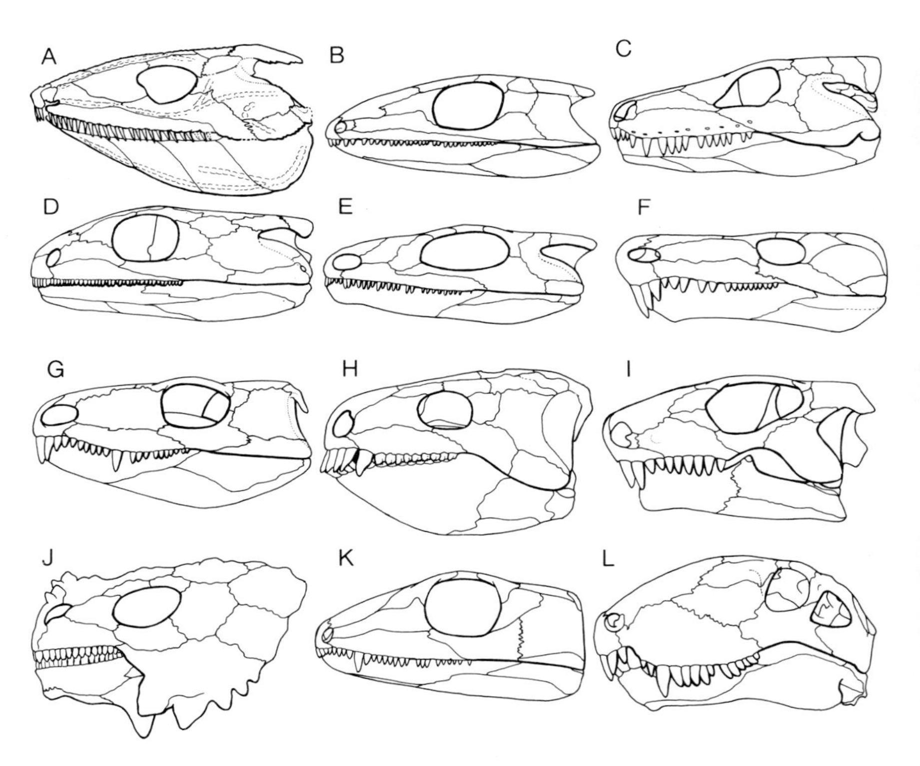

Fig. 3. Skulls in left lateral aspect. (A) The anthracosaur *Palaeoherpeton*; (B) the anthracosaur *Gephyrostegus*; (C) the seymouriamorph cotylosaur *Seymouria*; (D) the seymouriamorph *Kotlassia*; (E) the seymouriamorph *Nycteroleter*; (F) the diadectomorph cotylosaur *Limnoscelis*; (G) the diadectomorph *Tseajaia*; (H) the diadectomorph *Diadectes*; (I) the procolophonid reptile *Procolophon*; (J) the pareiasaur *Scutosaurus*; (K) the captorhinomorph *Protorothyris*; (L) the pelycosaur *Dimetrodon*. Not to scale.

system is of little use taxonomically, except to indicate that the Cotylosauria were not reptiles.

Except for the aberrant late Permian form *Nycteroleter*, all batrachosaurs have a lachrymal extending from the orbit to the external naris, a condition that may safely be regarded as primitive. *Nycteroleter* (Efremov, 1938, 1940; Chudinov, 1957) has been described as having a greatly reduced lachrymal that does not extend to the external naris (Fig. 3). Efremov noted the difficulties in deciphering the structure of this small animal and reconstructed significant portions of it "schematically". Many of these interpretations were produced under the impression that *Nycteroleter* was a procolophonid reptile with a short lachrymal. Such a relationship is now considered extremely unlikely.

Among the Cotylosauria, the squamosal exhibits the most diverse structure of any of the dermal skull elements. The primitive pattern of squamosal development in cotylosaurs appears to be essentially similar to that of primitive temnospondyls (Holmes and Carroll, 1975; Carroll, Chapter 12 of this volume), anthracosaurs (Panchen, 1964, 1975), microsaurs (Carroll and Gaskill, 1978) and captorhinomorph reptiles (Carroll, 1964, 1969a,b,c, 1971) with a near-vertical posterior margin with a medially directed occipital flange (Figs 2 and 3). This primitive condition has been retained in only one cotylosaur, the primitive diadectomorph *Limnoscelis*. Two patterns of squamosal modification have occurred in cotylosaurs. Among the Seymouriamorpha a deep otic notch developed by embayment of the posterior margin of the squamosal. The occipital flange was not reduced but folded inward to sheath the posterior surface of the deeply emarginated quadrate. The presumed tympanum that was stretched across the notch was supported entirely by the squamosal. This is quite in contrast to the condition among the advanced members of the Diadectomorpha (Tseajaiidae and Diadectidae) in which the occipital flange of the squamosal is greatly reduced, thus exposing the emarginated quadrate posteriorly. The tympanum that bridged the otic notch, and which was frequently preserved as an ossified membrane in *Diadectes*, was supported solely by the quadrate (Olson, 1947, 1950).

The palates of cotylosaurs and primitive reptiles are virtually identical and are highly derived relative to temnospondyl or anthracosaurian amphibians (Fig. 4). *Seymouria* is the only known

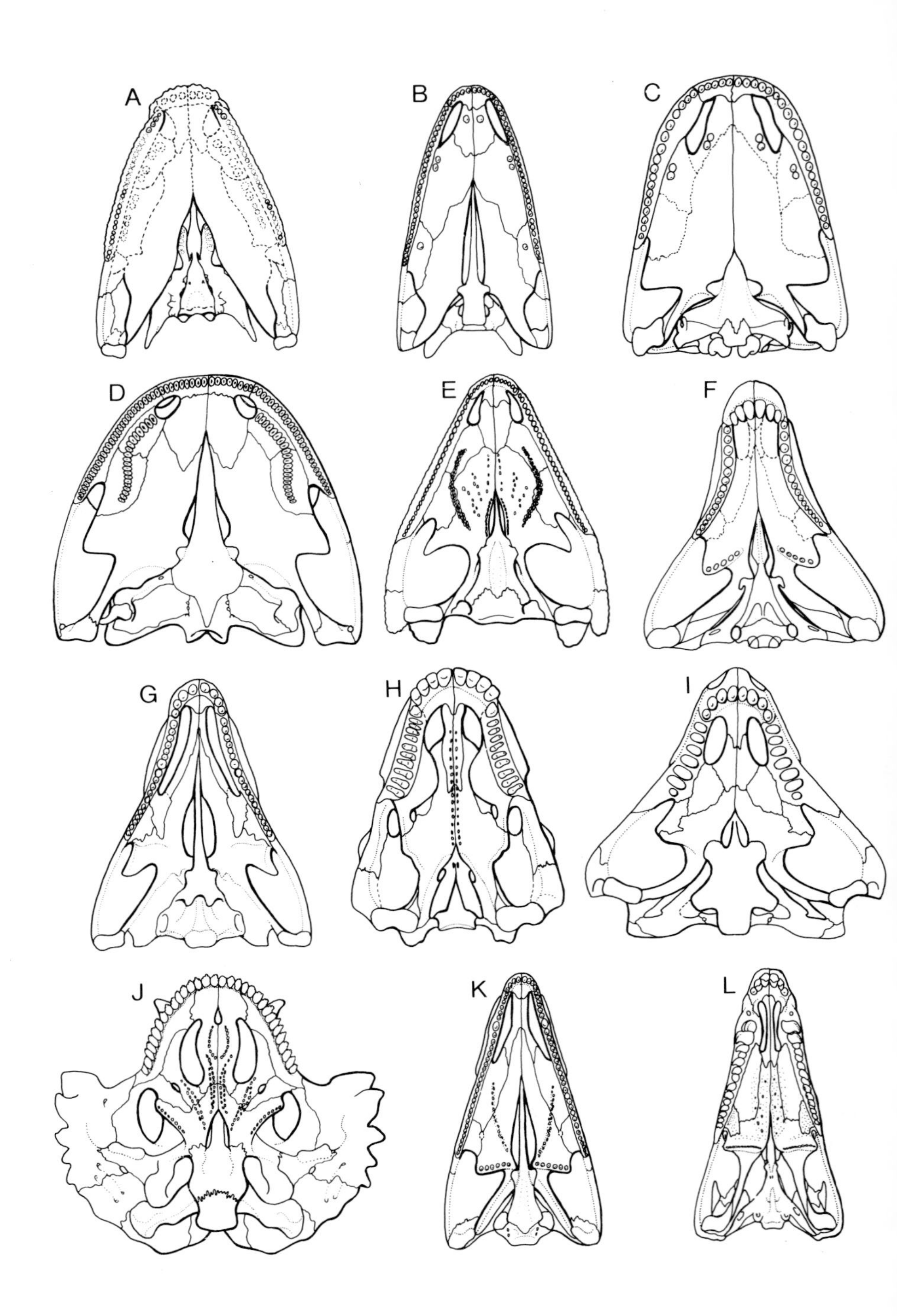
A
B
C
D
E
F
G
H
I
J
K
L

form that has retained the primitive rhipidistian palatal tusks. All other cotylosaurs have, at the most, rows of relatively small teeth (not tusks) or fields of small denticles on the surface of the palate. The types of denticulation of the palate in cotylosaurs are so diverse and so closely controlled by feeding habits that they are of little taxonomic use.

The presence of a transverse flange on the pterygoid has typically been regarded as a unique reptilian character (Carroll, 1969a,b). Instead, this must be interpreted as a shared derived character of all batrachosaurs as well as their reptilian derivatives. Nothing approaching the transverse flange of the pterygoid is present in temnospondyl or anthracosaurian labyrinthodonts or in "lepospondyl" amphibians. In the advanced seymouriamorphs, *Seymouria* and *Kotlassia*, the transverse flange is typically broad with a rounded posterolateral extremity (White, 1939; Bystrow 1944), whereas in *Nycteroleter* (Efremov, 1940) and in diadectomorphs it is more noticeably pointed and extended farther posteriorly. Carroll (1969c) has commented on the role of the transverse flange in the differentiation of a primitive adductor mass into anterior and posterior segments by the constriction of the subtemporal fenestra (presumably occupied by the posterior adductor musculature) by the flange. This is an optical illusion formed by projecting the outline of the transverse flange onto a flat surface. In fact, the transverse flange sheathed the anterior surface of the adductor (principally the M. pterygoideus) musculature and at no time restricted or impinged upon the region occupied by the adductor musculature. Recent investigation of the musculature of both recent and fossil amphibians and reptiles indicates that, if anything, the primitive condition in amphibians, including early batrachosaurs, was the existence of a highly differentiated multi-segmented adductor musculature. This further reinforces the thesis that the development

---

Fig. 4. Skull in palatal aspect. (A) The anthracosaur *Palaeoherpeton*; (B) the anthracosaur *Gephyrostegus*; (C) the seymouriamorph cotylosaur *Seymouria*; (D) the seymouriamorph *Kotlassia*; (E) the seymouriamorph *Nycteroleter*; (F) the diadectomorph cotylosaur *Limnoscelis*; (G) the diadectomorph *Tseajaia*; (H) the diadectomorph *Diadectes*; (I) the procolophonid reptile *Procolophon*; (J) the pareiasaur *Scutosaurus*; (K) the captorhinomorph *Protorothyris*; (L) the pelycosaur *Dimetrodon*. Not to scale.

of the transverse flange of the pterygoid was not related to differentiation of the adductor musculature. Terrestrial animals did not have the buoyancy and viscosity of the external medium that aquatic forms had opposing the action of gravitational forces by which items of prey could be held nearly stationary while the bite was shifted during feeding. The transverse flange of the pterygoid probably originated as an aid in inertial feeding wherein prey was ratcheted down the oesophagus or held while the mouth was opened to shift the bite.

In the seymouriamorph cotylosaurs the maxilla is curved laterally as in many wide-skulled labyrinthodont amphibians. Sharp conical teeth are present in *Nycteroleter* (Efremov, 1940) and in *Seymouria* (White, 1939) while in *Kotlassia* the teeth are compressed antero-posteriorly (Bystrow, 1944) (Figs 3 and 4). Labyrinthine infolding of the tooth dentine is common to most seymouriamorphs and to limnoscelid diadectomorphs. This was a retained primitive character which indicates that these animals all had a common rhipidistian ancestor with labyrinthine plication. This character is therefore of no significance in determining less remote relationships between the seymouriamorphs and the limnoscelids. The advanced diadectomorphs, the Tseajaiidae and Diadectidae, have lost the infolding of the tooth enamel. While this character seems to unite the members of this group, character loss is notoriously prone to development of convergences. Several of the advanced diadectomorphs develop distinctive dental structures. *Tseajaia* has a well-developed caniniform tooth that is situated at about the midpoint of the maxilla. The only other cotylosaur that possesses a maxillary caniniform tooth is the diadectid *Diadectes*. In this genus the caniniform is variable in size from large to quite small. It was always the first maxillary tooth and bore no auxillary cusps. Smaller and earlier diadectids do not appear to have had enlarged caniniform teeth. Diadectids are peculiar since they are the only Palaeozoic amphibians that developed massive, multi-cusped, grinding cheek teeth for chewing plant material. The postcaniniform maxillary and posterior dentary teeth were antero-posteriorly compressed with, typically, the development of two or three labiolingually arranged cusps. Diadectids are the only amphibians to have developed a secondary palate. This was accomplished by the medial development of prominent palatal flanges of the maxilla ventral to the primary palate (Fig. 4).

Although a full, hard palate never developed as in turtles, crocodiles, therapsids or mammals, a fleshy continuation of the secondary palate to form a floored nasal chamber is probable. Such development of a means of conducting inhaled air to the back of the pharynx to allow breathing during mastication is a convergent feature developed by many reptiles with specialized chewing mechanisms. However, the possession of these peculiar specializations serves to unite all members of the family Diadectidae.

Perhaps the most significant character separating batrachosaurs and their presumed descendants from other Palaeozoic tetrapods was the structure of the braincase. In crossopterygian rhipidistians and the most primitive known amphibians, the ichthyostegids, there is separate ossification of, and the possibility of relative movement between, the sphenethmoid and otico-occipital regions of the braincase anterior to the dorsum sella. This flexion of the braincase is known as neurokinesis. It was frequently accompanied by flexion between the parietals and postparietals in crossopterygians, a characteristic known as mesokinesis. In crossopterygians each braincase segment was attached to the overlying skull-roof segment thus permitting no movement between corresponding braincase and skull-roof regions. Primitively, tetrapods, such as the ichthyostegids, which had lost the rhipidistian mesokinetic joint, apparently retained a modified form of neurokinesis known as metakinesis that permitted movement between the otico-occipital moiety and the skull-roof by means of flexion in the lateral wall of the otico-occipital region of the braincase.

Two different forms of braincase architecture typical of more advanced, non-lepospondyl amphibians are exhibited by primitive labyrinthodonts and by batrachosaurs (Fig. 5). The early labyrinthodont braincase was a heavily ossified tube of conservative structure. It differed from the rhipidistian braincase in such advanced features as loss of the neurokinesis (and associated mesokinesis); progressive reduction of the embryonic fossa bridgei by dorsal migration of the paroccipital processes; and reduction or loss of the supraoccipital. The sphenethmoid region of the braincase was roofed by an extensive bony sheet, ventral projections from which separated this region into several longitudinal tubes or chambers (Panchen, 1964, 1972b). This prevented the sphenethmoid from being compressed dorsoventrally. The otico-occipital region was an osseous

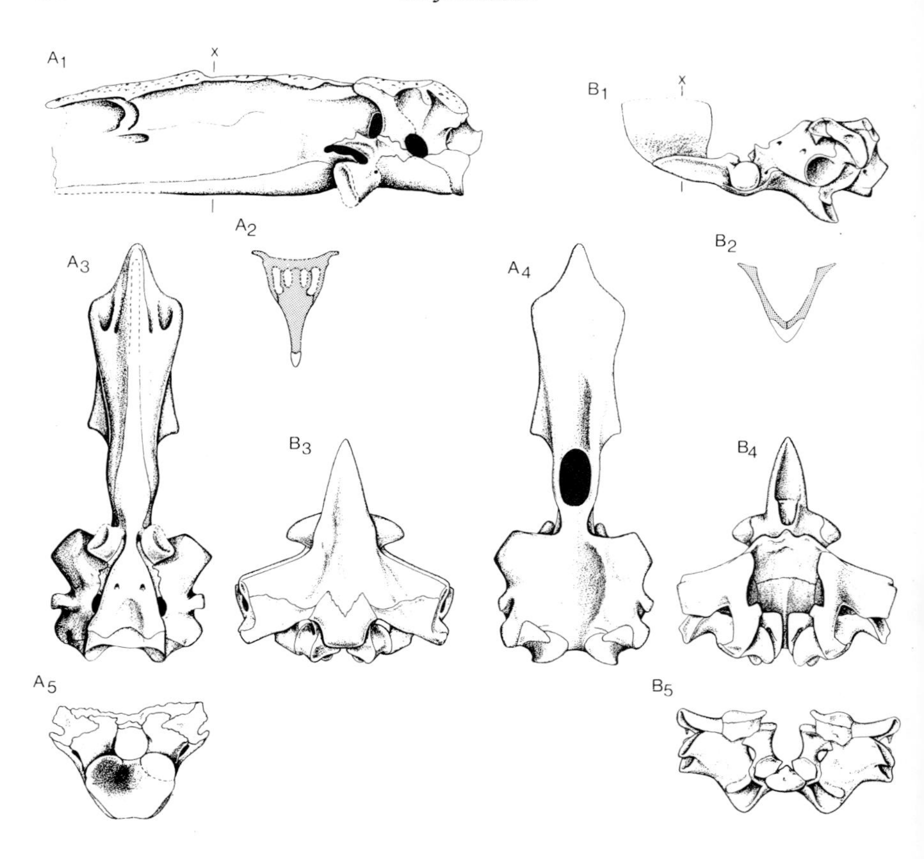

Fig. 5. Braincases. (A) The anthracosaur *Eogyrinus*; (B) the cotylosaurian batrachosaur *Seymouria*. (1) Left lateral aspect; (2) cross-section at point X; (3) ventral aspect; (4) dorsal aspect; (5) occipital aspect. Otico-occipital lengths equal. (A) After Panchen, (B) after White.

box with an extensive bony roof joining the otic capsules. As a unit, the whole braincase of labyrinthodonts, including both primitive temnospondyls – some advanced temnospondyls with braincases co-ossified with the skull-roof exhibit some reduction of the braincase (Watson, 1919; Romer, 1947) – and anthracosaurs, was sheathed dorsally by bone except for a pineal fontanelle (Sawin, 1941; Romer and Witter, 1942; Panchen, 1964).

The batrachosaurian braincase differs significantly from that of labyrinthodonts in that posteriorly the only dorsal roofing element is, at most, a narrow, posterior supraoccipital and even that may be lost as in *Seymouria*. The sphenethmoid region is typically open dorsally to produce a structure with a simple "Y"- or "V"-shaped cross-section which, when unossified, is easily compressed to allow some slight movement. *Kotlassia* (Bystrow, 1944) presents a slight variant to this pattern as a result of its relatively large size and heavy ossification. The sphenethmoid region appears to have been a "V"-shaped structure, as in other batrachosaurs, that has become thickened and more heavily ossified than normal since it is composed of a single ossification not a double one (dorsal and ventral) as Panchen (1964) has illustrated in embolomeres. Although almost totally ossified between the sola supraseptales, the sphenethmoid of *Kotlassia* has only a single paired canal for the olfactory nerve, just as in other batrachosaurs and reptiles (Bystrow, 1944), rather than double, paired canals for olfactory, vomeronasal and deep ophthalmic nerves as Panchen has described in embolomeres. Based on the information available, the sphenethmoid of *Kotlassia*, although heavily ossified, is batrachosaurian in character not anthracosaurian. In batrachosaurs there is no wide dorsal plate covering the dorsal surface of the braincase between the otic capsules, although a narrow supraoccipital is usually present. Only seymouriids appear to have lost the supraoccipital. In the absence of a large dorsal roof to the otico-occipital region of the braincase, and with distinctly separated sphenethmoid and otic-occipital regions, the braincase of batrachosaurs may be seen to be much different in structure to that of labyrinthodonts. The batrachosaurian braincase is typical of both cotylosaurs and reptiles and is a strong feature uniting the two groups. As Bock (1963) has suggested, it is unlikely that reptilian metakinesis is a neomorphic feature since loss of original neurokinesis has invariably led to incorporation of the braincase into the skull as a structural support element. Some form of flexion must have been retained in the lateral wall of the braincase to permit neurometakinesis to survive the loss of rhipidistian mesokinesis if Bock is correct. It would, therefore, seem to be impossible to develop a metakinetic batrachosaur–reptile braincase from that of an animal with a solidly walled braincase that did not permit dorsoventral flexion. It is instructive to examine the distribution of braincase characters particularly in relation to lateral braincase flexure within early tetrapods, but first some problems of terminology must be solved.

Some disagreement exists in the assignment of names to the bones bridging the dorsal surface of the otico-occipital region of the brain-case. In reptiles there is a narrow element, the supraoccipital, that is said to be formed from the embryonic tectum synoticum that unites the otic capsules (Gaupp, 1900, 1905; DeBeer, 1926, 1930, 1937; Romer, 1956). It is situated dorsal to the exoccipitals, each of which is derived from the pila ascendens (embryonic neural arch element) and touching the posterior edge of the prootic (otic capsule derivative). There is a dissenting opinion presented by Rice (1920) and Goodrich (1930) that identifies the supraoccipital as an ossification of the tectum posterior that unites the pilae ascendens. The pila ascendens (occipital arch) is separated from the otic capsule during early ontogenetic stages by the fissura metotica which later closes. The otic capsules primitively were united by a dorsal, but not necessarily separate, cartilage, the tectum synoticum, that is to be seen in the calcified chondrocrania of such large rhipidistian crossopterygians as *Eusthenopteron* (Jarvik, 1954, 1975). This lay anterior to the open fissura metotica and was thus separated by it from the discrete tectum posterior which is also well known as a separate calcified cartilage in rhipidistians. It is apparent that the primitive character state for both crossopterygians and their presumed tetrapod descendants was the possession of both an anterior tectum synoticum and a posterior tectum posterior. Clearly, as Rice (1920) and Goodrich (1930) assumed, only the tectum posterior ossification or supraoccipital, which is formed posterior to the fissura metotica, is present in reptiles. This is true also for the reptiles' batrachosaur ancestors and their close relatives, the cotylosaurs, except in the case of seymouriids where the supraoccipital has been lost. No batrachosaur has been found to have possessed an ossification of the tectum synoticum.

Labyrinthodont amphibians, both temnospondyl and anthra-cosaurian, primitively had a major dorsal ossification of the tectum synoticum. This ossification can be identified as such since a few labyrinthodonts, most notably the embolomerous anthracosaur *Palaeoherpeton* (Panchen, 1964), retain an easily identified supra-occipital that is separated from it by a suture. The anterior ossification between the otic capsules should be referred to as the synotic not the supraoccipital. Since the possession of both a synotic and a supraoccipital by *Palaeoherpeton* is a retained primitive character, it cannot serve to indicate relationships between anthracosaurs and batrachosaurs. Many labyrinthodonts show reduction and often the complete loss of the supraoccipital while retaining the synotic, a derived character that serves to separate these advanced members from the Batrachosauria. The cotylosaurs and reptiles both exhibit the same synapomorphic character state, the loss of the synotic, thus suggesting that they are related as synapomorphic sister-groups of the Batrachosauria. In cases where one or the other of the tectal elements is missing, it is often difficult to determine whether a synotic or a supraoccipital is present. Usually a supraoccipital exhibits a visible lateral suture along its union with the prootic and exoccipital while such a suture apparently is not present between the synotic and the prootic. It may be possible to argue that the synotic is actually a greatly expanded dorsal extension of the prootic (and embryonic otic capsule) that roofs the braincase, since, in the absence of any modern animal with a well-formed tectum synoticum, the embryonic origin of this cartilage still remains in doubt. Apparently a poorly ossified tectum synoticum but no tectum posterior (supraoccipital) is present in modern hylid and some leptodactylid frogs (Baldauf, 1963). The structure of the braincase of labyrinthodonts and batrachosaurs is here considered to be so different that neither group can be derived from or even closely related to each other. Cotylosaurs and their reptilian relatives (collectively the Batrachosauria and their derivatives) are in no way closely related to any known labyrinthodont, least of all to any anthracosaur as Romer (1947, 1956, 1966, 1968), Carroll (1969a,b) and Panchen (Chapter 13 of this volume) among others, have suggested.

The braincases of "lepospondyl" amphibians are poorly known except in the two microsaur genera *Pantylus* (Romer, 1969) and *Goniorhynchus* (Carroll and Currie, 1975). Many characters such as the dorsally open sphenethmoid region and anterior otico-occipital region are distinctly batrachosaurian in character. A broad plate unites the posterior region of the otic capsules and may be a homologue of the labyrinthodont synotic, a wide supraoccipital, or a fusion of both. It is not possible, on the basis of information presently available, to tell whether the microsaurian and batrachosaurian braincases are exhibiting convergent or homologous reduction of dorsal ossification.

The stimulus toward development of the different types of braincase structure came from the development of two different types of feeding strategies. The labyrinthodonts specialized originally in the large aquatic piscivore role in contrast to the original batrachosaurian terrestrial or semi-terrestrial insectivore feeding strategy. The aquatic piscivore feeding mode required the development of relatively large size and, as Olson (1961) has discussed, a kinetic-inertial jaw adduction system requiring great muscular force and skull strength. In contrast, batrachosaurs seem to have arisen as relatively small, light, terrestrial forms with emphasis on a slower speed, but with a more precise and adaptable static-pressure jaw adduction system. In batrachosaurs there appears to have been no early development of large size and cranial akinesis. Primitive neurokinesis, as was typical of rhipidistians, occurred through movement of the sphenethmoid about a fulcrum (basicranial articulation), relative to the otico-occipital segment. This movement was facilitated by an intracranial joint. During the rhipidistian–amphibian transition two things happened; the rhipidistian mesokinesis was lost and the parasphenoid expanded posteriorly to bring the basicranial articulation beneath the anterior extremity of the otico-occipital region of the braincase. This backward expansion has not occurred in *Ichthyostega* and a line of junction is visible between the two braincase regions (Jarvik, 1955) indicating not only that the cartilaginous otico-occipital segment could still move relative to the sphenethmoid region (neurokinesis), but also that movement between the skull-roof and the otico-occipital regions was possible by distortion of the soft braincase cartilages (metakinesis). Two separate courses of development are possible from this primitive tetrapod pattern. The labyrinthodont pattern, as seen in primitive temnospondyls and anthracosaurs, developed through heavy co-ossification of the sphenethmoid and otic-occipital regions with the loss of both neurokinesis and metakinesis. Carroll (Chapter 12 of this volume) has suggested that primitive neurometakinesis may have been retained by the primitive temnospondyl *Greererpeton*. The primitive batrachosaur pattern probably resembled that of modern *Sphenodon* and lizards in which neurokinesis may have been reduced by fusion of the parasphenoid to the ventral surface of the otico-occipital region (basioccipital when ossified). Slight neurokinesis may have been retained in small animals by flexion of the thin parasphenoid. Metakinesis remained because the thin, unossifed cartilages, and membranes of the chondrocranium were easily deformed by the jaw musculature. A braincase that lacked dorsal cartilages and ossifications was much more amenable to dorsoventral flexion than was one that was heavily roofed. All batrachosaurs have a braincase without a dorsal roof except for the narrow supraoccipital which frequently serves as a metakinetic axis or fulcrum about which the otico-occipital moiety rotates. The batrachosaur lineages diverged into two major groups, one remaining small or even becoming smaller still, and refining the primitive metakinesis into typical reptilian metakinesis, the other increasing in size and becoming akinetic. This latter group is the Cotylosauria. Although akinetic, the cotylosaurs exhibit evidence of their metakinetic ancestry in the structure of the dorsally open or incomplete braincase (Fig. 5).

Early batrachosaur metakinesis and its later more refined reptilian form probably operated in essentially the same manner. With an unossified dorsal and dorsolateral region, except for the narrow supraoccipital, the braincase could be compressed and flexed slightly dorsoventrally. This produced a slight amount of rotation of the skull-roof relative to the braincase. The compression of the anterior region of the braincase by slight flexion of the cartilaginous and membranous solum supraseptale and dorsal taeniae, just as in modern lizards, was accomplished by contraction of the small muscles of the M. constrictor dorsalis group. Large skulls (more than about 60 mm in length) were too heavy to be lifted by these muscles so that terrestrial tetrapods with skulls larger than this invariably lacked neuro- or metakinesis. Jaw adduction in neurokinetic or metakinetic tetrapods was accomplished most effectively by a multi-segmented M. adductor mandibulae (Fig. 6). In modern reptiles this is divided into the M. adductor mandibulae pars externus, pars internus (M. pterygoideus and M. pseudotemporalis) and pars posterior. The M. adductor mandibulae externus is further divided into partes superficialis, media and profunda, the latter two of which originate from the otico-occipital region of the braincase. The M. adductor mandibulae externus medius arises from the supraoccipital and the pars profunda from the anterior surface of the paroccipital process. The former bulges dorsally into the posterodorsal region of the adductor chamber while the latter, when contracted, bulges posteriorly through the enlarged post-temporal fenestrae (embryonic

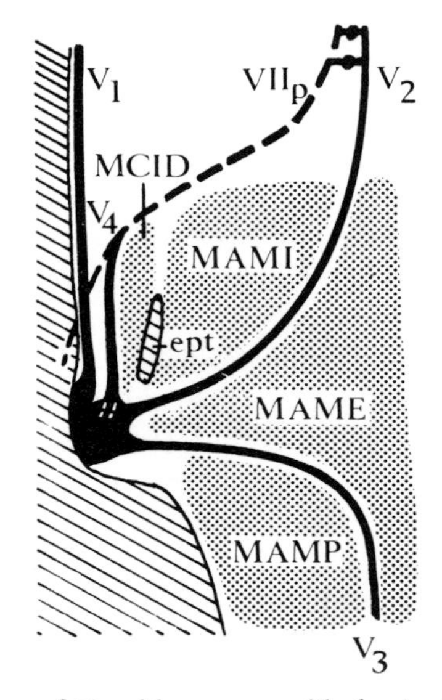

Fig. 6. Schematic diagram of M. adductor mandibulae in dorsal aspect.
Key to abbreviations. az, anterior zygapophysis; e, episphene; ec, ectepicondyle; en, entepicondyle; enf, entepicondylar foramen, ept, epipterygoid; h, hyposphene; hyc, hyostylar cartilage; ic, intercentrum, ls, lateral shelf of ilium; MAME, M. adductor mandibulae externus; MAMI, M. adductor mandibulae internus; MAMP, M. adductor mandibulae posterior; MCID, M. constrictor (internus) dorsalis; opr, opercular process; pc, pleurocentrum; pop, paroccipital process; pz, posterior zygapophysis; q, quadrate; qpr, quadrate process; sp, supinator process; V_1, ophthalmic ramus of trigeminal nerve; V_2, maxillary ramus of trigeminal nerve; V_3, mandibular ramus of trigeminal nerve; V_4, M. constrictor (internus) dorsalis ramus of trigeminal nerve; VII_p, lateral palatine ramus of facial nerve.

fossa bridgei). In this way a primitive reptile (and presumably a primitive, non-cotylosaurian batrachosaur) can be identified, although the latter has not yet been recognized in the fossil record. Typically, the occiput is vertical, the paroccipital processes are narrow and directed laterally, the supraoccipital is narrow and the post-temporal fenestrae are large. By this definition, procolophonids and pareiasaurs, often termed cotylosaurs, are either true reptiles or non-cotylosaurian batrachosaurs very close to the origin of reptiles (Figs 2 and 3). Much additional study is needed to establish their exact systematic position although they will probably be found to be reptiles, albeit very primitive ones, even more primitive than captorhinomorphs which are widely regarded as the most primitive reptiles (Carroll, 1964).

The Cotylosauria was a diverse group of large terrestrial or semi-terrestrial batrachosaurs. Because of their large size and massive skulls, they had lost the primitive batrachosaurian metakinesis, since the small constrictor dorsalis group muscles seem to have been unable to elevate the heavy skull-roof. Consequently, the cotylosaurian sphenethmoid region was always ossified to strengthen its attachment to the skull-roof but retained the "V" or "Y"-shaped batrachosaurian septum, except as modified in *Kotlassia* as noted above, rather than the multi-channelled labyrinthodont tube (Fig. 5). Loss of metakinesis eliminated much of the mechanical requirement for a segmented adductor musculature while the robust build of the mandible required the maintenance of a large muscle mass. In cotylosaurs it appears that the M. adductor mandibulae externus profundus segment, which bulges posteriorly through the post-temporal fenestra in reptiles, had been lost. This led to the expansion of the supraoccipital laterally, eventually to obliterate the post-temporal fenestra in diadectomorph cotylosaurs. In seymouriamorphs, the supraoccipital remained small, as in *Kotlassia*, or was unossified and the lateral ends of the paroccipital processes migrated dorsally to restrict the post-temporal fenestrae. Typically, the cotylosaurian occiput is oriented vertically to accommodate the massive M. adductor mandibulae externus medius and probably the M. pseudotemporalis. Only in the Diadectidae is the occipital plate inclined anterodorsally, probably in response to development of massive occipital musculature related to increased mobility of the head (Olson, 1936, 1947).

The tetrapod occipital condyle is typically formed of the basi-occipital and the paired exoccipitals. In labyrinthodonts it forms a concave surface against which the axis intercentrum, a large element, articulates (Figs 2 and 5). Typically there are no proatlases anterior to the paired atlas neural arches and hence no proatlas articulating surfaces on the exoccipitals in labyrinthodonts either. The occipital articulation of batrachosaurs is distinctly different. In both cotylosaurs and the batrachosaurian derivatives, the Reptilia, the occipital condyle is noticeably convex although it usually shows a central notochordal concavity. Although the atlas–axis complex in cotylosaurs has not been studied in detail, it appears that the basioccipital typically articulates with a small, anteriorly concave intercentrum and the condylar portion of the exoccipitals with the concave anterodorsal surface of the atlas neural arches just as in reptiles. Prominent proatlases are present that articulate anteriorly with the ascending pillar of the exoccipital lateral to the foramen magnum and posteriorly with the atlas neural arches. The convex, ball-shaped occipital condyle has, in the past, often been interpreted, as has the transverse flange of the pterygoid, as a distinctly reptilian character. This has caused systematic problems since it is characteristic of both seymouriamorph and diadectomorph cotylosaurs and, hence, must now be regarded as a shared derived character of batrachosaurs and their derivatives.

The structure of the middle ear region of two advanced cotylosaurs, *Seymouria* (Watson, 1916; White, 1939) and *Diadectes* (Watson, 1916, 1954; Olson, 1947, 1965, 1966a,b) has been studied extensively. These two forms have aberrant otic structures unlike those of more primitive cotylosaurs. *Limnoscelis* has a braincase and enclosed otic capsule essentially similar to that of primitive reptiles except that, because the stapes had been released from its supportive role in the primitive metakinetic mechanical system with the development of large size and cranial akinesis, there is no wide, circular articulating surface on the prootic, basioccipital and basisphenoid–parasphenoid together with which it forms the anteroventral corner of the rim of the fenestra ovalis. There is no extension of the parasphenoid lateral to the cristae ventrolaterales. The braincase of *Nycteroleter* is essentially similar in gross features (Efremov, 1940); little is known of its finer details. In advanced seymouriamorphs (Seymouriidae and Kotlassiidae), the otic capsule has become highly modified into a tubular lateral otic canal. In labyrinthodonts such as *Eryops* (Sawin, 1941), *Palaeoherpeton* (Panchen, 1964), and *Eogyrinus* (Panchen, 1972), as well as reptiles (Heaton, 1979), the ventral rim of the fenestra ovalis is formed by the parasphenoid and often the basisphenoid and basioccipital which it sheathes ventrally. On the basis of commonality among these diverse forms, this is considered to be the primitive condition. Typically the remainder of the rim of the fenstra ovalis is formed by the prootic anteriorly and dorsally, and the opisthotic posteriorly.

The lateral otic canal of *Seymouria* preserves these primitive relationships around the fenestra ovalis but with the basisphenoid and basioccipital not expanded laterally as far as the parasphenoid (White, 1939). *Kotlassia* exhibits a derived condition in which the parasphenoid has been excluded from the fenestra ovalis by the joining of the prootic and opisthotic ventrally. The condition of the lateral otic canal apparent in *Kotlassia* appears to be derivable from a form similar to that of *Seymouria*. In *Kotlassia* the prootic and opisthotic are sutured to the supraoccipital. This bone is unossified in *Seymouria*. In either case, a small post-temporal fenestra is located dorsal to the paroccipital process. The topography of the ventral surface of the parasphenoid in seymouriamorphs is subdued with no prominent cristae ventrolaterales.

In advanced diadectomorphs (Tseajaiidae and Diadectidae) a lateral otic canal of somewhat different structure is present. That of *Tseajaia* (Moss, 1972) is less well developed than in the more advanced form *Diadectes* (Watson, 1916, 1954; Olson, 1947, 1965, 1966a,b). In these animals the supraoccipital and the opisthotics are indistinguishably fused and sutured to the skull-roof. They form the dorsal and posterior walls of the lateral otic canal. Anteriorly the canal is almost completely surrounded by the prootic (periotic) with the parasphenoid making only a small contribution to the rim of the fenestra ovalis. In this character state, diadectomorphs exhibit the primitive pattern and thus this character neither supports nor opposes a close relationship between these animals. The more massively constructed lateral otic canal of diadectids with a separate, blind, lateral canal (Olson, 1947) and markedly different stapes that transmitted acoustic vibrations to the fenestra ovalis and inner ear (see below) suggests that the lateral otic canals of seymouriamorphs and advanced diadectomorphs may have been independently derived. In addition, the parasphenoid of diadectomorphs shows considerable development of ventral ridges. In *Limnoscelis* the cristae ventrolaterales are well developed as is a median, ventral, parasphenoid crest. In *Tseajaia* the cristae ventrolaterales have expanded posteroventrally over the anteroventral surface of the M. rectus capitis anterior insertion to form two pointed parasphenoidal wings. The median parasphenoid crest is subdued in the only known braincase of *Tseajaia* but this may have been a result of the extremely difficult preparational problems. In *Diadectes* the parasphenoidal wings are greatly expanded posteriorly and are often supported by a robust median parasphenoid crest, although the size of the latter is highly variable (Olson, 1966b).

The stapes is well known in both seymouriamorph and diadectomorph cotylosaurs, where two distinctive types are present as divergent developments from an ancestral rhipidistian hyomandibular or a stapes closely resembling it (Fig. 7). The rhipidistian hyomandibular has two distal processes; the opercular process dorsally and the quadrate or hyostylar process ventrally. The opercular process is now recognized as the antecedent and thus homologue of the extrastapes of labyrinthodont amphibians, and

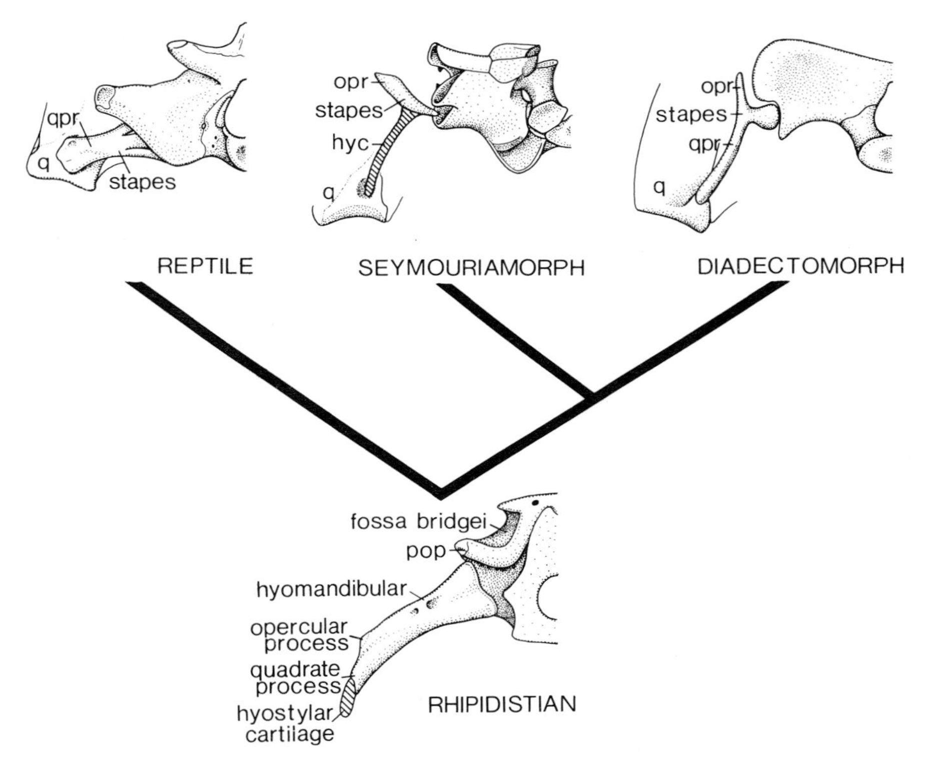

Fig. 7. Probable course of middle ear development in reptiles and cotylosaurs. Cross-hatching shows hyostylar cartilage continuation of quadrate process of stapes. See key to abbreviations in legend to Fig. 6.

the quadrate process as the antecedent of the extrastapes of reptiles (Shishkin, 1975; Lombard and Bolt, 1979). It appears that a stapes with both primitive processes remained in the earliest batrachosaurs. In the lineage leading to reptiles, the opercular process degenerated leaving a posteroventrolaterally directed stapes, the distal end of which articulated with a deep columellar recess in the posteromedial surface of the quadrate. In cotylosaurs both processes apparently were retained; the quadrate process articulated with the quadrate while the opercular process contacted the lateral wall of the tympanic cavity. The loss of neurokinesis in cotylosaurs released the stapes from its primitive supportive function and led to its gradual reduction in ossification. As it became lighter, it began to become sensitive to sound-induced vibration whereupon it became incorporated and modified as a hearing ossicle. This appears to have occurred in two different ways. In the Seymouriamorpha a labyrinthodont-like otic notch developed that apparently was bridged by a small tympanum located anterodorsally in the deep notch. In *Seymouria* a small rod-like stapes extends from the fenestra ovalis anterodorsally to the otic notch. This is similar to the form of stapes present in labyrinthodonts and, as with them, has an extrastapes formed from the primitive opercular process. White (1939) illustrated a small, crescentic protuberance on the posterior surface of the quadrate that probably served as an articulation with a cartilaginous quadrate process of the stapes. The stapes of *Kotlassia* is heavily ossified. There is no evidence either in the structure of the stapes or in the configuration of the quadrate that a quadrate process existed in this form. Thus, there appears to have been a trend in seymouriamorphs toward the development of the opercular (tympanic) process and a reduction of the quadrate process. The stapes is unknown in *Nycteroleter*.

In diadectomorphs the stapes, when preserved, exhibits a large, ossified quadrate process that articulates with the lateral tip of the paroccipital process above the fenestra ovalis. The proximal head of the stapes is greatly reduced or unossified where it enters the fenestra ovalis. The extrastapes is robust but extends laterally only a very short distance. In *Tseajaia* a short process extends dorsolaterally into the anterodorsal corner of the otic notch. There is a prominent, wide, ventrolaterally directed process that articulates with a pronounced crescentic process on the posterior surface of the quadrate. This process appears to be the homologue of the rhipidistian quadrate process. The unusual width of the process appears to be a result of expansion of the process to form an ossified manubrium in the mechanical sound-conducting system of the middle ear. It is too thick to have been an ossified tympanum. In diadectids the stapes is similar except that there appears to be a wide, thin, ossified tympanum attached to the lateral surface of the thick quadrate process. This latter process articulates with a massive circular tubercle on the posterior surface of the quadrate. The stapes of *Limnoscelis* is unknown. The two significantly different types of stapedial structure in seymouriamorphs and diadectomorphs appear to indicate separate evolutionary histories of both the stapes and the inner ear (including the superficially similar lateral otic canals).

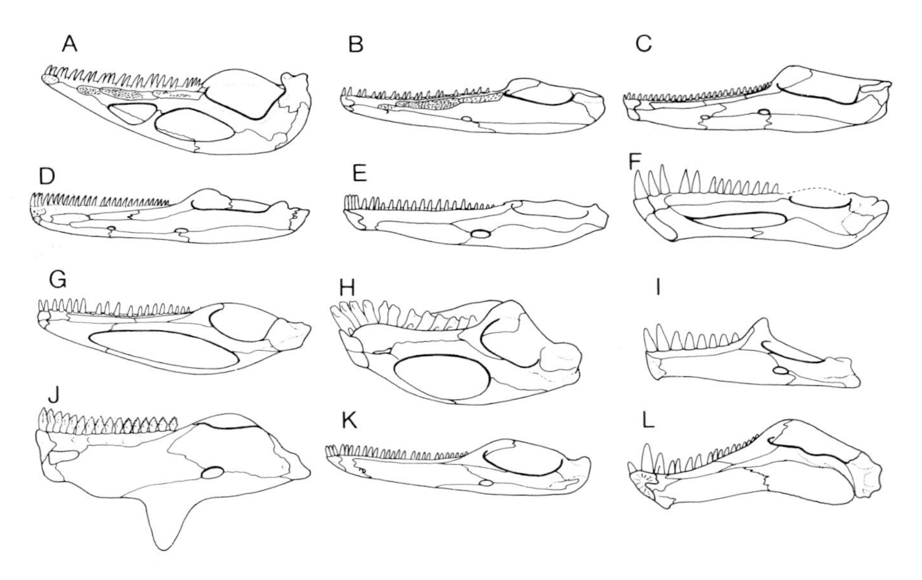

Fig. 8. Mandibles in internal aspect. (A) The anthracosaur *Eogyrinus*; (B) the anthracosaur *Gephyrostegus*; (C) the seymouriamorph cotylosaur *Seymouria*; (D) the seymouriamorph *Kotlassia*; (E) the seymouriamorph *Nycteroleter*; (F) the diadectomorph cotylosaur *Limnoscelis*; (G) the diadectomorph *Tseajaia*; (H) the diadectomorph *Diadectes*; (I) the procolophonid reptile *Procolophon*; (J) the pareiasaur *Scutosaurus*; (K) the captorhinomorph *Protorothyris*; (L) the pelycosaur *Dimetrodon*. Not to scale.

COTYLOSAURIAN MANDIBLE

The mandible of most cotylosaurs is well known (Figs 3 and 8). With the exception of dental characteristics, which generally resemble those of the corresponding maxillary dentition, only two major derived trends are consistently noticeable. The first trend is toward a general reduction in the number of bones in the mandible. Typically the postsplenial and first two coronoids are reduced or lost. This trend proceeded in both cotylosaurian suborders with considerable convergence and hence is of little taxonomic significance. The second trend is of far greater use.

Seymouriamorphs typically have two small foramina on the medial surface of the mandible, a posterior foramen intermandibularis caudalis (frequently referred to as the meckelian or inframeckelian foramen) and an anterior foramen intermandibularis oralis. This is the primitive condition present in rhipidistians, temnospondyls, some anthracosaurs (gephyrostegids) and many reptiles. In all of these animals the mandibles are relatively shallow with the meckelian canal apparently having been occupied by the meckelian cartilage and numerous nerves and blood vessels. The diadectomorph cotylosaurs present a contrast to the primitive form for the mandible is massively built with a large meckelian canal. A single, large, foramen intermandibularis communis perforates the medial surface of the mandible to give access to the meckelian canal. It is believed that both of these features are correlated with the existence of a large M. intramandibularis, a ventral lobe of the M. pseudotemporalis of the M. adductor mandibulae internus group (Heaton, 1978, 1979). The presence of an M. intramandibularis is thought to have been a primitive condition in batrachosaurs for it still exists in modern turtles and crocodilians. Only in diadectomorphs, crocodilians and possibly ophiacodont pelycosaurs has it developed into a major muscle mass. In these cases some means of accommodating the swelling of the contracted muscle was necessary. In diadectomorph cotylosaurs this was accomplished by expansion of the foramen intermandibularis caudalis to excavate the whole of the medial surface of the mandible and coalesce with the foramen intermandibularis oralis to form the foramen intermandibularis communis. In both crocodilians and ophiacodonts, bulging of the M. intermandibularis has been accommodated by the development of a large external mandibular foramen.

COTYLOSAURIAN AXIAL SKELETON

As far as is known, the primitive tetrapod vertebral column was of a multi-segmented notochordal form similar to that present in rhipidistian crossopterygian fish (Watson, 1919; Romer, 1947; Andrews and Westoll, 1970). The vertebral centrum consisted of two elements, an anterior intercentrum (hypocentrum) and a posterior pleurocentrum, and was capped by the neural arch. Two basic patterns occurred, one the rhachitomous pattern with a large intercentrum and a reduced pleurocentrum formed of paired dorsolateral ossifications, and the other pattern, often termed protoreptilian, in which the intercentrum might be large, small or absent but the pleurocentrum was always a large single ossification. It has been the custom to assume that because the vertebrae of the best known rhipidistian crossopterygians and the extremely successful early temnospondyl amphibians, including the earliest known amphibian, *Ichthyostega*, all had rhachitomous vertebral patterns, that the rhachitomous pattern was the primitive pattern for all tetrapods (Romer, 1947, 1956, 1966). Recently it has been discovered that, except for the seemingly aberrant ichthyostegid amphibians, at least some of the earliest members of each major tetrapod group, including temnospondyls, had protoreptilian vertebrae (Holmes and Carroll, 1977), prompting Panchen (1977a) to reiterate his suggestion that it might be the rhachitomous rather than the protoreptilian vertebral pattern that is derived. This is similar to the proposal first suggested by Watson (1919). If this is true, as current evidence suggests, then lepospondyls, primitive anthracosaurs, cotylosaurs and reptiles all possess vertebrae of a basic primitive character. The comparative sizes of the intercentrum and pleurocentrum should, therefore, not be used as a character to unite any of these groups.

There are relatively few advanced (derived) characters that are useful in separating the Cotylosauria from other amphibians on the basis of vertebral structure. Typically, the vertebral centrum (pleurocentrum) of cotylosaurs and other batrachosaurs and their descendants is greatly shortened along the axis of the vertebral column, with the neural spine and posterior zygapophyses normally projecting far posteriorly. The neural arch is typically low with massive dorsolateral swellings that displace the zygapophyses far laterally and orient their articular surfaces into a near-horizontal

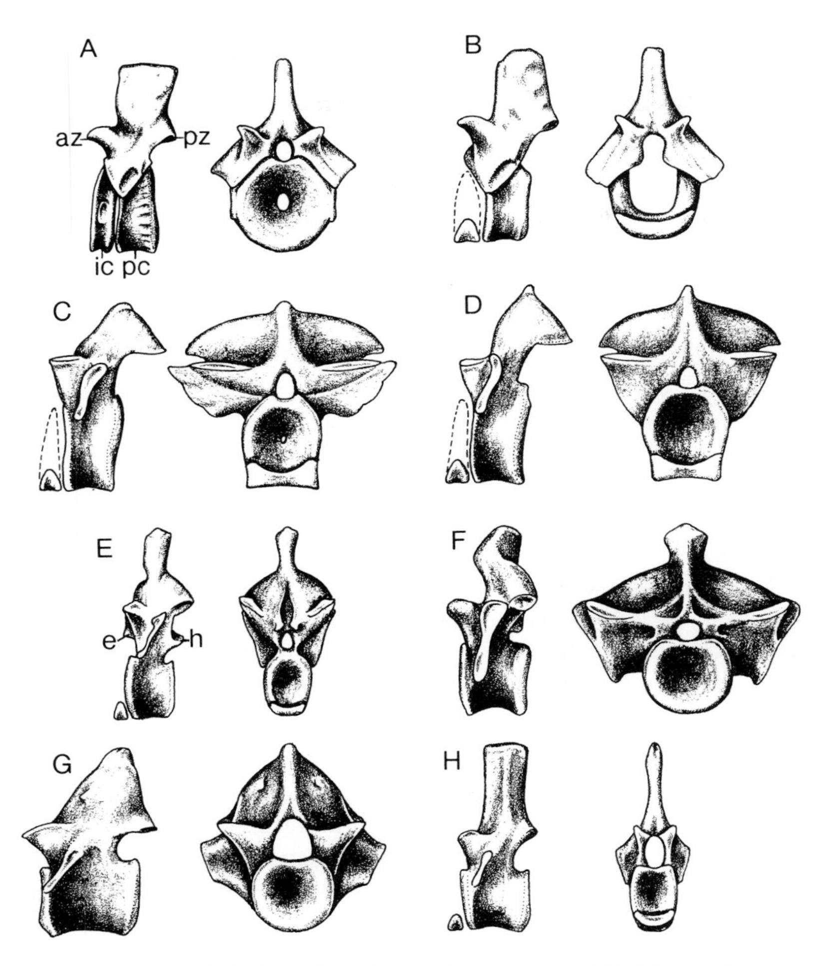

Fig. 9. Vertebrae in left lateral and anterior aspects. (A) The anthracosaur *Eogyrinus*; (B) the anthracosaur *Gephyrostegus*; (C) the seymouriamorph cotylosaur *Seymouria*; (D) the diadectomorph cotylosaur *Limnoscelis*; (E) the diadectomorph *Diadectes*; (F) the pareiasaur *Bradysaurus*; (G) the captorhinomorph *Eocaptorhinus*; (H) the captorhinomorph *Protorothyris*. (A) After Panchen, (B) after Carroll, (F) after Romer, (G) after Heaton and Reisz, (H) after Heaton and Reisz (from Carroll). Not to scale.

attitude. In *Seymouria* and *Diadectes* the diameter of the pleurocentrum of a mid-dorsal vertebra is about 35% of the width of the expanded neural arch (Romer, 1956), while in *Tseajaia* it is about

25% (Moss, 1972) and in *Limnoscelis* about 40% (Romer, 1946, 1956). In adults, and possibly juveniles, the neural arches are firmly attached to the pleurocentrum with no evidence of a suture between the elements (Fig. 9).

While the neural arches of the dorsal vertebrae of batrachosaurs are massively swollen, both the sacral and caudal vertebrae and the anterior cervical vertebrae have neural arches in which the swelling is greatly reduced or absent. The most anterior cervical elements of the atlas-axis complex are poorly known in cotylosaurs. Watson (1918) described the atlas–axis complex of *Seymouria* briefly. White (1939), in his thorough re-study of *Seymouria*, relied entirely on Watson's description. There is still some question as to whether Watson's description is accurate since at least one isolated cervical column attributed to *Seymouria* (Harvard Univ. MCZ 1646 – White, 1939) shows a distinctly different form. Watson did not describe a proatlas or articulating surfaces for it on the atlas neural arch in his specimen, while White noted a "well developed facet for the proatlas" in MCZ 1646. Clearly much further work must be done on the atlas–axis complex of *Seymouria*. Unfortunately only the atlas neural arch has been preserved and that in relatively poor condition, in *Kotlassia* (Bystrow, 1944). The atlas–axis complex of *Nycteroleter*, although apparently preserved in at least one specimen, has not been described fully. The atlas–axis complexes of diadectomorph cotylosaurs, although represented in numerous specimens, have not been prepared, illustrated and described as fully as would be necessary for a full discussion here. The general impression is that the atlas–axis complex of diadectomorphs is quite similar to that of primitive reptiles such as *Paleothyris* (Carroll, 1969a) with a lozenge-shaped axis neural spine and low atlas neural spine rather than like the tall, accuminate atlas and axis neural spines of Watson's (1928) specimen of *Seymouria*. It appears that a paired proatlas is a necessary element in the braincase–atlas articulation of diadectomorph cotylosaurs as it is in primitive reptiles, although this point must be confirmed.

The existence of wide dorsal neural arches in all batrachosaurs is a unique shared derived character of this group not possessed by any other group of amphibians. The microsaurian amphibian *Pantylus*, as noted by Carroll (1968) and Carroll and Gaskill (1978), has commonly been thought to have swollen neural arches; however, the neural arches "are little modified and the zygapophyses, although not as close to the midline as those of pelycosaurs and rhachitomes, extend barely as far laterally as the width of the centrum". Swelling of the neural arches occurs in most primitive reptilian lineages, including captorhinids, araeoscelids, bolosaurs, mesosaurs, procolophonids, pareiasaurs, placodonts and in many snakes. It is suspected, but not as yet proved, that, except in the case of the snakes and possibly the placodonts, swollen neural arches may be a primitive character of early reptiles. This would be expected as they are apparently batrachosaur derivatives. The swollen neural arches of the dorsal vertebrae of cotylosaurs typically extend laterally at least as far as the ends of the diapophyses (transverse processes), whereas in captorhinids (Fox and Bowman, 1966), araeoscelids (Vaughn, 1956; Reisz, 1975), petrolacosaurids (Reisz, 1975), placodonts (Romer, 1956) and snakes (Romer, 1956) they do not. The apophysis of diadectomorph cotylosaurs extends ventromedially as a web from the diapophysis to the centrum (pleurocentrum) and forms a long, single articulation (synapophysis) with the holocephalous proximal end of the dorsal rib. This contrasts with the form present in seymouriamorphs where the diapophysis is shorter and is articulated with the tuberculum of the dichocephalous proximal end of the dorsal rib while the capitulum articulates with the parapophysis of the intercentrum. It appears that the dichocephalous rib articulation of seymouriamorph cotylosaurs is primitive (Romer, 1966) since it is common to such diverse groups as ichthyostegalians (Jarvik, 1952), temnospondyls (Holmes and Carroll, 1977), anthracosaurians (Panchen, 1966, 1970) and discosauriscids (Špinar 1952) among labyrinthodonts; to lepospondyls (Carroll, 1968; Carroll and Gaskill, 1978); and to reptiles, including procolophonids (Colbert, 1946), pareiasaurians (first five cervical ribs only – Romer, 1956), primitive captorhinomorphs (Carroll and Baird, 1972) and pelycosaurians (Romer and Price, 1940). The holocephalous rib articulation of diadectomorph cotylosaurs appears to be a derived character shared by all members of that group and developed by the extension of a thin web of bone between the tuberculum and capitulum. Holocephalous ribs are present in both captorhinids and pareiasaurians (except cervical ribs) where they have developed independently, at least partially as a result of the reduction in size of the intercentrum.

COTYLOSAURIAN APPENDICULAR SKELETON

The primary pectoral girdle of rhipidistian crossopterygian fishes (Andrews and Westoll, 1969), and temnospondyl (Miner, 1925), anthracosaurian (Romer, 1957; Panchen, 1970) and microsaurian amphibians (Carroll, 1968; Carroll and Gaskill, 1978), is generally formed as a single ossification usually regarded as being homologous with the scapula of mammals. This is the primitive tetrapod condition. Cotylosaurs have a primary pectoral girdle composed of two endochondral ossifications, a primitive scapula and a neomorphic coracoid. This derived character has been reported in the seymouriamorph *Seymouria* (White, 1939), as well as in the diadectopmorphs *Limnoscelis* (Romer, 1946, 1956), *Tseajaia* (Moss, 1972) and *Diadectes* (Case, 1911; Olson, 1947). Primitive reptiles typically possess two distinct coracoids as reported in procolophonids (Colbert, 1946; Romer, 1956), pareiasaurs (Haughton and Boonstra, 1930; Boonstra, 1932, 1933; Romer, 1922, 1956), captorhinomorphs (Fox and Bowman, 1966; Carroll and Baird, 1972; Holmes, 1977), araeoscelids (Vaughn, 1955), petrolacosaurids (Reisz, 1975) and pelycosaurs (Romer and Price, 1940). This is an homologous shared derived character of primitive reptiles. It has not been established whether the coracoid of cotylosaurs is homologous with the anterior coracoid (procoracoid) only of reptiles or with both the anterior and posterior coracoids. A single element is found in millerosaurs (Watson, 1957; Gow, 1975), turtles (Romer, 1956), lizards (Romer, 1956) and archosaurs (Romer, 1956) where it is thought to represent a secondary reduction. Among the Cotylosauria there is also an exception to the pattern presented here. The seymouriamorph *Kotlassia* has been described by Bystrow (1944) as having only a single ossification of the pectoral girdle. This is thought to be a result of co-ossification of the scapula and coracoid in an adult animal rather than a retention of a primitive character. This illustrates one of the problems of using the

Fig. 10. Pelvic girdles in left lateral aspect. (A) The anthracosaur *Archeria*; (B) the anthracosaur *Gephyrostegus*; (C) the seymouriamorph cotylosaur *Seymouria*; (D) the diadectomorph cotylosaur *Limnoscelis*; (E) the diadectomorph *Tseajaia*; (F) the diadectomorph *Diasparactus*; (G) the captorhinomorph *Captorhinus*; (H) the pelycosaur *Dimetrodon*. (A) After Romer, (B) after Carroll, (C) after White, (D) after Romer, (E) after Moss, (F) after Case and Williston, (H) after Romer and Price. Not to scale. ls, lateral shelf of ilium.

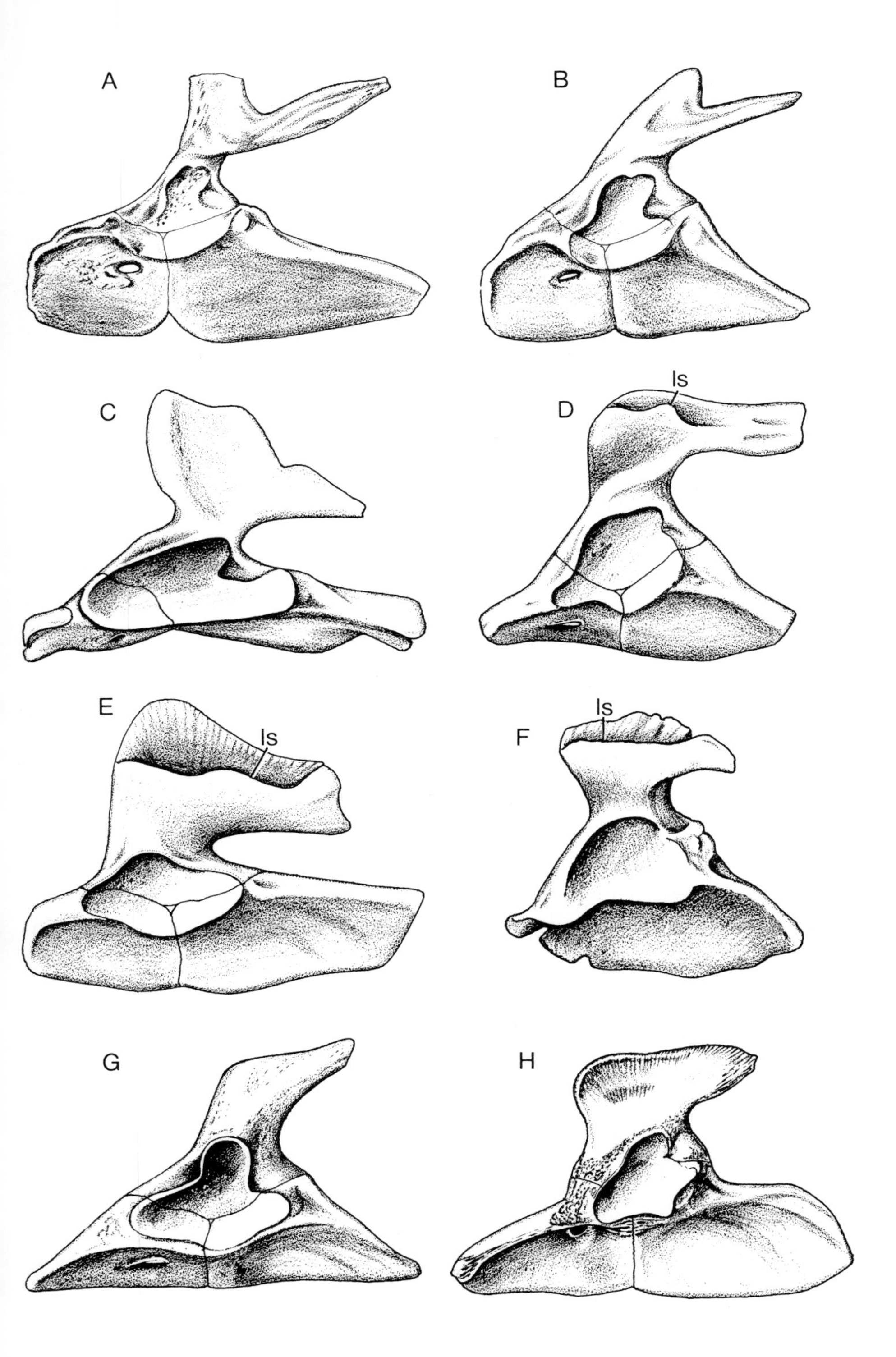

A
B
C
D
ls
E
ls
F
ls
G
H

structure of the endochondral shoulder girdle as a character, since co-ossification of the elements in adults is a common or even typical occurrence.

The pelvic girdles of archaic tetrapods are all of simple construction with few easily definable common characters (Fig. 10). The pelvic girdle consists of three endochondral ossifications, a dorsal ilium, an anteroventral pubis and a posteroventral ischium, the latter two forming the pubo-ischiadic plate. The few identifiable characters associated with the pelvic girdle are almost exclusively confined to the ilium. Although the ilia of archaic tetrapods vary considerably, one primitive feature has remained in most lineages. In temnospondyl (Miner, 1925), anthracosaurian (Romer, 1957) and microsaurian amphibians (Carroll, 1968; Carroll and Gaskill, 1978), and in procolophonid (Colbert, 1946), pareiasaurian (Gregory, 1946), captorhinomorph (Fox and Bowman, 1964; Carroll and Baird, 1972) and pelycosaurian reptiles (Romer and Price, 1940), the posteroventral edge of the ilium is steeply inclined posterodorsally at an angle of at least 20° to the base of the pubo-ischiadic plate. In both seymouriamorph cotylosaurs including *Seymouria* (White, 1939) and *Kotlassia* (Bystrow, 1944), and in diadectomorph cotylosaurs, the posteroventral margin of the ilium extends posteriorly parallel to the base of the pubo-ischiadic plate. While this is not a particularly good character to use since it is easily subject to distortion during preservation, it does appear to be consistent.

The two cotylosaurian suborders are clearly distinguishable since the diadectomorphs all share a distinctive derived character in the ilium. In limnoscelids, tseajaiids and diadectids, a prominent horizontal, laterally projecting shelf has developed on the external face of the ilium. Olson (1936) described this shelf as having developed to restrict the M. iliocostalis to the dorsal portion of the ilium to allow greater expansion of the M. iliofemoralis onto the iliac blade. This seems to be a possible interpretation since there would seem to be a selective advantage to having a large area of origin of this latter muscle in a large, heavy, terrestrial animal. Whether reptiles passed through an intermediate stage such as this, as Romer (1922, 1956) has suggested, is open to question for there is no osteological evidence to support this view. If they did not, as now seems probable, then Romer's interpretation of the "invasion of the iliac surface in reptiles" by the M. iliofemoralis must be abandoned. This suggests that although Olson may have been correct in his assessment of why the lateral iliac shelf developed in cotylosaurs, the manner in which it developed is still unknown. Further studies of the musculature of the pelvic girdle and hindlimb of primitive tetrapods are needed.

The humeri of all cotylosaurs are similar in form (Fig. 11). They possess the basic tetrahedral shape described by Romer (1922) with only a very short shaft separating the proximal and distal expansions. The proximal head is wide in dorsal (proximal dorsal) aspect with a prominent deltoid crest directed anteriorly and a separate longer pectoralis crest situated just posterior, ventral and distal to it. The distal extremity of the cotylosaurian humerus is formed by a broad expansion oriented at about 80° to the plane of the proximal head. The postaxial (distal dorsal) surface of this expansion is composed of a large, rounded entepicondyle that extends ventrally and somewhat distally. The entepicondyle is perforated proximo-axially by a large entepicondylar foramen. A prominent distodorsally oriented ectepicondyle extends distally beyond the capitellar articulation with the radius. A prominent dorsally directed supinator process arises from the humeral shaft and is separated from the ectepicondyle by a deep ectepicondylar notch and groove.

The humeri of non-cotylosaurian primitive amphibians are poorly known, thus making it difficult to evaluate the primitive and advanced characters of the humerus of cotylosaurs. The humeral construction of the earliest known amphibians, the ichthyostegids, appears to be grossly similar to that of the osteolepiform rhipidistian *Eusthenopteron* (Andrews and Westoll, 1970). In *Ichthyostega* there appears to have been no identifiable shaft between the proximal and distal heads of the humerus since the preaxial portion of the humerus was developed as a massive supinator–ectepicondylar plate surrounding an enclosed ectepicondylar foramen (Jarvik, 1955). No information is available concerning the structure of the entepicondylar region. All subsequent tetrapod groups have modified the ectepicondylar region of the humerus to some degree.

The anthracosaurian humerus retains a broad preaxial plate that tapers disto-axially so that no supinator process is developed nor is there a prominent ectepicondylar groove. The anthracosaurian humerus typically has a well-developed entepicondyle, pierced dorso-axially by a prominent entepicondylar foramen, giving it a typical

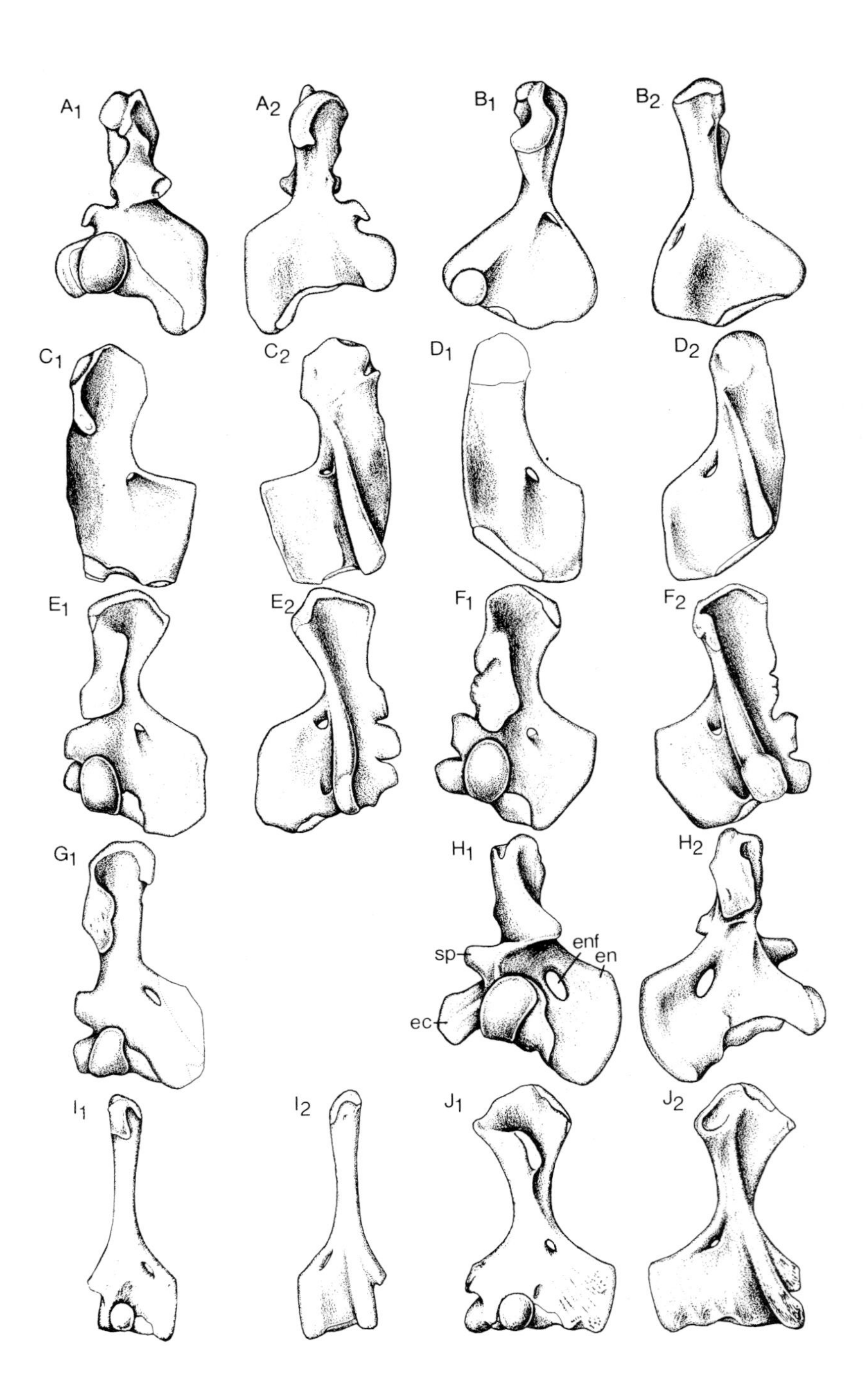

A1
A2
B1
B2
C1
C2
D1
D2
E1
E2
F1
F2
G1
H1
H2
sp
enf
en
ec
I1
I2
J1
J2

"L"-shaped form. This configuration is present in embolomere (Romer, 1957; Panchen, 1970), herpetospondyl (Holmes, Chapter 14 of this volume) and gephyrostegid (Carroll, 1969a,b, 1970; Boy and Bandel, 1973) anthracosaurs, as well as in such advanced aquatic reptile groups as the tangasaurid eosuchians (R. L. Carroll and P. J. Currie, personal communications), nothosaurs and plesiosaurs, where fore limb supination was not a significant locomotor function. All cotylosaurs have humeri highly modified beyond the primitive rhipidistian–ichthyostegid pattern and of a greatly different form from that typical of anthracosaurs. While much of the supinator–ectepicondylar plate has been reduced in cotylosaurs, with the subsequent reduction of the ectepicondylar foramen to a deep notch, the supinator process has remained large. This appears to be related to the effect of gravitational forces on the antebrachium of these predominantly terrestrial amphibians, thus requiring the presence of a long supinator process to produce a sufficient mechanical advantage to allow the fore limb to be rotated and extended forward at the beginning of a stride. Both the structural and functional histories of the humeri in cotylosaurs and anthracosaurs were completely different.

The humeri of temnospondyl labyrinthodont amphibians are not well known except in *Eryops* (Romer, 1922; Miner, 1925) and *Trematops* (Romer, 1922). In general, they more nearly resemble the humeri of cotylosaurs than they do the humeri of anthracosaurian labyrinthodonts. Both the supinator process and the ectepicondyle are well developed with a large, open ectepicondylar notch separating them. The entepicondyle is broad and rounded just as in cotylosaurs but differs from them in that the entepicondylar foramen is absent, either through closure of the foramen or by its exclusion from the dorsal edge of the entepicondylar blade.

Fig. 11. Humeri in distal ventral (1) and distal dorsal (2) aspects. (A) The temnospondyl *Eryops*; (B) the microsaur *Pantylus*; (C) the anthracosaur *Archeria*; (D) the anthracosaur *Gephyrostegus*; (E) the seymouriamorph cotylosaur *Seymouria*; (F) the diadectomorph cotylosaur *Limnoscelis*; (G) the diadectomorph *Tseajaia*; (H) the diadectomorph *Diadectes*; (I) the captorhinomorph *Paleothyris*; (J) the pelycosaur *Dimetrodon*. (A) After Miner, (B) after Carroll and Gaskill, (C) after Romer, (D) after Carroll, (I) after Heaton and Reisz, (J) after Romer and Price. Not to scale. See key to abbreviations in legend to Fig. 6.

The humerus in primitive reptiles is generally similar to that of cotylosaurs except that the reptilian humerus is more lightly built with a proportionately longer shaft. The supinator process is typically extended distodorsally close to the long ectepicondyle: between the two lie a shallow ectepicondylar notch and a long ectepicondylar groove. The entepicondyle is broad and is perforated by a prominent entepicondylar foramen. The humerus of the earliest known reptile, *Hylonomus*, a primitive captorhinormorph, is now believed to have a well-developed supinator process (formerly part of the type specimen of *Protoclepsydrops* – Carroll, 1964) which is in agreement with all indications that a supinator process is a primitive character in reptiles. An enclosed ectepicondylar foramen appears to have developed convergently in pareiasaurs, araeoscelids, lepidosaurs and edaphosaurian pelycosaurs.

The humeri of limbed lepospondyl amphibians are poorly known. Only in the microsaurs is there any reliable information available. In general, the microsaurian humerus resembles that of primitive reptiles in the development of a thin shaft and the great reduction or absence of the supinator process (Gregory *et al.*, 1956; Carroll, 1968; Carroll and Gaskill, 1978) but this is at least as likely a result of convergent development of characters associated with small size and, in some instances advanced terrestriality, as with any phylogenetic relationship.

Other features of the appendicular skeletons of archaic tetrapods are either poorly known or lacking in significant features upon which widespread comparisons can be made.

SYSTEMATICS

Most work done to date on what are recognized here as the Cotylosauria has been strictly descriptive in nature. Only Olson (1947, 1950, 1965, 1966a,b) has tried to analyse the relationships of members of this group in any detail. He concluded that the Seymouriamorpha and Diadectomorpha are closely related and quite separate from the Captorhinomorpha which have long been considered to be cotylosaurs. There can be little argument with these conclusions. What has been lacking has been the establishment of diagnoses of the different suborders and families of cotylosaurs or determination of the systematic position of the Cotylosauria.

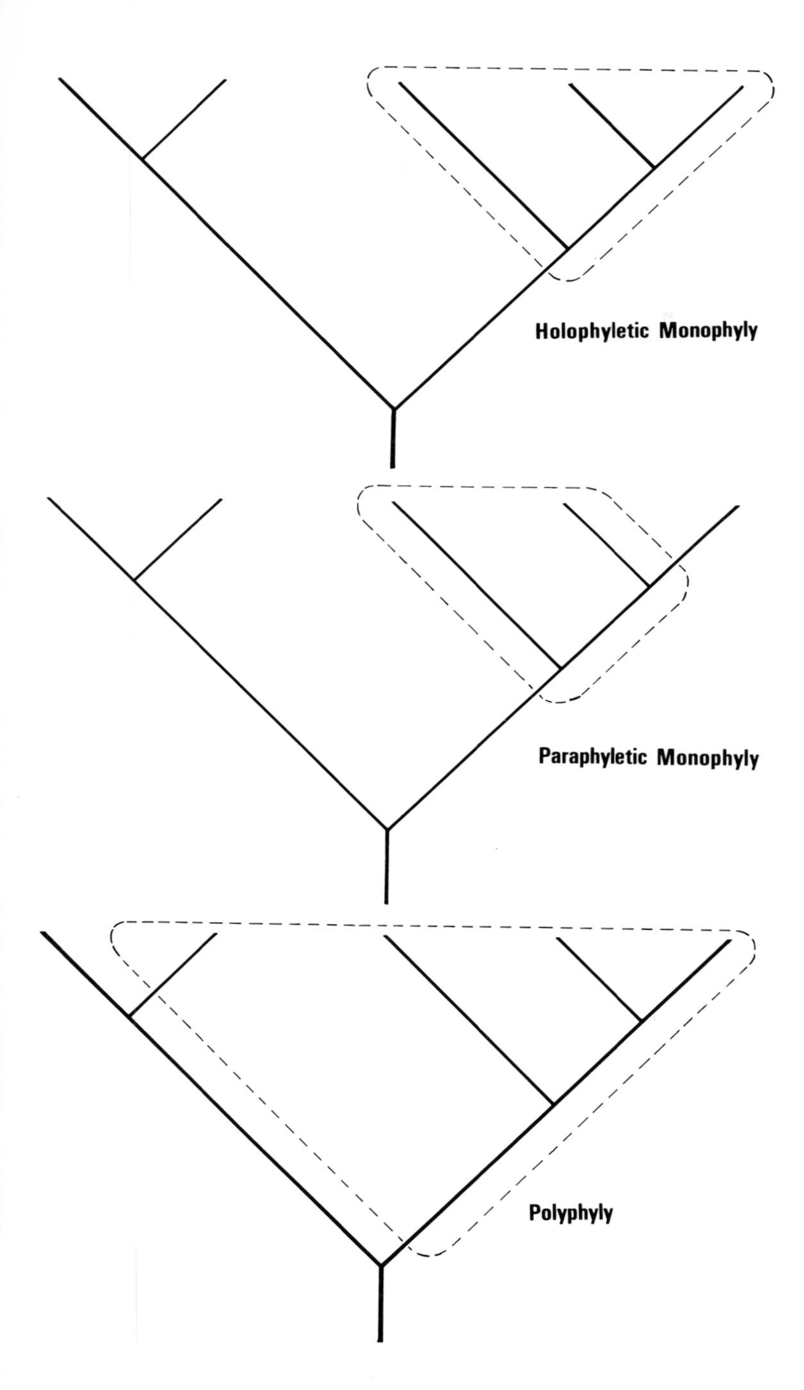

Fig. 12. Schematic models of phylogenetic systems.

Diagnoses of the Order Cotylosauria and the suborders Seymouriamorph and Diadectomorpha as well as the reptilian suborder Captorhinomorpha have been provided by Case (1911) and Romer (1956), although the latter reference is more accurately a short description of each rather than a diagnosis. All were phenetic analyses and included many primitive characters of little taxonomic value. While it must not be assumed that primitive characters have no significance in systematics, they are much more limited than are advanced or derived characters in determining close relationships. Such primitive characters as "skull generally sculptured superficially . . . normally no temporal fenestra or emargination of skull roof" when used to characterize the Cotylosauria (Romer, 1956) do not serve to differentiate this group from earlier labyrinthodont or lepospondyl amphibians, or even rhipidistian crossopterygian fish, and hence should not be used in a diagnosis. An analysis of the relationships of particular groups of organisms based solely on shared derived characters is the preferred systematic methodology.

An attempt has been made to define groups that are monophyletic, that is that contain a single ancestral species and its offspring (Fig. 12). Whereas monophyletic groups containing the ancestor and all of its offspring are used at low taxonomic levels (as in the case of the Order Cotylosauria and its constituent families), at higher taxonomic levels groups become unmanageably large and one or more monophyletic groups or descendents may be removed from a monophyletic assemblage and raised in rank to simplify the terminology. The remaining members of the monophyletic series form a paraphyletic assemblage. Nearly all higher taxa are paraphyletic, including the Amphibia (the Reptilia and their descendants have been removed) and the Reptilia (the Aves and Mammalia have been removed). There is nothing biologically or philosphically inconsistent in the use of paraphyletic groups in a taxonomic system, although it does make diagnoses more difficult since, in addition to stating the shared derived characters of all descendants of the hypothetical ancestor, one or more derived characters identifying the group to be removed must also be noted. Such is the case with the Subclass Batrachosauria from which the Reptilia and their descendants have been removed on the basis of the attainment of a reproductive mode employing an amniotic agg. The Batrachosauria is thus proposed as a paraphyletic taxon without apologies. The only important point biologically is

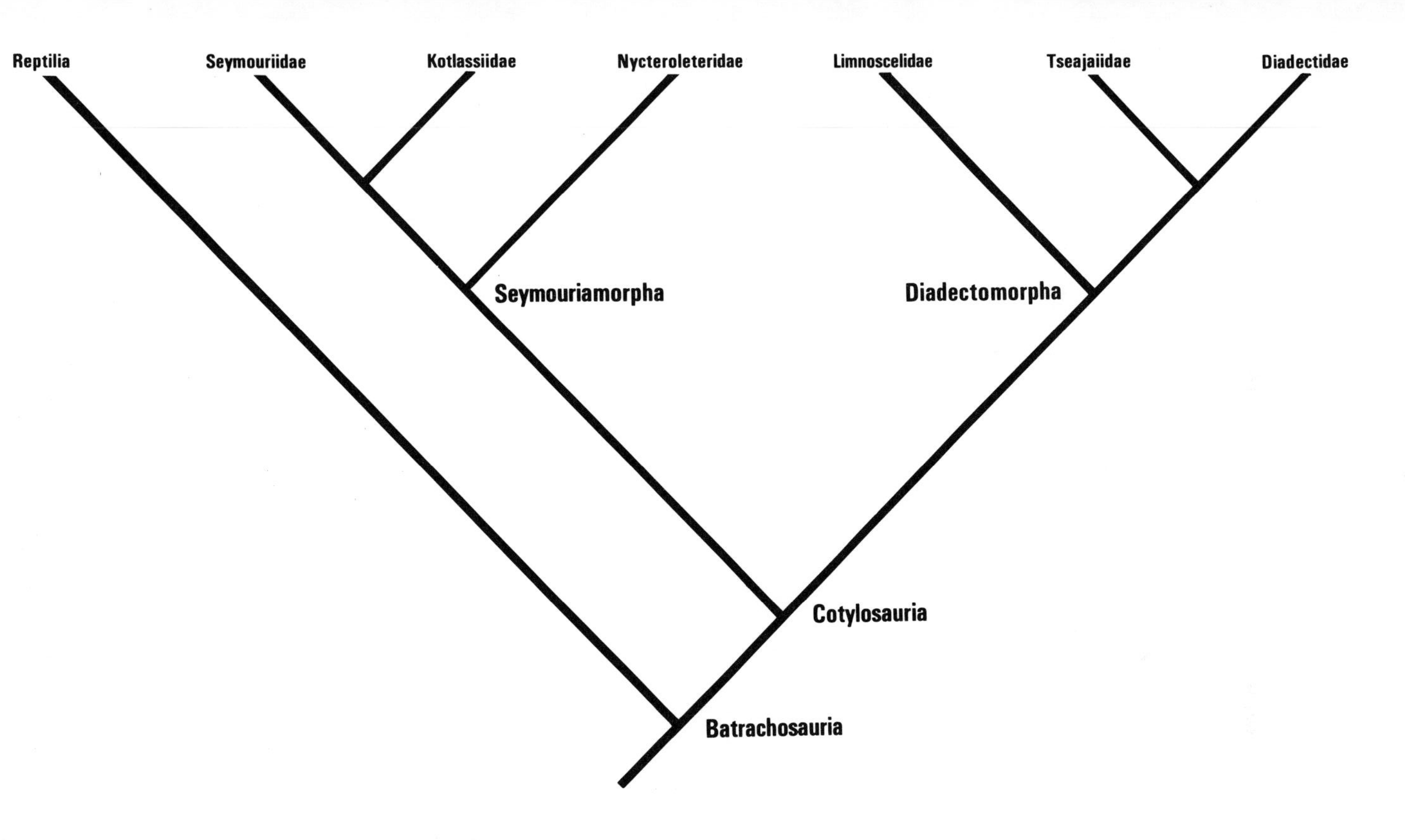

Fig. 13. Phylogeny and relationships of the Cotylosauria.

that it and its descendants are monophyletic rather than polyphyletic. This is basically "old fashioned" vertical classification.

An effort has been made to preserve as much of the existing terminology as possible. This is a preliminary study intended to stimulate renewed study of the members of this important taxonomic group. Until these animals have been studied in detail, it is thought unwise to erect any new major taxa. As a result, no superfamilies have been included in this classification although some have previously been proposed and some appear obvious from the cladogram presented here (Fig. 13). An additional reason for wishing to maintain the terminology has been to avoid adding another series of names to a group of animals whose nomenclature is already cluttered with abandoned names.

Class AMPHIBIA Linnaeus, 1758

Diagnosis. Tetrapod gnathostome vertebrates that have not developed extra-embryonic (amniotic) membranes in the egg.

Taxonomic status. Paraphyletic. Reptilian derivatives and their descendants have been removed. This class has usually been regarded as monophyletic although there have been notable dissenters from this opinion.

Basis for original diagnosis. Modern amphibians specifically the Anura (Salientia) and Urodela (Caudata) although initially some reptiles and fish were also included by Linnaeus.

Subclass BATRACHOSAURIA Efremov, 1946

Diagnosis. Amphibians with neurokinetic braincase (but no mesokinesis) facilitated by a dorsally open braincase (no tectum synoticum ossification); "V"- or "Y"-shaped sphenethmoid; convex single occipital condyle. Transverse flange on pterygoid. Pleurocentrum-based vertebrae with expanded neural arches in dorsal series. Separate coracoid ossification. Humerus with both a large supinator process and an entepicondylar foramen.

Taxonomic status. Paraphyletic. Reptilian derivatives and their descendants have been removed. Anthracosaurs which have at times been included in this subclass do not share any of these derived characters and hence are excluded to avoid polyphyly.

Basis for original diagnosis. The late Permian seymouriamorph cotylosaur *Kotlassia prima.*

Comment. These characters are primitive for the Batrachosauria and may be lost or modified in more derived descendants.

Class REPTILIA Linnaeus, 1758 [Order—taxon elevated to class status]

Diagnosis. Tetrapods (Batrachosaurs) that produce amniotic eggs but which have not developed feathers or mammary glands.

Additional derived characters (secondary diagnosis). Primitively small size. Skull metakinetic; large post-temporal fenestrae. Two coracoid ossifications.

Taxonomic status. Paraphyletic. Batrachosaur derivatives with birds and mammals excluded. Probably monophyletic ancestry. Raised from ordinal to class rank on the basis of the existence of an amniotic egg.

Basis for original diagnosis. Modern lepidosaurs, turtles and crocodilians.

Comments. The additional derived characters of reptiles developed at about the same time as the amniotic egg. Since it is not possible to state unequivocally which developed before, and are hence advanced amphibian characters, and which at the same time as or later than the development of the amniotic egg and are hence truly derived reptilian characters, they cannot be used in the primary diagnosis. For the purposes of palaeontological research where the existence of an amniotic egg cannot be established, these additional osteological characters must suffice. These characters are primitive for reptiles and may be lost or modified in more derived descendants.

Order COTYLOSAURIA Cope, 1880

Diagnosis. Batrachosaurs that have akinetic skulls with ossified sphenethmoid regions and reduced (or lost) post-temporal fenestrae. Apparent loss of M. adductor mandibulae externus profundus. 26 presacral and 2 sacral vertebrae. Single coracoid ossification.

Taxonomic status. Holophyletic (truly monophyletic).

Basis for original diagnosis. The diadectomorph cotylosaur *Diadectes sideropelicus (Empedocles molaris)*.

Comment. Even though in recent years (since 1965) the term Cotylosauria has come to be a virtual synonym of Captorhinomorpha, it rightfully belongs with *Diadectes* and its kin as was the case from 1880 to 1965. The characters presented in the diagnosis are primitive for the Cotylosauria and may be lost or modified in more derived cotylosaurs.

Suborder SEYMOURIAMORPHA Watson, 1917

Diagnosis. Cotylosaurs with a deep otic notch and squamosal tympanic support; slender rod-like ("opercular") stapes; reduced or absent cristae ventrolaterales of parasphenoid; wide, dorsolaterally directed, paroccipital processes; reduced or unossified supraoccipital.

Taxonomic status. Holophyletic.

Basis for original diagnosis. The seymouriid *Seymouria baylorensis*.

Comments. The postcranial skeletons of most primitive (non-diadectid) cotylosaurs are virtually indistinguishable except in the most complete and well preserved specimens. The skulls provide the most significant features for comparison. The occiput is particularly diagnostic since its dorsolaterally directed paroccipital processes give it a very labyrinthodont-like appearance, as do the deep otic notch and tabular horns.

Family NYCTEROLETERIDAE Efremov, 1940

Diagnosis. Seymouriamorphs with short lachrymal excluded from external naris; intertemporal lost. Transverse flange of pterygoid bearing large conical teeth. Limbs elongated.

Taxonomic status. Monogeneric.

Basis for original diagnosis. Nycteroleter bashkiricus.

Known occurrence. Early late Permian, U.S.S.R.

Comments. The exact taxonomic position of the Nycteroleteridae is equivocal although it seems best to include it within the Seymouriamorpha. While several specimens of *Nycteroleter* have been described by Efremov (1938, 1940) and Chudinov (1956), none has been done in sufficient detail to settle the question of its taxonomic position. This family seems to represent the primitive seymouriamorph condition before the development of the extended lateral otic canals typical of the Seymouriidae and the Kotlassiidae. It is because of the uncertainty of the exact phylogenetic position of this family that the Seymouriamorpha, and in parallel the Diadectomorpha, have not been divided into superfamilies.

Family SEYMOURIIDAE Williston, 1911

Diagnosis. Seymouriamorphs with extended lateral otic canal; stapes without quadrate process. Slight development of caniniform tooth. Reduced vertebral column with 23 or 24 presacral vertebrae.

Taxonomic status. Holophyletic.

Basis for original diagnosis. Seymouria baylorensis

Known occurrence. Middle early Permian of north-central Texas, south-eastern Utah and Europe.

Comments. The Seymouriidae are the most primitive of the advanced seymouriamorphs. They exhibit few derived characters that separate them from the more derived Kotlassiidae. The more generalized ancestors of the Kotlassiidae, if they are ever found, could probably be included in this family with minor alteration of the diagnosis. The current definition of the family Seymouriidae is essentially the same as that for the genus *Seymouria.*

Family KOTLASSIIDAE Romer, 1947

Synonym. Karpinskiosauridae Sushkin 1925.

Diagnosis. Seymouriamorphs with extended lateral otic canal; stapes without quadrate process. Skull low and broad; antorbital region shortened; marginal teeth compressed distomesially; palatine and ectopterygoid bearing sigmoidal row of distomesially compressed teeth. Limbs reduced; supinator process reduced; carpal and tarsal elements poorly ossified (possibly paedomorphic). Extensive dorsal armour.

Taxonomic status. Holophyletic.

Basis for original diagnosis. Kotlassia prima = "Karpinskiosaurus secundus".

Known occurrence. Late late Permian, North Dvina River Basin, U.S.S.R.

Comments. Further investigation will probably show that the Seymouriidae and Kotlassiidae should be grouped together in a single superfamily on the basis of the elongated lateral otic canal. Until more than just this one character is defined and the exact taxonomic position of *Nycteroleter* is determined, it is unwise to erect any new taxa.

Suborder DIADECTOMORPHA Watson, 1917

Diagnosis. Cotylosaurs with a vertical occipital plate formed from a wide supraoccipital and laterally directed paroccipital processes; post-temporal fenestrae greatly reduced or lost. Single splenial; single large foramen intermandibularis communis. Ilium with prominent lateral iliac shelf.

Taxonomic status. Holophyletic.

Basis for original diagnosis. The diadectid *Diadectes sideropelicus*.

Family LIMNOSCELIDAE Williston, 1911

Diagnosis. Diadectomorphs that have lost the intertemporal with the development of a postorbital–supratemporal contact.

Taxonomic status. Holophyletic.

Basis for original diagnosis. Limnoscelis paludis.

Known occurrence. Middle Pennsylvanian of Nova Scotia, Canada; late Pennsylvanian of central Colorado and north-central New Mexico, U.S.A.

Comments. Limnoscelis has long been regarded as a captorhinomorph reptile. Most of the characters thought to unite these two groups as reptiles are primitive and only indicate a distant batrachosaur ancestry of both. Until recently, the absence of an otic notch in both limnoscelids and such primitive reptiles as captorhinomorphs and pelycosaurs was considered to be a shared derived character resulting from the loss of a labyrinthodont otic notch. It now appears that the absence of an otic notch is a primitive character of amphibians (Carroll, Chapter 12 of this volume), thus removing the last seemingly significant character from consideration. The identification of three diagnostic features of diadectomorphs, the vertical plate-like occiput (except as highly modified in diadectids), the large foramen intermandibularis communis and the lateral iliac shelf in the pelvis, confirm the position of the Limnoscelidae within the Diadectomorpha. Since these features have been well known since Williston's original work (1911), it is surprising that it has taken so long to recognize the obvious similarity.

Family TSEAJAIIDAE Vaughn, 1964

Diagnosis. Diadectomorphs with elongated lateral otic canal; small parasphenoidal wings formed from cristae ventrolaterales. Stapes with heavy base, well-developed opercular process and long, slightly expanded, quadrate (hyostylar) process that articulates with a prominent notched boss on posterior surface of quadrate. Shallow otic notch with quadrate support for tympanum. Intertemporal lost, replaced by lateral lappet of parietal. Prominent caniniform tooth.

Taxonomic status. Monogeneric.

Basis for original diagnosis. Tseajaia campi.

Known occurrence. Early Permian of south-eastern Utah, U.S.A.

Comments. The Tseajaiidae has previously been considered to be a seymouriamorph family (Romer, 1966; Moss, 1972) principally as a result of Romer's (1964, 1966) decision to include the diadectomorphs within the Seymouriamorpha. Tseajaiids show little in common with seymouriamorphs save primitive characters. Some, such as the structure of the vertebral column, are striking in their similarity but are only indicative of a more ancient batrachosaur ancestry. It appears likely that an animal very close in structure to *Tseajaia* and probably assignable to this family gave rise to the Diadectidae.

Family DIADECTIDAE Cope, 1880

Diagnosis. Diadectomorphs with elongated lateral otic canal; complete loss of post-temporal fenestrae; large parasphenoidal wings formed from cristae ventrolaterales. Stapes with heavy base, well-developed opercular process and long, wide quadrate (hyostylar) process that articulates with a large round knob on posterior surface of quadrate. Shallow otic notch with quadrate support for tympanum. Intertemporal lost or fused to lateral edge of parietal. Medial expansion of maxilla to form partial hard secondary palate. Teeth reduced in number; elongate labiolingually with prominent cusps on crowns. Vertebral column reduced to 22 or 23 presacral vertebrae; vertebrae tall and shortened anteroposteriorly; accessory articulations (episphenes and hyposphenes) developed. Humerus with massive ectepicondyle; long, narrow, rounded supinator process. Ilium shortened anteroposteriorly.

Taxonomic status. Holophyletic.

Basis for original diagnosis. Diadectes sideropelicus = Empedocles molaris.

Known occurrence. Late Pennsylvanian of central Europe; late Pennsylvanian to late early Permian of eastern, central, south-central and south-western U.S.A.

Comments. The Diadectidae is a compact, easily diagnosed group. It was, from 1880 to 1965, always included within the Order Cotylosauria. Since the "type" cotylosaur is also the "type" of the Diadectidae, this family must form the core of the Cotylosauria. It is the one family that must not be removed from it.

The Diadectomorpha and its constituent members are better understood than are the Seymouriamorpha. It is probable that eventually the Tseajaiidae and

Diadectidae will be grouped together as a superfamily for which Diadectoidea (Romer, 1956) is the valid available name. This has not been done here for the sake of taxonomic symmetry. If one superfamily was to be erected, the other three implied should also be proposed. As mentioned previously, there is insufficient evidence to proceed with this at this time. In addition, some superfamily names have been proposed (e.g. Limnosceloidea, as an infraorder – Vaughn, 1955) without either a diagnosis or indication of included taxa. This raises numerous questions of nomenclatorial validity that have not been addressed as yet.

FORMS INCORRECTLY ASSIGNED TO THE COTYLOSAURIA

DISCOSAURISCIDAE

This group of early Permian amphibians was assigned to the Seymouriamorpha by Romer (1947) on the basis of a protoreptilian vertebral structure and a skull-roof exhibiting a contact between the tabular and parietal. As previously discussed, neither of these characters is sufficient to suggest close relationship with any cotylosaur. The discosauriscids do not have the transverse flange on the pterygoid, the distinct supinator process or the expanded neural arches (Špinar, 1952) that would identify them as batrachosaurs; the braincases are not sufficiently well known to be of any use. In most characters they appear most similar to the gephyrostegid anthracosaurs.

CAPTORHINOMORPHA

The Suborder Captorhinomorpha was assigned to the Cotylosauria by Watson (1917). We now recognize this as incorrect since the Cotylosauria is rightfully an amphibian order. The only characters shared by the Captorhinomorpha and the Cotylosauria are primitive characters indicating only a distant batrachosaur ancestry of both groups. The reptilian status of the Captorhinomorpha has never been challenged since members of this group share all of the derived characters noted in the secondary diagnosis of the Reptilia. In many characters such as the reduction of the bones of the lateral temporal series of the skull, the loss of the swelling of the neural arches, and the slimming and elongation of the limb elements, especially in the family Protorothyrididae, the Captorhinomorpha show considerable advances over much more primitive reptile groups such as the pareiasaurs and the procolophonids. The Captorhinormorpha should not be thought of as the most primitive reptilian suborder.

PROCOLOPHONIA

This suborder is currently thought to contain the superfamilies Procolophonoidea, Pareiasauroidea and Millerosauroidea (Romer, 1966). The Pareiasauroidea, and to a somewhat lesser degree the Procolophoroidea, have been considered to be closely related to the Diadectidae Olson (1947, 1950) even went as far as to include these latter three groups within a new subclass Parareptilia. The only features that are present in all members of these groups appear to be primitive for batrachosaurs (swollen dorsal neural arches) or to be convergences resulting from similar herbivorous feeding strategies (larger size and shortened vertebral column). Even these characters are not totally consistent since procolophonoids never developed to the very large size of pareiasaurs and some of the later diadectids. Both pareiasaurs and procolophonoids have skulls with vertical occiputs with narrow supraoccipitals, narrow laterally directed paroccipital processes and large post-temporal fenestrae. Because of their large size, the skulls of pareiasaurs are akinetic while those of small, primitive procolophonids are metakinetic (Ivakhnenko, 1972). Millerosaurs also satisfy this diagnosis. All of these characters accord well with the additional derived characters that may be used as a basis for an osteological diagnosis of the Reptilia. The Pareiasauroidea, Procolophonoidea and Millerosauroidea are all true, albeit very primitive, reptiles not closely related, except as distant batrachosaur descendants, to the amphibian Cotylosauria.

ACKNOWLEDGEMENTS

I am greatly indebted to the directors and personnel of the Museum of Comparative Zoology, Harvard University, the Peabody Museum of Natural History, Yale University, and the University of California Museum of Paleontology for their assistance and permission to study specimens under their care. Michael Fracasso of Yale University, New Haven, has freely discussed his research on the type specimen of *Limnoscelis* and Dr Peter P. Vaughn of the University of California, Los Angeles, has contributed greatly of his knowledge of cotylosaurs from the south-west U.S.A. To both of these gentlemen I am truly grateful. My greatest debt is owed to Dr Robert R. Reisz of Erindale College, University of Toronto, Mississauga, who has been confidant, editor and valued critic throughout this project.

REFERENCES

Andrews, S.M. and Westoll, T. S. (1970). The postcranial skeleton of *Eusthenopteron foordi* Whiteaves. *Trans. R. Soc. Edinb.* 68, 207–329.

Baldauf, R. J. (1963). Cartilages of the cranial roof in modern anurans. *Am. Zool.* 3, 497–498.

Bock, W. J. (1963). The evolution of cranial kinesis in early tetrapods. *Am. Zool.* 3, 487.

Boy, J. A. and Bandel, K. (1973). *Bruckterepeton fiebigi* n. gen. n. sp. (Amphibia: Gephyrostegida) der erste Tetrapode aus dem Rheinisch-Westfälischen Karbon (Namur B; W. Deutschland). *Palaeontographica* **145A**, 39–77.

Broili, F. (1904). Permische Stegocephalen und Reptilien aus Texas. *Palaeontographica* **51**, 1–120.
Broom, R. and Haughton, S. H. (1913). On the skeleton of a new pareiasaur (*Pareiasuchus peringueyi* g. et sp. nov.). *Ann. S. Afr. Mus.* **12**, 17–25.
Bystrow, A. P. (1944) *Kotlassia prima* Amalitzky. *Bull. geol. Soc. Am.* **55**, 37–416.
Carroll, R. L. (1964). The earliest reptiles. *J. Linn. Soc.* (*Zool.*) **45**, 61–83.
Carroll, R. L. (1967). Labyrinthodonts from the Joggins Formation. *J. Paleont.* **41**, 111–142.
Carroll, R. L. (1968). The postcranial skeleton of the Permian microsaur *Pantylus*. *Can. J. Zool.* **46**, 1175–1192.
Carroll, R. L. (1969a). A middle Pennsylvanian captorhinomorph and the interrelationships of primitive reptiles. *J. Paleont.* **43**, 151–170.
Carroll, R. L. (1969b). The origin of the reptiles. *In* "Biology of the Reptilia" (A.d'A. Bellairs *et al.*, eds), Vol. 1, pp. 1–44. Academic Press, London and New York.
Carroll, R. L. (1969c). Problems of the origin of reptiles. *Biol. Rev.* **44**, 151–170.
Carroll, R. L. (1971). Quantitative aspects of the amphibian–reptilian transition. *Forma et Functio* **3**, 165–178
Carroll, R. L. and Baird, D. (1972). Carboniferous stem-reptiles of the family Romeriidae. *Bull. Mus. comp. Zool. Harv.* **143**, 321–364.
Carroll, R. L. and Currie, P. J. (1975). Microsaurs as possible apodan ancestors. *Zool. J. Linn. Soc.* **57**, 229–247.
Carroll, R. L. and Gaskill, P. (1978). The Order Microsauria. *Mem. Am. Phil. Soc.* No. 126, 1–211.
Case, E. C. (1911). A revision of the Cotylosauria of North America. *Publs Carnegie Instn* **145**, 1–122.
Case, E. C. and Williston, S. W. (1913). Description of a nearly complete skeleton of *Diasparactus zenos* Case. *Publs Carnegie Instn* **181**, 17–35.
Chudinov, P. (1957). Cotylosaurs from the Upper Permian Redbeds deposits of the Preurals. *Trudy̆ Paleont. Inst. Akad. Nauk SSSR* **48**, 19–87. (In Russian.)
Clark, J. and Carroll, R. L. (1973). Romeriid reptiles from the Lower Permian. *Bull. Mus. comp. Zool. Harv.* **144**, 353–407.
Colbert, E. H. (1946). *Hypsognathus*, a Triassic reptile from New Jersey. *Bull. Am. Mus. nat. Hist.* **86**, 229–274.
Cope, E. D. (1878). Descriptions of extinct Batrachia and Reptilia from the Permian formations of Texas. *Proc. Am. phil. Soc.* **17**, 505–530.
Cope, E. D. (1880). Second contribution to the history of the vertebrates of the Permian formation of Texas. *Proc. Am. phil. Soc.* **19**, 38–58.
Cope, E. D. (1883). Fourth contribution to the history of the vertebrates of the Permian formation of Texas. *Proc. Am. phil. Soc.* **20**, 628–636.
Cope, E. D. (1889). Synopsis of the families of vertebrata. *Am. Nat.* **23**, 849–877.
DeBeer, G. R. (1926). Studies on the vertebrate head. II. The orbito-temporal region of the skull. *Q. J. Microsc. Sci.* **70**, 263–370.
DeBeer, G. R. (1930). The early development of the chondrocranium of the lizard. *Q. J. Microsc. Sci.* **73**, 707–739.

DeBeer, G. R. (1937). "The Development of the Vertebrate Skull." Oxford University Press.
Efremov, I. A. (1938). On some new Permian reptiles of the U.S.S.R. *Dokl. Acad. Sci. U.S.S.R.* **19**, 379–466.
Efremov, I. A. (1940). Die Mesen-Fauna der permischen Reptilien. *Neues Jb. Min.* **84**, 379–466.
Efremov, I. A. (1946). On the subclass Batrachosauria – an intermediary group between amphibians and reptiles. *Izv. Acad. Nauk. S.S.S.R.* (*Biol.*) 615–638.
Fox, R. C. and Bowman, M. C. (1966). Osteology and relationships of *Captorhinus aguti* (Cope) (Reptilia: Captorhinomorpha). *Paleont. Contr. Univ. Kans. Vertebr.* **11**, 1–79.
Frazzetta, T. H. (1962). A functional consideration of cranial kinesis in lizards. *J. Morph.* **111**, 287–320.
Gaupp, E. (1900). Das Chondrocarnium von *Lacerta agilis*, Ein Beitrag zum Verständnis der Amniotenschädels. *Anat. Heft., Wiesbaden* **15**, 433–595.
Gaupp, E. (1905). Die Entwicklung des Kopfskelets. *Hertwigs Handb. Entwickl. lehre* **3**, 573–875.
Goodrich, E. S. (1930). "Studies on the Structure and Development of Vertebrates." Macmillan, London
Gow, C. E. (1972). The osteology and relationships of the Millerettidae (Reptilia: Cotylosauria). *J. Zool., Lond.* **167**, 219–264.
Gregory, W. K. (1946). Pareiasaurs versus placodonts as near ancestors to the turtles. *Bull. Am. Mus. nat. Hist.* **86**, 275–326.
Gregory, J. T., Peabody, F. E. and Price, L. I. (1956). Revision of the Gymnarthridae, American Permian microsaurs. *Bull. Peabody Mus. nat. Hist.* **10**, 1–77.
Heaton, M. J. (1978). "The Cranial Soft Anatomy and Functional Morphology of a Primitive Captorhinid Reptile." Ph.D. Thesis, McGill University, Montreal.
Heaton, M. J. (1979). The cranial anatomy of primitive captorhinid reptiles from the Pennsylvanian and Permian of Oklahoma and Texas. *Bull. Okla. geol. Surv.* **127**, 1–84.
Holmes, R. B. (1977). The osteology and musculature of the pectoral limb of small captorhinids. *J. Morph.* **152**, 101–140.
Holmes, R. B. and Carroll, R. L. (1977). A temnospondyl amphibian from the Mississippian of Scotland. *Bull. Mus. comp. Zool. Harv.* **147**, 489–511.
Ivakhnenko, M. F. (1973). Skull structure in the early Triassic procolophonian *Tichvinskia vjatkensis*. *Paleont. Zh.* 1973, 511–518.
Jaeger, E. C. (1955). "A Source-book of Biological Names and Terms", 3rd Edn. C. S. Thomas, Springfield, Illinois.
Jarvik, E. (1944). On the dermal bones, sensory canals and pit lines of the skulls in *Eusthenopteron foordi* Whiteaves, with some remarks on *E. säve-söderberghi* Jarvik. *K. svenska VetenskAkad. Handl.* (3)**21**(3), 3–48.
Jarvik, E. (1954). On the visceral skeleton in *Eusthenopteron* with a discussion of the parasphenoid and palatoquadrate in fish. *K. svenska VetenskAkad. Handl.* (4)5(1), 1–104.

Jarvik, E. (1955). The oldest tetrapods and their forerunners. *Scient. Mon. N.Y.* **80**, 141–154.
Jarvik, E. (1975). On the *saccus endolymphaticus* and adjacent structures in osteolepiforms, anurans, and urodeles. *Probl. Act. Paléontol. (Évol. Vert.) Colloq. int. C.n.R.S.* No. 218, 191–211.
Jollie, M. (1978). Book review of "Morphology and Biology of Reptiles" (A. d'A. Bellairs and C. Cox, eds), *Evolution* **32**, 221–223.
Konjukova, E. D. (1953). Terrestrial vertebrate fauna of the Lower Permian of the northern pre-Urals (River Inta basin). *Dokl. Akad. Nauk S.S.S.R.* **89**, 723–726. (In Russian.)
Lombard, E. R. and Bolt, J. R. (1979). Evolution of the tetrapod ear; an analysis and reinterpretation. *Biol. J. Linn. Soc., Lond.* **11**, 19–76.
Miner, R. W. (1925). The pectoral limb of *Eryops* and other primitive tetrapods. *Bull. Am. Mus. nat. Hist.* **51**, 145–312.
Moss, J. L. (1972). The morphology and phylogenetic relationships of the lower Permian tetrapod *Tseajaia campi* Vaughn (Amphibia: Seymouriamorpha). *Univ. Calif. Publ. geol. Sci.* **98**, 1–63.
Olson, E. C. (1936). The dorsal axial musculature of certain primitive Permian tetrapods. *J. Morph.* **59**, 265–311.
Olson, E. C. (1947). The family Diadectidae and its bearing on the classification of reptiles. *Fieldiana: Geology* **11**, 1–53.
Olson, E. C. (1950). The temporal region of the Permian reptile *Diadectes*. *Fieldiana: Geology* **10**, 63–67.
Olson, E. C. (1961). Jaw mechanisms: rhipidistians, amphibians, reptiles. *Am. Zool.* **1**, 205–215.
Olson, E. C. (1965). Relationships of *Seymouria, Diadectes* and *Chelonia*. *Am. Zool.* **5**, 295–307.
Olson, E. C. (1966a). The middle ear – morphological types in amphibians and reptiles. *Am. Zool.* **6**, 399–419.
Olson, E. C. (1966b). Relationships of *Diadectes*. *Fieldiana*: *Geology* **14**, 199–227.
Olson, E. C. (1971). "Vertebrate Paleozoology." Wiley, New York.
Olson, E. C. (1977). The exploitation of land by early tetrapods. *In* "Morphology and Biology of Reptiles" (A. d'A. Bellairs and C. B. Cox, eds), pp. 1–30. Academic Press, London and New York.
Panchen, A. L. (1964). The cranial anatomy of two Coal Measures anthracosaurs. *Phil. Trans. R. Soc.* **B247**, 593–637.
Panchen, A. L. (1966). The axial skeleton of the labyrinthodont *Eogyrinus attheyi*. *J. Zool., Lond.* **150**, 199–222.
Panchen, A. L. (1970). Teil 5a Anthracosauria. "Handbuch der Paläoherpetologie." Fischer, Stuttgart.
Panchen, A. L. (1972a). The interrelationships of the earliest tetrapods. *In* "Studies in Vertebrate Evolution—Essays presented to Dr. F. R. Parrington, F.R.S." (K. A. Joysey and T. S. Kemp, eds), pp. 65–87. Oliver and Boyd, Edinburgh.
Panchen, A. L. (1972b). The skull and skeleton of *Eogyrinus attheyi* Watson

(Amphibia: Labyrinthodontia). *Phil. Trans. R. Soc.* **B263**, 279–326.

Panchen, A. L. (1975). A new genus and species of anthracosaur amphibian from the Lower Carboniferous of Scotland and the status of *Pholidogaster pisciformis* Huxley. *Phil. Trans. R. Soc.* **B269**, 581–640.

Panchen, A. L. (1977a). The origin and early evolution of tetrapod vertebrae. *In* "Problems in Vertebrate Evolution" (S. M. Andrews, R. S. Miles and A. D. Walker, eds), pp. 289–318. Academic Press, London and New York.

Panchen, A. L. (1977b). On *Anthracosaurus russelli* Huxley (Amphibia: Labyrinthodontia) and the family Anthracosauridae. *Phil. Trans. R. Soc.* **B279**, 447–512.

Reisz, R. R. (1975). "*Petrolacosaurus kansensis* Lane: the Earliest Known Diapsid Reptile." Ph.d. Thesis, McGill University, Montreal.

Reisz, R. R. (1977). *Petrolacosaurus kansensis*, the oldest known diapsid reptile. *Science* **196**, 1091–1093.

Rice, E. L. (1920). The development of the skull in the skink, *Eumeces quinquelineatus* L. I. The chondrocranium. *J. Morph.* **34**, 119–220.

Romer, A. S. (1922). The locomotor apparatus of certain primitive and mammal-like reptiles. *Bull. Am. Mus. nat. Hist.* **46**, 517–606.

Romer, A. S. (1946). The Permian reptile *Limnoscelis* restudied. *Am. J. Sci.* **244**, 149–188.

Romer, A. S. (1947). Review of the Labyrinthodontia. *Bull. Mus. comp. Zool. Harv.* **99**, 1–368.

Romer, A. S. (1956). "Osteology of the Reptiles." University Press, Chicago.

Romer, A. S. (1957). The appendicular skeleton of the Permian embolomerous amphibian, *Archeria. Contr. Mus. Geol. Univ. Mich.* **13**, 103–159.

Romer, A. S. (1964). *Diadectes* an amphibian? *Copeia* 1964, 718–719.

Romer, A. S. (1966). "Vertebrate Paleontology", 3rd Edn. University Press, Chicago.

Romer, A. S. (1968). "Notes and Comments on Vertebrate Paleontology." University Press, Chicago.

Romer, A. S. (1969). The cranial anatomy of the Permian amphibian *Pantylus. Breviora* No. 314, 1–37.

Romer, A. S. and Price, L. I. (1940). The review of the Pelycosauria. *Spec. Pap. geol. Soc. Am.* **28**, 1–538.

Romer, A. S. and Witter, R. V. (1942). *Edops*, a primitive rhachitomous amphibian from the Texas redbeds. *J. Geol.* **50**, 925–960.

Säve-Söderbergh, G. (1945). Notes on the trigeminal musculature in non-mammalian tetrapods. *Nova Acta Regia Soc. Sci. Upp.* (IV), **13**(7).

Sawin, H. J. (1941). The cranial anatomy of *Eryops megacephalus. Bull. Mus. comp. Zool. Harv.* **88**, 407–463.

Shishkin, M. A. (1975). Labyrinthodont middle ear and some problems of amniote evolution. *Probl. Act. Paléont. (Évol. Vert.) Colloques int. C.n.R.S.* No. 128, 337–348.

Špinar, Z.V. (1952). Revise některých moravských Diskosauriscidů (Labyrinthodontia). *Rozpr. ústred. Ustavu geol.* **15**, 1–160.

Sushkin, P. P. (1925). On the representatives of the Seymouriamorpha, supposed primitive reptiles, from the Upper Permian of Russia, and on their phylogenetic relations. *Occ. Pap. nat. Hist. Soc. Boston.* **51**, 179–181.

Vaughn, P. P. (1955). The Permian reptile *Araeoscelis* restudied. *Bull. Mus. comp. Zool. Harv.* **113**, 305–467.

Watson, D. M. S. (1916). On the structure of the braincase in certain lower Permian tetrapods. *Bull. Am. Mus. nat. Hist.* **35**, 611–636.

Watson, D. M. S. (1917). A sketch classification of the pre-Jurassic tetrapod vertebrates. *Proc. zool. Soc. Lond.* 1917, 167–186.

Watson, D. M. S. (1918). On *Seymouria*, the most primitive known reptile. *Proc. zool. Soc. Lond.* 1918, 267–301.

Watson, D. M. S. (1919). The structure, evolution and origin of the Amphibia – the orders Rhachitomi and Stereosponyli. *Phil. Trans. R. Soc.* **B209**, 1–73.

Watson, D. M. S. (1954). On *Bolosaurus* and the origin and classification of reptiles. *Bull. Mus. comp. Zool. Harv.* **111**, 297–449.

Watson, D. M. S. (1957). On *Millerosaurus* and the early history of the sauropsid reptiles. *Phil. Trans. R. Soc.* **B240**, 335–440.

White, T. E. (1939). Osteology of *Seymouria baylorensis*. *Bull. comp. Zool. Harv.* **85**, 323–410.

Williston, S. W. (1911). A new family of reptiles from the Permian of New Mexico. *Am. J. Sci.* (4)**31**, 378–398.

19 | The Pelycosauria: A Review of Phylogenetic Relationships

R. R. REISZ

Department of Biology, Erindale College, University of Toronto, Mississauga, Ontario, Canada

Abstract: Shared derived characters of the skeleton form the basis for the re-evaluation of the widely accepted theories of relationships of the pelycosaurian reptiles. The pelycosaurs, together with their descendants, form a natural group and possess such derived characters as broad anteriorly tilted occipital plate formed by the paroccipital process of the opisthotic and the lateral process of the supraoccipital, reduced post-temporal fenestrae, single median postparietal, and septomaxilla composed of a broad base and a dorsal process. The presence of the inferior temporal fenestra cannot be considered to be a shared derived character of this group, because diapsids also have this feature. The association of the six valid pelycosaurian families within three suborders, the Ophiacodontia, Sphenacodontia and Edaphosauria, and the implied phylogeny, is based almost exclusively on untenable assumptions and shared primitive characters. These taxa are therefore rejected and a new pattern of relationships is offered and tested with shared derived characters. A detailed review of the pelycosaurs, family by family, is needed before a new classification and the required erection of new taxa based on monophyletic groups is attempted.

Systematics Association Special Volume No. 15, "The Terrestrial Environment and the Origin of Land Vertebrates", edited by A. L. Panchen, 1980, pp. 553-592, Academic Press, London and New York.

INTRODUCTION

The origin and early radiation of amphibians during the Palaeozoic is a remarkable phenomenon. But even more remarkable in many ways was the early evolution of synapsid reptiles. The fossil record indicates that the synapsid pelycosaurs was the first group of exclusively terrestrial tetrapods to dominate the scene from the Middle Pennsylvanian into the upper part of the Lower Permian. In the Pennsylvanian more than 50% of the known reptile genera were pelycosaurs; in the Lower Permian over 70% were the large terrestrial carnivorous and herbivorous pelycosaurs; in the Upper Permian the pelycosaurs' therapsid descendants were the dominant terrestrial vertebrates. The pelycosaurs not only include some of the oldest known reptiles and the largest Pennsylvanian and Lower Permian terrestrial vertebrates, but also represent the earliest successful adaptation of terrestrial tetrapods to the herbivorous mode of life.

Pelycosaurian reptiles have been the subject of study for over a century. The first major stage in the study of pelycosaurs, initiated by Cope, culminated in the publication in 1907 of the "Revision of the Pelycosauria of North America" by E.C. Case. Most of the important materials then known were illustrated, some of the taxonomy was corrected and a classification was provided. This classification (Case, 1907, pp. 17–75) into three families, based mainly on dental characters, formed the nucleus of later patterns of classification. The great amount of materials found after Case's work was only partially reflected in the description of the pelycosaurs *Varanops* and *Casea* by Williston (1911, 1913), of *Ophiacodon mirus* by Williston and Case (1913), of *Pantelosaurus* by von Huene (1925) and of *Eothyris* and *Secondontosaurus* by Romer (1937). A second stage in the study of this group culminated in the publication in 1940 of the "Review of the Pelycosauria", a monumental work by Romer and Price. In this study three pelycosaurs, *Ophiacodon, Dimetrodon* and *Edaphosaurus*, representatives of three highly divergent types, were described in great detail. Sketches of various bones and skeletal reconstructions, when warranted by the fossil remains, were also presented. In the classification adopted by Romer and Price the main pattern of the previous arrangements were followed, with most of the differences being the result of taxonomic and morphological reviews:

Order Pelycosauria Cope 1878
Suborder Ophiacodontia Nopsca 1923
Family Ophiacondontidae Nopsca 1923
Family Eothyrididae Romer and Price 1940
Suborder Sphenacodontia Marsh 1878
Family Varanopsidae Romer and Price 1940
Family Sphenacodontidae Marsh 1878
Suborder Edaphosauria Cope 1882
Family Nitosauridae Romer and Price 1940
Family Lupeosauridae Romer and Price 1940
Family Edaphosauridae Cope 1882
Family Caseidae Williston 1912

The three suborders are mainly enlargements of the families presented by Case (1907). The above classification and a phylogeny also provided by Romer and Price (1940, Fig. 2) represent a theory of relationships. Although many papers on pelycosaurs have been published in the last four decades and many new species have been described, no new general treatment of the group has been undertaken, and the classification and phylogeny adopted by Romer and Price has been retained with little change. Olson, for example, published extensively on caseid pelycosaurs (Olson, 1954, 1962, 1968) but has, despite certain misgivings, retained Romer and Price's association of caseids and edaphosaurs. Both Vaughn (1958b) and Langston (1965) argued that eothyridids could be closely related to the caseids, but associated the former group with the ophiacodonts. Langston did suggest in his discussion of pelycosaur relationships that edaphosaurs and caseids should be placed in different suborders because they were derived from different groups of primitive pelycosaurs, but was largely ignored because of the strong criticisms of this theory by Stovall *et al.* (1966).

The purpose of the present paper is to re-evaluate the relationships of pelycosaurian reptiles. Past hypotheses of relationships of the members of this group have been presented mainly as statements, without attempts at testing those hypotheses. Diagnoses of the Order Pelycosauria, of the Suborders Ophiacodontia, Sphenacodontia and Edaphosauria and of the member families have been provided by Romer and Price (1940), and by Romer (1956) in his "Osteology of the Reptiles". These have all been basically short descriptions rather than diagnoses, without distinction of primitive and derived characters. Although it is incorrect to assume that primitive characters have no significance in systematics, they are more limited in value in determining relationships than are advanced or derived characters. The preferred systematic methodology is the one where the hypotheses of relationships are tested using shared derived characters. I therefore propose to test the hypotheses of relationships of pelycosaurs presented by Romer and Price and adopted in subsequent papers. This is a preliminary study, part of a general review of the pelycosaurs. Until this group has been re-studied in detail it is unwise to erect new taxa. It is for this reason that no revised classification is offered, but a theory of relationships is presented in the form of a cladogram (Fig. 17 – the figures appear at the end of this paper). It is hoped that this approach to the study of pelycosaurs will pinpoint the problems that need to be confronted in a review of the group.

ORDER PELYCOSAURIA

Romer and Price (1940), in their discussion of pelycosaur relationships, came to the conclusion that the presence of the lateral temporal opening was the only valid diagnostic feature of the group. The margins of the lateral temporal opening in pelycosaurs are formed posteriorly by the squamosal, dorsally by the posterior process of the postorbital, anteriorly by the ventral process of the postorbital and the dorsal process of the jugal, and ventrally by the posterior process of the jugal, the squamosal or the quadratojugal. The presence of lateral temporal openings of this configuration is not restricted to pelycosaurs and their descendants, but is also found in diapsid reptiles. It is therefore fortunate that we can distinguish pelycosaurs and their descendants from all other amniotes on the basis of a series of shared derived characters restricted to this group. Probably the most readily recognizable of these characters is the presence of a broad anteriorly tilted occipital plate. This plate is formed by the large supraoccipital and by the paroccipital process of the opisthotic (Figs 2 and 4). In addition, the lateral processes of these bones greatly restrict the post-temporal fenestrae. In all pelycosaurs the postparietal is a single median occipitally oriented bone. No other primitive reptile possesses these features. In captorhinomorphs, for example, the paroccipital process extends laterally either in cartilage or bone, but the supraoccipital does not extend laterally to form a plate, the post-temporal fenestrae are relatively large and the post-parietal is paired. In all pelycosaurs where the septomaxilla is preserved it is highly distinctive. It is a vertical sheet of bone in the posterior portion of the external narial opening, consisting of a broad base, which straddles the premaxilla and maxilla, and a dorsal process. In other primitive reptiles, when preserved, the septomaxilla is a relatively simple curved sheet of bone. Two other cranial features may be shared derived characters for pelycosaurs, but greater knowledge of the palate and braincase of other primitive reptiles is needed before the status of these characters is established. In all pelycosaurs where the appropriate portions of the braincase and palate are preserved, firstly the dorsum sellae is formed by the prootic and secondly the medial process of the jugal, sutured to the pterygoid, is covered ventrally by the ectopterygoid. In *Captorhinus* where the skull is well known the dorsum sellae is formed by the basisphenoid, and the jugal–pterygoid suture is present but not covered by the ectopterygoid. In modern diapsids the dorsum sellae is also formed by the basisphenoid but the jugal–pterygoid suture is not present. Instead the ectopterygoid extends between the jugal and the pterygoid.

In the postcranial skeleton primitive morphological patterns persist in large numbers and there are no recognizable synapomorphies for the group. Most pelycosaurs in which an accurate presacral vertebral count is available possess 27 presacral vertebrae, but some of the caseids appear to have as few as 24 (Olson, 1968). In captorhinomorphs the number varies from 25 presacral vertebrae in captorhinids to 30 in protothyridids. In the early eosuchian *Petrolacosaurus* there are 26 presacral vertebrae. The vertebrae (Fig. 5) not only vary too much in shape and size between pelycosaurian families to provide a shared derived character for the group, but even within the sphenacodont pelycosaurs the neural spines vary in size and shape from short, normal-width spines in *Haptodus* and tall, slightly widened spines in *Sphenacodon*, to very tall, greatly expanded spines in *Ctenospondylus* and very tall, narrow spines in *Dimetrodon* and *Secodontosaurus*. There are from five (primitive amniote condition) to seven cervical vertebrae, and from two (primitive amniote condition) to four sacral vertebrae.

The appendicular skeleton of pelycosaurs is built upon the fundamental tetrapod plan seen in most primitive reptiles. However, some advances do seem to occur among pelycosaurs. In ophiacodonts, for example, the pelvic girdle is similar to that in captorhinomorphs and primitive diapsids. The iliac blade has a narrow, long, posterior process, the symphysis is greatly thickened on the pubis and in most forms there is a lateral pubic tubercle. In sphenacodonts, edaphosaurs and caseids the pelvic girdle is advanced in the following features: the iliac blade is broad, the symphysis is not greatly thickened on the pubis but is of uniform thickness along the medial margin of the pelvis and there is no lateral pubic tubercle. These advanced features, which appear to have developed independently in the three pelycosaurian groups, are probably related to the development of an expanded vertebral-pelvic attachment formed by three or four sacral ribs. In varanopsids, where two sacral ribs are present, the primitive pelvic structure prevails (Fig. 12). Caseids and edaphosaurs, the herbivorous pelycosaurs, retain the primitive limb proportions; whereas the carnivorous pelycosaurs, the ophiacodonts, varanopsids and especially the sphenacodonts, develop relatively slender, long, distal limb elements and hence higher posture and a longer stride.

Although pelycosaurs appear to form a natural group, readily distinguishable from other contemporary amniotes on the basis of a series of cranial shared derived characters, their phylogenetic relationships and taxonomic position are not clear. This is partly because the anatomy of most primitive reptiles is poorly known and partly because of geological and geographical patterns of distribution. It is generally accepted that pelycosaurs and captorhinomorphs are closely related. However, study of their osteology reveals only two synapomorphies (shared derived characters). In both groups the tabulars are restricted to the occiput and in early captorhinomorphs and all pelycosaurs the neural arches are unswollen. Even these synapomorphies are questionable. It is quite possible, for example, that the narrow neural arches may have evolved independently in the two groups as adaptations to a rapid mode of locomotion. The tabular in the pelycosaurs *Eothyris* and *Dimetrodon* is not completely restricted to the occiput, but is partially visible in lateral view (Fig. 2). What most authors have neglected to emphasize is that these two groups of early reptiles differ in a number of significant features. Captorhinomorphs typically have lost the contact between the postorbital and the supratemporal, reduced or eliminated the tabular, and have a small, narrow supratemporal and a single pair of coronoids in the lower jaw. In the captorhinomorph tarsus the lateral centrale is either very small or is lost. These characters represent synapomorphies for captorhinomorphs and for diapsids. Pelycosaurs, on the other hand, are primitive in that they retain the contact between the postorbital and supratemporal, the supratemporal is large (a broad sheet in some pelycosaurs, a long narrow strip in others), the tabular is also a large sheet of bone, and there are two pairs of coronoids in the lower jaws. In the tarsus the lateral centrale is equal in size to the medial centrale in primitive pelycosaurs. It therefore appears that the pelycosaurs and their descendants are a sister-group of all other true reptiles (captorhinomorphs, diapsids and their descendants) as indicated in Fig. 17.

Well over a century ago reptilian remains were discovered in the Upper Permian and Triassic deposits of South Africa. These forms, clearly related to mammals, are now considered to constitute the Order Therapsida. As already indicated, Cope and subsequent workers described from the Lower Permian American deposits the more primitive reptiles that now constitute the Order Pelycosauria. It is now generally agreed that the Pelycosauria represent an initial stage in the evolution towards mammals, antecedent to the Therapsida. These two orders together form the Subclass Synapsida. The division of the Synapsida into these two groups has been based on occurrence, the pelycosaurs being until recently restricted to the Pennsylvanian and Lower Permian of North America and Western Europe, and the therapsids flourishing from the Upper Permian onward in Africa and Russia. Recent discoveries have broken down both the temporal and geographical separation of the two groups, making them into a single, continuous, major assemblage (Olson, 1962, Fig. 69). The morphological similarities between advanced carnivorous pelycosaurs and primitive therapsids are so great that the two categories can be maintained only by purely arbitrary means. Olson found, for example, that some fossils from both North America and Russia are intermediate between the better known members of the two orders. He found that in order to maintain the integrity of the Pelycosauria the intermediate forms had to be arbitrarily placed within the Therapsida (Olson, 1962).

The monophyly of the synapsids is supported by the fact that the shared derived characters cited for pelycosaurs are really synapomorphies for all synapsids. Pelycosaurs therefore represent a paraphyletic grade and cannot be defined on the basis of shared derived characters that do not apply to therapsids and mammals as well.

Romer and Price (1940) subdivided the order into eight families that were placed into three suborders, a reflection of their belief that the pelycosaurs consisted of three major adaptive radiations. The Suborder Ophiacodontia included, according to them, the most primitive pelycosaurs, the fish-eating ophiacodonts and the poorly known eothyridids. In the second major group, the Suborder Sphenacodontia, Romer and Price included for the first time varanopsids and sphenacodontids, pelycosaurs that have adapted to a truly carnivorous mode of life. In the third group, the Edaphosauria, a number of primitive forms, the nitosaurids, a poorly known form *Lupeosaurus* and the better know edaphosaurids and caseids were included. The last two groups represent adaptation to the herbivorous mode of life. These suborders, because of their composition, are not monophyletic groups and are either based almost exclusively on primitive features, or characters that were probably developed independently within several member families.

SUBORDER OPHIACODONTIA

The diagnostic features of the Suborder Ophiacodontia, presented by Romer and Price (1940) and adopted by Romer (1956), fall into three categories: primitive reptilian characters, primitive pelycosaurian characters (features found in all pelycosaurs) and shared derived characters restricted to one of the families of this suborder.

Primitive reptilian cranial features cited include the presence of concave dorsal orbital margins, lack of lateral projection of frontal or postorbital, extension of the lacrimal to the naris, nearly straight ventral skull margin, presence of well-developed pterygoid flanges, jaw articulation almost in line with the tooth row, slender lower jaws with symphysis restricted to the dentary. Primitive reptilian vertebral features include subcircular centrosphenes, short transverse processes, unexcavated neural arches, two sacral vertebrae. Primitive reptilian appendicular characters include the shape of the iliac blade, concentration of the pelvic symphysis on the pubis, short limbs, presence of widely divergent supinator process on the humerus, well-developed medial centrale of the pes, presence of gastralia.

Cranial features cited in the diagnosis of the suborder, but actually found in all pelycosaurs, include the presence of an anteriorly slanted occipital plate and large tabulars.

Shared derived characters restricted to the ophiacodontids but cited as diagnostic for the suborder include elongated slender snout, and small clavicular and interclavicular heads. These features are synapomorphies of the Family Ophiacodontidae and are not found in the Family Eothyrididae, placed by Romer and Price (1940) into this suborder. No synapomorphies for this suborder can be recognized and this group has no taxonomic or phylogenetic validity. The two families formerly included in this taxon are diagnosable on the basis of shared derived characters and will be discussed in a later section.

SUBORDER SPHENACODONTIA

The diagnostic features of the Suborder Sphenacodontia, introduced by Romer and Price (1940) and adopted by Romer (1956), also fall into three categories as in the case of the Suborder Ophiacodontia: primitive reptilian characters, shared derived characters not seen in the herbivorous pelycosaurs and in eothyridids, but found in ophiacodontids, sphenacodontids and varanopsids, and shared derived characters restricted to the families of pelycosaurs included in this suborder and to their descendants.

Primitive reptilian postcranial features cited as part of the diagnosis of this suborder are the presence of large ventral clavicular plate and large interclavicular head. These features were considered advanced by Romer and Price (1940) because they incorrectly assumed that the pattern seen in ophiacodontids (small clavicular and interclavicular heads) is primitive. All non-synapsid Palaeozoic reptiles have large-headed clavicles and interclavicles, indicating that this condition is primitive for reptiles.

Shared derived characters found in varanopsids and sphenacodontids, that were included in the diagnosis of this suborder, but are also seen in ophiacodontid pelycosaurs, are the presence of an elongated facial region, sharply demarcated dorsal and lateral skull surfaces, jaw articulation well posterior to the skull condyle, teeth sharply pointed and recurved.

Shared derived characters of varanopsids, sphenacodontids and their therapsid descendants include the presence of a well-developed retroarticular process on the lower jaw, vertebrae with excavated neural arches, ventral keels on cervical, anterior dorsal and mid-dorsal centra. Where preserved, the atlantal centrum reaches the ventral surface of the column and is not sutured to the axial intercentrum. In the appendicular skeleton the scapulae are tall and have an anteroposteriorly narrow base, the adductor crest and the fourth trochanter of the femur are represented by rugosities rather than by crests and the lateral centrale is lost or excluded from contact with the astragalus. The lack of good cranial diagnostic features is a reflection of our poor knowledge of the anatomy of varanopsids. Although Romer and Price (1940) noted that both varanopsids and sphenacodontids have well-developed retroarticular processes, the structural pattern of this process in varanopsids is intermediate between that seen in ophiacodontids and sphenacodontids. Work now in progress on the anatomy of *Aerosaurus* should help establish the exact phylogenetic relationships of these groups of pelycosaurs. Some of the other shared derived characters cited above are also questionable. For example, the excavation of the neural spines and the ventral keel of varanopsids are less well developed than those in sphenacodontids. It has been generally assumed that the primitive condition among reptiles in the atlas–axis complex is the one found in varanopsids and sphenacodontids. Recent studies on primitive reptiles indicate, however, that in captorhinomorphs, millerosaurs and diapsid reptiles, as well as in ophiacodonts and caseids, the atlantal centrum does not reach the ventral surface of the vertebral column and the axial intercentrum is suturally attached or fused to it ventrally. This suggests that the latter pattern represents the primitive reptilian condition and that the sphenacodontid and varanopsid condition is derived.

Although the Suborder Sphenacodontia is diagnosable on the basis of shared derived characters, its taxonomic status is paraphyletic, because the therapsid reptiles have been arbitrarily removed from this group. The two families included in this taxon are monophyletic groups and will be discussed in a later section.

SUBORDER EDAPHOSAURIA

Romer and Price (1940) included in this taxon not only the caseids and edaphosaurs, formerly believed to represent very different types of pelycosaurs, but also placed two other families here, the newly erected Nitosauridae and Lupeosauridae. This association was unfortunately based on a combination of primitive reptilian characters, and a series of derived characters that probably evolved independently in two groups of animals adapting to a herbivorous mode of life.

The primitive reptilian features cited as part of the diagnosis of this suborder include low skull, short antorbital region, smoothly rounded dermal skull-roof, extension of the lacrimal to the naris, low occiput, short paroccipital process of the opisthotic. In the postcranial skeleton the following primitive features were also included in the diagnosis: lack of ventral keel on the centra, lack of excavation on the neural arches, poorly developed rib tubercula, large ventral clavicular blade, short scapula, broad anterior coracoid plate, poorly developed triceps process on the posterior coracoid.

The diagnosis of this suborder is also based on a series of derived characters, seen in both caseids and edaphosaurids, that are clearly related to adaptations to a herbivorous diet: small head, jaw articulation below the level of the tooth row, heavy lower jaw with deep symphysis, loss of caniniform teeth. In the postcranial skeleton the cervical vertebrae are small. Although these derived characters shared between caseids and edaphosaurids appear to suggest close phylogenetic relationships, detailed examination of the characters indicates that they represent convergence. Although both caseids and edaphosaurids have relatively small heads, the skull to trunk ratio in caseids is distinctly smaller than in edaphosaurids (Romer and Price, 1940). Although the jaw articulation is below the level of the tooth row in both, this is achieved in different ways in the two groups. In edaphosaurids the ventral edge of the jugal and quadratojugal is strongly concave and most of lateral surface of the squamosal extends below the lateral temporal opening. In caseids, on the other hand, there is only a slight obtuse angle formed at the end of the maxilla, between the tooth row and the rest of the ventral margin of the cheek. Both the jugal and quadratojugal have straight ventral margins and the lateral surface of the squamosal extends only behind the lateral temporal opening. Although the lower jaw is massive in both caseids and edaphosaurids, the caseid mandible is nearly rectangular in lateral aspect, with deep anterior and posterior ends and little dorsal expansion at the middle. In edaphosaurids, on the other hand, the mandible is greatly expanded both dorsoventrally and mediolaterally at the level of the posterior coronoid. Both anteriorly and posteriorly from this region the edaphosaurid mandible narrows considerably (Fig. 2). Deep mandibular symphyses are found not only in caseids and edaphosaurids, but also in many sphenacodontids. Although both caseids and edaphosaurids lack caniniform teeth, their marginal dentition is quite different. The marginal dentition in edaphosaurids is homodont and the teeth are simple peg-like structures. In caseids, on the other hand, the marginal dentition is heterodont and the posterior maxillary and mandibular teeth are cusped. The small size of the cervical vertebrae is obviously related to the small size of the skull.

Of the four families included in this suborder by Romer and Price (1940), the Caseidae and Edaphosauridae are monophyletic groups and will be discussed in a later section, but the Nitosauridae and Lupeosauridae appear to be invalid taxa. The family Nitosauridae was erected (Romer and Price, 1940) for the reception of *Nitosaurus* and *Mycterosaurus*. *Nitosaurus* is based on a series of poorly preserved and completely unassociated skeletal elements that include limb bones, partial pelvic girdles and some vertebrae, all of which are difficult to identify, plus a poorly preserved skull fragment similar to *Oedaleops* (Langston, 1965; Stovall *et al.*, 1966) which includes an incomplete premaxilla and maxilla and an impression of part of the dentary. The fragments of the pelvis and the humeri and femora correspond closely both in size and shape to those in varanopsid pelycosaurs. These postcranial elements are too large to be associated with the skull fragment, even if the small-headed caseids are used for comparison. Fragments of a varanopsid pelycosaur *Aerosaurus* (Romer, 1937) have been found in the same locality as those of *Nitosaurus*, that is, El Cobre Canyon, New Mexico. The diagnosis of this species is therefore based on elements that probably belonged to two different types of pelycosaur. The diagnosis of *Mycterosaurus*, on the other hand, is based not only upon two partial pelycosaurian skulls and some postcranial fragments, but also upon the pelvic girdle and limb elements of a small temnospondylous amphibian. Study of this material, now in progress, indicates that the name *Mycterosaurus* may be applicable to a small varanopsid pelycosaur.

The family Lupeosauridae was erected (Romer and Price, 1940) for the reception of a single species, *Lupeosaurus kayi*. This species is based on fragmentary postcranial remains. The only features that bar this form from inclusion in the Family Edaphosauridae, according to Romer and Price (1940), are primitive: the apparent absence of an ectepicondylar foramen on the humerus, the absence of cross-bars on the neural spines and the presence of a posterior process of the iliac blade. What is more significant is that almost all of the morphological features cited in the diagnosis of this animal and of the family are primitive features. The only distinguishing features of this species are the presence of tall, narrow, neural spines on the dorsal vertebrae, slightly expanded rugose neural spines on the proximal caudal vertebrae and a supraglenoid foramen on the supraglenoid buttress. These features are not sufficient for the proper diagnosis of a pelycosaurian family and this species must be considered a pelycosaur *incertae sedis*.

The groups in the classification of pelycosaurian reptiles adopted by Romer and Price (1940) and subsequent workers have now been reduced to six valid taxa. These six families will be discussed in a particular order, one which corresponds to the sequence in which they are arranged on the cladogram presented in Fig. 17. This sequence reflects the theory of pelycosaurian relationships proposed in this paper. Revised diagnoses of these families are provided and the derived characters of each family are discussed. Those primitive features in one family that appear in derived form in other families are also listed.

FAMILY EOTHYRIDIDAE Romer and Price 1940

Diagnosis. Small pelycosaurs with the jugals excluded from the ventral margin of the cheek by a long anterior process of the quadratojugal; rounded posterior border of the squamosal, providing poor separation between the cheek and occipital surfaces of this bone.

Taxonomic status. Monophyletic.

Basis of original diagnosis. Eothyris parkeyi Romer.

Discussion. This family was erected as a provisional group by Romer and Price (1940) not only for the reception of the small pelycosaur *Eothyris* known from a single skull, but also for the fragmentary remains of three pelycosaurs of large size, *Stereophallodon ciscoensis* Romer, *Stereorhachis dominans* Gaudry and *Baldwinonus trux* Romer and Price. These pelycosaurs were included within this family because they are "ophiacodont pelycosaurs, primitive in most known regards but paralleling the higher sphenacodonts in the development of much enlarged canines and showing a tendency toward elongation of the vertebral column " (Romer and Price, 1940, pp. 246–247). Both the presence of caniniform teeth and of long vertebrae represent primitive conditions for both pelycosaurs and captorhinomorphs, and the dental and vertebral patterns seen in the advanced ophiacodont *Ophiacodon* are derived conditions. Both *Stereophallodon* and *Stereorhachis* are, as indicated by Romer and Price (1940), ophiacodont pelycosaurs that have retained the primitive dental and vertebral patterns also seen in the early ophiacodont *Archeothyris* (Reisz, 1972, 1975), and are therefore included in the Family Ophiacodontidae. *Baldwinonus trux*, on the other hand, appears to present a more difficult taxonomic problem because the type specimen includes fragments of ophiacodont vertebrae, sphenacodont neural spines and the maxilla of a sphenacodont pelycosaur. The study of this material is now in progress, but it is certain that none of the materials associated with this species is an eothyridid pelycosaur.

The specialized, pointed snout of *Eothyris* was not noticed in either the original or subsequent descriptions because the premaxilla was pushed into the narial opening *post mortem* and the dorsal process was broken at the anterior tip. In his description of another small eothyridid, *Oedaleops campi,* Langston (1965) indicated not only that this species is remarkably similar to *Eothyris*, but also that these pelycosaurs are closely related to the caseids. *Oedaleops* has smaller caniniform teeth than *Eothyris*, similar in size to those found in early captorhinomorphs, but the former eothyridid is peculiar in that the supratemporal projects slightly beyond the posterior edge of the cheek. The exact configuration of the posterior end of the supratemporal in *Eothyris* is not known because of overdevelopment. This posterior projection of the supratemporal may represent another shared derived character, but further materials are needed before this can be verified.

Eothyridids, represented by two species, *Eothyris parkeyi* and *Oedaleops campi*, are closely related to the caseid pelycosaurs as indicated by a series of shared derived characters listed in Table 1 (2) and shown in the cladogram (Fig. 17). Significant shared primitive characters found in both caseids and eothyridids that appear in derived form in all other pelycosaurs are worth noting: the width of the skull is greater than the height, even in the region of the snout, the frontal either does not extend to the orbit (*Eothyris*) or the orbital margin of the frontal is very short (*Oedaleops* and all caseids), and the supratemporal bone is large and broad. In captorhinomorphs, millerosaurs and procolophonids the skull retains the primitive cotylosaur pattern of a low-profiled skull. This primitive condition persists only in caseid and eothyridid pelycosaurs. In other pelycosaurs either the snout (in the edaphosaurs) or the whole skull (in ophiacodonts, varanopsids and sphenacodonts) becomes narrow and deep. Only in cotylosaurs (*Limnoscelis, Diadectes, Seymouria*) and the caseid and eothyridid pelycosaurs is the primitive pattern of the frontal retained: in other pelycosaurs and in early captorhinomorphs one-third of the dorsal orbital margin is formed by the frontal. The broad supratemporal, a primitive reptilian and cotylosaurian character, is modified in captorhinomorphs, diapsids and advanced pelycosaurs, but persists as a broad sheet in caseids and eothyridids.

FAMILY CASEIDAE Williston 1912

Diagnosis. Small to very large pelycosaurs with greatly enlarged external narial openings; anteroposteriorly narrow squamosals; large pineal foramen anterior to the midpoint of the interparietal suture; anterior premaxillary and mandibular teeth the largest in the jaw; small terminal cusps on posterior marginal dentition; maxillary and mandibular dentition reduced in number; jaw articulation slightly below the level of the tooth row. Dorsolaterally expanded ribs; reduced phalangeal formulae ranging from 2–3–4–4–3 to 2–2–2–3–2.

Taxonomic status. Monophyletic.

Basis of original diagnosis. Casea broilii Williston.

Discussion. This family has been reviewed by Olson (1968). He has been able to show not only that such primitive reptilian features as large ventral clavicular blade, large head of the interclavicle, humerus with relatively short shaft and large ends, femur with large internal trochanter, broad epipodials much shorter than the propodials, are present in early small caseids, but also that these features are exaggerated in the later, large forms. In addition, larger caseids are specialized in that they have increased the breadth of the scapula, and have fully formed ectepicondylar foramen on the humerus, and shortened, massive, metapodials and phalanges. The unique construction of the ilium of *Casea*, with a narrow neck and large posterior and anterior flares, may be either the primitive caseid condition, or a specialization of this genus. In later, larger caseids there is either a slight or no anterior flare and only a moderate posterior flare on the iliac blade. Many of the advanced caseid features, peculiar to the larger species, are also found in edaphosaurs. This clearly represents convergence.

FAMILY EDAPHOSAURIDAE Cope 1882

Diagnosis. Small to very large herbivorous pelycosaurs with homodont, peg-like marginal dentition reduced in number; massive crushing dentition on palate and mandible; broad frontals and postfrontals overhanging the orbits; posterior ramus of the postorbital very slender; relatively large lower temporal fenestrae. The jaw articulation far below the level of the tooth row as a result of the posteroventral projection of the squamosal and quadratojugal. Greatly elongated neural spines with cross-bars on presacral vertebrae: slightly rugose neural spines on anterior caudal vertebrae.

Taxonomic status. Monogeneric.

Basis of original diagnosis. Edaphosaurus cruciger (Cope).

Discussion. Edaphosaurid pelycosaurs are advanced over caseids and eothyridids in a number of features shared with ophiacodontids, varanopsids and sphenacodontids, in addition to the diagnostic features listed above. These shared derived characters are listed in Table I (3). The only questionable feature listed there is the position of the pineal foramen. In all caseids the pineal foramen is anterior to the midpoint of the interparietal suture. In eothyridids, however, the pineal foramen appears to be at the midpoint in *Eothyris* and slightly posteriorly of the midpoint in *Oedaleops*. Both of these species are known from single specimens, where the salient portions of the skull-table were overprepared in the former and fractured in the latter. More specimens are therefore needed before this possible character contradiction is resolved.

Significant shared primitive characters found in edaphosaurids are worth noting: as in caseids and eothyridids the nasals are short, roughly equal in length to the parietals, the skull-table is gently convex in cross-section and continuous with the cheeks in the region of the snout, and the jaw articulation is at the level of the occipital condyles.

FAMILY OPHIACODONTIDAE Nopsca 1923

Diagnosis. Small to large carnivorous pelycosaurs with greatly elongated nasals and lacrimals. Axis neural spine hatchet-shaped, with a large anterior process; height of axis neural spine equal to anteroposterior length; neural spine of dorsal vertebrae broad anteroposteriorly, nearly equal to the length of the centrum; transverse processes of dorsal vertebrae webbed; ventral plate of clavicle narrow, head of interclavicle small; lateral pubic tubercle present; epipodials nearly equal in length to the propodials.

Taxonomic status. Monophyletic.

Basis for original diagnosis. Ophiacodon mirus Marsh

Discussion. Many of the diagnostic features of the family are modified or exaggerated in advanced, large ophiacodonts. In *Ophiacodon retroversus*, for example, the nasals, lacrimals and frontals have become more than 300% longer than the parietals. The dorsal vertebrae have become compressed anteroposteriorly in all large species of *Ophiacodon* exaggerating the skull to trunk ratio. (The skull to trunk ratio thus increases from 34% in *Archeothyris* to 64% in *Ophiacodon uniformis*.) The propodials have become massive with broad ends: for example, the distal end of the humerus is formed mostly by the entepicondyle. In the pelvic girdle the lateral pubic tubercle is large in small primitive forms, but becomes reduced in the larger ophiacodonts; it is lost in the largest known ophiacodont, *Ophiacodon retroversus.*

Although Romer and Price (1940) considered ophiacodonts to be the most primitive pelycosaurs, specialized to a piscivorous mode of life, this study indicates that they are advanced over eothyridids, caseids and edaphosaurids in a number of features shared with varanopsid and sphenacodontid pelycosaurs, in addition to the diagnostic features listed above. These shared derived characters, listed in Table I (4), confirm the phylogenetic relationships indicated in the cladogram (Fig. 17). Significant shared primitive characters seen in opiacodontids are worth noting: as in eothyridids, caseids and edaphosaurids, the neural arches are not excavated, the ventral edges of the centra in dorsal vertebrae are rounded, the ventral margin of the scapula is long, the fore limbs are shorter than the hind limbs, and the lateral centrale is equal in size to the medial centrale and articulates with the astragalus.

FAMILY VARANOPSIDAE Romer and Price 1940

Diagnosis. Carnivorous pelycosaurs of moderate size with skull-table deeply incised above the orbits; orbital margins of the frontals formed without lateral extension; posterior process of the maxilla extended beyond the posterior margin of the orbit; ventral margin of the inferior temporal fenestra long and straight, roughly parallel to the ventral margin of the skull.

Taxonomic status. Monophyletic.

Basis of original diagnosis. Varanops brevirostris Williston.

Discussion. Only three genera can be included in this family with confidence: *Varanops, Aerosaurus* and *Varanodon*. All other genera tentatively included by Romer and Price (1940) are either lost to science, are not pelycosaurs, are not varanopsids or are not diagnosable. Varanopsid pelycosaurs are advanced over eothyridids, caseids, edaphosaurids and ophiacodontids in a number of postcranial features (Table 1 (5)). The lack of shared derived cranial characters between varanopsids and sphenacodontids is a reflection of our poor knowledge of the anatomy of the former. Work now in progress on the anatomy of *Aerosaurus* should greatly increase our understanding of this group. Primitive features of varanopsids that appear in derived form in sphenacodontids include the presence of the jugal–quadratojugal suture, contribution of the quadratojugal to the ventral margin of the lateral temporal opening and the lack of a reflected lamina on the angular bone.

FAMILY SPHENACODONTIDAE Marsh 1878

Diagnosis. Small to very large carnivorous pelycosaurs with very large, plate-like dorsal process on the stapes, and anteroposteriorly flattened columella; septomaxilla has both dorsal and posterior processes extending either to the nasal and lacrimal respectively, or to the anterior and ventral processes of the nasal. Epipodials long and slender; hind limbs equal in length to the fore limbs, pes equal in size to manus.

Taxonomic status. Monophyletic.

Basis for original diagnosis. Sphenacodon ferox Marsh.

Discussion. Sphenacodontid pelycosaurs are advanced over all other pelycosaurs in a number of significant cranial features shared with therapsid reptiles, in addition to the diagnostic features listed above. These shared derived characters, shown in Table I (6), indicate that the sphenacodont pelycosaurs are a sister-group to the therapsid reptiles and their descendants, the mammals. Significant share primitive characters of sphenacodontids that appear in derived form in therapsids include the presence of a well-developed dorsal process of the septomaxilla, the presence of large, well-developed basicranial articulating facets and the presence of relatively large interpterygoid vacuities.

ACKNOWLEDGEMENTS

I wish to express my thanks to Ms Diane Scott who prepared the illustrations presented in this paper. These drawings are based mainly on the works of L. Price and the late A.S. Romer whose great contributions provide an indispensable background to the study of pelycosaurs. This work was supported by a grant from the National Research Council of Canada.

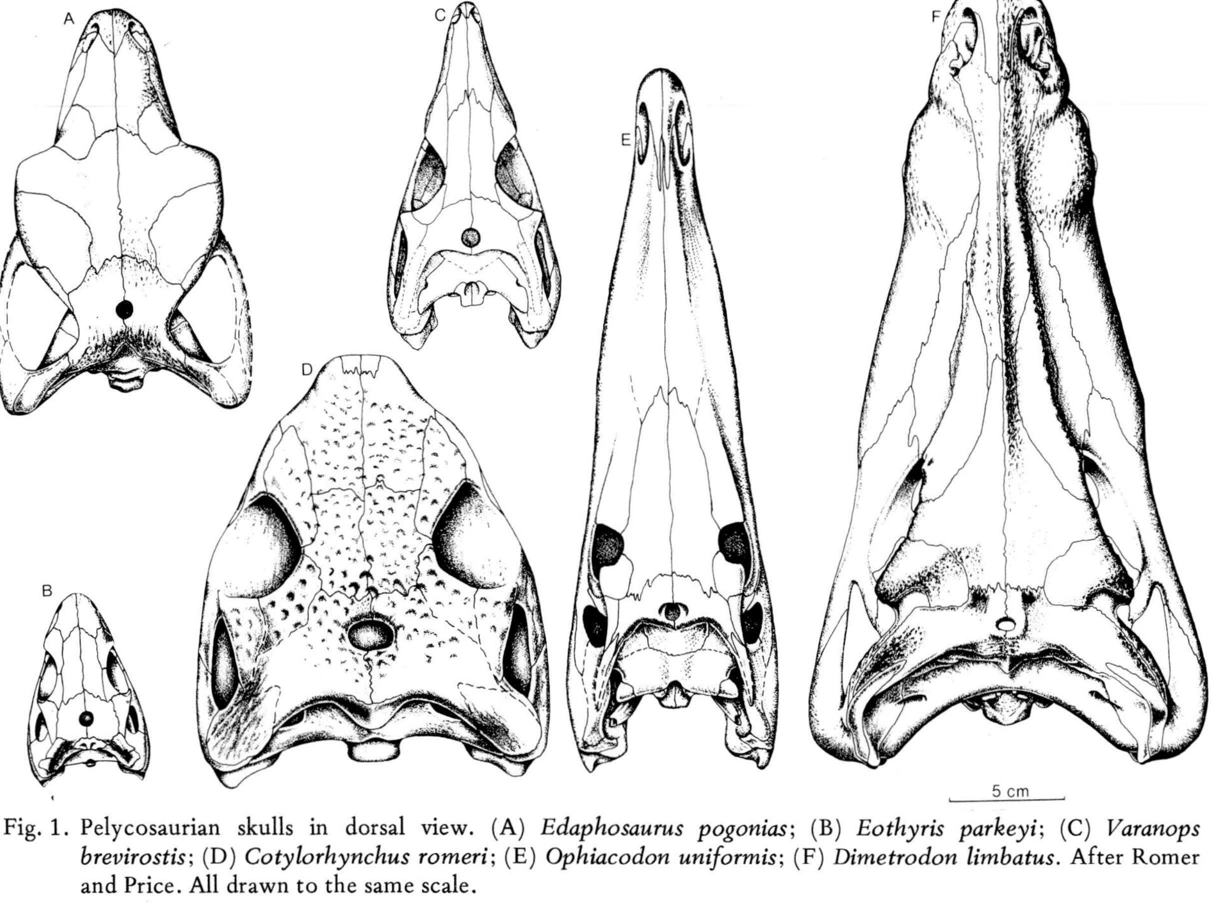

Fig. 1. Pelycosaurian skulls in dorsal view. (A) *Edaphosaurus pogonias*; (B) *Eothyris parkeyi*; (C) *Varanops brevirostis*; (D) *Cotylorhynchus romeri*; (E) *Ophiacodon uniformis*; (F) *Dimetrodon limbatus*. After Romer and Price. All drawn to the same scale.

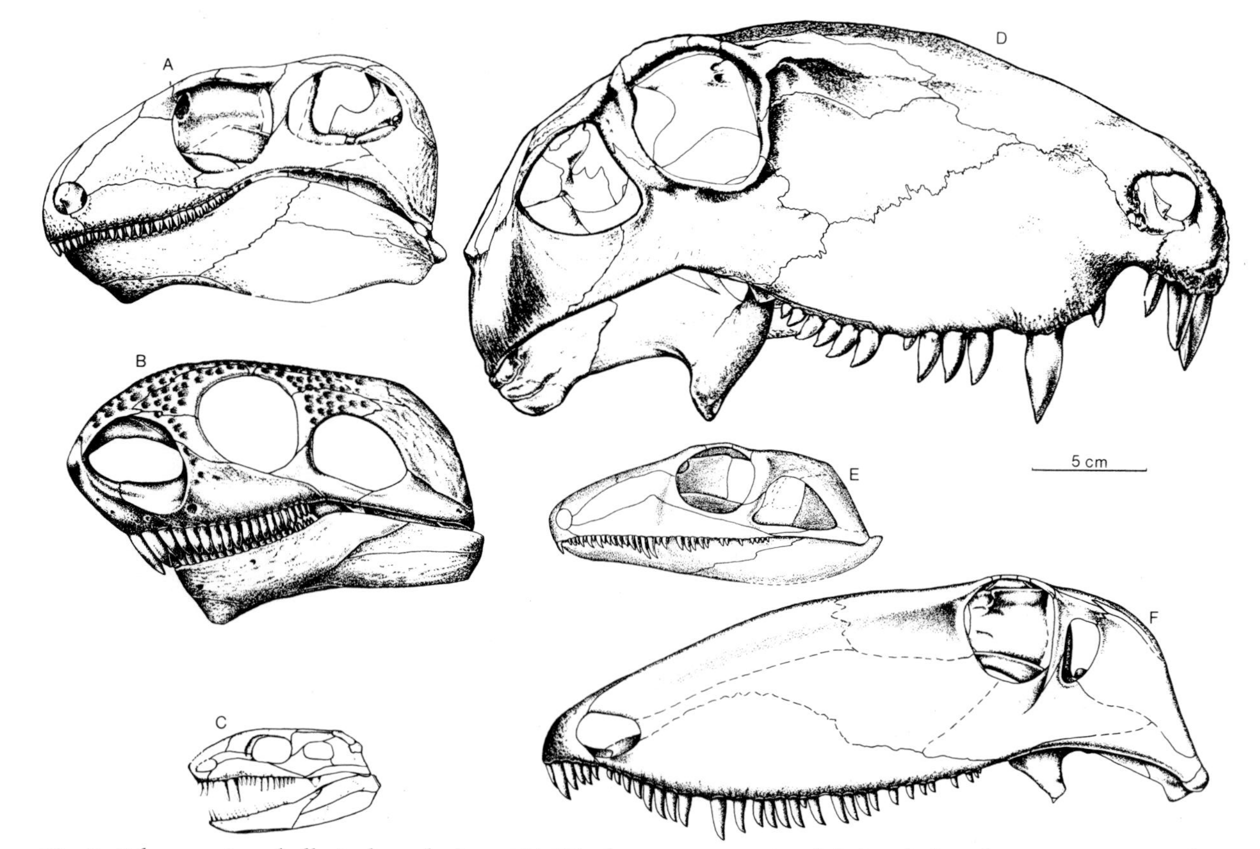

Fig. 2. Pelycosaurian skulls in lateral view. (A) *Edaphosaurus pogonias*; (B) *Cotylorhynchus romeri*; (C) *Eothyris parkeyi*; (D) *Dimetrodon limbatus*; (E) *Varanops brevirostris*; (F) *Ophiacodon uniformis*. After Romer and Price. All drawn to the same scale.

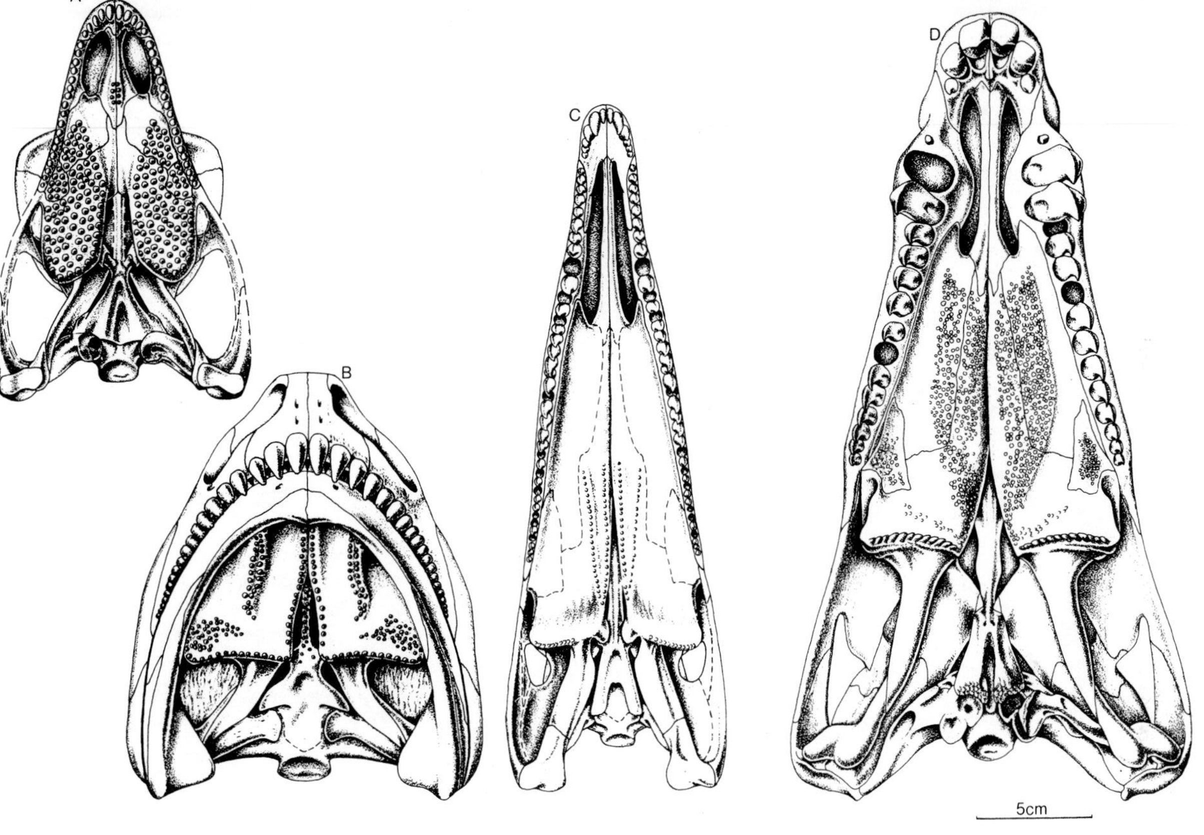

Fig. 3. Pelycosaurian skulls in palatal view. (A) *Edaphosaurus pogonias*; (B) *Cotylorhynchus romeri*; (C) *Ophiacodon uniformis*; (D) *Dimetrodon limbatus*. After Romer and Price. All drawn to the same scale.

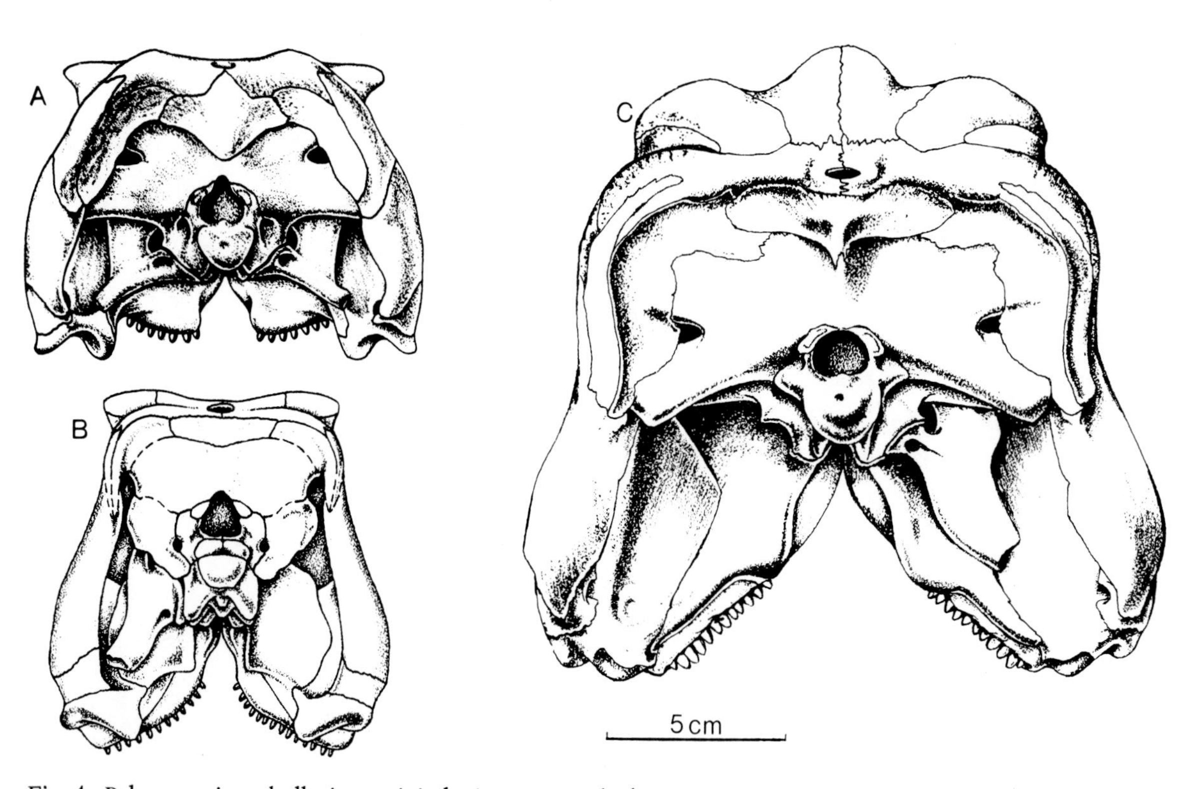

Fig. 4. Pelycosaurian skulls in occipital view. (A) *Edaphosaurus pogonias*; (B) *Ophiacodon uniformis*; (C) *Dimetrodon limbatus*. After Romer and Price. All drawn to the same scale.

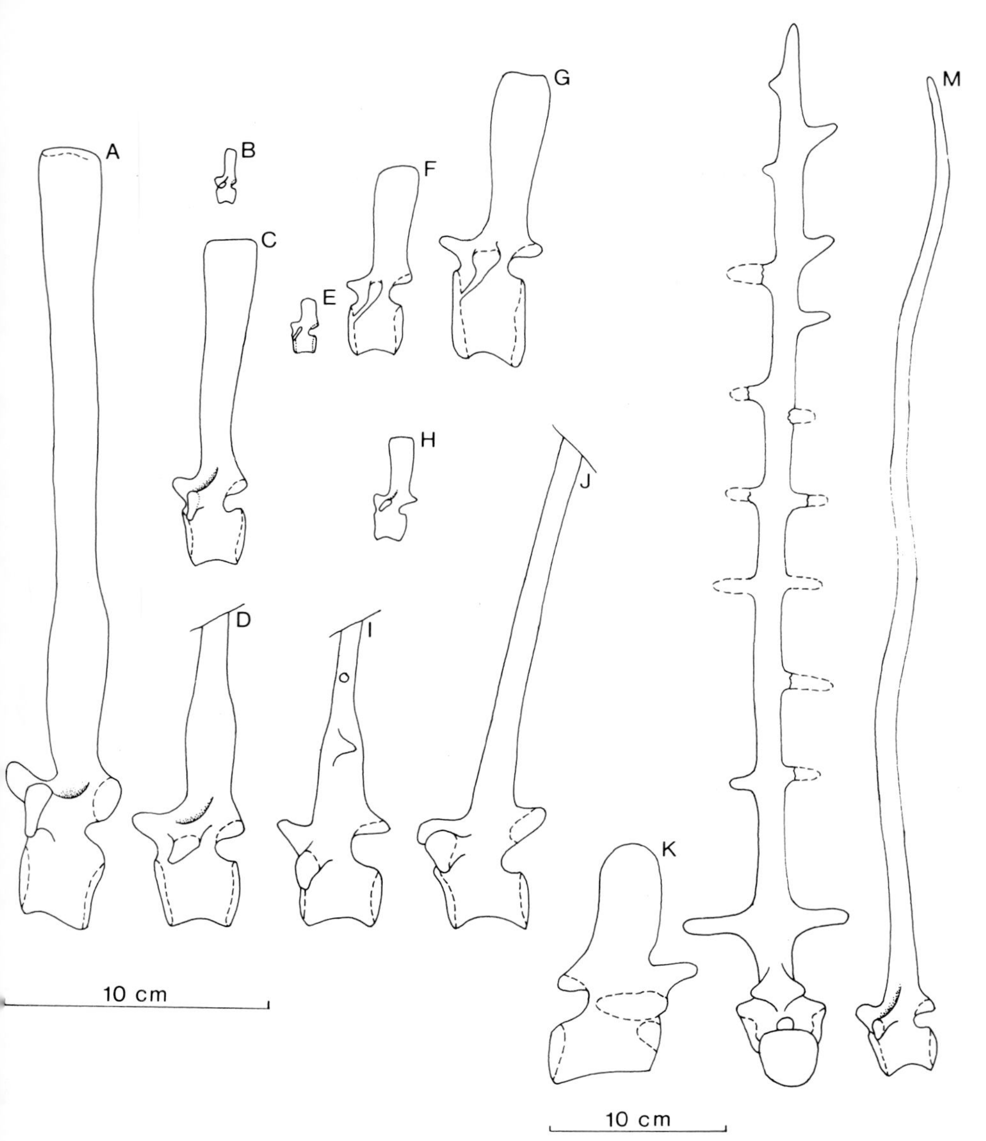

Fig. 5. Pelycosaurian dorsal vertebrae in lateral view. (A) *Ctenospondylus casei;* (B) *Haptodus garnettensis;* (C) *Sphenacodon ferox;* (D) *Dimetrodon limbatus;* (E) *Archeothyris florensis;* (F) *Ophiacodon mirus;* (G) *Ophiacodon retroversus;* (H) *Varanops brevirostris;* (I) *Edaphosaurus boanerges;* (J) *Lupeosaurus kayi;* (K) *Cotylorhynchus romeri;* (L) *Edaphosaurus cruciger (anterior);* (M) *Dimetrodon macrospondylus.* (A) After Romer and Price, (B) after Currie, (C and D) after Romer and Price, (E) after Reisz, (F–J) after Romer and Price, (K) after Stovall, Price and Romer, (L and M) after Case.

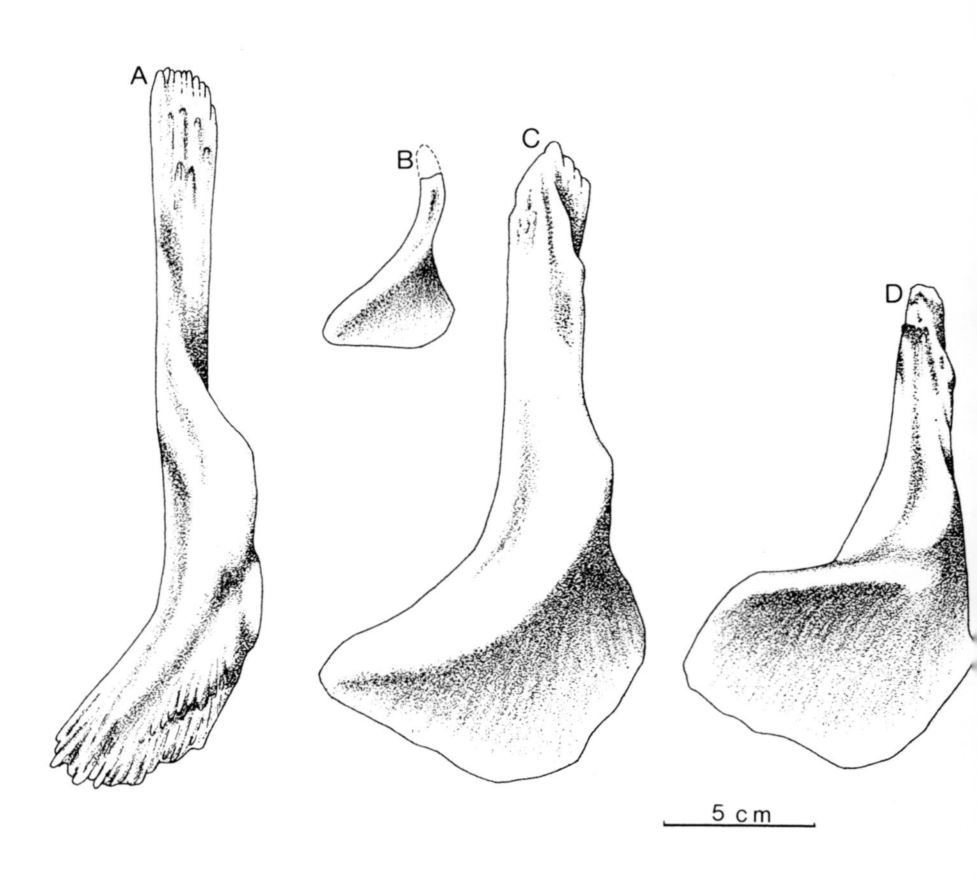

Fig. 6. Pelycosaurian clavicles in outer aspect. (A) *Ophiacodon retroversus;* (B) *Varanops brevirostris;* (C) *Dimetrodon limbatus;* (D) *Edaphosaurus boanerges.* After Romer and Price. All drawn to the same scale.

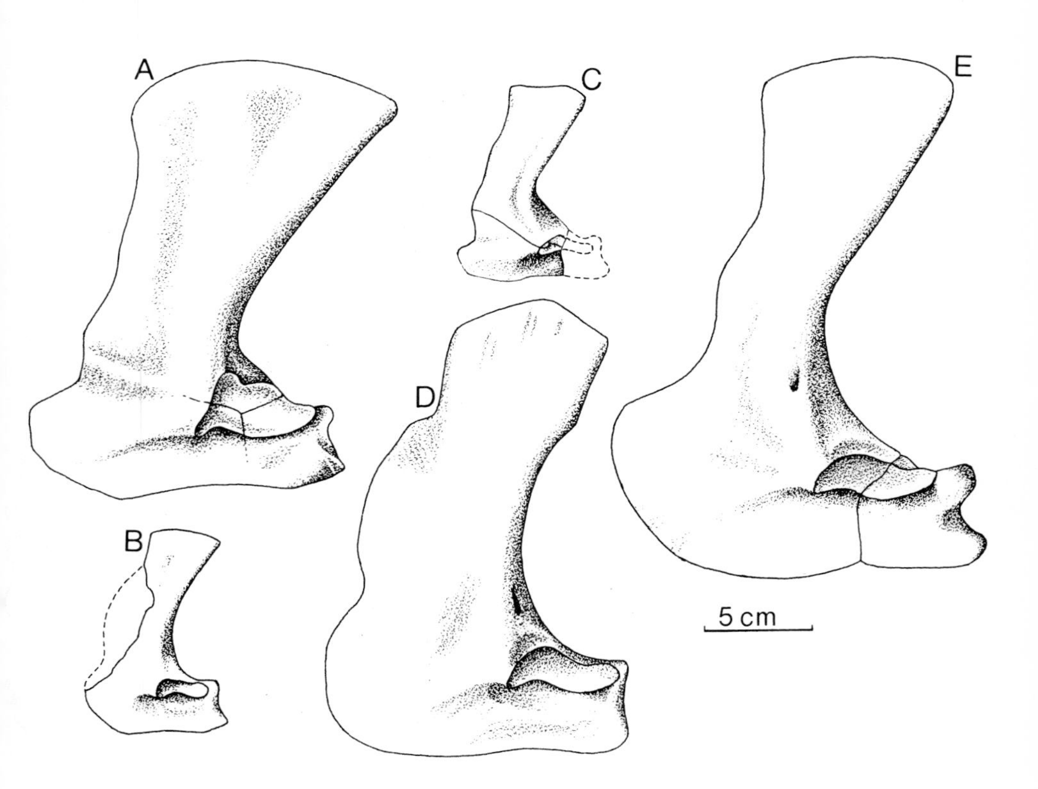

Fig. 7. Pelycosaurian scapulocoracoids in lateral view. (A) *Ophiacodon retroversus;* (B) *Casea broilii;* (C) *Varanops brevirostris;* (D) *Edaphosaurus boanerges;* (E) *Dimetrodon limbatus.* After Romer and Price. All drawn to the same scale.

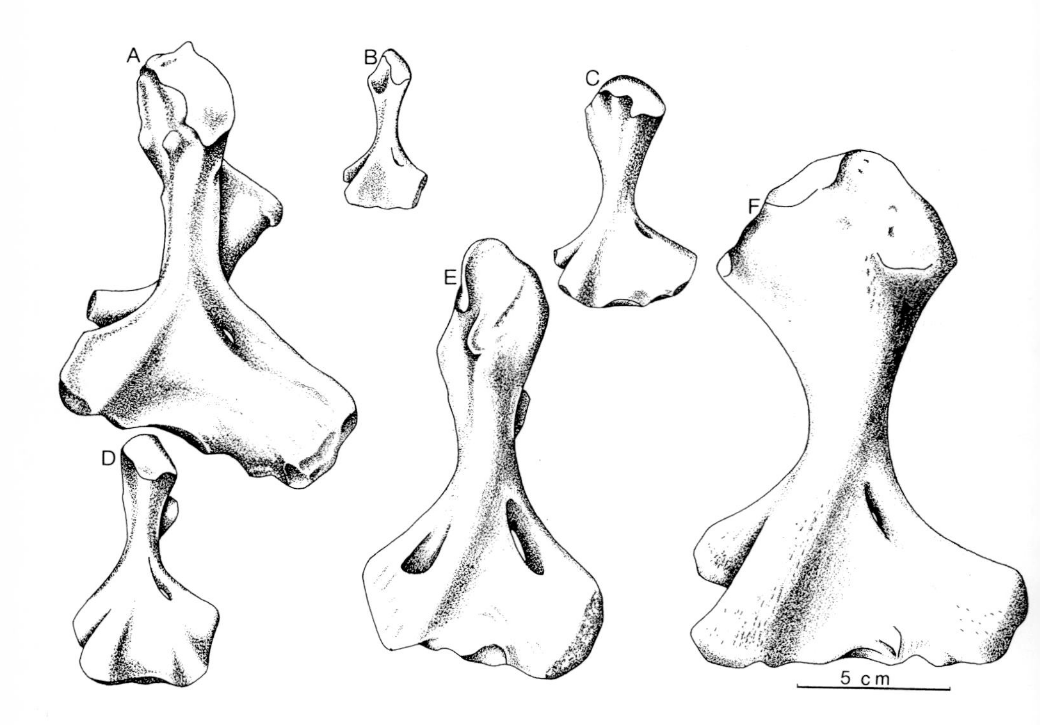

Fig. 8. Pelycosaurian humeri in dorsal view. (A) *Ophiacodon retroversus;* (B) *Oedaleops campi;* (C) *Varanops brevirostris;* (D) *Casea broilii;* (E) *Edaphosaurus boanerges;* (F) *Dimetrodon limbatus.* (A) After Romer and Price, (B) after Langston, (C–F) after Romer and Price. All drawn to the same scale.

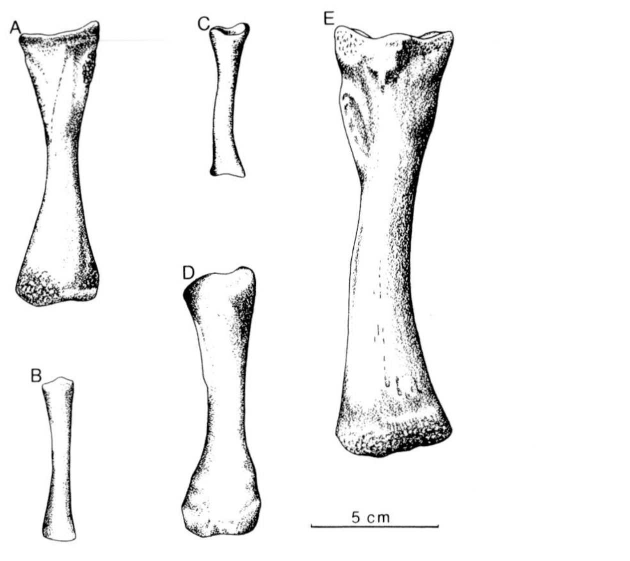

Fig. 9. Pelycosaurian radii in dorsal view. (A) *Ophiacodon retroversus*; (B) *Casea broilii*; (C) *Varanops brevirostris*; (D) *Edaphosaurus boanerges*; (E) *Dimetrodon limbatus*. After Romer and Price. All drawn to the same scale.

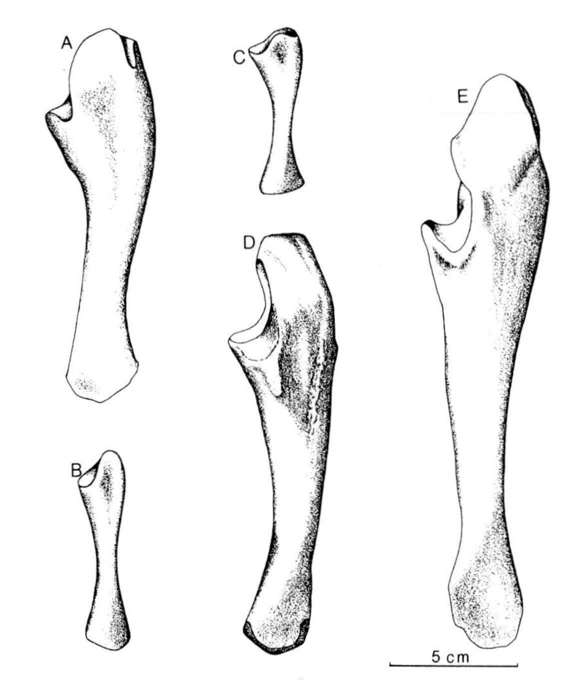

Fig. 10. Pelycosaurian ulnae in dorsal view. (A) *Ophiacodon retroversus*; (B) *Casea broilii*; (C) *Varanops brevirostris*; (D) *Edaphosaurus boanerges*; (E) *Dimetrodon limbatus*. After Romer and Price. All drawn to the same scale.

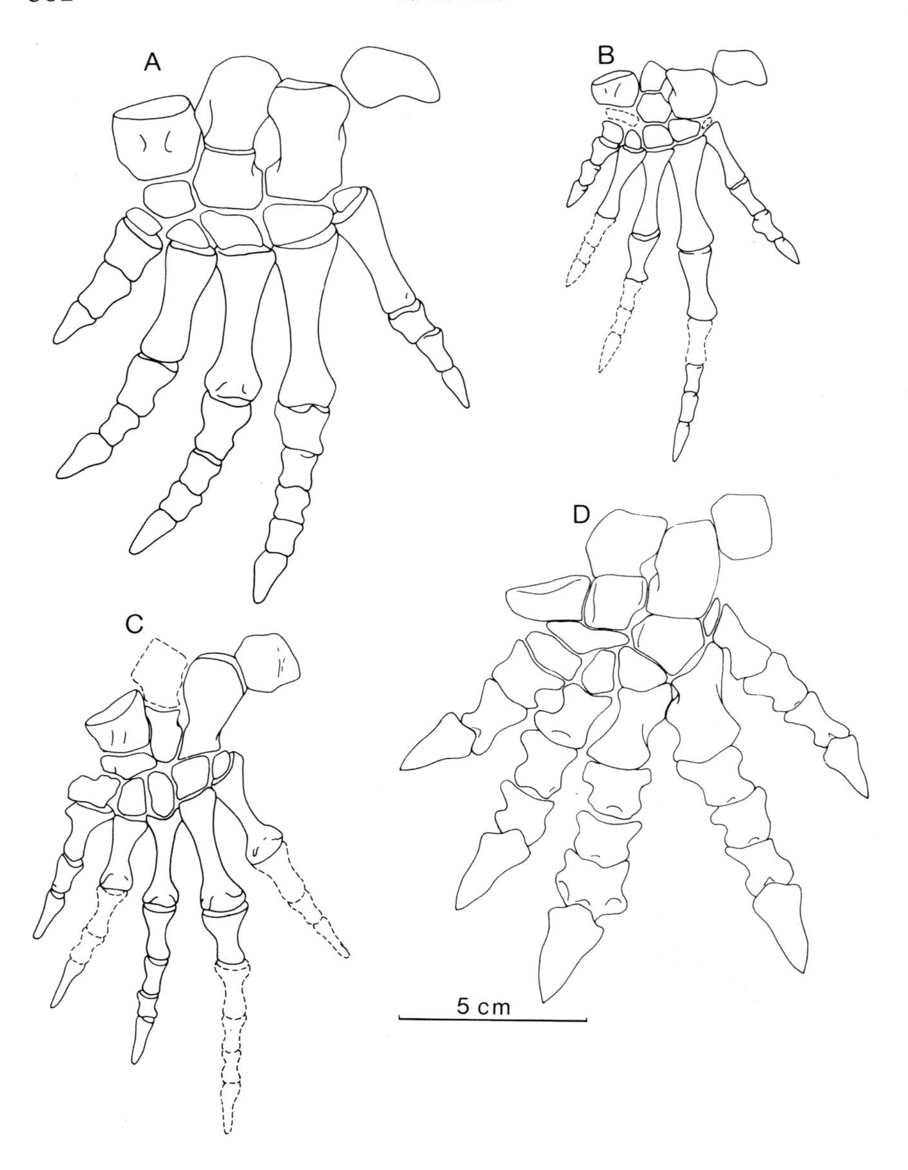

Fig. 11. Pelycosaurian manus in dorsal view. (A) *Ophiacodon retroversus;* (B) *Varanops brevirostris;* (C) *Dimetrodon milleri;* (D) *Casea rutena.* (A–C) After Romer and Price, (D) after Sigogneau-Russell and Russell. All drawn to the same scale.

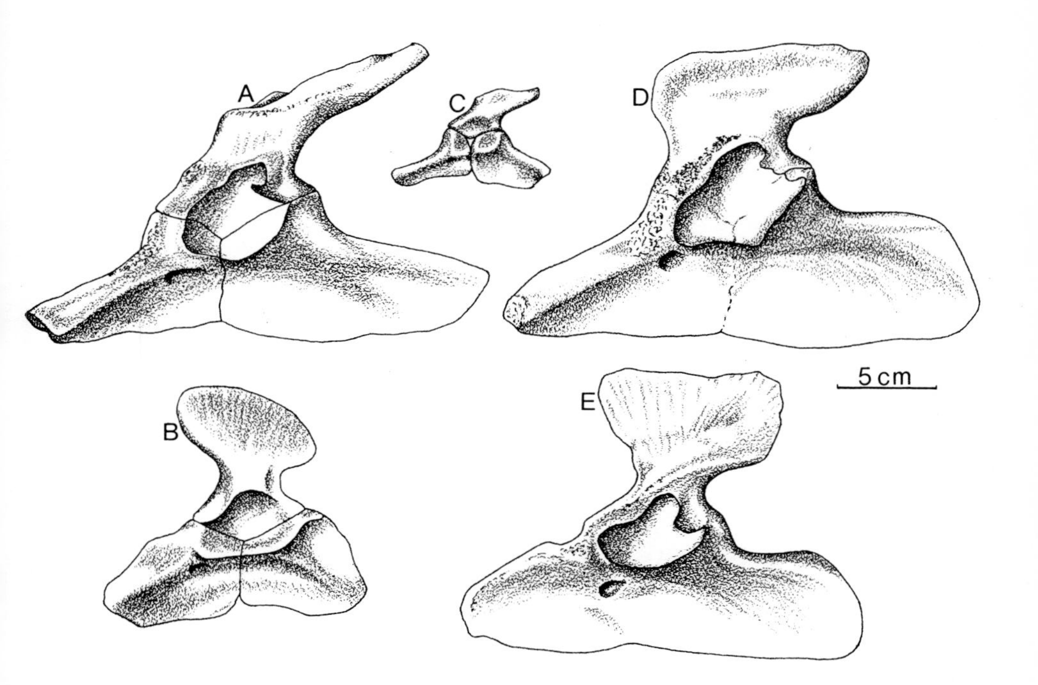

Fig. 12. Pelycosaurian pelves in lateral view. (A) *Ophiacodon retroversus;* (B) *Casea broilii;* (C) *Varanops brevirostris;* (D) *Dimetrodon limbatus;* (E) *Edaphosaurus boanerges.* After Romer and Price. All drawn to the same scale.

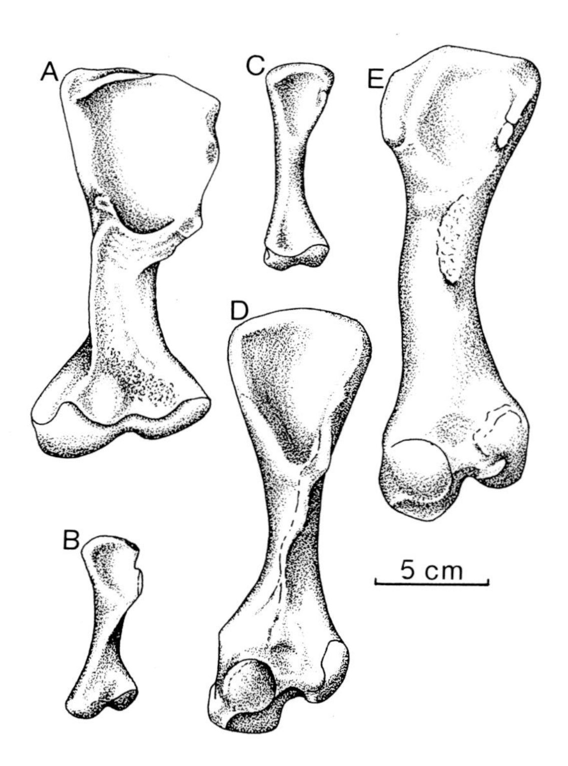

Fig. 13. Pelycosaurian femora in ventral view. (A) *Ophiacodon retroversus;* (B) *Casea broilii;* (C) *Varanops brevirostris;* (D) *Edaphosaurus boanerges;* (E) *Dimetrodon limbatus*. After Romer and Price. All drawn to the same scale.

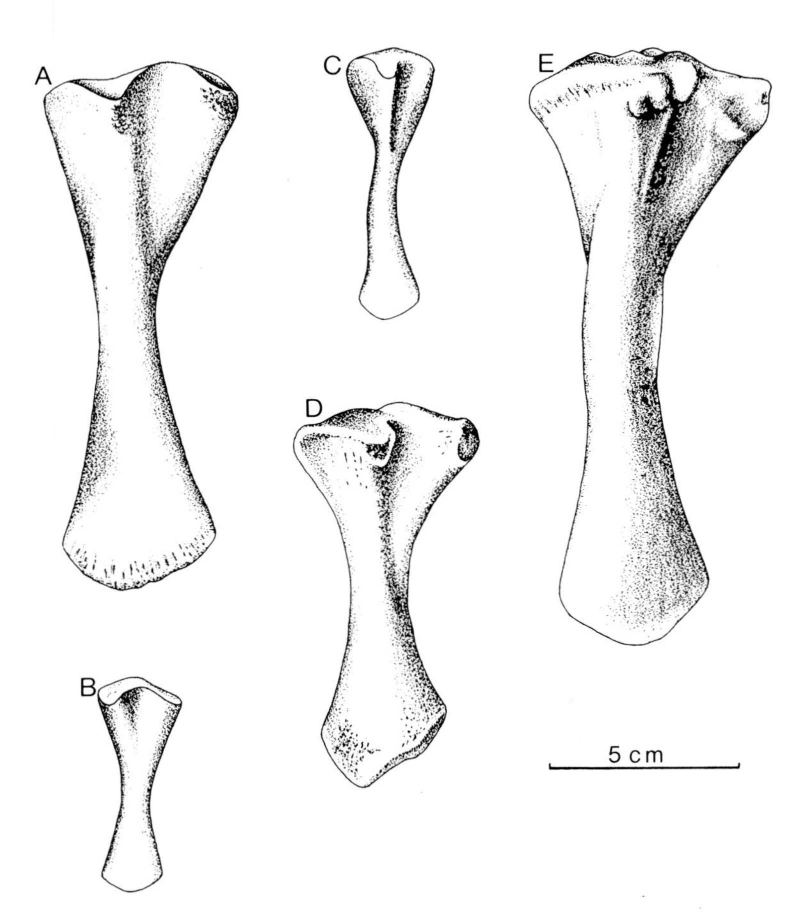

Fig. 14. Pelycosaurian tibiae in dorsal view. (A) *Ophiacodon retroversus;* (B) *Casea broilii;* (C) *Varanops brevirostris;* (D) *Edaphosaurus boanerges;* (E) *Dimetrodon limbatus.* After Romer and Price. All drawn to the same scale.

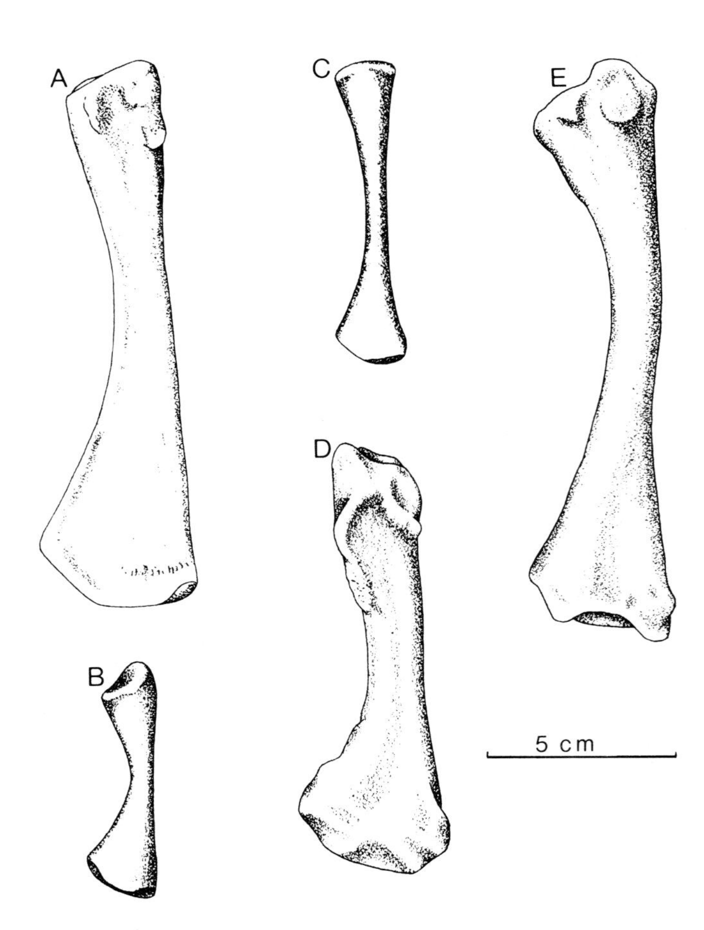

Fig. 15. Pelycosaurian fibulae in dorsal view. (A) *Ophiacodon retroversus;* (B) *Casea broilii;* (C) *Varanops brevirostris;* (D) *Edaphosaurus boanerges;* (E) *Dimetrodon limbatus.* After Romer and Price. All drawn to the same scale.

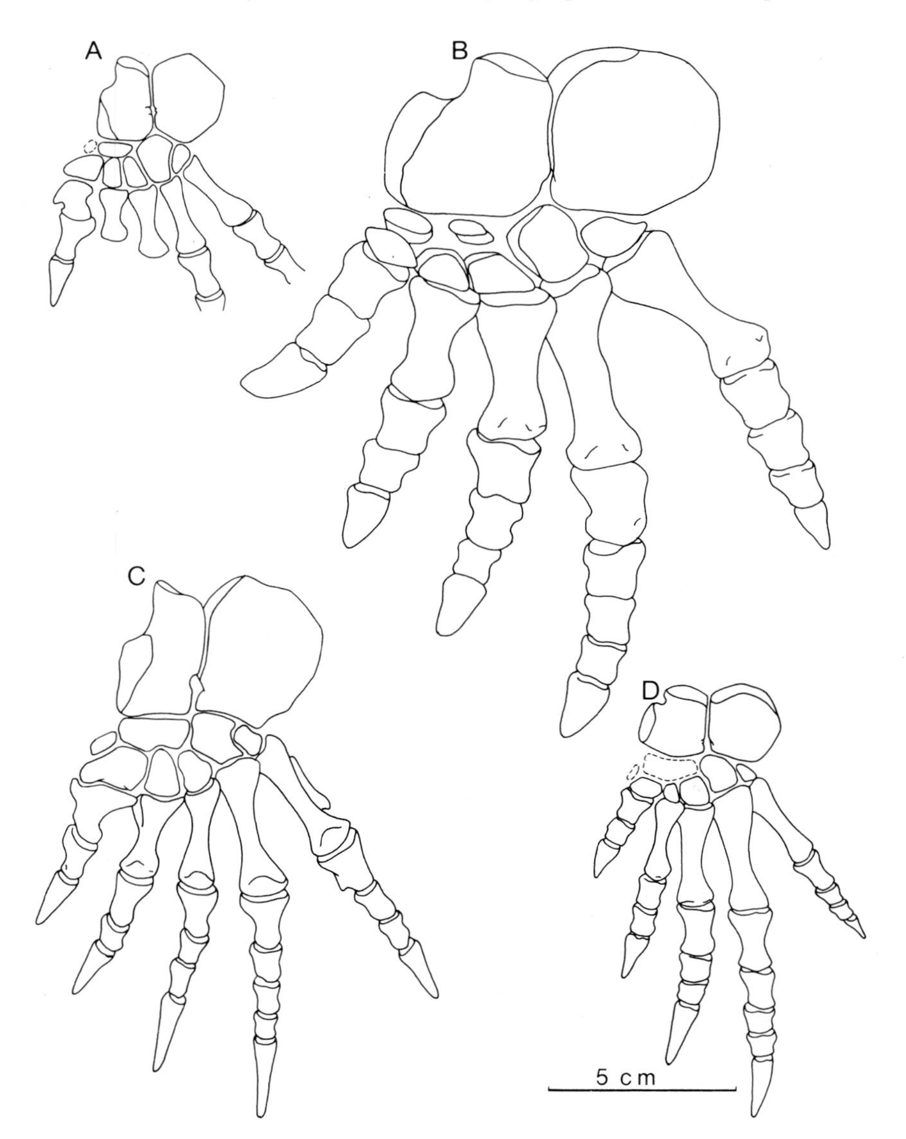

Fig. 16. Pelycosaurian pedes in dorsal view. (A) *Casea broilii;* (B) *Ophiacodon retroversus;* (C) *Dimetrodon milleri;* (D) *Varanops brevirostris.* After Romer and Price. All drawn to the same scale.

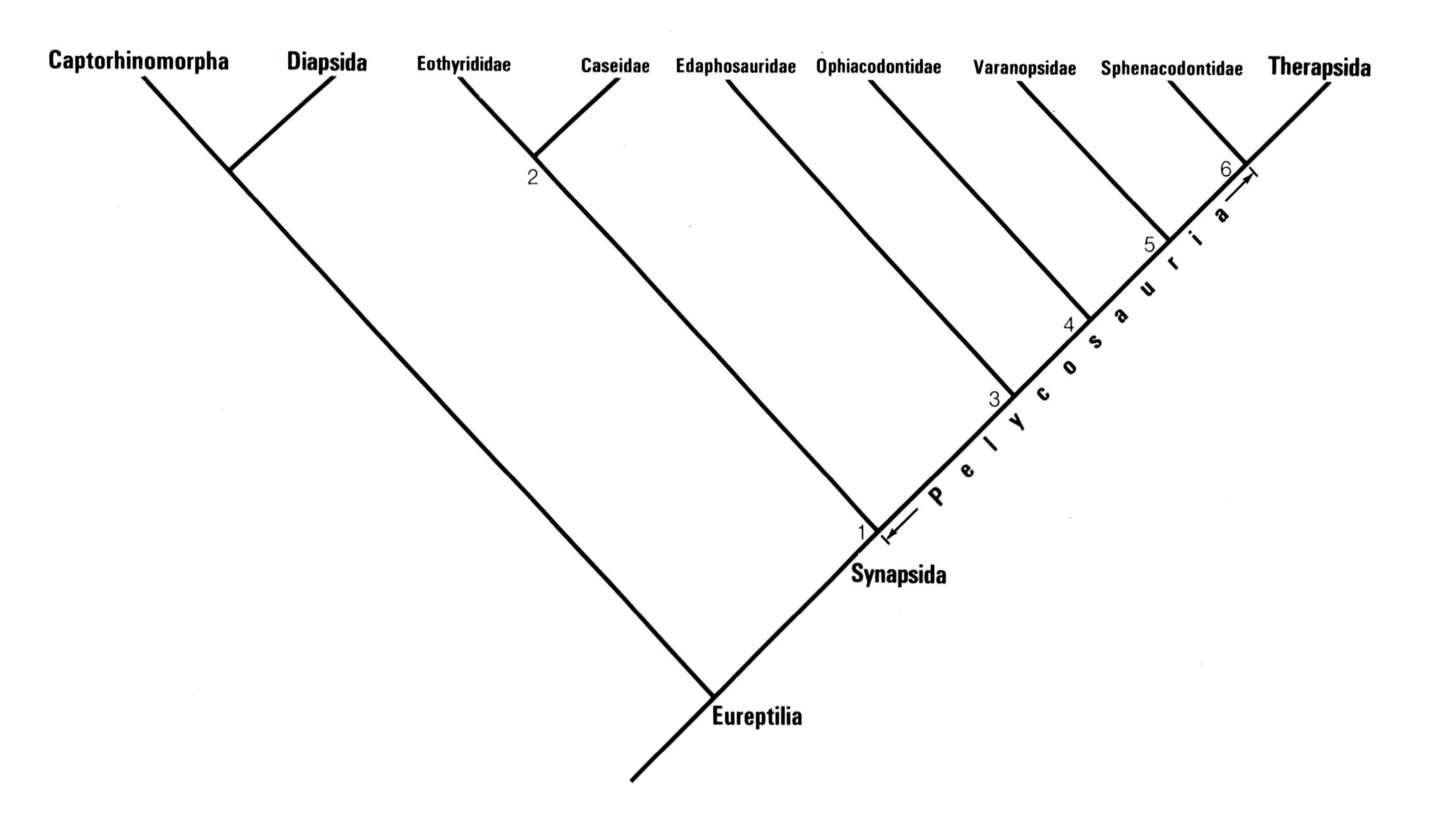

Fig. 17. A theory of relationships for pelycosaurian reptiles. The numbers refer to shared derived characters listed in Table I.

Table I. Shared derived characters for the theory of relationships shown in Fig. 17. The numbers correspond to the nodes numbered in the cladogram.

1. Broad anteriorly tilted occipital plate
   Small post-temporal fenestrae bound by paroccipital process of opisthotic, lateral process of supraoccipital and tabular
   Septomaxilla comprised of a broad base and a dorsal process
   Single median postparietal
   Medial process of the jugal sutured to pterygoid but eptopterygoid present
   (Inferior temporal fenestra)

2. Pointed rostrum formed by anteriorly tilted dorsal process of premaxilla
   Anteroposteriorly elongated external narial opening
   Ventral orbital margin partially formed by maxilla

3. Height of skull greater than the width in the region of the snout
   At least one-third of the dorsal orbital margin formed by the frontal
   Narrow, long supratemporal
   (Pineal posterior to the midpoint of the interparietal suture)

4. Nasal longer than the parietal at the midline
   Skull table sharply separated from the cheek in the region of the snout
   Lateral extension of the skull-table behind the orbits formed by the postfrontal and postorbital
   Jaw articulation far behind the occipital condyle
   Strongly recurved marginal dentition

5. Excavated neural arches
   Ventral keel on dorsal vertebrae
   Supraglenoid foramen on scapular surface
   Tall narrow necked scapula

6. Small quadratojugal
   Reflected lamina of the angular
   Long ventrolaterally directed paroccipital processes of the opisthotic

REFERENCES

Broom, R. (1927). On a new type of mammal-like reptile from the South African Karroo beds *(Anningia megalops)*. *Proc. zoo.. Soc. Lond.* 1927, 227–232.

Broom, R. (1930). On a new primitive theromorph *(Eumatthevia bolli)*. *Am. Mus. Novit.* No. 446, 1–4.

Carroll, R.L. (1964). The earliest reptiles. *J. Linn. Soc. (Zool.)* 45, 61–83.

Carroll, R.L. and Baird, D. (1972). Carboniferous stem reptiles of the family *Romeriidae*. *Bull. Mus. comp. Zool. Harv.* **143**, 321–364.

Case, E.C. (1903). The structure and relationship of the American Pelycosauria. *Am. Nat.* **37**, 85–102.

Case, E.C. (1907). Revision of the Pelycosauria of North America. *Publs Carnegie Instn* **55**, 1–176.

Currie, P.J. (1977). A new Haptodontine sphenacodont (Reptilia: Pelycosauria) from the Upper Pennsylvanian of North America. *J. Paleont.* **51**, 927–943.

Fox, R. C. (1962). Two new pelycosaurs from the Lower Permian of Oklahoma. *Univ. Kansas Publ. Mus. nat. Hist.* **12**, 297–307.

Gow, C.E. (1972). The osteology and relationships of the Millerettidae (Reptilia: Cotylosauria). *J. Zool. Lond.* **167**, 219–264.

Huene, F., von (1902). Uebersicht über die Reptilien der Trias. *Geol. Pal. Abh. Jena.* **6**, 1–84.

Huene, F., von (1908). Neue und verkannte Polycosaurier-Reste aus Europa. *Zentbl. Min. Geol., Pal.* 1908, 431–434.

Huene, F., von (1913). The skull elements of the Permian Tetrapoda in the American Museum of Natural History, New York. *Bull. Am. Mus. nat. Hist.* **32**, 315–386.

Heune, F., von (1925). Ein neuer Pelycosaurier aus der unteren Permformation Sachsens. *Geol. Pal. Abh. Jena.* **18** (N.F. **14**), 215–264.

Langston, W. (1965). *Oedaleops campi* (Reptilia: Pelycosauria). A new genus and species from the Lower Permian of New Mexico and the family Eothyrididae. *Bull. Texas mem. Mus.* **9**, 1–46.

Leidy, J. (1854). On *Bathygnathus borealis*, an extinct saurian of the New Red Sandstone of Prince Edward's Island. *J. Philad. Acad. nat. Sci.* (2)**2**, 327–330.

Matthew, W.D. (1908). A four-horned pelycosaurian from the Permian of Texas. *Bull. Am. Mus. nat. Hist.* **24**, 183–185.

Olson, E. C. (1954). Fauna of the Vale and Choza: 7. Pelycosauria: Family Caseidae. *Fieldiana: Geology* **10**, 193–204.

Olson, E.C. (1962). Late Permian terrestrial vertebrates, U.S.A. and U.S.S.R. *Trans. Am. phil. Soc. (N.S.)* **52**, 3–224.

Olson, E.C. (1965). New Permian vertebrates from the Chikasha formation in Oklahoma. *Circ. Okla. geol. Surv.* **70**, 3–70.

Olson, E.C. (1968). The family Caseidae. *Fieldiana: Geology* **17**, 223–349.

Olson, E.C. and Barghusen, H. (1962). Permian vertebrates from Oklahoma and Texas. *Circ. Okla. geol. Surv.* **59**, 3–58.

Olson, E.C. and Beerbower, J.R. (1953). The San Angelo formation, Permian of Texas, and its vertebrates. *J. Geol.* **61**, 389–423.

Reisz, R. (1972). Pelycosaurian reptiles from the Middle Pennsylvanian of North America. *Bull. Mus. comp. Zool. Harv.* **144**, 27–62.
Reisz, R. (1975). Pennsylvanian pelycosaurs from Linton, Ohio and Nyrany, Czechoslovakia. *J. Paleont.* **49**, 522–527.
Romer, A.S. (1937). New genera and species of pelycosaurian reptiles. *Proc. New Engl. zool. Club* **16**, 89–96.
Romer, A.S. (1945). The late Carboniferous vertebrate fauna of Kounova (Bohemia) compared with that of the Texas redbeds. *Am. J. Sci.* **243**, 417–442.
Romer, A.S. (1956). "Osteology of the Reptiles." University Press, Chicago.
Romer, A.S. (1961). A large ophiacodont pelycosaur from the Pennsylvanian of the Pittsburgh region. *Breviora* No. 144, 1–7.
Romer, A.S. and Price, L.I. (1940). Review of the Pelycosauria. *Geol. Soc. Am. spec. Pap.* No. 28, 538pp.
Sigogneau-Russell, D. and Russell, D.E. (1974). Étude du premier Caside (Reptilia, Pelycosauria) d'Europe occidentale. *Bull. Mus. natn. Hist. nat., Paris* **230**, 145–216.
Stovall, J.W. (1937). *Cotylorhynchus romeri*, a new genus and species of pelycosaurian reptile from Oklahoma. *Am. J. Sci.* **34**, 308–313.
Stovall, J. W., Price, L. I. and Homer, A. S. (1966). The postcranial skeleton of the giant Permian pelycosaur *Cotylorhynchus romeri*. *Bull. Mus. comp. Zool. Harv.* **135**, 1–30.
Vaughn, P.P. (1958a). A pelycosaur with subsphenoidal teeth from the Lower Permian of Oklahoma. *J. Wash. Acad. Sci.* **48**, 44–47.
Vaughn, P.P. (1958b). On a new pelycosaur from the Lower Permian of Oklahoma, and on the origin of the family Caseidae. *J. Paleont.* **32**, 981–991.
Watson, D.M.S. (1914). Notes on *Varanosaurus acutirostris* Broili. *Ann. Mag. nat. Hist.* **13**, 297–310.
Watson, D.M.S. (1916). On the structure of the brain-case in certain Lower Permian tetrapods. *Bull. Am. Mus. nat. Hist.* **35**, 611–636.
Williston, S. W. (1911). "American Permian Vertebrates." University Press, Chicago.
Williston, S.W. (1913). The skulls of *Araeoscelis* and *Casea*, Permian reptiles. *J. Geol.* **21**, 743–747.
Williston, S.W. (1915). A new genus and species of American Theromorpha, *Mycterosaurus longiceps*. *J. Geol.* **23**, 554–559.
Williston, S.W. (1917). *Sphenacodon* Marsh, a Permocarboniferous theromorph reptile from New Mexico. *Proc. nat. Acad. Sci. U.S.A.* **2**, 650–654.
Williston, S.W. and Case, E.C. (1912). The Permo-Carboniferous of northern New Mexico. *J. Geol.* **20**, 1–12.
Williston, S.W. and Case, E.C. (1913a). A description of *Edaphosaurus* Cope. *Publs Carnegie Instn* **181**, 71–81.
Williston, S.W. and Case, E.C. (1913a). Description of a nearly complete skeleton of *Ophiacodon* Marsh. *Publs Carnegie Instn* **181**, 37–59.
Williston, S.W. and Case, E.C. (1913c). A description of *Scoliomus puercensis*, new genus and species. *Publs Carnegie Instn* **181**, p. 60.

20 | Phylogenetic Relationships of the Major Groups of Amniotes

EUGENE S. GAFFNEY

Department of Vertebrate Paleontology, The American Museum of Natural History, New York, U.S.A.

Abstract: Despite the apparent agreement about amniote phylogenies and classifications in most textbooks, hypotheses on this subject are poorly tested and often not even stated explicitly. The most often seen classification is based on temporal fenestrae, a character condemned by most current workers. However, efforts have been made to produce explicit hypotheses, notably those of Goodrich (1916, 1930). For clarity, I have interpreted the problem as an attempt to resolve a three-taxon statement consisting of Testudines, Diapsida and Synapsida, using shared derived characters. These taxa are hypothesized as being monophyletic, with hypotheses of relationship of contained taxa tested by shared derived characters. Goodrich's hypothesis is that Testudines + Diapsida are the monophyletic sister-group of Synapsida, tested by aortic arch and heart morphology. Recent workers (notably Holmes, 1975) have argued that the aortic arch pattern common to Testudines + Diapsida is primitive for all amniotes and therefore incompetent to test this hypothesis. Similarly, Watson's (1954) ear characters are also consistent with the other alternatives. The second alternative using these three taxa is that Testudines is the sister-taxon to a monophyletic Synapsida + Diapsida. Temporal fenestration versus non-fenestration is the classic character consistent with this hypothesis, and the presence of a lower fenestra is common to Synapsida and Diapsida. However,

---

Systematics Association Special Volume No. 15, "The Terrestrial Environment and the Origin of Land Vertebrates", edited by A. L. Panchen, 1980, pp. 593–610, Academic Press, London and New York.

there are contradictions and it is only the presence of another character, Jacobson's organ, that persuades me to prefer this phylogeny over the preceding one.

Parsons (1970) has developed the hypothesis that Jacobson's organ occurs only in non-testudinate amniotes. Even in amniotes that have lost Jacobson's organ in the adult, such as birds and crocodiles, the definitive features are identifiable during development, while in turtles there is no indication of these features embryologically. The third alternative, that Diapsida is the sister-group to Testudines + Synapsida, has not been seriously developed and is ignored here. Therefore, although the second alternative is preferred, it is apparent that both are viable and work is needed to seek out further relevant characters.

INTRODUCTION

In an earlier paper (Gaffney, 1979a) I presented a hypothesis of tetrapod monophyly that I thought could be successfully tested using a cladistic methodology, that is shared derived characters. In the present paper I will try to test some phylogenetic ideas that are closer to God's noblest creature, the turtle. The relationships of turtles to other tetrapods have been the subject of numerous studies but, rather than present a historical review of these ideas, let me simply state that I think there are only two viable alternatives at the present time: namely that turtles are the sister-group of all other amniotes or that synapsids are the sister-taxon to turtles + remaining amniotes.

My reason for choosing these alternatives is discussed below, but it is important at this point to emphasize the difficulties involved. First of all, I have studied turtles for some years now and still feel that hypotheses about the primitive morphotype (in the sense of Schaeffer *et al.*, 1972) for that group have not been adequately tested. Secondly, the adequate testing of hypotheses regarding turtle relationships requires as much knowledge of other amniotes as it does of turtles, and I do not have it. Reliance on the literature is satisfactory to a point, but I sincerely hope that persons well versed in some of the problems presented here will be stimulated to add their expertise to this question.

What I have tried to do in this paper is set up the primary problem in dealing with turtle relationships (i.e. a resolution of the three taxon statements indicated below) and present some criticisms of previous tests of these hypotheses as well as a new test. I have barely scratched the surface of potential areas for new tests of these hypotheses and I do not offer any pretence of completeness. Rather, I am trying to indicate a point of view and a direction for future work that should be fruitful.

The question of turtle relationships has generally been asked in the classic evolutionists' form: "How and from what group did the turtles originate?" or "What are the ancestors of turtles?" Questions of this sort can never be answered adequately, particularly if we insist upon some sort of testability (i.e. falsifiability in the sense of Popper – see Platnick and Gaffney, 1977, 1978a,b) in our hypotheses. Rather the question should be phrased: "What is the closest known relative of turtles?" Questions of this sort can be dealt with more objectively by formulating hypotheses and testing them with distributions of unique characters (for further discussion of this methodology see Gaffney, 1979a,b). I have therefore interpreted the question of turtle relationships as one of resolving a three-taxon statement involving Testudines, Synapsida and Diapsida. Although euryapsids (plesiosaurs, ichthyosaurs and their presumed relatives) are excluded from this discussion, I do not think that the study is seriously weakened. There have been arguments in favour of a close relationship between placodonts and turtles (cf. Jaekel, 1911), but these do not seem to merit much attention at present. In any case, the absence of euryapsids would not alter the validity of an argument including the three taxa mentioned above.

The three possible cladograms involving Testudines, Synapsida and Diapsida are shown in Fig. 1. Of these three, only the first two have been developed to any extent in the literature; I am not aware of any serious argument for the third possibility and will not deal with it further. The first two, however, figure very importantly in amniote relationships and one or the other is implicit (or explicit) in nearly all discussions of "the origin of reptiles". The first hypothesis (A) is the classic arrangement based on temporal fenestrae in which the "anapsid" or fenestra-less forms are seen to be primitive with respect to fenestrated forms. To characterize this hypothesis I have chosen Olson's (1947) term "Eureptilia", even though his argument was quite independent of fenestrae (and later retracted). The monophyly of Testudines + Diapsida (Fig. 1B) is the work of Goodrich (1916, 1930) who adopted Huxley's term "Sauropsida" for the combined taxon. In this paper I will use the terms Eureptilia and Sauropsida solely to refer to respective combinations of taxa without any implications of characters or arguments by the authors of the taxa. I shall begin by discussing monophyly of the major groups

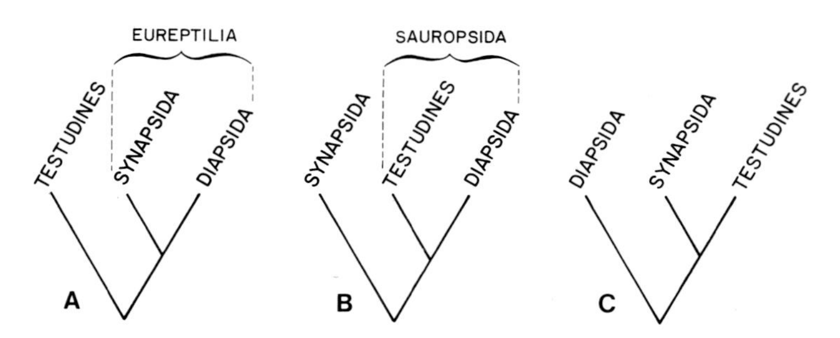

Fig. 1. The three alternative hypotheses (three taxon statements) relating Synapsida, Diapsida and Testudines. Hypotheses A and B are discussed in the text, the hypothesis A is preferred by the Author.

involved, Amniota, Testudines, Synapsida and Diapsida, and will then discuss character distributions which test the two alternative hypotheses.

Among the living tetrapods, the presence of an amniotic egg unites the Mammalia, Aves, Crocodilia, Squamata, *Sphenodon* and Testudines. The strict monophyly of this assemblage has rarely been seriously questioned although some have suggested that either turtles or mammals have "split off" at the amphibian (non-amniote) level, but I have interpreted these arguments as being consistent with amniote monophyly in my sense.

The choice of an outgroup for testing character distibutions for uniqueness with regard to Amniota is a difficult one. As I have argued elsewhere (Gaffney, 1979b) the "outgroup" is really all other taxa except the one being tested for monophyly, but, in order to make these comparisons easier, some assumptions about relationships should be made. As far as amniotes are concerned, comparisons among the living tetrapods are relatively limited; the Lissamphibia is the only extant group of non-amniote tetrapods available. Fossil forms, however, provide a wealth of potential near relatives and for the outgroup we must use all non-amniote tetrapods ("Amphibia").

TESTUDINES

Elsewhere (Gaffney, 1975) I have presented an argument for turtle monophyly as well as a cladogram for included taxa. I do not know of any author arguing against chelonian monophyly and I accept this hypothesis here. However, it is easy to argue for turtle monophyly because they possess so many unique features, particularly in the postcranium. It is harder to find characters in common with other amniotes, particularly advanced ones that are not simply amniote characters.

Although it has been suggested (Gaffney and McKenna, 1979) that turtles may be closely related to captorhinids, whether or not this is the case does not affect the resolution of the three-taxon statement discussed here. Uniquely among amniotes, captorhinids share with turtles the absence of the ectopterygoid but this is the only feature with this distribution that I have found. (Although mammals lack an ectopterygoid other synapsids do not and its presence must be considered a primitive synapsid feature.) Turtles and captorhinids also have relatively large post-temporal fenestrae and well-developed medial jugal processes, and both lack the tabular, but these features occur elsewhere in the Amniota. However, if we do assume that captorhinids are the sister-taxon to turtles, and modify the three-taxon statement to captorhinids, diapsids and synapsids, how might this be resolved? Do captorhinids and diapsids have any advanced features in common? Both groups have relatively well developed post-temporal fenestrae in contrast to synapsids, but a comparison of the diapsid *Petrolacosaurus* (Reisz, 1977) with various "romeriids" (Carroll and Baird, 1972) does not show an unusually great difference in occipital morphology. I am not aware of any other features that have this distribution. Similarly, I do not know of any advanced features common to captorhinids and synapsids. Therefore, for the present, inclusion or exclusion of the Captorhinidae as the sister-taxon to Testudines does not affect the resolution of the three taxon statements indicated above.

DIAPSIDA

The monophyly of this group (see Fig. 2) is classically based on the presence of two temporal fenestrae, the dorsal one being surrounded by the parietal, postorbital and squamosal, and the lower one by the postorbital, squamosal, jugal and quadratojugal. There are some difficulties in the use of this character, the principal one being the possibility that it is really two characters. The lower fenestra also occurs throughout the Synapsida and is often considered the defining character of that group. Euryapsids have the upper fenestra without

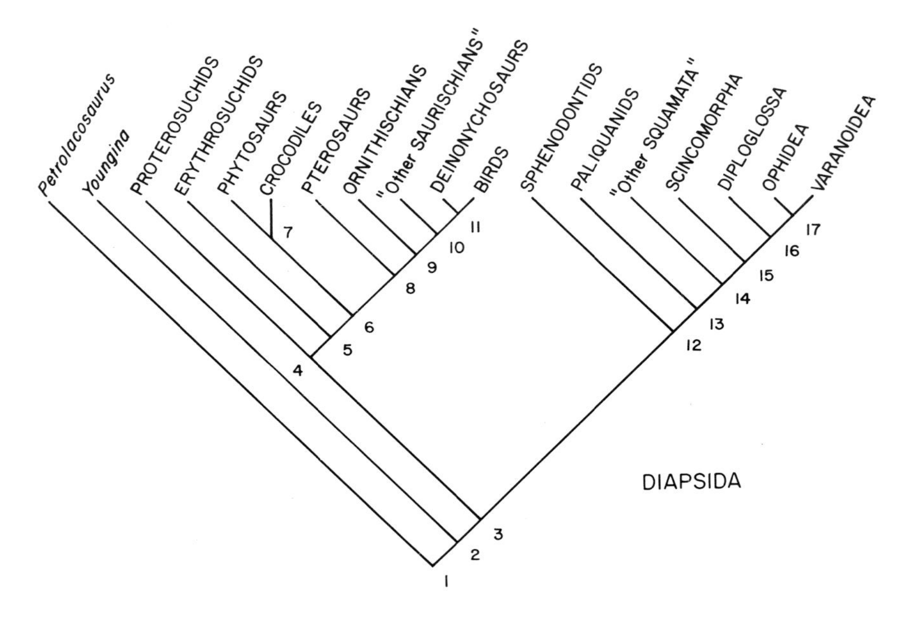

Fig. 2. A hypothesis of relationships for selected taxa of Diapsida. Not all known diapsid taxa are indicated. The numbers refer to characters listed in Table I and are advanced here to show that a testable scheme for diapsid relationships exists. It is not intended to represent a thoroughly tested phylogeny.

the lower one (although there have been suggestions to the contrary – R. L. Carroll, personal communication; see also Kuhn-Schnyder, 1963), the result being that if either fenestra is used as a synapomorphy, diapsids are included in both groups, but neither fenestra defines an exclusive group with regard to the other fenestra. The problem might be that we are looking at the hole rather than the doughnut following Romer's (1956, p. 467) suggestion. The diapsids were originally referred to as "two-arched reptiles" and the presence of the upper arch is unique to this group. That is, a temporal arch formed by a triradiate postorbital and a triradiate squamosal is unique to diapsids.

Another diapsid derived character is the presence of paired suborbital fenestrae formed by the palatines, maxillae and ectopterygoids. These are found in all groups having the diapsid temporal condition. Again, it is the structure of the surrounding

Table I. Synapomorphies hypothesized for a cladogram of selected Diapsida (numbers correspond with those in Fig. 2)

1. Upper and lower temporal arch formed by triradiate postorbital and triradiate squamosal; suborbital fenestrae formed by palatines, maxillae and ectopterygoids (also found in *Araeoscelis*, which has only an upper temporal fenestra) (Reisz, 1977).
2. Quadrate partially exposed posteriorly (Carroll, 1977).
3. Quadrate completely exposed posteriorly and to some extent laterally; hooked fifth metatarsal; ossified sternum (Carroll, 1977).
4. Antorbital fenestra.
5. Mandibular fenestra; tarsus with two distal and two proximal elements; calcaneum with moderate tuber (Cruickshank, 1972).
6. Ossified laterosphenoid.
7. "Crocodyloid tarsus": socket on calcaneum, ball on astragalus (Sill, 1974; Chatterjee, 1978).
8. Astragalus and calcaneum rounded cap fixed on tibia and fibula (Bakker and Galton, 1974; Wellnhofer, 1978).
9. Perforate acetabulum; femoral head with distinct neck; manus digits II and III elongate, IV and V small (Bakker and Galton, 1974).
10. The "Saurischia" is recognized only on the basis of characters that are primitive for Dinosauria, but in the absence of alternatives, I will use it here.
11. Manus digit III small.
12. Ossified epiphyses in pectoral girdle and ulna–ulnare; determinate growth; mesotarsal joint; fused astragalus and calcaneum (but latter is absent in paliguanids) (Carroll, 1977).
13. Fenestrated pectoral girdle; quadratojugal absent or extremely small; quadrate free from lower temporal arch (Carroll, 1977).
14. Postparietal, tabular absent; lacrimal small or absent; supratemporal large; parietals fused and with well-developed descending processes (Carroll, 1977).
15. Divided tongue tip; complex body musculature (Camp, 1923).
16. Divided tongue with sheath (Camp, 1923).
17. Intramandibular hinge; posterior prolongation of nares.

bones that is the character, not the hole *per se*. The coincidence, however, is not complete. *Araeoscelis* (Vaughn, 1955) has typical suborbital fenestrae but only the upper temporal opening.

Within the diapsids, Squamata + Archosauria + Rhynchocephalia may be considered as a monophyletic taxon on the basis of the ear region. The quadrate in these forms is exposed posteriorly for its entire height and is usually somewhat concave anteriorly. The stapes is thin and light, and articulates with a tympanic membrane posterior to the quadrate. In what is here hypothesized as the primitive condition, seen in Romeriidae, Captorhinidae and pelycosaurs, the squamosal has an occipital flange that covers most of the quadrate

Table II. Synapomorphies hypothesized for a cladogram of selected Synapsida (numbers correspond with those in Fig. 3)

1. Single postparietal on occipital plate which is tilted anteriorly; dorsum sellae consists of prootic; septomaxilla has broad base and dorsal process (Reisz, Chapter 19 of this volume).
2. Reflected lamina on mandible (Kemp, 1972a,b); lacrimal does not enter nares.
3. Fused basipterygoid joint; medial flanges of pterygoid sutured to braincase; adductor musculature attachment area on external surface of skull; supratemporal absent; reflected lamina relatively large in comparison to (2); no lacrimal–nasal contact (Boonstra, 1963).
4. Relatively large adductor musculature attachment sites in comparison to (3); zygomatic arch bulges laterally.
5. Relatively small quadrate lying in recess on squamosal; relatively small quadratojugal; all in comparison to (4).
6. Large suborbital (palatal) fenestrae; narrow intertemporal region caused by well-developed adductor musculature attachment areas.
7. Postfrontal absent; palatal teeth absent; well-developed palatal vault (Kemp, 1972a,b).
8. Quadratojugal in a slit in squamosal; palatal fenestrae closed; large epipterygoid with trigeminal notch (Kemp, 1972a,b).
9. Divided occipital condyles; well-developed secondary palate; coronoid process with masseter attachment area on lateral surface; multi-cusped teeth (Barghusen, 1968; Hopson and Crompton, 1969).
10. Angle on dentary; surangular–squamosal articulation; postdentary elements rod-like.
11. Occipital portion of squamosal emarginated dorsally.

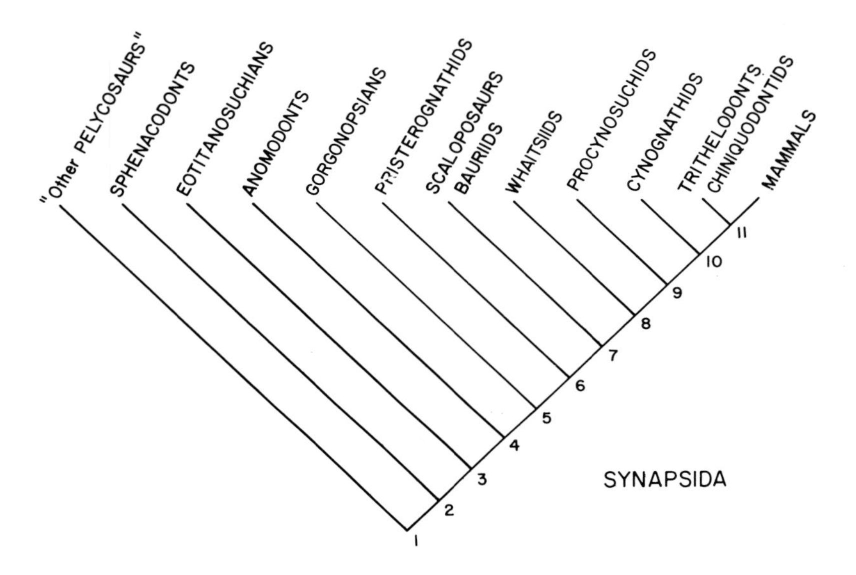

Fig. 3. A hypothesis of relationships for selected taxa of Synapsida. Not all known synapsid taxa are indicated. The numbers refer to characters listed in Table II and are advanced here to show that a testable scheme for synapsid relationships exists. It is not intended to represent a thoroughly tested phylogeny.

posteriorly and the stapes is a heavy element that articulates with a depression on the quadrate. *Petrolacosaurus* (Reisz, 1977) has the diapsid temporal condition, characteristic suborbital fenestrae and an occipital flange on the squamosal as well as a massive stapes. In *Youngina*, also a diapsid with suborbital fenestrae, the occipital flange on the squamosal is small but still present dorsally (the stapes is unknown). Another feature absent in *Petrolacosaurus* and uniting Rhynchocephalia, Squamata and Archosauria is the development of ventral processes on the parietals (Reisz, 1977).

SYNAPSIDA

The synapsid temporal condition is a single fenestra primitively formed by the postorbital, squamosal and jugal bones and is morphologically the same as the lower temporal opening in diapsids. If one uses the presence of a lower temporal fenestra as a synapomorphy it is consistent only with Diapsida + Synapsida (including all pelycosaurs) and does not uniquely define Synapsida. However, Reisz (Chapter 19 of this volume) has suggested three other characters (Table II) which he argues are synapomorphies for Synapsida, and I will accept them here.

ALTERNATIVE 1: MONOPHYLY OF DIAPSIPA + TESTUDINES (SAUROPSIDA)

1. Ear Region

Watson (1954) (see also Vaughn, 1955) advanced the idea that the structure of the ear region in amniotes supported Goodrich's Sauropsida–Theropsida hypothesis. The sauropsid ear region would be characterized by a high quadrate, open posteriorly by the absence of a squamosal flange, and concave anteriorly. The stapes is thin and articulates with a tympanic membrane posterior to the quadrate. The theropsid condition is that seen in captorhinomorphs and pelycosaurs in which the quadrate is relatively low, covered by an occipital flange of the squamosal and has a pit for the reception of the distal end of a massive stapes. Watson argued that the sauropsid condition was primitive because it retained the "otic notch" of anthracosaurs while the theropsid condition was more advanced. Much of this argument involved the presence or absence of the tympanic membrane, but if we direct our attention to the preserved morphology and set aside the unresolvable question of characters that are not preserved in fossils, the alternatives seem clearer. A comparison of "amphibian" and amniote quadrates suggests that the sauropsid condition is derived and that the theropsid condition is primitive. The otic notch of anthracosaurs and temnospondyls is formed by the squamosal, supratemporal and tabular, and the quadrate is not involved. It is only in *Diadectes*, a form of dubious relationships, that the quadrate has sauropsid features. Furthermore, if we accept my interpretation of Reisz' (1977) hypothesis, that is, that *Petrolacosaurus* is the sister-taxon to remaining diapsids, then the "sauropsid" ear character does not delimit diapsids. *Petrolacosaurus* has a typical romeriid-captorhinid quadrate and stapes morphology suggesting that this condition, the "theropsid" ear type of Watson, is primitive for diapsids and that the posteriorly open quadrate of turtles and more advanced diapsids was independently derived.

The ear morphology, then, I interpret as follows: the "theropsid" condition is primitive for amniotes, the "sauropsid" condition is a parallelism acquired separately by turtles and diapsids.

2. Aortic Arches

Goodrich (1916) presented evidence concerning the heart and aortic arches of tetrapods that supported the monophyly of Testudines + Diapsida (Sauropsida). Goodrich argued that in the Sauropsida the carotid arteries branch off the right fourth aortic arch while in the Theropsida the carotids branch off the left fourth aortic arch. Neither condition could be considered primitive with respect to the other, and Goodrich suggested that the common ancestor would have both right and left aortic arches present. Although Goodrich's (1916, p. 271) argument is essentially derivationist:

> . . . it does not seem possible for a heart which has once started, so to speak, to evolve along the Sauropsidan line to change its course and revert to the Theropsidan . . .

his characters are nonetheless readily interpreted as good shared derived characters for each group.

More recent work on heart morphology disputes Goodrich's conclusions with regard to heart and aortic arch phylogeny. Holmes (1975, p. 212) has provided the most direct criticism:

> The reptilian heart is a much misinterpreted organ. It has been inaccurately described in many comparative anatomy textbooks . . . This misconception appears to have resulted primarily from the influence of the usually reliable E. S. Goodrich, who published a wholly inaccurate diagram of the reptilian heart . . .

This criticism of Goodrich's figure (reproduced here as Fig. 4C) is justified; the heart morphology is incorrect. This does not mean that Goodrich was ignorant of vertebrate heart morphology; the remainder of Chapter X in his 1930 *magnum opus* shows his expertise and understanding of this topic.

Holmes' most important criticism (1975, p. 225), however, is with regard to Goodrich's statement of the theropsid aortic arch condition.

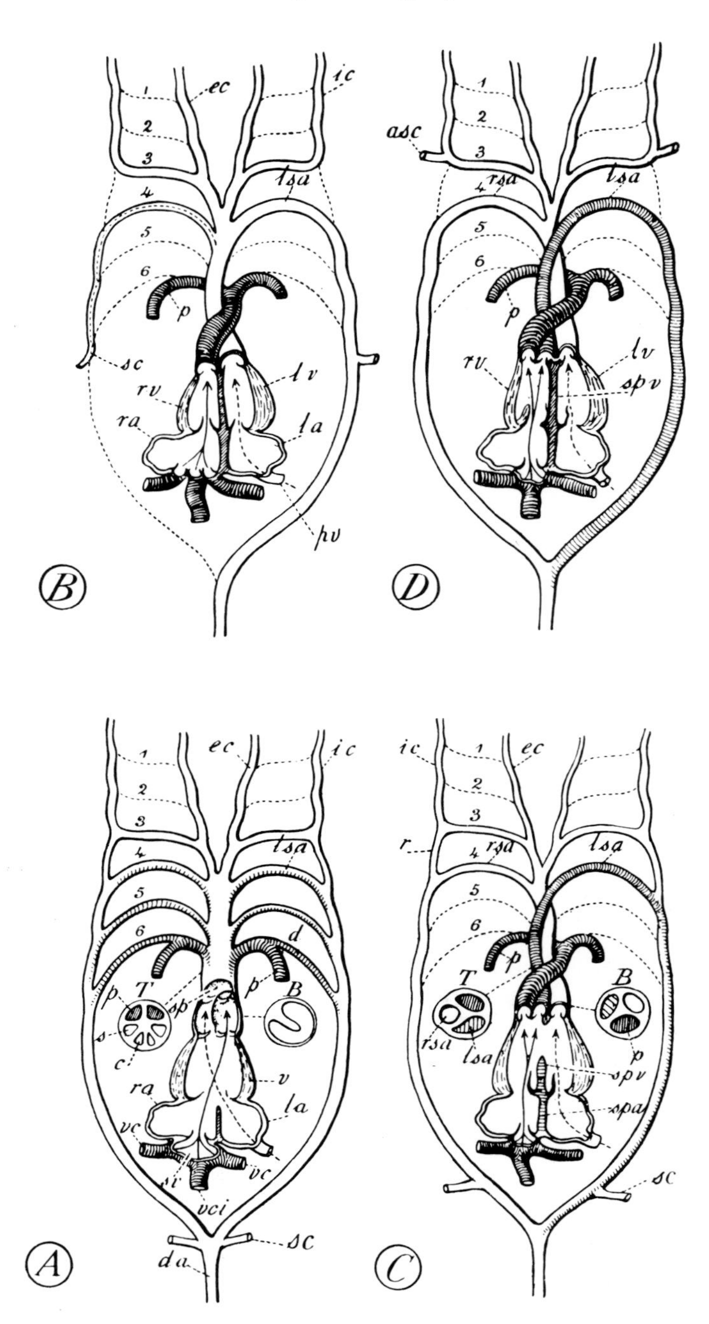
ec
ic
1
2
3
4
5
6
lsa
p
sc
rv
lv
ra
la
pv
B
asc
rsa
spv
D
T
sp
s
c
v
vc
si
vci
d
da
A
r
spa
C

> It is claimed that since the carotids of mammals arise from an aorta that curves to the *left*, this arrangement cannot have evolved from the reptilian one . . . This argument is invalid on every count . . . The common carotids of mammals arise embryologically as anterior extensions of the ventral aorta (as is the case in all tetrapods); no change in position is required at all. The only modification necessary is elimination of the subdivision of the ventral aorta into right and left aortas.

Anurans already have a partial subdivision of the aorta into right and left halves (Goodrich, 1930; Holmes, 1975) suggesting that the presence of the septum is the primitive amniote condition and that its absence is derived. In this interpretation, the theropsid condition would be a synapomorphy for that group while the sauropsid condition would be synapomorphic only at the level of Amniota and would be plesiomorphic within the Amniota. The aortic arches, then, would appear to be *hors de combat* in the amniote arguments; their structure can be considered consistent with both alternatives that I have presented here.

ALTERNATIVE 2: MONOPHYLY OF SYNAPSIDA + DIAPSIDA (EUREPTILIA)

The argument, typified by Williston in various papers (Williston, 1925) that turtles "truly" lack temporal fenestrae and were therefore to be allied with "stem reptiles" or cotylosaurs, has the implicit hypothesis that fenestrae are a derived feature and characterize a monophyletic group. Unfortunately, as discussed above, there is no specific fenestrated condition that can be stated as being typical of this group. An argument can be made, however, that the lower fenestra is such a feature because it occurs throughout Diapsida and Synapsida. In this case we must then accept the hypothesis that

---

Fig. 4. Diagram of the heart and aortic arches in (A) an amphibian, (B) a mammal, (C) a non-crocodilian reptile and (D) a crocodile, as presented by Goodrich (1930). See text for discussion.

Key to abbreviations. asc. Anterior subclavian; d, ductus Botalli; ec, external carotid; ic, internal carotid; la, left auricle; lsa, left systemic arch; lv, left ventricle; p, pulmonary artery; r, portion of lateral aorta remaining open only in *Sphenodon* and some Lacertilia as ductus arteriosus; ra, right auricle; rsa, right systemic arch; rv, right ventricle; spa, interauricular septum; spv, interventricular septum; sv, sinus venosus; v, ventricle; vc, vena cava superior; vci, vena cava inferior.

euryapsids are modified diapsids, that is, that they have lost the lower opening. *Araeoscelis* also must be cited as a character reversal as it only has the upper fenestra. Nonetheless, if we are to use a phylogenetic hypothesis involving temporal fenestrae, then this one is the most parsimonious.

JACOBSON'S ORGAN

Parsons' extensive work on reptilian nasal anatomy and development (1959a,b, 1967, 1970) has resulted in an interesting character distribution: the definitive Jacobson's organ occurs only in non-testudinate amniotes.

> In all amniotes with the exception of turtles . . . Jacobson's organ appears as a ventromedial outpocketing of the early embryonic nasal cavity and, at an almost equally early stage, a single lateral concha is formed. (Parsons, 1967, p.411.)

The interesting thing about the Jacobson's organ is that it is absent or greatly reduced in adult crocodiles, birds and some mammals, but, even in the forms that entirely lack it as adults, the ventromedial pocket with distinctive histology and an associated concha is formed in the embryo. Turtles, however, have no indication of these structures during development.

> It is possible that turtles are descended from a stock possessing the normal amniote nasal characters and that they have lost the concha and secondarily replaced Jacobson's organ with a simpler structure, but I find this hard to believe. (Parsons, 1967, p. 411.)

The taxonomic distribution of amniotes in which nasal development is known is relatively extensive and it is hard to impute this feature on that basis. Although lissamphibians are sometimes considered to have a "Jacobson's organ", Parsons has restricted the use of the term to the above morphology only.

I have put together the following summary of Jacobson's organ morphology from Parsons' work. In the adult, Jacobson's organ is usually a ventromedial outpocketing of the nasal cavity that lacks Bowman's glands in its sensory epithelium (Bowman's glands are found in the remaining olfactory epithelium of nearly all tetrapods) and has olfactory nerve fibres that tend to reach the accessory olfactory bulb rather than the main olfactory bulb. These last two

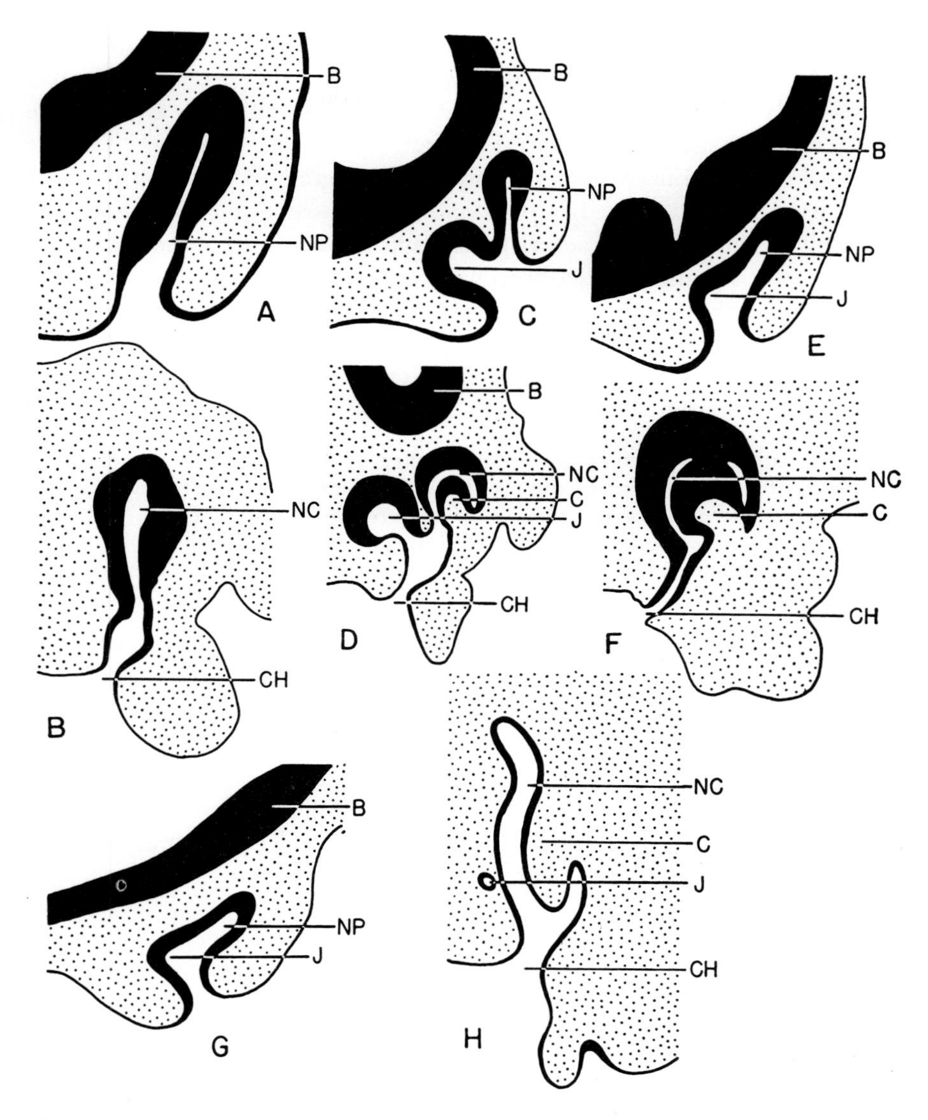

Fig. 5. Transverse sections through the nasal cavities of amniote embryos. (A and B) *Chrysemys;* (C and D) *Thamnophis*; (E and F) *Crocodylus*; (G and H) *Homo*. (A),(C),(E) and (G) show a stage shortly before the fusion of the nasal processes, and (B),(D),(F) and (H) a stage shortly after their fusion. After Parsons (1959a).

Key to abbreviations. B, Brain; C, concha; CH, choana; J, Jacobson's organ; NC, nasal cavity; NP, nasal pit.

features also characterize a region of tissue found in Lissamphibia and Testudines that has sometimes been called Jacobson's organ, but it is only in Eureptilia that these are associated with a definitive structure consisting of a ventral pocket.

Within the Eureptilia, the size and degree of development of Jacobson's organ vary. In *Sphenodon* and most Squamata the organ is well developed, characteristically with a complex concha, and in most adult squamates Jacobson's organ loses its nasal connection and opens into the palate. Living archosaurs lack a Jacobson's organ as well as the accessory olfactory bulb in the adults, but in the embryo of both crocodiles and birds the ventral outpocketing is clearly present. In mammals

> ...Jacobson's organ is generally retained as a small digitiform organ connected to the roof of the mouth ... In some mammals it may retain its primitive position as a pocket off the ventromedian nasal wall, and in others it is lost completely in the adult. An accessory olfactory bulb is present whenever Jacobson's organ persists in the adult ... (Parsons, 1959a, p. 180.)

This character seems to hold up under scrutiny; the presence of the feature in the embryos of Eureptilia that have lost it as adults being a particularly good argument for its never being present in turtles.

ACKNOWLEDGEMENTS

I would like to thank Dr E. Bruce Holmes and Dr Thomas Parsons for offering constructive criticism of this paper.

REFERENCES

Bakker, R. T. and Galton, P. M. (1974). Dinosaur monophyly and a new class of vertebrates. *Nature, Lond.* **248**, 168–172.

Barghusen, H. R. (1968). The lower jaw of cynodonts (Reptilia, Therapsida) and the evolutionary origin of mammal-like adductor jaw musculature. *Postilla* No. 116, 1–49.

Boonstra, L. D. (1963). Early dichotomies in the therapsids. *S. Afr. J. Sci.* **59**, 176–195.

Camp, C. L. (1923). Classification of the lizards. *Bull. Am. Mus. nat. Hist.* **48**, 289–481.

Carroll, R. L. (1977). The origin of lizards. *In* "Problems in Vertebrate Evolution" (S. M. Andrews, R. S. Miles and A. D. Walker, eds), pp. 359–396. Academic Press, London and New York.

Carroll, R. L. and Baird, D. (1972). Carboniferous stem-reptiles of the Family Romeriidae. *Bull. Mus. comp. Zool. Harv.* **143**, 321–364.

Chatterjee, S. (1978). A primitive parasuchid (phytosaur) reptile from the Upper Triassic Maleri Formation of India. *Palaeontology* **21**, 83–127.

Cruickshank, A. R. I. (1972). The proterosuchian thecodonts. *In* "Studies in Vertebrate Evolution" (K. A. Joysey and T. S. Kemp, eds), pp. 89–119. Winchester Press, New York.

Gaffney, E. S. (1975). A phylogeny and classification of the higher category of turtles. *Bull. Am. nat. Hist.* **155**, 387–436.

Gaffney, E. S. (1979a). Tetrapod monophyly: A phylogenetic analysis. *Bull. Carnegie Mus. nat. Hist.* No. 13, pp. 92–105.

Gaffney, E. S. (1979b). "An introduction to the Logic of Phylogeny Reconstruction." Columbia University Press.

Gaffney, E. S. and McKenna, M. C. (1979). A Late Permian captorhinid from Rhodesia. *Am. Mus. Novit.* No. 2688, 1–15.

Goodrich, E. S. (1916). On the classification of the Reptilia. *Proc. R. Soc. Lond.* **89**, 261–276.

Goodrich, E. S. (1930). "Studies on the Structure and Development of Vertebrates." Macmillan, London.

Holmes, E. B. (1975). A reconsideration of the phylogeny of the tetrapod heart. *J. Morph.* **147**, 209–228.

Hopson, J. A. and Crompton, A. W. (1969). Origin of mammals. *In* "Evolutionary Biology" (T. Dobzhansky, M. Hecht and W. C. Steere, eds), Vol. 3, pp.15–72. Appleton-Century-Crofts, New York.

Jaekel, O. (1911). Placechelys placodonts aus der Obertrias des Bakony. Resultate der wissenschaftlichen Erforschung des Balatonsees, *1*, Anhang. "Palaeontologie der Umgebung des Balatonsees", 1–90.

Kemp, T. S. (1972a). The jaw articulation and musculature of the whaitsiid Therocephalia. *In* "Studies in Vertebrate Evolution" (K. A. Joysey and T. S. Kemp, eds), pp. 213–230. Winchester Press, New York.

Kemp, T. S. (1972b). Whatsiid Therocephalia and the origin of cynodonts. *Phil. Trans. R. Soc.* **264**, 1–54.

Kuhn-Schnyder, E. (1963). Wege der Reptiliensystematik. *Paläont. Z.* **37**, 61–87.

Olson, E. C. (1947). The Family Diadectidae and its bearing on the classification of the reptiles. *Fieldiana:Geology* **11**, 1–53.

Parsons, T. H. (1959a). Nasal anatomy and the phylogeny of reptiles. *Evolution* **13**, 175–187.

Parsons, T. S. (1959b). Studies on the comparative embryology of the reptilian nose. *Bull. Mus. comp. Zool. Harv.* **120**, 101–277.

Parsons, T. S. (1967). Evolution of the nasal structure in the lower tetrapods. *Am. Zool.* **7**, 397–413.

Parsons, T. S. (1970). The nose and Jacobson's organ. *In* "Biology of the Reptilia" (C. Gans and T. S. Parsons, eds), Vol. 2. Academic Press, London and New York.

Platnick, N. I. and Gaffney, E. S. (1977). Systematics: A Popperian perspective. [Reviews of] The logic of scientific discovery, by K. R. Popper; Conjectures and refutations: The growth of scientific knowledge, by K. R. Popper. *Syst. Zool.* **26**, 360–365.

Platnick, N. I. and Gaffney, E. S. (1978a). Evolutionary biology: A Popperian perspective. [Reviews of] The poverty of historicism, by K. R. Popper; Objective knowledge: An evolutionary approach, by K. R. Popper; Unended quest: An intellectual autobiography, by K. R. Popper. *Syst. Zool.* **27**, 137–141.

Platnick, N. I. and Gaffney, E. S. (1978b). Systematics and the Popperian paradigm. [Reviews of] The philosophy of Karl Popper, by P. A. Schulpp; Criticism and the growth of knowledge, I. Lakatos and A. Musgrave, eds; The philosophy of Karl Popper, by R. J. Ackermann; Karl Popper, by B. Magee. *Syst. Zool.* **27**, 381–388.

Reisz, R. (1977). *Petrolacosaurus*, the oldest known diapsid reptile. *Science* **196**, 1091–1093.

Romer, A. S. (1956). "The Osteology of the Reptiles." University Press, Chicago.

Schaeffer, B., Hecht, M. K. and Eldredge, N. (1972). Paleontology and phylogeny. *In* "Evolutionary Biology" (T. Dobzhansky, M. Hecht and W. C. Steere, eds), Vol. 6. Appleton-Century-Crofts, New York.

Sill, W. D. (1974). The anatomy of *Saurosuchus galilei* and the relationships of the rauisuchid thecodonts. *Bull. Mus. comp. Zool. Harv.* **146**, 317–362.

Vaughn, P. P. (1955). The Permian reptile *Araeoscelis* restudied. *Bull. Mus. comp. Zool. Harv.* **113**, 305–367.

Watson, D. M. S. (1954). On *Bolosaurus* and the origin and classification of the reptiles. *Bull. Mus. comp. Zool. Harv.* **111**, 295–449.

Wellnhofer, P. (1978). Pterosauria. *In* "Encyclopedia of Paleoherpetology" (P. Wellnhofer, ed.), Part 19, pp. 1–82. Fischer, Stuttgart.

Williston, S. W. (1925). "The Osteology of the Reptiles." Harvard University Press.

Author Index

S

Subject Index

The numbers in *italics* refer to figures or tables

M